A Themed Issue Dedicated to Professor John B. Goodenough on the Occasion of His 100th Birthday Anniversary

A Themed Issue Dedicated to Professor John B. Goodenough on the Occasion of His 100th Birthday Anniversary

Editors

Jean Etourneau
Claude Delmas
Stéphane Jobic
Myung-Hwan Whangbo

MDPI • Basel • Beijing • Wuhan • Barcelona • Belgrade • Manchester • Tokyo • Cluj • Tianjin

Editors

Jean Etourneau
ICMCB Institut de la Matière
Condensée de Bordeaux
Pessac
France

Claude Delmas
ICMCB Institut de la Matière
Condensée de Bordeaux
Pessac
France

Stéphane Jobic
Institut des Matériaux
Jean Rouxel
Nantes
France

Myung-Hwan Whangbo
Department of Chemistry
North Carolina State University
Raleigh
USA

Editorial Office
MDPI
St. Alban-Anlage 66
4052 Basel, Switzerland

This is a reprint of articles from the Special Issue published online in the open access journal *Molecules* (ISSN 1420-3049) (available at: www.mdpi.com/journal/molecules/special_issues/molecules_Professor_John_Goodenough).

For citation purposes, cite each article independently as indicated on the article page online and as indicated below:

LastName, A.A.; LastName, B.B.; LastName, C.C. Article Title. *Journal Name* **Year**, *Volume Number*, Page Range.

ISBN 978-3-0365-3195-3 (Hbk)
ISBN 978-3-0365-3194-6 (PDF)

© 2022 by the authors. Articles in this book are Open Access and distributed under the Creative Commons Attribution (CC BY) license, which allows users to download, copy and build upon published articles, as long as the author and publisher are properly credited, which ensures maximum dissemination and a wider impact of our publications.

The book as a whole is distributed by MDPI under the terms and conditions of the Creative Commons license CC BY-NC-ND.

Contents

Jean Etourneau, Claude Delmas, Stéphane Jobic and Myung-Hwan Whangbo
In Honor of John Bannister Goodenough, an Outstanding Visionary
Reprinted from: *Molecules* **2021**, *26*, 6624, doi:10.3390/molecules26216624 1

Atsuo Yamada
High-Voltage Polyanion Positive Electrode Materials
Reprinted from: *Molecules* **2021**, *26*, 5143, doi:10.3390/molecules26175143 7

Jinhua Hong, Shunsuke Kobayashi, Akihide Kuwabara, Yumi H. Ikuhara, Yasuyuki Fujiwara and Yuichi Ikuhara
Defect Engineering and Anisotropic Modulation of Ionic Transport in Perovskite Solid Electrolyte $Li_xLa_{(1-x)/3}NbO_3$
Reprinted from: *Molecules* **2021**, *26*, 3559, doi:10.3390/molecules26123559 13

Julia H. Yang, Haegyeom Kim and Gerbrand Ceder
Insights into Layered Oxide Cathodes for Rechargeable Batteries
Reprinted from: *Molecules* **2021**, *26*, 3173, doi:10.3390/molecules26113173 25

Margaud Lécuyer, Marc Deschamps, Dominique Guyomard, Joël Gaubicher and Philippe Poizot
Electrochemical Assessment of Indigo Carmine Dye in Lithium Metal Polymer Technology
Reprinted from: *Molecules* **2021**, *26*, 3079, doi:10.3390/molecules26113079 37

Jan L. Allen, Bria A. Crear, Rishav Choudhury, Michael J. Wang, Dat T. Tran and Lin Ma et al.
Fast Li-Ion Conduction in Spinel-Structured Solids
Reprinted from: *Molecules* **2021**, *26*, 2625, doi:10.3390/molecules26092625 51

Tanusri Saha-Dasgupta
The Fascinating World of Low-Dimensional Quantum Spin Systems: Ab Initio Modeling
Reprinted from: *Molecules* **2021**, *26*, 1522, doi:10.3390/molecules26061522 69

Edouard Boivin, Jean-Noël Chotard, Christian Masquelier and Laurence Croguennec
Towards Reversible High-Voltage Multi-Electron Reactions in Alkali-Ion Batteries Using Vanadium Phosphate Positive Electrode Materials
Reprinted from: *Molecules* **2021**, *26*, 1428, doi:10.3390/molecules26051428 97

Hyun-Joo Koo, Reinhard Kremer and Myung-Hwan Whangbo
Unusual Spin Exchanges Mediated by the Molecular Anion $P_2S_6^{4-}$: Theoretical Analyses of the Magnetic Ground States, Magnetic Anisotropy and Spin Exchanges of MPS_3 (M = Mn, Fe, Co, Ni)
Reprinted from: *Molecules* **2021**, *26*, 1410, doi:10.3390/molecules26051410 119

Xueyang Li, Hongyu Yu, Feng Lou, Junsheng Feng, Myung-Hwan Whangbo and Hongjun Xiang
Spin Hamiltonians in Magnets: Theories and Computations
Reprinted from: *Molecules* **2021**, *26*, 803, doi:10.3390/molecules26040803 137

Myung-Hwan Whangbo, Hyun-Joo Koo and Reinhard K. Kremer
Spin Exchanges between Transition Metal Ions Governed by the Ligand p-Orbitals in Their Magnetic Orbitals
Reprinted from: *Molecules* **2021**, *26*, 531, doi:10.3390/molecules26030531 163

Sang-Kyu Lee, Hun Kim, Sangin Bang, Seung-Taek Myung and Yang-Kook Sun
WO_3 Nanowire/Carbon Nanotube Interlayer as a Chemical Adsorption Mediator for High-Performance Lithium-Sulfur Batteries
Reprinted from: *Molecules* **2021**, *26*, 377, doi:10.3390/molecules26020377 . . . . **189**

Seona Kim, Guntae Kim and Arumugam Manthiram
A Bifunctional Hybrid Electrocatalyst for Oxygen Reduction and Oxygen Evolution Reactions: Nano-Co_3O_4-Deposited $La_{0.5}Sr_{0.5}MnO_3$ via Infiltration
Reprinted from: *Molecules* **2021**, *26*, 277, doi:10.3390/molecules26020277 . . . . **203**

Guowei Zhao, Kota Suzuki, Masaaki Hirayama and Ryoji Kanno
Syntheses and Characterization of Novel Perovskite-Type $LaScO_3$-Based Lithium Ionic Conductors
Reprinted from: *Molecules* **2021**, *26*, 299, doi:10.3390/molecules26020299 . . . . **213**

Samir F. Matar and Jean Etourneau
Electronic and Magnetic Structures of New Interstitial Boron Sub-Oxides $B_{12}O_2$:X (X = B, C, N, O)
Reprinted from: *Molecules* **2020**, *26*, 123, doi:10.3390/molecules26010123 . . . . **225**

Michel Pouchard
John B. Goodenough's Role in Solid State Chemistry Community: A Thrilling Scientific Tale Told by a French Chemist
Reprinted from: *Molecules* **2020**, *25*, 6040, doi:10.3390/molecules25246040 . . . . **239**

Editorial

In Honor of John Bannister Goodenough, an Outstanding Visionary

Jean Etourneau [1],*, Claude Delmas [1], Stéphane Jobic [2] and Myung-Hwan Whangbo [3],*

1. ICMCB-CNRS, University of Bordeaux, 33600 Pessac, France; delmas@icmcb-bordeaux.cnrs.fr
2. Institut des Matériaux Jean Rouxel, Université de Nantes, CNRS, IMN, 44000 Nantes, France; stephane.jobic@cnrs-imn.fr
3. Department of Chemistry, North Carolina State University, Raleigh, NC 27695, USA
* Correspondence: jean.etourneau@icmcb-bordeaux.cnrs.fr (J.E.); mike_whangbo@ncsu.edu (M.-H.W.)

Citation: Etourneau, J.; Delmas, C.; Jobic, S.; Whangbo, M.-H. In Honor of John Bannister Goodenough, an Outstanding Visionary. *Molecules* **2021**, *26*, 6624. https://doi.org/10.3390/molecules26216624

Received: 22 September 2021
Accepted: 25 October 2021
Published: 1 November 2021

Publisher's Note: MDPI stays neutral with regard to jurisdictional claims in published maps and institutional affiliations.

Copyright: © 2021 by the authors. Licensee MDPI, Basel, Switzerland. This article is an open access article distributed under the terms and conditions of the Creative Commons Attribution (CC BY) license (https://creativecommons.org/licenses/by/4.0/).

John B. Goodenough won the Nobel Prize in Chemistry in 2019 with Stanley Wittingham and Akira Yoshino for their fundamental contributions to the development of lithium-ion batteries. Calls for recognition for John's pioneering work on lithium-ion batteries had been launched at the Nobel Committee for many years and finally the calls were answered! This was wonderful, though long overdue, news for the communities of solid-state chemistry and materials sciences. John impacted these areas of research greatly via his constant desire to account for the physical properties of solids through the discussion of their local crystal structures and their chemical bonding. Let us notice that, on this special occasion, John also set the record for being the oldest person, at 97, to receive this prestigious honor!

John is a world-renowned outstanding scientist who won the Nobel Prize not only for his pioneering work on lithium-ion batteries, but also for his profound influence on solid-state chemistry, paving the way into the 21st century toward the modern approach to materials sciences where physics and chemistry are intimately related. Very early in the sixties of the last century, he demystified solid-state physics for solid-state chemists by developing qualitative tools with which to discuss and predict in two areas of transition-metal oxides: (i) he showed how to understand the magnetism of transition-metal oxides in terms of spin exchanges between transition-metal magnetic ions bridged by a common main-group ligand atom, providing the qualitative rules known as the Goodenough–Kanamori rules, and (ii) he related the metal versus insulator behaviors of transition-metal oxides to whether the interactions between adjacent transition-metal atoms dominate over those involving individual transition-metal atoms. In both areas of research, John showed how the seemingly complex problems can be reduced to the level of chemical bonding and local structure–property relationships, thereby strongly inspiring generations of scientists!

During the International Conference organized by Paul Hagenmuller in Bordeaux in 1964, which gathered the most influential solid-state physicists and chemists from Europe and North America involved in the study of transition-metal oxides, John clearly demonstrated the importance of solid-state physics in understanding the physical properties of solids with a view to using them for applications. From that moment on, John became a mentor and a guide for many of us in the discipline of solid-state chemistry. Throughout his stays in Europe and Bordeaux in particular, John inspired people around him by his vision and his fruitful intuition. John was a pioneer not only in lithium-ion batteries but also in developing phenomenological mechanisms useful in explaining a variety of physical properties of transition-metal oxides. For further discussion, see the article of M. Pouchard [1].

John Goodenough is one of those rare scientists whose impact on our daily lives is not only obvious, but also essential. His contributions have appreciably changed the way we live and, in a world always in search of new and better energy solutions, point towards promising directions for the future. Even at the age of 99, he still continues to develop new polymers and battery concepts with researchers in his laboratory. He is now largely focused on developing fully solid-state batteries with many challenging requirements such as low cost, long cycle life, high volumetric density and fast rates of charge and discharge.

John is probably one of first to establish a strong link between basic and applied research, which is of prime importance for our society. John showed an insatiable desire to interpret observations of extended solid-state compounds by developing conceptual tools based on local bonding and local structures. With this Special Issue dedicated to John Goodenough, we salute him for his long and illustrious career, significantly advancing the discipline of solid-state chemistry and materials sciences. For all of us, John is clearly an example to follow.

John Goodenough has a warm personality and he often said that, to live one's life to the fullest, one should be able to interact with people who want to interact with you. All our discussions with John, professional or private (see in Figure 1), were extremely rich in a friendly and relaxed atmosphere punctuated by his legendary and contagious laughter! John enjoyed traveling all over the world and sharing scientific discussions with people from almost every country in the world.

Figure 1. John Goodenough and Jean Etourneau at Jean's in the summer of 1998.

1. Special Issue Dedicated to Professor John B. Goodenough

After the introduction by Michel Pouchard [1] that spotlights the outstanding contributions of Professor John B. Goodenough to solid-state chemistry, this Special Issue consists of 14 contributions representing two areas of research John worked on, namely, nine contributions on lithium-ion batteries and five on magnetic properties.

From a theory describing the physical properties of a system, we often demand two things that are often incompatible with each other. One is to demand a conceptual framework with which to organize what is observed and predict qualitative trends. The other is to demand a quantitative tool with which to describe with numerical accuracy. The Goodenough–Kanamori rules, formulated in the late 1950s, provided a conceptual framework with which to discuss whether spin exchange interactions between transition-metal magnetic ions M present in M-L-M bridges, formed by sharing a common main group ligand atom L, are antiferromagnetic or ferromagnetic. The conceptual picture given by the Goodenough–Kanamori rules has greatly influenced the thinking of chemists and physicists for many decades.

The 2019 Nobel Prize in chemistry to John B. Goodenough, underlines another important scientific field of interest of John, namely, the development of lithium-ion batteries that are now indispensable in our lives (smartphones, laptops, electric vehicles, etc.). Goodenough was particularly credited for the choice of Li_xCoO_2 and then Li_xNiO_2 materials as positive electrodes in batteries, which truly revolutionized the field of electrochemistry. He also showed that the potential delivered by a battery can be increased by replacing oxide anions by polyanions (LiFePO$_4$), and continues to develop solid electrolytes with high conductivity of ionic species. All these materials or their derivatives are now used in commercial batteries.

In what follows, we briefly comment on the main points of each contribution.

1.1. Contributions in the Area of Spin Exchanges and Magnetism

In the following, the first four papers are invited contributions.

(1) In the late 1990s and early 2000s, it became increasingly clear that spin exchange interactions between magnetic ions do occur even if they do not share a common ligand. They are M-L... L-M and M-L... A... L-M type interactions with A as a d^0 (i.e., S = 0) cation. They reflect that the magnetic orbitals of a cation M forming an ML$_n$ polyhedron with surrounding ligands are given by d-states of ML$_n$, in which the p-orbitals of L are combined out-of-phase with the d-orbitals of M, and that all types of spin exchanges are governed largely by the ligand p-orbitals of the magnetic orbitals. The qualitative aspects of spin exchanges were reviewed by Myung-Hwan Whangbo, Hyun-Joo Koo and Reinhard K. Kremer [2], in which they discuss how the qualitative trends in spin exchanges are related to the arrangements of the magnetic orbitals, providing the structure–property relations with which to understand the magnetic properties of complex magnetic solids.

(2) Interactions between spins of a magnetic system are very weak and are very small in energy scale. This necessitates the use of a spin Hamiltonian defined in terms of several phenomenological parameters (e.g., spin exchanges, Dzyaloshinskii–Moriya interactions, etc.) to have quantitative predictions on magnetic properties. How to estimate the numerical values of such parameters systematically and accurately has been a challenge for a long time, which eventually led to an energy-mapping analysis based on first-principles DFT+U and DFT+hybrid calculations in the early to mid-2000s. To determine the parameters defining a model spin Hamiltonian, one analyzes the relative energies of a set of broken-symmetry ordered spin states using two different Hamiltonians; one generates the relative energies by using the model Hamiltonian made up of the parameters to determine and also by performing DFT+U or DFT+hybrid numerical calculations. Then, the two energy spectra are mapped to determine the numerical values of the desired parameters. This energy-mapping analysis is generalized in the review written by Xueyang Li, Hongyu Yu, Feng Lou, Junsheng Feng, Myung-Hwan Whangbo and Hongjun Xiang [3]. This review examines the origin of various possible interactions and shows how to compute the values of these interactions. Having such a quantitative tool is crucial because the purpose of using a model spin Hamiltonian is not to include all possible phenomenological parameters, but to include the minimal number of parameters needed to describe the observed physical properties. The quantitative tool allows one to sort out which parameters are essential.

(3) The spin exchange parameters can be estimated by using a method different from the energy-mapping method. For instance, when non-spin-polarized DFT calculations are employed to describe a magnetic solid of transition-metal magnetic ions, there occur partially filled d-state band(s) so that the solid is predicted to be metallic instead of magnetic insulating. Nevertheless, one can determine the parameters of a tight binding approximation that can reproduce the calculated d-state electronic band structure. From such parameters, the spin exchange parameters necessary for discussing the magnetic properties can be determined. Tanusri Saha-Dasgupta analyzed the d-state band structures in terms of the N-th order muffin orbital-downfolding technique to determine the spin lattices appropriate for a number of insulating magnetic oxides. Saha-Dasgupta provided a review on her studies based on this method in [4].

(4) In each layer of a layered compound MPS$_3$ (M = Mn, Fe, Co, Ni), the metal ions M^{2+} form a honeycomb arrangement with a molecular anion P$_2$S$_6^{4-}$ located at the center of every hexagon of M^{2+} ions. Thus, the spin exchange between M^{2+} ions is mediated by the symmetry-adapted group orbitals of P$_2$S$_6^{4-}$. Hyun-Joo Koo, Reinhard K. Kremer and Myung-Hwan Whangbo [5] probed the spin exchanges of MPS$_3$ (M = Mn, Fe, Co, Ni)

by DFT+U calculations and analyzed their trends to find several unusual types of spin exchanges unknown from other types of spin exchanges known so far.

(5) The contribution by Samir Matar and Jean Etourneau [6] examined the chemical bonding and electronic structures of interstitial boron suboxides $Bi_{12}O_2X$ (X = B, C, N, O) using first-principles DFT calculations to find that $Bi_{12}O_2X$ has unpaired electrons on X for X = C and N with moments of 1.9 and 1 μ_B, respectively, in the ferromagnetic ground state.

1.2. Contributions in the Area of Lithium-Ion Batteries

In the following, the first eight papers are invited contributions.

(1) The article by Julia H. Yang, Haegyeom Kim and Gerbrand Ceder [7] perfectly illustrates the philosophy of Goodenough concerning his pursuit to understand the structure–property relationship, rationalize experimental observations and improve the characteristics of a studied material or device. On the basis of DFT calculations, the authors evidence how the topology of layered structures impacts the electrochemical performances. In a more general way, the article demonstrates how crucial first-principles calculations are in accounting for the properties of materials and how they help to select new promising candidates for electrodes.

(2) Atsuo Yamada [8] provided a short review on polyanion positive electrode materials, which can generate high-voltage generation batteries with high-density energy. This paper gives a general overview of cell voltage monitoring vs. transition element electronic configuration. Currently, a challenging problem is to find cations other than those derived from Ni, Co and V, three transition elements that are still commonly used but are too expensive. As alternative possibilities, investigations were carried out with Fe^{3+}/Fe^{2+} and Cr^{4+}/Cr^{3+} redox couples, and sulfate and phosphate as the framework structure with Li^+ or Na^+ as mobile species.

(3) High-voltage multi-electron reactions in alkali-ion batteries using vanadium phosphate positive electrodes are described by Edouard Boivin, Jean-Noël Chotard, Christian Masquelier and Laurence Croguennec [9]. They clearly showed how the vanadyl distortion affects the reversibility of the intercalation/de-intercalation mechanism, the voltage of the battery and the number of exchange electrons per transition metal, and how these parameters can be modified by controlling the local chemical environment of vanadium, i.e., by an appropriate change in the composition (e.g., fluorine/oxygen exchange) of the material.

(4) Lithium–sulfur batteries can provide a higher energy density than classical Li-ion batteries at a lower cost, and they are environmentally friendly. Here, Sang-Kyu Lee, Hun Kim, Sangin Bang, Seung-Taek Myung and Yang-Kook Sun [10] show how the addition of WO_3 nanowires mixed with carbon nanotubes at the separator/cathode interface may improve the performance of such a device.

(5) Until recently, batteries and specifically positive electrodes were made from inorganic materials. Margaud Lécuyer, Marc Deschamps, Dominique Guyomard, Joël Gaubicher and Philippe Poizot [11] report the electrochemical performance of Li-based metal batteries involving indigo carmine, an insoluble organic salt, as the positive electrode. This opens the possibility toward inexpensive and renewable materials, a desired goal that future batteries must achieve.

(6) In this article devoted to a perovskite-type Li-ion conductor, Guowei Zhao, Kota Suzuki, Masaaki Hirayama and Ryoji Kanno [12] focus on a $(Li_xLa_{1-x/3})ScO_3$ solid solution prepared under high pressure and doped with Ce^{4+} and Zr^{4+} or Nb^{5+} to achieve a high ionic conductivity.

(7) This is another article devoted to a perovskite-type Li-ion conductor. Jinhua Hong, Shunsuke Kobayashi, Akihide Kuwabara, Yumi H. Ikuhara, Yasuyuki Fujiwara and Yuichi Ikuhara [13] studied $(Li_xLa_{1-x/3})NbO_3$ to examine the structural defects, such as point defects and grain boundaries, that may significantly perturb the migration of Li^+ ions. Controlling such structural defects is essential in constructing solid electrolyte batteries

and their all solid-state variants. This study presents new very-high-resolution electron microcopy techniques.

(8) This contribution by Seona Kim, Guntae Kim and Arumugam Manthiram [14] is concerned with rechargeable metal–air batteries. As an example of cost-effective electrocatalysts with effective bifunctional activity for both oxygen reduction and oxygen evolution, they discussed a hybrid catalyst, Co_3O_4-infiltrated $La_{0.5}Sr_{0.5}MnO_{3-\delta}$, to demonstrate that hybrid catalysts are a promising approach for oxygen electrocatalysts for renewable and sustainable energy devices.

(9) In the last contribution [15], Jan L. Allen, Bria A. Crear, Rishav Choudhury, Michael J. Wang, Dat T. Tran, Lin Ma, Philip M. Piccoli, Jeff Sakamoto and Jeff Wolfenstine examined Li-stuffed spinels with a high conductivity of Li as a material for a fully structured all solid battery. Their work provides a useful concept for enabling a single-phase fully solid electrode without interphase impedance.

Funding: This research received no external funding.

Conflicts of Interest: The authors declare no conflict of interest.

References

1. Pouchard, M. John B. Goodenough's Role in Solid State Chemistry Community: A Thrilling Scientific Tale Told by a French Chemist. *Molecules* **2020**, *25*, 6040. [CrossRef] [PubMed]
2. Whangbo, M.-H.; Koo, H.-J.; Kremer, R.K. Spin Exchanges between Transition Metal Ions Governed by the Ligand p-Orbitals in Their Magnetic Orbitals. *Molecules* **2021**, *26*, 531. [CrossRef] [PubMed]
3. Li, X.; Yu, H.; Lou, F.; Feng, J.; Whangbo, M.-H.; Xiang, H. Spin Hamiltonians in Magnets: Theories and Computations. *Molecules* **2021**, *26*, 803. [CrossRef] [PubMed]
4. Saha-Dasgupta, T. The Fascinating World of Low-Dimensional Quantum Spin Systems: Ab Initio Modeling. *Molecules* **2021**, *26*, 1522. [CrossRef] [PubMed]
5. Koo, H.-J.; Kremer, R.; Whangbo, M.-H. Unusual Spin Exchanges Mediated by the Molecular Anion $P_2S_6^{4-}$: Theoretical Analyses of the Magnetic Ground States, Magnetic Anisotropy and Spin Exchanges of MPS_3 (M = Mn, Fe, Co, Ni). *Molecules* **2021**, *26*, 1410. [CrossRef] [PubMed]
6. Matar, S.F.; Etourneau, J. Electronic and Magnetic Structures of New Interstitial Boron Sub-Oxides $B_{12}O_2$:X (X = B, C, N, O). *Molecules* **2021**, *26*, 123. [CrossRef] [PubMed]
7. Yang, J.H.; Kim, H.; Ceder, G. Insights into Layered Oxide Cathodes for Rechargeable Batteries. *Molecules* **2021**, *26*, 3173. [CrossRef] [PubMed]
8. Yamada, A. High-Voltage Polyanion Positive Electrode Materials. *Molecules* **2021**, *26*, 5143. [CrossRef] [PubMed]
9. Boivin, E.; Chotard, J.-N.; Masquelier, C.; Croguennec, L. Towards Reversible High-Voltage Multi-Electron Reactions in Alkali-Ion Batteries Using Vanadium Phosphate Positive Electrode Materials. *Molecules* **2021**, *26*, 1428. [CrossRef] [PubMed]
10. Lee, S.-K.; Kim, H.; Bang, S.; Myung, S.-T.; Sun, Y.-K. WO_3 Nanowire/Carbon Nanotube Interlayer as a Chemical Adsorption Mediator for High-Performance Lithium-Sulfur Batteries. *Molecules* **2021**, *26*, 377. [CrossRef] [PubMed]
11. Lécuyer, M.; Deschamps, M.; Guyomard, D.; Gaubicher, J.; Poizot, P. Electrochemical Assessment of Indigo Carmine Dye in Lithium Metal Polymer Technology. *Molecules* **2021**, *26*, 3079. [CrossRef] [PubMed]
12. Zhao, G.; Suzuki, K.; Hirayama, M.; Kanno, R. Syntheses and Characterization of Novel Perovskite-Type $LaScO_3$-Based Lithium Ionic Conductors. *Molecules* **2021**, *26*, 299. [CrossRef] [PubMed]
13. Hong, J.; Kobayashi, S.; Kuwabara, A.; Ikuhara, Y.H.; Fujiwara, Y.; Ikuhara, Y. Defect Engineering and Anisotropic Modulation of Ionic Transport in Perovskite Solid Electrolyte $Li_xLa_{(1-x)/3}NbO_3$. *Molecules* **2021**, *26*, 3559. [CrossRef] [PubMed]
14. Kim, S.; Kim, G.; Manthiram, A. A Bifunctional Hybrid Electrocatalyst for Oxygen Reduction and Oxygen Evolution Reactions: Nano-Co_3O_4-Deposited $La_{0.5}Sr_{0.5}MnO_3$ via Infiltration. *Molecules* **2021**, *26*, 277. [CrossRef] [PubMed]
15. Allen, J.L.; Crear, B.A.; Choudhury, R.; Wang, M.J.; Tran, D.T.; Ma, L.; Piccoli, P.M.; Sakamoto, J.; Wolfenstine, J. Fast Li-Ion Conduction in Spinel-Structured Solids. *Molecules* **2021**, *26*, 2625. [CrossRef] [PubMed]

High-Voltage Polyanion Positive Electrode Materials

Atsuo Yamada

Department of Chemical System Engineering, The University of Tokyo, Tokyo 113-8656, Japan; yamada@chemsys.t.u-tokyo.ac.jp

Abstract: High-voltage generation (over 4 V versus Li^+/Li) of polyanion-positive electrode materials is usually achieved by Ni^{3+}/Ni^{2+}, Co^{3+}/Co^{2+}, or V^{4+}/V^{3+} redox couples, all of which, however, encounter cost and toxicity issues. In this short review, our recent efforts to utilize alternative abundant and less toxic Fe^{3+}/Fe^{2+} and Cr^{4+}/Cr^{3+} redox couples are summarized. Most successful examples are alluaudite $Na_2Fe_2(SO_4)_3$ (3.8 V versus sodium and hence 4.1 V versus lithium) and β_1-$Na_3Al_2(PO_4)_2F_3$-type $Na_3Cr_2(PO_4)_2F_3$ (4.7 V versus sodium and hence 5.0 V versus lithium), where maximizing ΔG by edge-sharing Fe^{3+}-Fe^{3+} Coulombic repulsion and the use of the $3d^2/3d^3$ configuration of Cr^{4+}/Cr^{3+} are essential for each case. Possible exploration of new high-voltage cathode materials is also discussed.

Keywords: cathode; polyanion; high-voltage

1. Introduction

Polyanion-positive electrode material for lithium batteries was identified by Delmas, Goodenough, and their co-workers for the NASICON $M_2(XO_4)_3$ framework in the 1980s [1–3]. Later on, Padhi, Nanjundaswamy, and Goodenough discovered a very promising positive electrode material, $LiFePO_4$ [4], which is now widely commercialized for stationary use or a power source for electric vehicles. A common advantage of polyanion-type electrodes is their long-term stability of operation due to the rigid structural framework. Additional advantages inherent to $LiFePO_4$ that have led to its commercial application are (i) lithium can be extracted at the first charge and functions as a charge carrier, moving back and forward upon charge/discharge, (ii) it can withstand self-decomposition to guarantee a high level of safety, and (iii) it has a suitable operating voltage of 3.4 V versus lithium, which is not so high that it decomposes electrolytes but not too low that energy density is sacrificed [5].

Toward higher voltages, Mn analogue $LiMnPO_4$ (4.1 V versus lithium) was investigated but its low electrochemical activity was not acceptable, and this negative feature is common for all Mn-based polyanion-positive electrode materials [4,6–8]. Vanadium-based compounds such as $LiVPO_4F$ [9] operate well at a reasonable voltage range around 4.2 V, but they have been excluded as a commercial option due to the element's toxicity and volume change during Li^+ de/intercalation [10,11]. For an even higher voltage, Co^{3+}/Co^{2+} and Ni^{3+}/Ni^{2+} redox couples show activity at >4.5 V [12–14], but their highly oxidizing nature induces several side reactions unless careful design is applied to both the electrolyte and electrode composite. For a sodium analogue, electrode operation at higher voltages is more important as the Na/Na^+ potential is ca. 0.3 V higher than the Li/Li^+ potential. Within the polyanionic materials, strategic design toward high-voltage operation is almost the same in the case of lithium, achieved by introducing V, Co, Ni as a redox center, as represented by $Na_4Co_3(PO_4)_2P_2O_7$ [15–17].

However, the use of V, Co, or Ni is a challenging option for battery engineers as they entail cost and toxicity issues. In particular, for a sodium battery system, a high-voltage system with more abundant and cheap elements would be ideal. In this short review article, after summarizing the influential factors dominating positive electrode voltage, our recent successful attempts to activate Fe^{3+}/Fe^{2+} and Cr^{4+}/Cr^{3+} redox couples, in $Na_2Fe_2(SO_4)_3$ (3.8 V versus sodium and hence 4.1 V versus lithium) [18] and β_1-$Na_3Al_2(PO_4)_2F_3$-type $Na_3Cr_2(PO_4)_2F_3$ (4.7 V versus sodium and hence 5.0 V versus lithium) [19], will be demonstrated.

2. Toward Higher Voltage

2.1. Inductive Effect in Polyanionic Compounds

The voltage trend of polyanion-based positive electrode materials roughly follows the formal charges of the central atoms in the polyanions, consisting of the idea of the inductive effect [20]. The presence of strong X-O covalency stabilizes the antibonding M^{3+}/M^{2+} state through an M-O-X inductive effect to generate an appropriately high voltage. A series of compounds including large polyanions $(XO_4)^{y-}$ (X = S, P, As, Mo, W, y = 2 or 3) were explored, and the use of $(PO_4)^{3-}$ and $(SO_4)^{2-}$ has been shown to stabilize the structure and lower the M^{3+}/M^{2+} redox energy to useful levels.

2.2. Thermodynamic Modification

In essence, the voltage is defined as the difference between the lithium chemical potential in the cathode and in the anode, leading to a simple thermodynamic definition, ignoring PV and TS terms: (P = pressure, V = volume, T = temperature, and S = entropy), $E = (G_{Li} + G_{charged} - G_{discharged})/nF$. where G_{Li}, $G_{charged}$, and $G_{discharged}$, are the Gibbs free energies of lithium metal, charged cathode, and discharged cathode, respectively; n is the number of electrons in the redox reaction, and F is the Faraday constant. The overall thermodynamic scheme for voltage generation is summarized in Figure 1 based on the Born–Haber cycle [21].

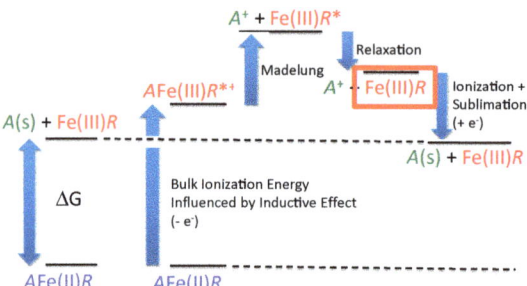

Figure 1. A Born–Haber cycle for the definition of voltage generation in cathode materials based on Fe^{3+}/Fe^{2+} redox couple. The notations R and R* represent relaxed and unrelaxed frameworks, respectively. Bulk ionization energy, which is closely related to the inductive effect, is an approximation (electronic part) but is not identical to ΔG.

2.3. Choice of Transition Metals

Figure 2 shows a schematic derivation of the operation voltages and d-band positions of 3d transition metal phosphates in sodium ion batteries. [19] In general, a transition metal ion M^{n+} with a higher atomic number has a deeper valence level owing to a larger effective nuclear charge, resulting in a higher $M^{(n+1)+}/M^{n+}$ redox potential. Naturally, phosphates with end representatives of the 3d series, such as Co^{2+} and Ni^{2+}, typically Na_2CoPO_4F and $Na_4M_3(PO_4)_2P_2O_7$ (M = Co^{2+} and Ni^{2+}), have been reported as high-voltage (4.3 V, 4.4 V, 4.8 V, respectively) cathode materials [16,17,22–24]. However, end representatives of the 3d series suffer from energy-level increments either by spin exchange penalty or crystal field splitting. On the other hand, the $3d^3$ electron of Cr^{3+} in the t_{2g} orbital is free from both spin exchange and crystal field splitting, which can be compensated for the smaller nuclear charge. Indeed, Cr^{4+}/Cr^{3+} redox couples in phosphates generate >4.5 V vs. Na/Na^+ (as presented below) [19], comparable to Co- or Ni-based phosphates.

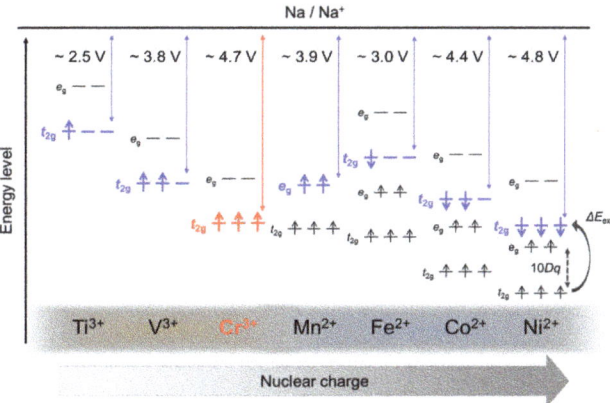

Figure 2. Schematic comparison of operating voltages and d-electron configurations for phosphate-based cathode materials containing 3d transition metals [19]. 10Dq and ΔE_{ex} indicate an octahedral crystal field splitting energy for 3d orbitals and exchange splitting energy, respectively. Orange and blue shading corresponds to valences of 3 and 2, respectively. Note that there is also a contribution of the Madelung and other energies to the cell voltage that is superimposed on the electronic contribution of the transition metal ions (see Figure 1). Permission is granted by Chemistry of Materials.

3. High-Voltage System with Fe^{3+}/Fe^{2+} and Cr^{4+}/Cr^{3+}

3.1. Pyrophosphates

Of particular interest is the triplite phase of LiFeSO4F [25] and metal-doped Li2FeP2O7 [26,27] possess edge-sharing FeO$_6$ octahedra to minimize the Fe-Fe distance, as distinguished from other, lower-voltage Fe-based polyanion electrodes with corner-sharing octahedra. A shorter Fe^{3+}–Fe^{3+} distance in the charged state is effective for enlarging $G_{charged}$, and hence the operating voltage E, while the influence of the discharged state $G_{discharged}$ with smaller charge Fe^{2+} can be subordinated in energetics.

The cell voltage for these two materials can reach as high as 3.9 V (vs. Li), which is higher than the value of 3.8 V calculated from the standard redox potentials. The latter has been suggested to be the highest achievable voltage for a Li ion battery utilizing the Fe^{3+}/Fe^{2+} redox couple in solid. As shown in Figure 3, the potential tunability for the Fe^{3+}/Fe^{2+} redox couple at the unusually high-voltage region of 3.5–3.9 V vs. lithium is similar for any metal M doping in the Li$_2$M$_x$Fe$_{1-x}$P$_2$O$_7$ system [27]. The phenomena include two aspects: (1) two redox reactions at different potentials are stabilized with the doping of foreign metal M, (2) with more dopant, both of the redox reactions upshift to higher potential, and one even approaches 4 V. Substitution of M into Fe sites may suppress the migration of Fe from the FeO$_5$ site upon charging, and the two original, distinct Fe sites become robust to stabilize the edge-sharing geometry of FeO$_5$ and FeO$_6$ polyhedrals with large Fe^{3+}-Fe^{3+} coulombic repulsion energy, leading to the two distinct redox reactions with inherently high potentials. The change in the relative energy of the intermediate compounds, which is induced by the unfavorable $V_{Li}{'}$-$M^{2+}(M_{Fe}{}^{\times})$ and/or Li$_{Li}{}^{\times}$-Fe^{3+}(Fe$_{Fe}$•) interaction in the doping case, may be a reason for the further potential upshifting. The classic inductive effect cannot explain the redox potential upshifting phenomenon in this case.

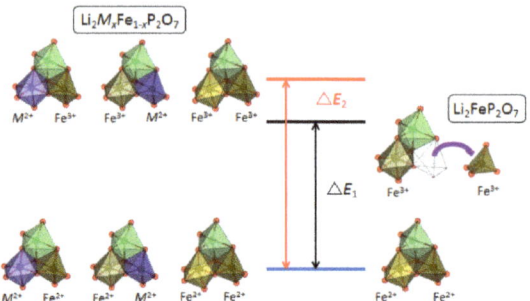

Figure 3. Schematic description of free energy difference between starting and delithiation materials. The right-hand portion is the pristine $Li_2FeP_2O_7$ system. The spontaneous structural rearrangement (Fe's migration) destroys the edge-sharing configuration and decreases the free energy of the delithiated state, which results in an energy difference of $\triangle E_1$. The left-hand portion is the doping system $Li_2M_xFe_{1-x}P_2O_7$. After full delithiation, the Li concentration is higher than that in the $Li_2FeP_2O_7$ case, because all of the M ions remain inert. The remaining Li can block the Fe migration and can stabilize the Fe's original local structure and the whole crystal structure, which means that the energy difference $\triangle E_2$ should be higher than $\triangle E_1$.

3.2. Alluaudites

Compaction of the MO_6 dimer can be more pronounced in an alluaudite framework, where two edge-shared MO_6 octahedra are bridged by a small XO_4 tetrahedron and the M-M distance becomes much shorter (Figure 4). During the search along the Na_2SO_4-$FeSO_4$ tie line, we discovered the first sulfate compound with an alluaudite-type framework [18]. Deviating sharply from most of the $A_xM_2(XO_4)_3$-type compounds adopting the NASICON-related structures, $Na_2Fe_2(SO_4)_3$ does not contain the lantern units $[M_2(XO_4)_3]$. It would be convenient to denote $AA'BM_2(XO_4)_3$ as general *alluaudite*-type compounds, where A = partially occupied Na(2), A' = partially occupied Na(3), B = Na(1), M = Fe^{2+}, and X = S in the present case.

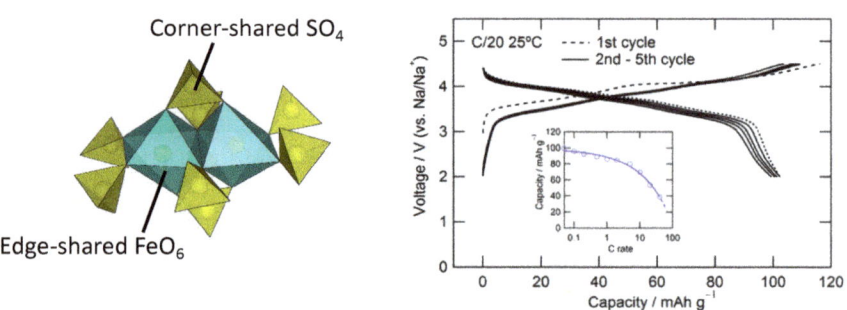

Figure 4. Local coordination structure and charge–discharge voltage profile of $Na_2Fe_2(SO_4)_3$.

The $Na_2Fe_2(SO_4)_3$ offers an average potential of 3.8 V (vs. Na/Na$^+$), with smooth, sloping charge–discharge profiles over a narrow voltage range of 3.3–4.3 V, which is the highest Fe^{3+}/Fe^{2+} redox potential obtained in any material environment (Figure 5) [18]. The abnormally high voltage can be explained by the thermodynamic definition of voltage explained in Section 2.2; the edge-sharing geometry of the Fe octahedra in $Na_2Fe_2(SO_4)_3$ will raise $G_{charged}$ due to the strong Fe^{3+}–Fe^{3+} repulsion, leading to high E. Additionally, it offers excellent rate kinetics and cycling stability without requiring any additional cathode optimization. It forms an open framework host for the efficient (de)intercalation of Na ions with very low activation energy.

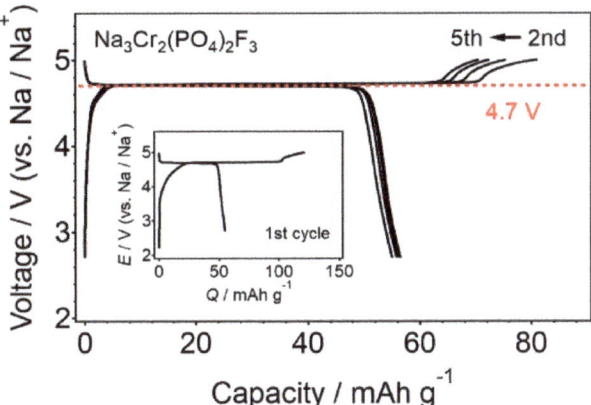

Figure 5. Galvanostatic charge–discharge curves of $Na_3Cr_2(PO_4)_2F_3$ electrode in Na half-cell at a rate of 0.1 C between 2.7 and 5.0 V vs. Na/Na^+ (1 C = 63.8 mA g^{-1}). Inset shows the first cycle.

An remarkable feature is that, now, the most commonly accessible redox Fe^{3+}/Fe^{2+} can, in principle, generate the high voltage of 3.8 V vs. sodium (and hence 4.1 V vs. lithium). However, the hygroscopicity of the sulphate compounds must be carefully managed.

3.3. β_1-$Na_3Al_2(PO_4)_2F_3$-Type Fluoride Phosphates

Considering the d-level considerations in Section 2.3, an extremely high operating potential of 4.7 V vs. Na/Na^+ was identified in $Na_{3-x}Cr_2(PO_4)_2F_3$ (0 < x < 1) on account of the Cr^{4+}/Cr^{3+} ($3d^2/3d^3$) redox couple, providing a promising design strategy for a high-voltage positive electrode material [19]. Whilst further Na^+ extraction (x > 1) to form $NaCr_2(PO_4)_2F_3$ above 5.0 V vs. Na/Na^+ remains elusive, optimization of the durable cell components for high-voltage operation may enable more activity and greater reversibility. Overall, utilizing low-cost Cr^{4+}/Cr^{3+} ($3d^2/3d^3$), instead of Co^{3+}/Co^{2+} ($3d^6/3d^7$) or Ni^{3+}/Ni^{2+} ($3d^7/3d^8$) as in previously reported polyanion compounds, is worthwhile for the realization of batteries with higher energy density.

4. Summary and Perspective

Initiated by Delmas, Goodenough, and co-workers in the 1980s, polyanion-type positive electrode materials now represent a large group of materials for reversible Li^+, Na^+, and K^+ insertion. With a suitable combination of transition metal and framework structure, the operating voltage can be tuned, leading sometimes to a suitable high-voltage range for practical application. Although $LiFePO_4$ is the only compound that has been widely applied for commercial use to date, continuous exploration is ongoing in the community toward better batteries with lower cost, high voltage, high safety, and a long calendar life. In addition to the widely examined redox couple based on Fe^{3+}/Fe^{2+}, Cr^{4+}/Cr^{3+} could be an inexpensive yet higher-voltage option for future material development in polyanion compounds.

Funding: This research was funded by Element Strategy Initiative of MEXT, Grant Number JP-MXP0112101003.

Conflicts of Interest: The author declares no conflict of interest.

Sample Availability: Samples are not available without a certain contract.

References

1. Nadiri, A.; Delmas, C.; Salmon, R.; Hagenmuller, P. Chemical and electrochemical alkali metal intercalation in the 3D framework of $Fe_2(MoO_4)_3$. *Rev. Chim. Minérale* **1984**, *21*, 537–544.
2. Manthiram, A.; Goodenough, J.B. Lithium insertion into $Fe_2(MO_4)_3$ frameworks: Comparison of M = W with M. *J. Solid State Chem.* **1987**, *71*, 349–360. [CrossRef]
3. Delmas, C.; Cherkaoui, F.; Nadiri, A.; Hagenmuller, P. A Nasicon-type phase as intercalation electrode: $NaTi_2(PO_4)_3$. *Mat. Res. Bull.* **1987**, *22*, 631–639. [CrossRef]
4. Padhi, A.K.; Nanjundaswamy, K.S.; Goodenough, J.B. Phospho-olivines as positive-electrode materials for rechargeable lithium batteries. *J. Electrochem. Soc.* **1997**, *144*, 1188–1194. [CrossRef]
5. Yamada, A.; Chung, S.-C.; Hinokuma, K. Optimized $LiFePO_4$ for lithium battery cathodes. *J. Electrochem. Soc.* **2001**, *148*, A224–A229. [CrossRef]
6. Yonemura, M.; Yamada, A.; Takei, Y.; Sonoyama, N.; Kanno, R. Comparative kinetic study of olivine $Li_x MPO_4$ (M=Fe, Mn). *J. Electrochem. Soc.* **2004**, *151*, A1352–A1356. [CrossRef]
7. Delacourt, C.; Laffont, L.; Bouchet, R.; Wurm, C.; Leriche, J.-B.; Morcrette, M.; Tarasco, J.-M.; Masqueliera, C. Toward understanding of electrical limitations (electronic, ionic) in $LiMPO_4$ (M = Fe, Mn) electrode materials. *J. Electrochem. Soc.* **2005**, *152*, A913–A921. [CrossRef]
8. Yamada, A.; Takei, Y.; Koizumi, H.; Sonoyama, N.; Kanno, R.; Ito, K.; Yonemura, M.; Kamiyama, T. Electrochemical, magnetic, and structural investigation of the $li_x(Mn_yFe_{1-y})PO_4$ olivine phases. *Chem. Mater.* **2006**, *18*, 804–813. [CrossRef]
9. Barker, J.; Saidi, M.Y.; Swoyer, J.L. Lithium Iron(II) Phospho-olivines prepared by a novel carbothermal reduction method. *Electrochem. Solid State Lett.* **2003**, *6*, A53–A55. [CrossRef]
10. Ellis, B.L.; Ramesh, T.N.; Davis, L.J.M.; Goward, G.R.; Nazar, L.F. Structure and electrochemistry of two-electron redox couples in lithium metal fluorophosphates based on the tavorite structure. *Chem. Mater.* **2011**, *23*, 5138–5148. [CrossRef]
11. Chayambuka, K.; Mulder, G.; Danilov, D.L.; Notten, P.H.L. From Li-ion batteries toward Na-ion chemistries: Challenges and opportunities. *Adv. Energy Mater.* **2020**, *10*, 1–11. [CrossRef]
12. Amine, K.; Yasuda, H.; Yamachi, M. Olivine $LiCoPO_4$ as 4.8 V electrode material for lithium batteries. *Electrochem. Solid State Lett.* **2000**, *3*, 178–179. [CrossRef]
13. Wolfenstine, J.; Allen, J. Ni^{3+}/Ni^{2+} redox potential in $LiNiPO_4$. *J. Power Sources* **2005**, *142*, 389–390. [CrossRef]
14. Masquelier, C.; Croguennec, L. Polyanionic (phosphates, silicates, sulfates) frameworks as electrode materials for rechargeable Li (or Na) batteries. *J. Am. Chem. Soc.* **2013**, *113*, 6552–6591.
15. Bianchini, M.; Fauth, F.; Brisset, N.; Weill, F.; Suard, E.; Masquelier, C.; Croguennec, L. Comprehensive investigation of the $Na_3V_2(PO_4)_2F_3$–$NaV_2(PO_4)_2F_3$ system by operando high resolution synchrotron X-ray diffraction. *J. Am. Chem. Soc.* **2015**, *27*, 3009–3020. [CrossRef]
16. Nose, M.; Nakayama, H.; Nobuhara, K.; Yamaguchi, H.; Nakanishi, S.; Iba, H. $Na_4Co_3(PO_4)_2P_2O_7$: A novel storage material for sodium-ion batteries. *J. Power Sources* **2013**, *234*, 175–179. [CrossRef]
17. Gezović, A.; Vujković, M.J.; Milović, M.; Grudić, V.; Dominko, R.; Mentus, S. Recent developments of $Na_4M_3(PO_4)_2(P_2O_7)$ as the cathode material for alkaline-ion rechargeable batteries: Challenges and outlook. *Energy Storage Mater.* **2021**, *37*, 243–273. [CrossRef]
18. Barpanda, P.; Oyama, G.; Nishimura, S.; Chung, S.-C.; Yamada, A. A 3.8-V earth-abundant sodium battery electrode. *Nat. Commun.* **2014**, *5*, 1–6. [CrossRef]
19. Kawai, K.; Asakura, D.; Nishimura, S.; Yamada, A. 4.7 V Operation of the Cr^{4+}/Cr^{3+} redox couple in $Na_3Cr_2(PO_4)_2F_3$. *Chem. Mater.* **2021**, *33*, 1373–1379. [CrossRef]
20. Padhi, A.K.; Nanjundaswamy, K.S.; Masquelier, C.; Okada, S.; Goodenough, J.B. Effect of structure on the Fe^{3+}/Fe^{2+} redox couple in iron phosphates. *J. Electrochem. Soc.* **1997**, *144*, 1609–1613. [CrossRef]
21. Yamada, A. Systematic studies on "abundant" battery materials: Identification and reaction mechanisms. *Electrochemistry* **2016**, *84*, 654–661. [CrossRef]
22. You, Y.; Manthiram, A. Progress in high-voltage cathode materials for rechargeable sodium-ion batteries. *Adv. Energy Mater.* **2018**, *8*, 1701785. [CrossRef]
23. Chayambuka, K.; Mulder, G.; Danilov, D.L.; Notten, P.H.L. Sodium-ion battery materials and electrochemical properties reviewed. *Adv. Energy Mater.* **2018**, *8*, 1–49. [CrossRef]
24. Kubota, K.; Yokoh, K.; Yabuuchi, N.; Komaba, S. Na_2CoPO_4F as a high-voltage electrode material for Na-ion batteries. *Electrochemistry* **2014**, *82*, 909–911. [CrossRef]
25. Barpanda, P.; Ati, M.; Melot, B.; Rousse, G. A 3.90V iron-based fluorosulphate material for lithium-ion batteries crystallizing in the triplite structure. *Nat. Mater.* **2011**, *10*, 772–779. [CrossRef] [PubMed]
26. Furuta, N.; Nishimura, S.; Barpanda, P.; Yamada, A. Fe^{3+}/Fe^{2+} couple approaching 4 V in $Li_{2-x}(Fe_{1-y}Mn_y)P_2O_7$ pyrophosphate cathodes. *Chem. Mater.* **2012**, *24*, 1055–1061. [CrossRef]
27. Ye, T.; Barpanda, P.; Nishimura, S.; Furuta, N.; Chung, S.-C.; Yamada, A. General observation of Fe^{3+}/Fe^{2+} redox couple close to 4 V in partially substituted $Li_2FeP_2O_7$ pyrophosphate solid-solution cathodes. *Chem. Mater.* **2013**, *25*, 3623–3629. [CrossRef]

Article

Defect Engineering and Anisotropic Modulation of Ionic Transport in Perovskite Solid Electrolyte $Li_xLa_{(1-x)/3}NbO_3$

Jinhua Hong [1], Shunsuke Kobayashi [1], Akihide Kuwabara [1], Yumi H. Ikuhara [1], Yasuyuki Fujiwara [2] and Yuichi Ikuhara [1,3,*]

[1] Nanostructures Research Laboratory, Japan Fine Ceramics Center, Nagoya 456-8587, Japan; jinhuahong436@gmail.com (J.H.); s_kobayashi@jfcc.or.jp (S.K.); kuwabara@jfcc.or.jp (A.K.); yumi@jfcc.or.jp (Y.H.I.)
[2] Faculty of Engineering, Shinshu University, Nagano 380-8553, Japan; yamato2010@shinshu-u.ac.jp
[3] Institute of Engineering Innovation, The University of Tokyo, Tokyo 113-8586, Japan
* Correspondence: ikuhara@sigma.t.u-tokyo.ac.jp

Citation: Hong, J.; Kobayashi, S.; Kuwabara, A.; Ikuhara, Y.H.; Fujiwara, Y.; Ikuhara, Y. Defect Engineering and Anisotropic Modulation of Ionic Transport in Perovskite Solid Electrolyte $Li_xLa_{(1-x)/3}NbO_3$. *Molecules* **2021**, *26*, 3559. https://doi.org/10.3390/molecules26123559

Academic Editors: Claude Delmas, Stephane Jobic and Myung-Hwan Whangbo

Received: 5 May 2021
Accepted: 7 June 2021
Published: 10 June 2021

Publisher's Note: MDPI stays neutral with regard to jurisdictional claims in published maps and institutional affiliations.

Copyright: © 2021 by the authors. Licensee MDPI, Basel, Switzerland. This article is an open access article distributed under the terms and conditions of the Creative Commons Attribution (CC BY) license (https://creativecommons.org/licenses/by/4.0/).

Abstract: Solid electrolytes, such as perovskite $Li_{3x}La_{2/1-x}TiO_3$, $Li_xLa_{(1-x)/3}NbO_3$ and garnet $Li_7La_3Zr_2O_{12}$ ceramic oxides, have attracted extensive attention in lithium-ion battery research due to their good chemical stability and the improvability of their ionic conductivity with great potential in solid electrolyte battery applications. These solid oxides eliminate safety issues and cycling instability, which are common challenges in the current commercial lithium-ion batteries based on organic liquid electrolytes. However, in practical applications, structural disorders such as point defects and grain boundaries play a dominating role in the ionic transport of these solid electrolytes, where defect engineering to tailor or improve the ionic conductive property is still seldom reported. Here, we demonstrate a defect engineering approach to alter the ionic conductive channels in $Li_xLa_{(1-x)/3}NbO_3$ ($x = 0.1 \sim 0.13$) electrolytes based on the rearrangements of La sites through a quenching process. The changes in the occupancy and interstitial defects of La ions lead to anisotropic modulation of ionic conductivity with the increase in quenching temperatures. Our trial in this work on the defect engineering of quenched electrolytes will offer opportunities to optimize ionic conductivity and benefit the solid electrolyte battery applications.

Keywords: defect engineering; perovskite electrolyte; lithium-ion battery; migration pathway; anisotropic response

1. Introduction

Commercial lithium-ion batteries have shaped the new era and people's daily lives, owing to their successful portable electronics applications in vehicles and mobile phones, etc. However, there are still common issues to be addressed in this industry, such as safety, electrochemical and mechanical stability, and cycling life, which are intrinsic disadvantages of the flammable organic liquid electrolytes mostly employed in the current commercial Li-ion batteries [1–4]. Solid electrolyte batteries [5–8] or all-solid-state batteries have long been considered as the future of battery technology to avoid safety issues such as leakage and explosion, and to minimize cycling instability due to side reactions or metal dendrite growth. Thus, solid electrolyte materials [9–12] have attracted extensive research interest in the advanced architecture design [13,14] of Li-ion batteries and Li-air, Li-S [15], and Li-Br$_2$ batteries [16,17] with exceptionally high energy density.

Among the available lithium-ion-conducting solid electrolytes, ceramic oxides ($10^{-5} \sim 10^{-3}$ S·cm^{-1}) such as perovskite materials $Li_{3x}La_{2/1-x}TiO_3$ (LLTO) [18–21] and $Li_xLa_{(1-x)/3}NbO_3$ (LLNO) [22–24], anti-perovskite Li_3OX (X = Cl, Br)[25–27] and garnet structured $Li_7La_3Zr_2O_{12}$ (LLZO) [28–31] have received much attention due to their good electrochemical stability and considerable potential to push the limit of ionic conductivity towards a desired level ($\sim 10^{-2}$ S·cm^{-1}) in the industrial application of batteries. Their ionic conductivity follows such a microscopic ion migration mechanism: low occupancy of the Li^+ on the vacancy sites; low migration energy barrier for ion hopping; and network-like available sites (vacancies) [32] to interconnect the migration pathways of the mobile ions in these solid electrolytes [1]. In practical battery applications, the presence of point defects, grain/domain boundaries [21,33] in polycrystalline materials, and electrolyte/electrode interface resistance [34,35] would play a dominant role in determining the Li ionic conductive performance.

Ma et al. [21,33] have demonstrated the atomic structure of grain boundaries in polycrystalline oxide electrolytes which severely degrade the ionic conductivity compared to the perfect bulk form. Very recently, Dawson et al. [36] utilized large-scale molecular dynamic simulations to show that the grain size distribution of the polycrystalline electrolytes is key to the overall ionic transport. The conductivity has a strong decrease [36] when the grain size is smaller than 100 nm, and it increases and converges to 85% of the conductivity of perfect bulk crystal when grain size exceeds 400 nm. The directional solidification method has been demonstrated to yield high-quality single-crystal electrolyte ingots at the decimetre scale [37], where the limited grain boundaries (if they exist between large-size domains) may not remarkably influence the macroscopic ionic conductivity. Herein, atomic structure characterization of other common structural disorders is also quite necessary to understand their effect on the ion diffusion. Furthermore, there have not been adequate experimental reports on how to utilize defect engineering to perfect or tailor the ionic conductivity at the atomic scale.

In this work, we will show the formation of various defects by the rearrangement of La sites in $Li_xLa_{(1-x)/3}NbO_3$ (LLNO, $x = 0.1$~0.13) electrolytes after a quenching process and that the resulted La interstitials/vacancies induce the anisotropic change in the ionic conductivity. Through aberration-corrected scanning transmission electron microscopy (STEM), we have identified the layered structure property of LLNO by atomic resolution annular dark/bright field (ADF/ABF) imaging, together with energy dispersive X-ray spectroscopy (EDX) mapping to visualize the layered chemical structure in atomic resolution. The experimental quenching process of single-crystal LLNO gives rise to vast La vacancy defects in the La, Li-coexisting A_1 layer and octahedral interstitial La atoms occupied in the "empty" O-containing A_2 layer. This La rearrangement mechanism leads to the anisotropic experimental findings that the in-A_1-plane conductivity increased, while the out-of-A_1-plane conductivity decreased when the quenching temperature elevated. Based on this microscopic diagram, the migration of La atoms along different kinetic pathways results in the formation of in-A_1-plane vacancies and in-A_2-plane interstitials which are responsible for the anisotropic modulation of ionic conductivity through defect engineering. Our trial in this work on the defect characterization of the quenched single-crystal LLNO will offer new opportunities to optimize the ionic conductivity and benefit in its potential applications in the new solid electrolyte batteries.

2. Results and Discussion

Solid electrolyte $Li_xLa_{(1-x)/3}NbO_3$ (LLNO) has a perovskite structure, as shown in Figure 1a, where the La atoms (green balls) are located at the vertex sites with 2/3 occupancy. This means 1/3 of the vertex La sites are inherent vacancies which accommodate Li ions. Hence, this special Li-containing plane is named as the A_1 layer [23,24] with rich vacancy networks, and the middle-plane gap between neighbouring A_1 layers is called the A_2 layer, which contains only O atoms (purple in Figure 1a). In the charging or discharging process of the Li ion battery, these layers could be the Li^+-conductive channels for ionic transport in the solid electrolyte.

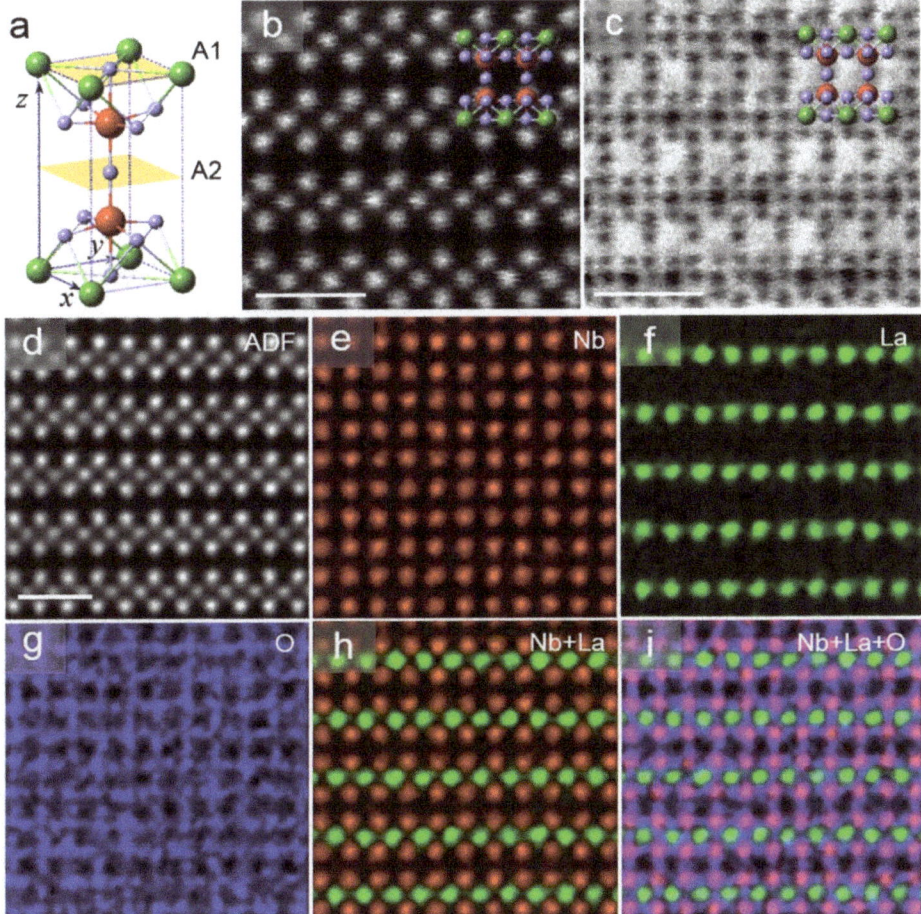

Figure 1. (**a**) Structure model of perovskite electrolyte LLNO. In the unit cell, green atoms stand for coexisting Li, La sites with an occupancy of 2/3, red for Nb with 100% occupancy, and blue for O with 100% occupancy. The A_1 layer marks the La, Li-coexisting layer, and A_2 layer is the "empty" mid-plane gap between A_1 layers. (**b**,**c**) Atomically resolved HAADF and ABF images in [100] direction, respectively. The atomic models superposed suggest the invisible O atoms can be clearly identified in the ABF image. (**d**–**i**) STEM-EDX mappings to reveal the chemical structure. Atomic resolution EDX mapping depicts the layered distribution of La in A_1 plane which is partly replaced by Li ions. Scale bars: 1 nm.

Figure 1b,c show the atomic resolution high angle annular dark/bright field scanning transmission electron microscopy (HAADF/ABF-STEM) images of the $Li_xLa_{(1-x)/3}NbO_3$ ($x \approx 0.13$) crystal in the [100] direction. Heavy atoms La and Nb are clearly presented in the HAADF image, while the O columns invisible in HAADF imaging can be unambiguously resolved in the ABF image in Figure 1c. These atomically resolved images demonstrate that there exist no interstitial La atoms in the A_2 layer of the pristine electrolyte crystal.

Furthermore, energy dispersive X-ray spectroscopy (EDX) mappings clearly show the atom-by-atom chemical identification of LLNO with distinguished layered or squared distribution of La, Nb and O lattices, as shown in Figure 1d–i. These EDX mappings show that there exist only La and O in the A_1 plane, and the layered La distribution (green in Figure 1f) also indicates the ionic conductive channel, since Li replaces part of the La atoms within the A_1 plane in the pristine LLNO. As a supplement to EDX detection, electron energy loss spectroscopy (EELS) demonstrated a weak signal around 60 eV that originated from the Li-K edge as shown in Figure S1, which indicates the low content of Li in the pristine samples.

In order to improve the ionic conductivity properties of the single-crystal electrolyte, material processes such as annealing and quenching were conducted to tailor the microstructures of the materials [22]. These high-temperature material processes result in ordered structure modulation [23], atom migration, new defects' formation, phase segregation or precipitation. Hence, defect engineering by materials annealing/quenching is to be explored to discover new atomic mechanisms for the enhancement of the ionic conductivity.

Figure 2a–c show the HAADF images of the electrolytes quenched at different temperatures. In contrast with the pristine and 700 °C quenched samples (almost the same) in Figure 2a, more defects such as interstitial atoms and novel complicated intermediate precipitates emerge in the samples quenched at the higher temperature of 1000 °C (Figure 2b). For crystals quenched at the even higher temperature of 1300 °C in Figure 2c, the interstitial atoms fully occupy the "empty" A_2 layer which should contain only HAADF-invisible O atoms. One can also see the more obvious differences in the HAADF images in Figures S2 and S3.

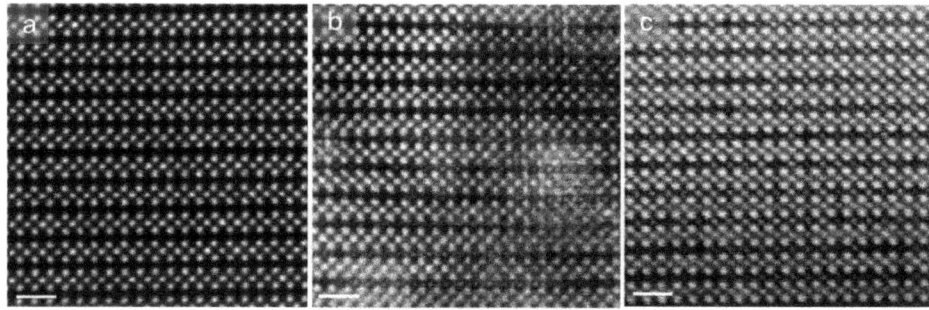

Figure 2. (**a**–**c**) Structure evolution of pristine and 700 °C-, 1000 °C-, and 1300 °C-quenched LLNO electrolytes, respectively, imaged along [100] direction. Scale bars: 1 nm. Defects appear with the increase in quenching temperatures. The "empty" O-containing A_2 layers in the pristine and 700 °C-quenched electrolytes in (**a**) underwent a complex transition with diverse defects precipitating in (**b**), and became fully occupied by interstitials in (**c**) at the high temperature of 1300 °C.

As shown in the HAADF images in the [001] direction in Figure 3, in contrast with the uniform square lattice of the pristine non-quenched electrolytes, vast vacancies appear in the 1300 °C quenched samples, highlighted by yellow circles. The simulated HAADF image in the Figure 3a inset shows the dark atoms correspond to the green La atoms in the superposed structural model. This confirms that vast La atoms move out of the original A1 layer and leave vacancies in the samples when quenched at high temperatures. Therefore, a rearrangement of La lattice atoms has occurred where La atoms within the A_1 plane migrate into the "empty" A_2 layer, behaving as the obvious "interstitial" defects in Figure 2c. From the serial images Figure 2a–c, it is also reasonable to propose that the regular "interstitial" defects (in Figure 2c) could be a metastable configuration in the La migration.

However, for the electrolytes quenched at the intermediate temperature 1000 °C (Figure 2b), much more diverse "precipitated" defects with unique atomic structures appear, shown in Figure 4a–c and also highlighted by ellipses in different colours. After analysing the diverse defects formed in the 1000 °C-quenched samples, only three types of defects are found to be the most common, highlighted by ellipses in colours standing for specific atomic structures. Initial interpretation of the emerging of diverse defects in the annealing/quenching process will also be discussed later.

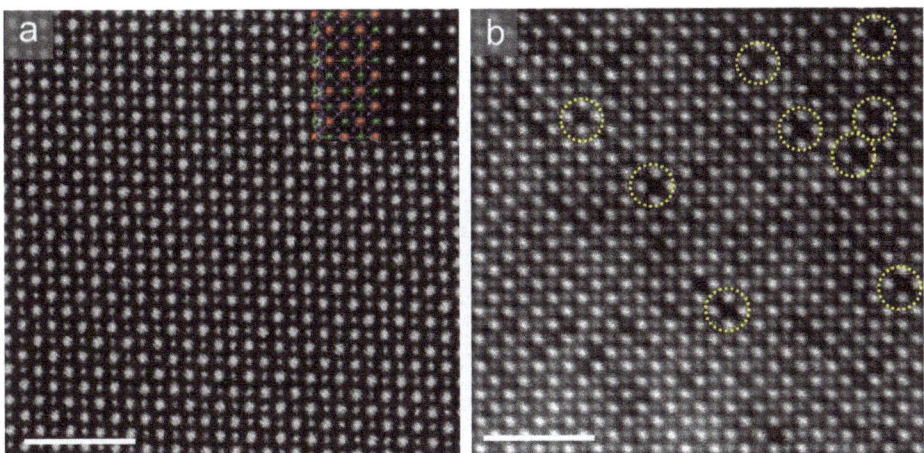

Figure 3. Atomically resolved HAADF images of pristine (**a**) and 1300 °C-quenched (**b**) LLNO electrolytes along [001] direction. Scale bars: 2 nm. La vacancies are not present in pristine non-quenched samples, where the inset simulated HAADF image in (**a**) agrees well with the experimental image. Dark atomic columns correspond to La in the inset of (**a**), as indicated by the green atoms (La) in the superposed structure model. Vast La vacancies appear in the 1300 °C-quenched sample, as highlighted by the yellow circles in (**b**).

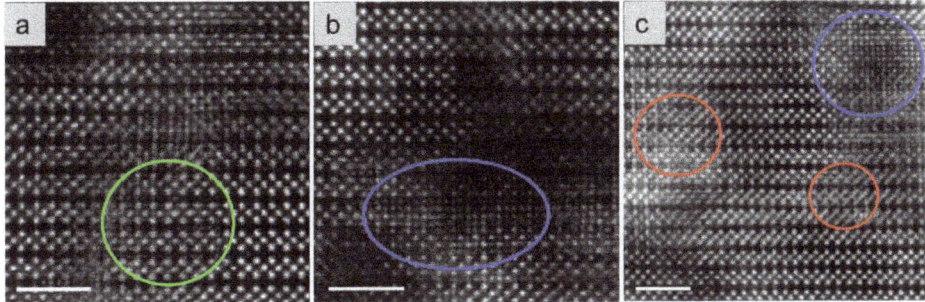

Figure 4. (**a**–**c**) The diverse defects formed in the 1000 °C-quenched electrolytes containing many intermediate defects, highlighted by ellipses in different colours. Scale bars: 2 nm.

To better illustrate the atomic structure, we present the models and experimental HAADF images of different defects systematically in Figure 5b–d,h,j,l. After careful comparison of these defects with the normal layered matrix lattice in Figure 5a,e,f, it can be inferred that the diverse occupancies of interstitial sites must stem from La atomic migration via different kinetic pathways. As the "empty" A_2 layer has adequate space to accommodate migrating La atoms activated by high temperature annealing/quenching, the $La_{A1} \rightarrow La_{A2}$ migration will easily occur along the proposed kinetic pathway, as marked by the yellow arrows in Figure 5b. This will lead to the most commonly observed metastable defects in Figure 2c. On the other hand, as depicted previously, 2/3 of La sites within the A_1 layer are occupied by La atoms and the left 1/3 La sites correspond to vacancies or Li in the pristine electrolyte LLNO. Hence, the nearest-neighbouring La-site hopping (indicated by arrows in Figure 5c) or second-nearest-neighbouring La-site hopping (arrows in Figure 5d) will take place via the vacancy mechanism.

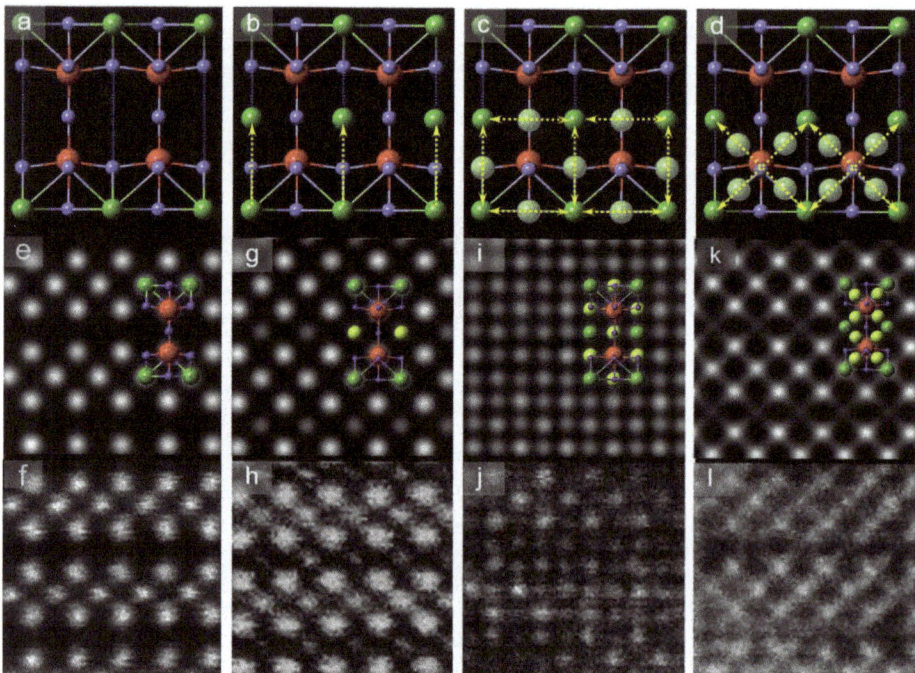

Figure 5. (**a,b**) Structure models of the ground state and metastable states. Note that the green La atoms have a probability of 1/3 to be a vacancy even in the pristine electrolyte. $La_{A1} \rightarrow La_{A2}$ migration leads to the metastable state in (**b**), where yellow arrows mark the kinetic pathways. Consistent with the previous models, red balls stand for Nb, green for La/Li and purple for O. (**c,d**) Nearest-neighbouring and second-nearest-neighbouring La-site migration, together with interstitially trapped La (light green within the yellow arrows). (**e,f**) Simulated and experimental HAADF images of ground-state pristine lattice. (**g,h**) Simulated and experimental HAADF images of metastable defects in (**b**). (**i–l**) Simulated and experimental HAADF images of diverse defects corresponding to different La migration pathways. Structure models are overlaid on the simulated images in (**e,g,i,k**), which are consistent with the experimental images.

Following these migration pathway hypotheses, the normal La_{A1} in Figure 5a can be regarded as the ground state configuration and La_{A2} in Figure 5b as the metastable state. Then, vacancies in these ground/metastable configurations offer considerable possibilities to the transitions between both states, leaving diverse intermediate defects (light green in Figure 5c,d) in the nearest- and second-nearest-neighbouring hopping pathways. Based on the structural models in Figure 5a,b, the simulated HAADF images of the ground-state/metastable configuration (Figure 5e,g, respectively) are consistent with the experimental results of pristine and quenched electrolytes in Figure 5f,h and Figure 2a,c. One can also see the effect of different La_{A2} occupancies on the quantitative intensity of the simulated HAADF image in Figure S4. Other diverse configurations trapped by quenching at the intermediate temperature in Figure 5c,d will contribute to abnormal lattice imaging, as simulated in Figure 5i,k, respectively, in accordance with the experimental observations in Figure 5j,l. Note that only a part of the La_{A1} and La_{A2} sites preserves vacancies, hence intermediate defects due to random migration of La should be superposed onto the perfect lattice without La vacancies in the image simulation, leading to a better matching between Figure 5i,k and Figure 5j,l. The qualitative experiment/simulation agreement confirms the diverse La-migration-induced defects between normal lattices.

To interpret the formation of these defects, we propose a temperature-dependent configuration diagram. The pristine unit cell configuration has the lowest formation energy, with all La situated in a close-packing polyhedron formed by 12 neighbouring O atoms

(Figure 1a), corresponding to the ground state. High-temperature thermal activation may break the La-O bonds and yield the in-A_1-plane, in-A_2-plane, and inter-plane $La_{A1} \to La_{A2}$ migrations when there exist plentiful vacancies or empty A_2 sites to accommodate La. A finite-time thermal process allows the mobile La to reach metastable sites with a lower coordination number (Figures 2c, 4 and 5) compared with the pristine La sites, yielding metastable phases with higher formation energies. The complicated kinetic pathways (Figure 5) for atomic migration in solids contribute to different metastable configurations after the finite-time thermal process, resulting in the formation of diverse defects. This resembles the Martensitic transformation of quenched steel in a diffusionless and military manner.

The $La_{A1} \to La_{A2}$ transition not only generates the emerging of the metastable configurations, but also creates a new chemical environment and coordination structure of the O ligands which can be detected by energy loss spectrum near-edge fine structures (ELNES). As shown in Figure 6, the K edges of O atoms (1s→final state) with double peaks, denoted as I_1 and I_2, demonstrate obvious variations in differently quenched samples. The double peaks in the O K-edge are mainly caused by t_{2g} and e_g splitting due to hybridization of unoccupied O 2p orbitals with Nb $3d^0$ orbitals [38] in the crystal field of octahedral O-ligands surrounding Nb. From the atomic structure characterization in Figures 2 and 4, the samples quenched at 1000 °C contain both perfect lattice and diverse precipitated defects, acting as an intermediate mixture of the pristine and 1300 °C-quenched samples. Although no obvious chemical shift occurs at the O K-edges, the intensity ratio of double peaks (I_1/I_2) increases with the quenching temperature. This indicates us that the metastable La_{A2} configuration contributes to the high I_1/I_2 ratio. As most O atoms reside outside of the A_2 planes and are less coordinated with La after the $La_{A1} \to La_{A2}$ transition, this indicates a higher unoccupied density of states of the hybridized e_g orbitals. Besides the ELNES of O, the $M_{2,3}$ edges of Nb also present variations when the quenching temperature increases (Figure S5), resulting from atomic migration and coordination rearrangement.

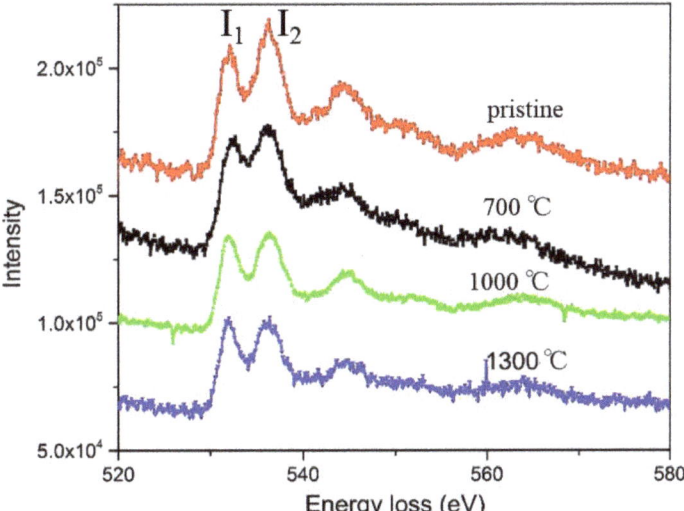

Figure 6. The O K-edges of differently quenched single-crystal electrolytes. The relative ratio of double peaks I_1/I_2 changes obviously with the quenching temperature, highly correlated with the metastable La_{A2} formation.

In the LLNO electrolytes, vast La vacancies within the A_1 layer emerge and the $La_{A1} \to La_{A2}$ transition occurs under high-temperature annealing/quenching. This will give rise to the enhancement of the in-A_1-plane ionic conductivity, since part of La_{A1} moves out of the Li migration pathways. In contrast, the out-of-A_1-plane (perpendicular to A_1 plane) Li ion transport would be partly blocked due to the presence of La_{A2} interstitial defects (Figure 2c). As seen in Figure 7, the conductivity in the [100] direction (in-A_1-plane) increases by almost one order of magnitude to 10^{-4} S·cm^{-1}, while the conductivity in the [001] direction (out-of-A_1-plane) decreases by three times as the quenching temperature increases. Note the measurements are all conducted on quenched single-crystal electrolytes. As a solid electrolyte, LLNO is an electronic insulator, and its macroscopic ionic conductivity is dominated by mobile ions with small radii such as Li$^+$. In our EELS measurement, we did not detect any local reduction of the valence state of Nb^{5+} in the single-crystal LLNO, which avoided the appearance of electronic conductivity. Local minor oxygen vacancies may exist and compensate the small amount of Li deficiency induced by the heat treatment. It is well known that oxide-ion conductivity usually appears at elevated temperatures [39]. We suppose that oxide ion conductions in the quenched LLNO samples hardly contribute to the total conductivities at room temperature.

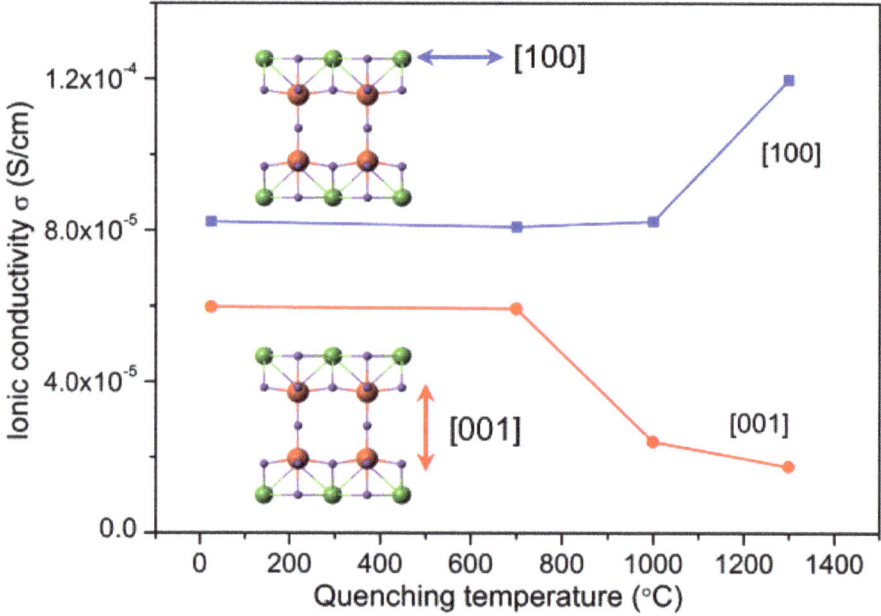

Figure 7. The anisotropic change of the ionic conductivity with the increasing quenching temperature. The in-A_1-plane conductivity (in [100] direction) increases due to the $La_{A1} \rightarrow La_{A2}$ transition, forming more vacancies and space for ionic diffusion. As a result, the presence of La_{A2} and other intermediate defects block the out-of-A_1-plane pathway (in [001] direction).

Defect engineering through atomic migration in the high-temperature quenching accounts for the anisotropic modulation of Li ionic transport (Figure 8). This can be interpreted by the anisotropic Li$^+$-migration energy barriers with the presence of La-vacancies in the A_1 layer and metastable La_{A2} interstitials in the "empty" A_2 layer. The former La vacancies will obviously decrease the energy barrier of the Li$^+$ in-A_1-plane migration due to the extra space freedom for Li ionic diffusion, and the latter interstitial La_{A2} atoms will increase the energy barrier of the Li$^+$ out-of-A_1-plane migration. Anisotropic modulation of ionic conductivity through this defect formation mechanism will also offer a possibility to tune the ionic conductive pathways in the future design of solid electrolytes in battery applications.

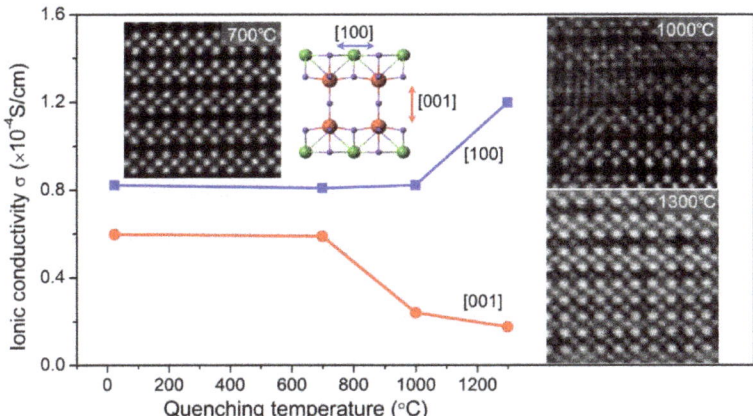

Figure 8. Quenching-induced defects and anisotropic changes in ionic conductivity.

3. Sample Preparation and Characterization

Single-phased $Li_xLa_{(1-x)/3}NbO_3$ sintered bodies were synthesized according to the following formula, where the desired Li composition was determined.

$xLiNbO_3 + (1-x)La_{1/3}NbO_3 \rightarrow Li_xLa_{(1-x)/3}NbO_3$

The $Li_xLa_{(1-x)/3}NbO_3$ sintered body placed in the platinum crucible was melted at 1362 °C, then solidified to grow the single crystal by crucible translation in the vertical temperature gradient furnace [37]. To maintain structural stability, the Li content should be kept as low as $x < 0.2$ to obtain high-quality single crystals. The prepared $Li_xLa_{(1-x)/3}NbO_3$ single crystals were annealed at 700 °C, 1000 °C and 1300 °C for ten minutes and then quenched to room temperature. The Li contents of the quenched samples were hardly changed in comparison to that of the pristine single crystal measured by inductive couple plasma (ICP) spectroscopy. The wafers for ionic conductivity measurement and microstructure observation were prepared by cutting the LLNO single crystals.

The ionic conductivity was measured by impedance spectroscopy in a 2-electrode setup using a Hewlett Packard 4192A impedance analyser, operated at 50 Hz~13 MHz, in air and at room temperature. The (001) and (100) surfaces of the samples were polished and then sputter-coated by Au to serve as the electrodes.

Through grinding the single-crystal wafer in ethanol, thin flakes of electrolytes were deposited onto a Cu TEM grid with holey carbon film for STEM observation. Samples were cleaned in a plasma cleaner (JIC-410, JEOL Ltd., Tokyo, Japan) to remove surface contamination or the amorphous layer before the TEM characterization. Atomically resolved HAADF and ABF imaging were conducted on an aberration-corrected TEM (JEM-2100F, JEOL Ltd., Tokyo, Japan) at 200 kV. The convergence angle of the incident electron probe was set to 25 mrad, the HAADF acceptance angle was 70–240 mrad, and the ABF detector 11–22 mrad. EEL spectra were obtained using an EEL spectrometer (Tridiem ERS, Gatan, Inc., Warrendale, PA, USA) attached to a Wien filter monochromated aberration-corrected STEM (JEM-2400FCS, JEOL Ltd., Tokyo, Japan) operated at 200 kV. EEL spectra were recorded in STEM mode, using 0.1 eV per channel and an energy resolution of 300 meV (full-width at half-maximum of zero-loss peak). The convergence and collection semi-angles were 33 and 43 mrad, respectively. Atomic resolution STEM-EDX mapping was performed on a JEM-ARM200CF (JEOL Ltd., Tokyo, Japan) at 200 kV, equipped with a double large-window silicon drift EDX detector forming a large effective collection solid angle of 1.7 Sr. HAADF image simulation was performed by the software QSTEM (C.T. Koch, Berlin, Germany) to match the experimental results.

4. Conclusions

In summary, the diverse defects emerging in the quenching process of LLNO were systematically characterized by atomically resolved HAADF-STEM imaging and were revealed as the ground/metastable configurations involved in different kinetic pathways of La atoms. This further suggests the diverse atomic processes and energy barriers of La migration via vacancy mechanisms. Meanwhile, the La vacancies' rearrangement in the $La_{A1} \rightarrow La_{A2}$ transition leads to the anisotropic response of the ionic conductivity to the increasing quenching temperature of the electrolytes. Our investigation on this perovskite electrolyte has provided the possibility of tuning ionic conductivity through specific defect engineering to promote its application in solid electrolyte batteries or all-solid-state batteries.

Supplementary Materials: Supporting information is available online, Figures S1 and S5: Core level EELS of solid electrolytes LLNO. Figures S2 and S3: Vacancies or interstitial defects in the quenched electrolytes LLNO. Figure S4: Simulated HAADF image of LLNO in [100] direction with different A_2-layer occupancies.

Author Contributions: Sample preparation and STEM observation, J.H., S.K. and Y.F.; Structure and composition analysis, Y.H.I.; Writing—original draft, J.H. and A.K.; Writing—review and editing, Y.I. All authors have read and agreed to the published version of the manuscript.

Funding: Part of this work was supported by Grant-in-Aid for Specially Promoted Research (Grant No. JP17H06094), Grant-in-Aid for Scientific Research on Innovative Areas (Grant No. JP19H05788) from the Japan Society for the Promotion of Science (JSPS), and "Nanotechnology Platform" (Project No. 12024046) of MEXT.

Institutional Review Board Statement: Not applicable.

Informed Consent Statement: Not applicable.

Data Availability Statement: The data presented in this study are available on request from the corresponding author.

Acknowledgments: We thank Keiichi Kohama and Hideki Iba in Toyota Motor Corporation for valuable discussions on this matter. We also thank R. Yoshida, T. Kato in JFCC and Feng Bin at the University of Tokyo for their help in TEM characterization.

Conflicts of Interest: The authors declare no competing financial interest.

Sample Availability: The samples in this study are available for only research purpose on request from the corresponding author.

References

1. Manthiram, A.; Yu, X.; Wang, S. Lithium battery chemistries enabled by solid-state electrolytes. *Nat. Rev. Mater.* **2017**, *2*, 16103. [CrossRef]
2. Quartarone, E.; Mustarelli, P. Electrolytes for solid-state lithium rechargeable batteries: Recent advances and perspectives. *Chem. Soc. Rev.* **2011**, *40*, 2525–2540. [CrossRef]
3. Takada, K. Progress and prospective of solid-state lithium batteries. *Acta Mater.* **2013**, *61*, 759–770. [CrossRef]
4. Wang, Y.; Richards, W.D.; Ong, S.P.; Miara, L.J.; Kim, J.C.; Mo, Y.; Ceder, G. Design principles for solid-state lithium superionic conductors. *Nat. Mater.* **2015**, *14*, 1026. [CrossRef]
5. Jiang, C.; Li, H.; Wang, C. Recent progress in solid-state electrolytes for alkali-ion batteries. *Sci. Bull.* **2017**, *62*, 1473–1490. [CrossRef]
6. Fan, L.; Wei, S.; Li, S.; Li, Q.; Lu, Y. Recent Progress of the Solid-State Electrolytes for High-Energy Metal-Based Batteries. *Adv. Energy Mater.* **2018**. [CrossRef]
7. Li, J.; Ma, C.; Chi, M.; Liang, C.; Dudney, N.J. Solid Electrolyte: The Key for High-Voltage Lithium Batteries. *Adv. Energy Mater.* **2014**, *5*. [CrossRef]
8. Famprikis, T.; Canepa, P.; Dawson, J.A.; Islam, M.S.; Masquelier, C. Fundamentals of inorganic solid-state electrolytes for batteries. *Nat. Mater.* **2019**, *18*, 1278–1291. [CrossRef]
9. Zhang, H.; Li, C.; Piszcz, M.; Coya, E.; Rojo, T.; Rodriguez-Martinez, L.M.; Armand, M.; Zhou, Z. Single lithium-ion conducting solid polymer electrolytes: Advances and perspectives. *Chem. Soc. Rev.* **2017**, *46*, 797–815. [CrossRef]

10. Zheng, Y.; Wang, J.; Yu, B.; Zhang, W.; Chen, J.; Qiao, J.; Zhang, J. A review of high temperature co-electrolysis of H2O and CO2 to produce sustainable fuels using solid oxide electrolysis cells (SOECs): Advanced materials and technology. *Chem. Soc. Rev.* **2017**, *46*, 1427–1463. [CrossRef]
11. Chen, L.; Li, Y.; Li, S.-P.; Fan, L.-Z.; Nan, C.-W.; Goodenough, J.B. PEO/Garnet Composite Electrolytes for Solid-State Lithium Batteries: From "Ceramic-in-Polymer" to "Polymer-in-Ceramic". *Nano Energy* **2018**, *46*, 176–184. [CrossRef]
12. Liu, B.; Gong, Y.; Fu, K.; Han, X.; Yao, Y.; Pastel, G.; Yang, C.; Xie, H.; Wachsman, E.D.; Hu, L. Garnet Solid Electrolyte Protected Li-Metal Batteries. *ACS Appl. Mater. Inter.* **2017**, *9*, 18809–18815. [CrossRef]
13. Zhang, Z.; Zhao, Y.; Chen, S.; Xie, D.; Yao, X.; Cui, P.; Xu, X. An advanced construction strategy of all-solid-state lithium batteries with excellent interfacial compatibility and ultralong cycle life. *J. Mater. Chem. A* **2017**, *5*, 16984–16993. [CrossRef]
14. Fu, K.K.; Gong, Y.; Hitz, G.T.; McOwen, D.W.; Li, Y.; Xu, S.; Wen, Y.; Zhang, L.; Wang, C.; Pastel, G. Three-dimensional bilayer garnet solid electrolyte based high energy density lithium metal-sulfur batteries. *Energy Environ. Sci.* **2017**, *10*, 1568–1575. [CrossRef]
15. Lei, D.; Shi, K.; Ye, H.; Wan, Z.; Wang, Y.; Shen, L.; Li, B.; Yang, Q.H.; Kang, F.; He, Y.B. Progress and Perspective of Solid-State Lithium-Sulfur Batteries. *Adv. Funct. Mater.* **2018**. [CrossRef]
16. Wang, L.; Wang, Y.; Xia, Y. A high performance lithium-ion sulfur battery based on a Li2S cathode using a dual-phase electrolyte. *Energy Environ. Sci.* **2015**, *8*, 1551–1558. [CrossRef]
17. Chang, Z.; Wang, X.; Yang, Y.; Gao, J.; Li, M.; Liu, L.; Wu, Y. Rechargeable Li//Br battery: A promising platform for post lithium ion batteries. *J. Mater. Chem. A* **2014**, *2*, 19444–19450. [CrossRef]
18. Gao, X.; Fisher, C.A.; Kimura, T.; Ikuhara, Y.H.; Kuwabara, A.; Moriwake, H.; Oki, H.; Tojigamori, T.; Kohama, K.; Ikuhara, Y. Domain boundary structures in lanthanum lithium titanates. *J. Mater. Chem. A* **2014**, *2*, 843–852. [CrossRef]
19. Gao, X.; Fisher, C.A.; Kimura, T.; Ikuhara, Y.H.; Moriwake, H.; Kuwabara, A.; Oki, H.; Tojigamori, T.; Huang, R.; Ikuhara, Y. Lithium atom and A-site vacancy distributions in lanthanum lithium titanate. *Chem. Mater.* **2013**, *25*, 1607–1614. [CrossRef]
20. Inaguma, Y.; Liquan, C.; Itoh, M.; Nakamura, T.; Uchida, T.; Ikuta, H.; Wakihara, M. High ionic conductivity in lithium lanthanum titanate. *Solid State Commun.* **1993**, *86*, 689–693. [CrossRef]
21. Ma, C.; Cheng, Y.; Chen, K.; Li, J.; Sumpter, B.G.; Nan, C.W.; More, K.L.; Dudney, N.J.; Chi, M. Mesoscopic framework enables facile ionic transport in solid electrolytes for Li batteries. *Adv. Energy Mater.* **2016**, *6*. [CrossRef]
22. Hu, X.; Fisher, C.A.; Kobayashi, S.; Ikuhara, Y.H.; Fujiwara, Y.; Hoshikawa, K.; Moriwake, H.; Kohama, K.; Iba, H.; Ikuhara, Y. Atomic scale imaging of structural changes in solid electrolyte lanthanum lithium niobate upon annealing. *Acta Mater.* **2017**, *127*, 211–219. [CrossRef]
23. Hu, X.; Kobayashi, S.; Ikuhara, Y.H.; Fisher, C.A.; Fujiwara, Y.; Hoshikawa, K.; Moriwake, H.; Kohama, K.; Iba, H.; Ikuhara, Y. Atomic scale imaging of structural variations in La(1−x)/3LixNbO3 (0 ≤ x ≤ 0.13) solid electrolytes. *Acta Mater.* **2017**, *123*, 167–176. [CrossRef]
24. Gao, X.; Fisher, C.A.; Ikuhara, Y.H.; Fujiwara, Y.; Kobayashi, S.; Moriwake, H.; Kuwabara, A.; Hoshikawa, K.; Kohama, K.; Iba, H. Cation ordering in A-site-deficient Li-ion conducting perovskites La(1−x)/3Li x NbO3. *J. Mater. Chem. A* **2015**, *3*, 3351–3359. [CrossRef]
25. Zhao, Y.; Daemen, L.L. Superionic conductivity in lithium-rich anti-perovskites. *J. Am. Chem. Soc.* **2012**, *134*, 15042–15047. [CrossRef] [PubMed]
26. Li, Y.; Zhou, W.; Xin, S.; Li, S.; Zhu, J.; Lü, X.; Cui, Z.; Jia, Q.; Zhou, J.; Zhao, Y. Fluorine-Doped Antiperovskite Electrolyte for All-Solid-State Lithium-Ion Batteries. *Angew. Chem. Int. Edit.* **2016**, *55*, 9965–9968. [CrossRef] [PubMed]
27. Hood, Z.D.; Wang, H.; Samuthira Pandian, A.; Keum, J.K.; Liang, C. Li2OHCl crystalline electrolyte for stable metallic lithium anodes. *J. Am. Chem. Soc.* **2016**, *138*, 1768–1771. [CrossRef] [PubMed]
28. Deviannapoorani, C.; Dhivya, L.; Ramakumar, S.; Murugan, R. Lithium ion transport properties of high conductive tellurium substituted Li7La3Zr2O12 cubic lithium garnets. *J. Power Sources* **2013**, *240*, 18–25. [CrossRef]
29. Ma, C.; Rangasamy, E.; Liang, C.; Sakamoto, J.; More, K.L.; Chi, M. Excellent Stability of a Lithium-Ion-Conducting Solid Electrolyte upon Reversible Li+/H+ Exchange in Aqueous Solutions. *Angew. Chem.* **2015**, *127*, 131–135. [CrossRef]
30. Kazyak, E.; Chen, K.-H.; Wood, K.N.; Davis, A.L.; Thompson, T.; Bielinski, A.R.; Sanchez, A.J.; Wang, X.; Wang, C.; Sakamoto, J. Atomic layer deposition of the solid electrolyte garnet Li7La3Zr2O12. *Chem. Mater.* **2017**, *29*, 3785–3792. [CrossRef]
31. Jiang, T.; He, P.; Wang, G.; Shen, Y.; Nan, C.-W.; Fan, L.-Z. Solvent-Free Synthesis of Thin, Flexible, Nonflammable Garnet-Based Composite Solid Electrolyte for All-Solid-State Lithium Batteries. *Adv. Energy Mater.* **2020**, *10*, 1903376. [CrossRef]
32. Duchardt, M.; Ruschewitz, U.; Adams, S.; Dehnen, S.; Roling, B. Vacancy-Controlled Na+ Superion Conduction in Na11Sn2PS12. *Angew. Chem. Int. Edit.* **2018**, *57*, 1351–1355. [CrossRef] [PubMed]
33. Ma, C.; Chen, K.; Liang, C.; Nan, C.-W.; Ishikawa, R.; More, K.; Chi, M. Atomic-scale origin of the large grain-boundary resistance in perovskite Li-ion-conducting solid electrolytes. *Energy Environ. Sci.* **2014**, *7*, 1638–1642. [CrossRef]
34. Ma, C.; Cheng, Y.; Yin, K.; Luo, J.; Sharafi, A.; Sakamoto, J.; Li, J.; More, K.L.; Dudney, N.J.; Chi, M. Interfacial Stability of Li Metal–Solid Electrolyte Elucidated via in Situ Electron Microscopy. *Nano Lett.* **2016**, *16*, 7030–7036. [CrossRef] [PubMed]
35. Liu, B.; Fu, K.; Gong, Y.; Yang, C.; Yao, Y.; Wang, Y.; Wang, C.; Kuang, Y.; Pastel, G.; Xie, H. Rapid Thermal Annealing of Cathode-Garnet Interface toward High-Temperature Solid State Batteries. *Nano Lett.* **2017**, *17*, 4917–4923. [CrossRef]
36. Dawson, J.A.; Canepa, P.; Famprikis, T.; Masquelier, C.; Islam, M.S. Atomic-Scale Influence of Grain Boundaries on Li-ion Conduction in Solid Electrolytes for All-Solid-State Batteries. *J. Am. Chem. Soc.* **2018**, *140*, 362–368. [CrossRef]

37. Fujiwara, Y.; Hoshikawa, K.; Kohama, K. Growth of solid electrolyte LixLa(1−x)/3NbO3 single crystals by the directional solidification method. *J. Cryst. Growth* **2016**, *433*, 48–53. [CrossRef]
38. Saitoh, M.; Gao, X.; Ogawa, T.; Ikuhara, Y.H.; Kobayashi, S.; Fisher, C.A.J.; Kuwabara, A.; Ikuhara, Y. Systematic analysis of electron energy-loss near-edge structures in Li-ion battery materials. *Phys. Chem. Chem. Phys.* **2018**, *20*, 25052–25061. [CrossRef]
39. Thangadurai, V.; Weppner, W. Recent progress in solid oxide and lithium ion conducting electrolytes research. *Ionics* **2006**, *12*, 81–92. [CrossRef]

Review

Insights into Layered Oxide Cathodes for Rechargeable Batteries

Julia H. Yang [1], Haegyeom Kim [2] and Gerbrand Ceder [1,2,*]

1. Department of Materials Science and Engineering, UC Berkeley, Berkeley, CA 94720, USA; juliayang@berkeley.edu
2. Materials Sciences Division, Lawrence Berkeley National Laboratory, Berkeley, CA 94720, USA; haegyumkim@lbl.gov
* Correspondence: gceder@berkeley.edu

Abstract: Layered intercalation compounds are the dominant cathode materials for rechargeable Li-ion batteries. In this article we summarize in a pedagogical way our work in understanding how the structure's topology, electronic structure, and chemistry interact to determine its electrochemical performance. We discuss how alkali–alkali interactions within the Li layer influence the voltage profile, the role of the transition metal electronic structure in dictating O3-structural stability, and the mechanism for alkali diffusion. We then briefly delve into emerging, next-generation Li-ion cathodes that move beyond layered intercalation hosts by discussing disordered rocksalt Li-excess structures, a class of materials which may be essential in circumventing impending resource limitations in our era of clean energy technology.

Keywords: layered oxide cathodes; alkali–alkali interactions; electronic structure; Li diffusion

Citation: Yang, J.H.; Kim, H.; Ceder, G. Insights into Layered Oxide Cathodes for Rechargeable Batteries. *Molecules* 2021, 26, 3173. https://doi.org/10.3390/molecules26113173

Academic Editors: Stephane Jobic, Claude Delmas and Myung-Hwan Whangbo

Received: 2 May 2021
Accepted: 24 May 2021
Published: 26 May 2021

Publisher's Note: MDPI stays neutral with regard to jurisdictional claims in published maps and institutional affiliations.

Copyright: © 2021 by the authors. Licensee MDPI, Basel, Switzerland. This article is an open access article distributed under the terms and conditions of the Creative Commons Attribution (CC BY) license (https://creativecommons.org/licenses/by/4.0/).

1. Introduction to the O3 Structure

Rechargeable Li-ion batteries have enabled a wireless revolution and are currently the dominant technology used to power electric vehicles and provide resilience to a grid powered by renewables. Research in the 1970s to create superconductors by modifying the carrier density of chalcogenides through intercalation [1] transitioned into energy storage when Whittingham demonstrated in 1976 a rechargeable battery using the layered TiS_2 cathode and Li metal anode [2]. Soon thereafter, Mizushima, Jones, Wiseman, and Goodenough demonstrated that a much higher voltage could be achieved by reversible Li de-intercalation from layered $LiCoO_2$ [3], energizing generations of rechargeable battery research. In this short review we revisit our work in understanding a few basic relationships between the structure, electronic structure, and properties of layered cathode materials.

A layered rocksalt cathode oxide adopts the general formula A_xMO_2 (A: alkali cation, M: metal cation, O: oxygen anion). The O anions form a face-centered cubic (FCC) framework with octahedral and tetrahedral sites. These two environments are face sharing and form a topologically connected network. When fully alkaliated such that x ~ 1, the compound consists of AO_2 and MO_2 edge-sharing octahedra. The layered structure, illustrated in Figure 1, is aptly named because AO_2/MO_2 octahedra form alternating (111) planes of the FCC oxygen lattice when fully lithiated. The A and M cations alternate in the abc repeat unit of the oxygen framework to form a−b_c−a_b−c_, stacking where the minus sign "−" indicates the location of M and the underscore "_" gives the position of the A ions. Because the oxygen stacking has a repeat unit of three and the metal layering repeats every two layers, periodicity is achieved after six oxygen layers. Under the structural classification by Delmas et al. for layered cathode oxides [4], the layered rocksalt cathode structure is commonly referred to as O3: O for the octahedral alkali ion environment (not to be confused with O for oxygen) and 3 for the number of MO_2 slabs in a repeat unit. The

O3 structure is equivalent to the structure of α-NaFeO$_2$ and the cation ordering is also known in metallic alloys as L1$_1$ (CuPt prototype) [5].

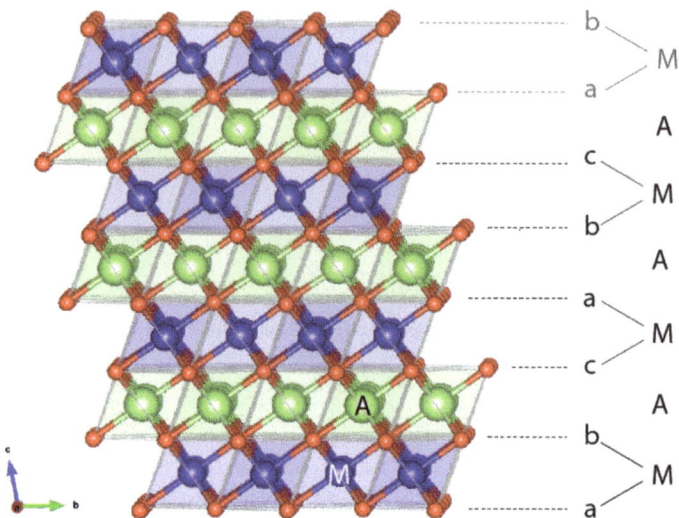

Figure 1. Representative O3 structure showing the abc stacking sequence of oxygen ions (red), thus creating various coordination environments for the alkali ion A (green) and metal ion M (blue). In an O3 repeat unit, M and A are coordinated below and above by oxygen layers. The beginning of another repeat unit with a-M-b stacking is in gray.

It is now well understood that the ordering of AO$_2$ and MO$_2$ in alternating layers is not the most favored cation ordering from an electrostatic perspective. Instead, the layered structure finds its stability in the size difference of A and M [6,7] as it allows A-O and M-O bond distances to relax independently of each other. This independent A-O and M-O bond accommodation explains how a larger A cation, such as Na$^+$, can form the layered structure with a wide range of M radii [8], whereas a smaller A, such as Li$^+$, only forms stable O3 compounds with a limited range of smaller M radii, namely Co^{3+} [9], V^{3+} [10], Ni^{3+} [11], and Cr^{3+} [10].

2. Evolution from LiCoO$_2$ to NMC

Today, O3 cathodes have evolved in several directions from LiCoO$_2$ [12] for use cases beyond portable electronics. Anticipating potential cost and resource problems with Co [13], research in the 1980s and 1990s mostly focused on substitutions of Co by Ni [14]. However, consideration of the low cost of Mn and the high stability of the Mn^{4+} charge state led the community towards layered LiMnO$_2$. Even though this structure is not the thermodynamically stable state of LiMnO$_2$ [15], Delmas [16] and Bruce [17] were able to synthesize it by ion exchange from the stable NaMnO$_2$. Unfortunately, the high mobility of Mn^{3+} [18] leads to a rapid transformation of the layered structure into the spinel structure upon cycling [19] because of its pronounced energetic preference at the Li$_{0.5}$MnO$_2$ composition [20]. Attempts to stabilize layered LiMnO$_2$ with Al [21] or Cr [22,23] substitution were only partially successful and led to the formation of a phase intermediate between layered and spinel [24]. Then, in 2001, several key papers were published that would pave the way for the highly successful Ni-Mn-Co (NMC) cathode series: Ohzuku showed very high capacity and cyclability in Li(Ni$_{1/3}$Mn$_{1/3}$Co$_{1/3}$)O$_2$ [25], known as NMC-111, and in Li(Ni$_{1/2}$Mn$_{1/2}$)O$_2$ [26]; Lu and Dahn published their work on the Li(Ni$_x$Co$_{1-2x}$Mn$_x$)O$_2$ [27] and its Co-free Li-excess version Li(Ni$_x$Li$_{1/3-2/x}$Mn$_{2/3-x/3}$)O$_2$ [28]. In these compounds Ni is valence +2 and Mn is +4 [29], thereby stabilizing the layered material against Mn migration and providing double redox from Ni^{2+}/Ni^{4+}. At this point the NMC cathode series was born. Since then, Ni-rich NMC cathodes have become of great interest to both academia and industry because they deliver a capacity approaching 200 mAh/g and demonstrate high energy density, good rate capability, and moderate cost [30–32].

In this short article, we summarize some general and fundamental understanding we have gained in layered oxide cathodes, without delving into issues with very specific compositions. We focus on the roles of the alkali–alkali interaction, electronic structure, and alkali diffusion, and illustrate how these fundamental features conspire to control the electrochemical behavior of O3-structured layered oxides.

3. Alkali–Alkali Interactions, Alkali/Vacancy Ordering, and Voltage Slope

The voltage of a cathode compound is set by the chemical potential of its alkali ions [33] which itself is the derivative of the free energy with respect to alkali concentration. This thermodynamic connection between voltage and free energy creates a direct relation between the voltage profile, the alkali–alkali interactions, and phase transformations as functions of alkali content. While Na_xMO_2 compounds show many changes in the stacking of the oxygen host layers when the Na content is changed, phase transitions in Li_xMO_2 materials are mostly driven by the Li-vacancy configurational free energy, resulting from Li^+-Li^+ interactions in the layer [20,34]. In layered compounds with a single transition metal, such as Li_xCoO_2 and Li_xNiO_2, such phase transitions are easily observed as voltage plateaus and steps in the electrochemical charge–discharge profiles as shown in Figure 2a. For a first-order phase transformation, for example from Phase I to Phase II, the Gibbs phase rule dictates that the Li chemical potential should be constant, hence the voltage remains constant while one phase transforms into the other. Phases in which the alkali ions are well-ordered usually display a rapid voltage change as the alkali content is changed, reflecting the high energy cost of trying to create off-stoichiometry in ordered phases. This is in contrast to solid solutions which have smoother voltage profiles as a function of alkali concentration. For example, both theory [35] and experiments [36,37] indicate that in Li_xCoO_2 a monoclinic phase appears with lithium and vacancies ordered in rows for $x \approx 0.5$ [36]. In Li_xNiO_2, Li-vacancy ordering is responsible for stable phases at x ~ 0.8, ~0.5, and ~0.25–0.3 [38–40]. When many transition metals are mixed, as in NMC cathodes, the Li^+-Li^+ interaction remains present, but Li-vacancy ordering is suppressed by the electrostatic and elastic perturbations on the Li site caused by the distribution of the Ni, Mn and Co in the transition metal layers.

The Li^+-Li^+ interaction is mostly electrostatic but is highly screened by the charge density on the oxygen ions, leading to a rather small effective interaction in layered Li_xMO_2 compounds and small voltage slope. This is a critical feature of Li_xMO_2 compounds that gives them high capacity in a relatively narrow voltage window compared to other alkali compounds, as explained below. The effective interaction between intercalating ions increases significantly when larger alkali ions (e.g., Na^+ and K^+) are used in the layered structure [41–43]. These larger alkali ions increase the oxygen slab distance, reducing the oxygen charge density available for screening within the alkali layer [20,42,43]. The larger effective repulsion between the Na^+ or K^+ ions affects the phase transition and electrochemistry in a very significant way as shown in Figure 2b. For example, Na_xCoO_2 has stronger Na-vacancy ordering and thus more pronounced voltage steps compared to Li_xCoO_2. This phenomenon [44,45] becomes even more significant in K_xCoO_2 [46,47]. The effect of the intercalant's size on the phase transitions and voltage steps is not just limited to Co-containing compounds but is also generally applicable to other transition metal systems as described in a recent review [43].

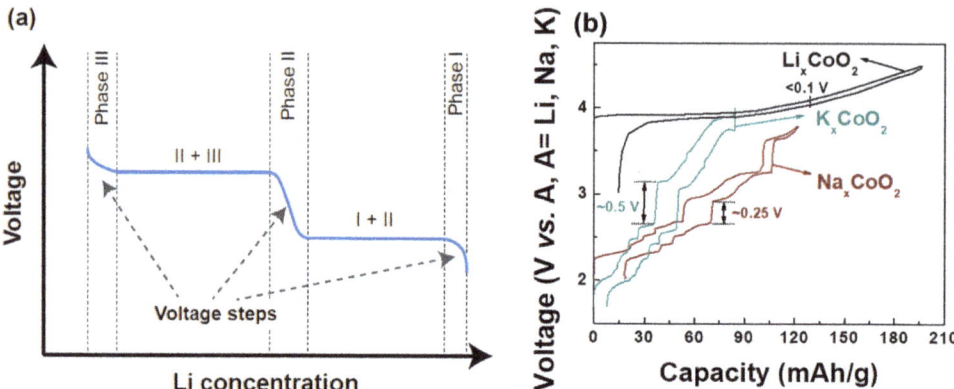

Figure 2. (a) Typical charge–discharge of intercalation-based cathode materials. A voltage step indicates new phase formation. (b) Charge–discharge comparison of O3-Li$_x$CoO$_2$, P2-Na$_x$CoO$_2$, and P2-K$_x$CoO$_2$ [44,47,48]. Voltage curves for P2-Na$_x$CoO$_2$, P2-K$_x$CoO$_2$, and O3-Li$_x$CoO$_2$ are reproduced with permissions from [44,47,48]. The voltage curve for P2-K$_x$CoO$_2$ is licensed under CC BY-NC 4.0 [47].

In practice, the larger effective interaction between alkali ions in the layered transition metal oxides is detrimental for their electrochemical performance. First, more phase transitions are likely to induce more mechanical stress in the cathode structure during charging and discharging, causing possible fracture of electrode particles. Second, a simple argument shows that the average voltage slope is proportional to the effective interaction: V(x) is equal to $-\mu_{Li}(x)$, and since $\mu_{Li}(x) = \frac{\partial G}{\partial x}$, $\frac{\partial V}{\partial x} = -\frac{\partial^2 G}{\partial x^2}$. In a simple regular solution model for mixing, this second derivative of the free energy is proportional to the effective interaction [41,43,49]. Hence, when the effective interaction is large, as in layered K$_x$MO$_2$ compounds, the voltage curve has a high slope, limiting the achievable capacity between fixed voltage limits. This analysis shows that the advantage of lithium systems in providing large capacity within reasonable voltage limits is in part due to the highly effective screening of the Li$^+$-Li$^+$ interaction by oxygen. For Na, and in particular for K-ion based intercalation energy storage, it may be more advantageous to search among poly-anion compounds for good cathodes [43].

4. Electronic Structure of LiCoO$_2$

The electronic structure of layered LiMO$_2$ oxides is well understood. Due to the large energy difference in electronic levels between Li and the transition metal (TM), their electronic states do not mix and the behavior of the compound is controlled by the (MO$_2$) complex within which the transition metal and oxygen hybridize. In the R3-m symmetry of the layered structure, the environment of the TM is pseudo-octahedral in that all TM-O bond lengths are of equal length, but O-TM-O angles have small deviations from those in a perfect octahedron. The pseudo-octahedral symmetry splits the otherwise degenerate TM-d orbitals in three (lower energy) t_{2g} and two (higher energy) e_g orbitals yielding an energy separation called the octahedral ligand-field splitting, abbreviated as Δ_0. A more complete schematic of an orbital diagram is given in Figure 3a. The t_{2g} orbitals are shown as "non-bonding" in this schematic though in reality some π-hybridization takes place between them and the oxygen p-orbitals [50]. In the most basic picture in which one considers the overlap of the TM-d orbitals with its ligand p-states, the t_{2g} orbitals are formed from the d_{xy}-type d-orbitals which point away from ligands. In contrast, the d_{z^2} and $d_{x^2-y^2}$ orbitals of the TM point toward the ligand creating σ-overlap. The e_g* orbitals are the anti-bonding component of this hybridization and are dominated by TM states, whereas the bonding components, e_g^b, sit deep in the oxygen-dominated part of the band structure. Hybridization of the oxygen 2p and the metal 4d and 4s make up the remaining part of the band structure. Because the e_g^* orbitals result from σ-overlap between TM d states and oxygen p-states their energy is most sensitive to the TM-O bond length. Inducing, for example, a Jahn–Teller distortion moves these levels considerably. One can recover the characteristics of this molecular orbital diagram in a more realistic band structure and density of states computed with Density Functional Theory using the meta-GGA SCAN density functional approximation [51] as shown in Figure 3b for $LiCoO_2$. The elemental contributions are indicated by the color of the bands with green being oxygen and red the Co $3d$ states. The two e_g^*-like bands above the Fermi level (solid line) have mixed Co and O contribution while the three t_{2g}-like bands below the Fermi level have more pure metal contribution. These three bands have a small bandwidth due to their non-bonding nature. The lowest six bands shown are the bands dominated by the oxygen states in green.

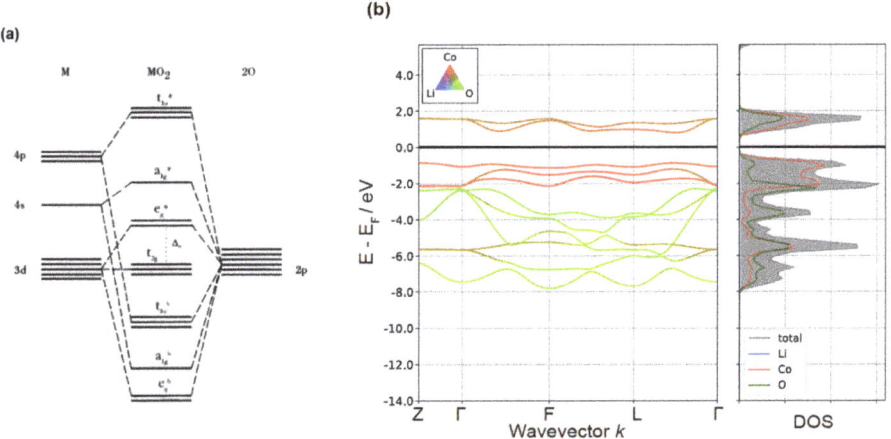

Figure 3. (a) The molecular orbital diagram for a $(MO_2)^-$ complex in octahedral environment. Reproduced with permission from [52]. (b) Calculated band structure and density of states with projections onto local orbitals for $LiCoO_2$, showing elemental contributions from Li (blue), Co (red), and O (green).

5. Electronic Structure Trends in Layered Li_xMO_2 Oxides

Because the ligand-induced splitting between transition metal d orbitals is fairly small, filling of states usually follows Hund's rule for creating high spin ions. Figure 4 illustrates this filling for $3d$ octahedral transition metal ions. Examples of this are Fe^{3+} (d^5) and Mn^{4+} (d^3). The high spin band filling implies that Mn^{3+} (d^4) and Fe^{4+} (d^4) with a single occupied e_g^* state are Jahn–Teller active ions [50,53]. The later transition metals form exceptions to the high-spin rule in that Co^{3+} and Ni^{4+} are low-spin d^6 with all electrons occupying t_{2g} states. We discuss below that this is a key reason for their predominance as redox-active materials in layered oxides. The lack of any filled antibonding states in Co^{3+} makes this cation also one of the smallest $3d$ TM ions, which is reflected in the very high crystal density of $LiCoO_2$ of 5.051 g/cm^3 [54] and the associated high energy density. This electronic structure-induced high density makes $LiCoO_2$ still the preferred cathode material for portable electronics where battery volume comes at a high premium.

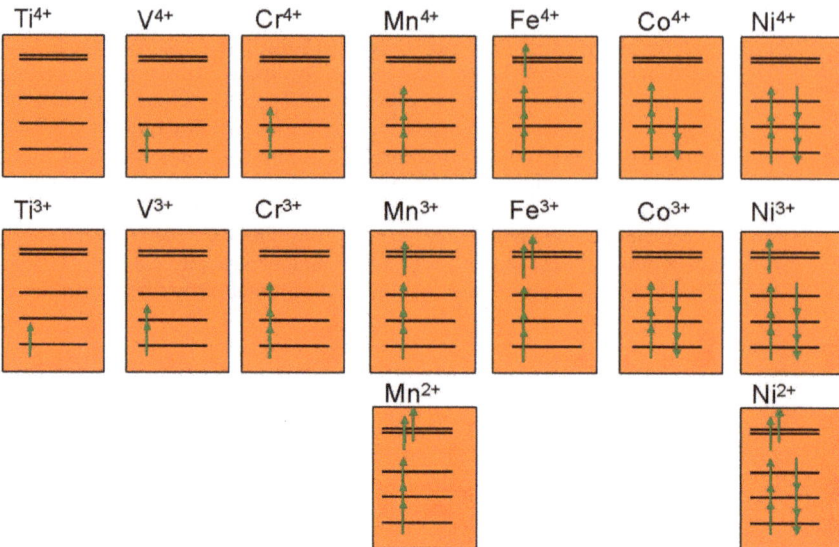

Figure 4. Expected transition metal band filling in the three t_{2g} and two e_g^* states for $3d$ transition metals in octahedral environments.

Upon Li removal the hybridization of the orbitals changes. As an electron is removed from TM states the remaining TM d electrons experience less intra-atomic Coulombic repulsion and their states move down in energy, bringing them closer to the oxygen p states, resulting in increased hybridization. There are two notable consequences from this rehybridization [51,55]: (1) As hybridization transfers (filled) oxygen states onto the metal it increases the electron density on the metal site. Somewhat counterintuitively, almost no electron density change occurs on the metal after delithiation, even though it is formally oxidized, something that had been recognized earlier outside of the battery field in various Mn-oxides. This rehybridization by the anion explains why the anion has an almost larger influence on the voltage than the choice of $3d$ TM ion in layered compounds [52]. Effectively, the flexible hybridization between the TM and the O ligand creates a charge density buffer on the TM. (2) Covalency increases upon charging, leading in some cases to the fully charged material taking on the O1 (octahedral alkali environment with a repeat unit of 1 [4]) structure which is typically found in more covalent materials such as CdI_2. The increased covalency also sharply contracts the Li slab spacing (distance between the oxygen layers around the Li-layer) and c-lattice parameter when most of the alkali is removed [56].

6. Implications of Electronic Structure on Layered Stability

The orbital filling of the transition metals also plays a critical role in the stability of the layered structure upon Li removal. Ions with filled t_{2g} levels are most stable in the octahedral environment and resist any migration into the Li layer [57]. Because the oxygen arrangement is topologically equivalent to an FCC lattice, octahedral cation sites edge-share with each other and face-share with a tetrahedral site. Since ion migration through a shared edge comes with a very high energy barrier, cation diffusion between octahedral sites requires passage through an intermediate tetrahedral site. For ions with filled t_{2g} states, this passage through the tetrahedral site and the shared anion face raises the energy substantially as the octahedral ligand field stabilization is lost, making these ions all but immobile. In contrast, ions with d^5-high-spin (e.g., Mn^{2+} or Fe^{3+}) and d^0 filling (e.g., Ti^{4+}, V^{5+}) tend to be much more ambiguous about their preferred anion coordination, and as a result, tend to migrate more easily [57–59]. The most stable octahedral cations are therefore low-spin-Co^{3+} (d^6) and low-spin-Ni^{4+} (d^6). In addition, high-spin-Mn^{4+} (d^3) also possesses very high octahedral stability as it adds Hund's rule coupling to the ligand field stabilization. These insights allow us to rationalize the prominence of the NMC class of layered oxides as cathodes in the Li-ion industry: Only Ni and Co have very high resistance against migration into the Li layer in the charged and discharged states. Mn^{4+} acts similarly, but cannot be used as a redox active element as its reduction or oxidation leads to an ion that is prone to migration [57,60,61]. Within the 3d-TM series there are unfortunately no other ions which can match the octahedral stability of the NMC chemistry, and layered oxides based on other 3d TM are unlikely to be practical. Hence, it is the basic electronic structure of the 3d transition metal ions which is the direct cause of the serious resource problem the Li-ion industry faces if it wants to scale to multiple TWh annual production with layered oxides [13,62].

7. Diffusion Mechanism

In understanding alkali transport in layered compounds, and more generally in closed-packed oxides, it is important to assess how structure and chemistry influence performance, and ultimately, how one can design novel dense cathodes that are not layered. In this section, we focus on Li diffusion. Even though it is likely that Na and K migrate through a similar pathway, much less work has been done to validate the transport mechanism of these larger alkalis. The octahedral-tetrahedral-octahedral topology introduced earlier determines the diffusion mechanism in layered materials. Lithium migrates along the minimum energy path between two stable sites via an activated state, with the activation barrier defined as the difference between the maximum energy point along the path and the initial equilibrium position of the ion [63]. In layered Li_xCoO_2 [35], migration between neighboring octahedral sites can in principle occur through the shared octahedral edge formed by an oxygen–oxygen dumbbell (a mechanism referred to as an octahedral dumbbell hop (ODH)), or via the tetrahedral site that faces-shares with the initial and final octahedron (a tetrahedral site hop (TSH)). Figure 5 illustrates the ODH and TSH mechanisms in Li_xCoO_2.

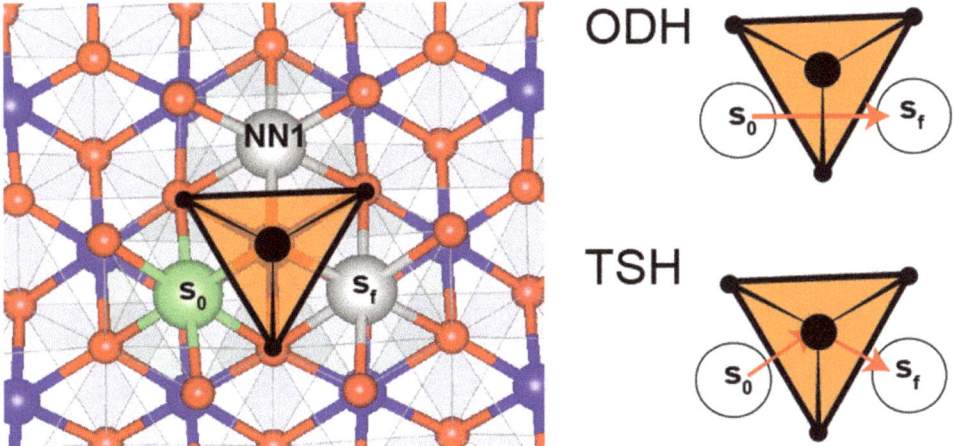

Figure 5. Li diffusion from site s_0 to s_f in layered materials. View of Li_xCoO_2 down the c-axis, with the top CoO_2 metal layer removed for clarity. The Li (green) layer and CoO_2 metal layer (Co: blue, O: red) below are shown. Black circles further indicate the O which coordinate the tetrahedral site colored in orange, showing how three of the tetrahedron faces are face-sharing with s_0, NN1, and s_f. The last face of the tetrahedron face-shares with Co in the metal layer below. Li in site s_0 can either diffuse from octahedral s_0 to octahedral s_f via the edge-sharing connection, thus completing an octahedral dumbbell hop (ODH) illustrated by the single red arrow, or through the empty tetrahedral site via the tetrahedral site hop (TSH) mechanism illustrated by the two red arrows. Site NN1 (white) is vacant.

A TSH requires the presence of a divacancy in the Li layer, which in Figure 5, implies that both s_f and NN1 need to be vacant. Van der Ven et al. used ab initio calculations to show that the ODH mechanism has a considerably higher activation barrier (~800 meV) than a TSH mechanism (230–600 meV) [63], establishing that Li diffusion in Li_xCoO_2 occurs predominantly by way of TSH for all practical lithiation levels. Even though the Li^+ in the activated state in the tetrahedron faces repulsion from a face-sharing $Co^{3+/4+}$ ion, the large Li slab spacing keeps the distance between Co and Li reasonable (Supplementary Materials).

The barrier for the TSH mechanism can vary from 230 to 600 meV due to local environment changes during lithiation. At x ~ 0.5, Li diffusion dips due to the ordering reaction [35] in agreement with experiment [36,64,65]. When a larger amount of Li (0.5 > x in Li_xCoO_2) is removed from the compound, a large decrease in the c-lattice parameter is observed experimentally and from first-principles [66–68]. Such lattice contraction increases the activation barrier significantly because it creates a smaller tetrahedron height and shorter distance between the activated Li^+ and the $Co^{3+/4+}$ ion. The larger positive charge on Co when the compound is more oxidized also contributes to an increase of the energy in the activated state. As a result, the activation barrier increases by hundreds of meV when delithiation increases past 0.5 Li [69]. While the precise behavior of the c-lattice parameter and slab spacing in NMC materials depends on the specific chemistry, the overall behavior is similar to $LiCoO_2$.

8. Beyond Layered Materials: DRX

It is now understood that the Li migration mechanism in layered oxides is a specific case of a more general framework for understanding ion transport in FCC close-packed oxides. Recent work [70,71] categorized the different environments that can occur around the tetrahedral activated state in close-packed oxides by the number of face-sharing transition metals it has. So-called nTM channels have n transition metals face-sharing, with the other face-sharing octahedral sites either occupied by Li or vacancies. Because minimally two Li (or vacant sites) are required to create a migration path, 4TM and 3TM channels do not participate in diffusion. Structures with only 2TM channels exist but display very poor Li mobility due to the large electrostatic repulsion Li^+ sees in the activated state from the two TM ions with which it face-shares. This theory explains why ordered γ-$LiFeO_2$, a compound with only 2TM channels, is not electrochemically active. Layered oxides contain 3TM and 1TM channels with Li diffusion occurring through the 1TM channels. However, the proximity of the TM to the Li^+ in a 1TM channel creates a strong dependence of the migration barrier on the size of the tetrahedron, which in layered materials is determined by the slab spacing. As shown extensively by Kang et al. [72,73], even small contractions of the slab spacing, caused by TM mixing in the Li layer, reduce Li mobility in a very substantial way. The "safest" migration paths are 0-TM channels: In the presence of a divacancy, a migrating Li^+ only electrostatically interacts with one other Li^+ ion making the activation energy rather insensitive to dimensional changes. Recent work has shown how to create cation-disordered materials in which transport occurs through these 0-TM channels [74]: When cations are fully disordered over the octahedral sites of a structure with FCC anion packing, all possible configurations around the activated state occur with some statistical probability, and percolation of the 0-TM channels into a macroscopic diffusion path occurs when more than 9% Li excess is present (i.e., x > 0.09 in $Li_{1+x}M_{1-x}O_2$). In reality, cation short-range order tends to reduce the amount of 0-TM channels from what would be in a random system [75] and a higher Li-excess content is needed or high-entropy ideas have to be applied to minimize short-range order [76]. Based on these insights, Li-excess rocksalt oxides with a disordered cation distribution have been shown to function as intercalation cathodes with high capacity [77]. Disordered Rocksalt Li-eXcess cathodes (DRX), also referred to as DRS (Disordered RockSalt) by some [78], do not require any specific ordering of the metal and Li cations, and can therefore be used with a broad range of redox active and non-redox active metals, alleviating the resource issue arising with

NMC layered cathodes. Indeed, almost all $3d$ and several $4d$ TM have been used as redox couples, including V^{4+}/V^{5+} and partial V^{3+}/V^{5+} [79], Mn^{2+}/Mn^{4+} [80], Mo^{3+}/Mo^{6+} [81], and partial Fe^{3+}/Fe^{4+} [82,83]. Redox-inactive d^0 TM elements, such as Ti^{4+}, V^{5+}, Nb^{5+}, and Mo^{6+} [77], play a particular role in DRX as they stabilize disorder [84], and their high valence compensates for excess Li content. Fluorination (anion substitution) is possible in DRX compounds which lowers the cation valence and extends cycle life by reducing oxygen redox [85]. Promising specific energies approaching 1000 Wh/kg (cathode only) have been achieved with Mn-Ti-based materials offering a possible low-cost, high-energy cathode solution for Li-ion.

9. Conclusions

The pioneering work of Professor Goodenough on $LiCoO_2$ [3] has led to a rich and widely used class of layered cathodes thereby transforming Li-ion into the leading energy storage technology for electronics, vehicles, and the grid. In this review, we discussed the topology of the layered structure and explain how the structure (1) sets the voltage slope trends among various alkali ions, (2) is critically limited to certain transition metals due to their electronic structure, and (3) controls the alkali diffusion mechanism. A 20-year effort to understand the phase stability, transport, and electronic structure in these compounds can now be broadened towards new high-energy density cathode materials, ensuring the future of Li-ion as an important contribution to clean energy technology.

Supplementary Materials: The following are available online. The supplemental information contains band structure calculation details.

Author Contributions: Writing—original draft preparation, J.H.Y., H.K., and G.C.; writing—reviewing and editing, J.H.Y., H.K. and G.C.; DFT-SCAN calculations, J.H.Y. and G.C. All authors have read and agreed to the published version of the manuscript.

Funding: This research received no additional funding.

Institutional Review Board Statement: Not applicable.

Informed Consent Statement: Not applicable.

Data Availability Statement: The $LiCoO_2$ structure used for calculating the band structure is freely available via the Materials Project database at https://materialsproject.org (accessed on 26 May 2021).

Acknowledgments: G.C. acknowledges long-term support from the Department of Energy, Office of Basic Energy Sciences, for the development of fundamental understanding of Li-intercalation cathodes. More recently, work on diffusion and on the development of DRX cathodes was supported by the Assistant Secretary for Energy Efficiency and Renewable Energy, Vehicle Technologies Office, under the Applied Battery Materials Program, of the U.S. Department of Energy under Contract No. DE-AC02-05CH11231, and under Contract No. DE-AC02-05CH11231, under the Advanced Battery Materials Research (BMR) Program.

Conflicts of Interest: The authors declare no conflict of interest.

References

1. Gamble, F.R.; Osiecki, J.H.; Cais, M.; Pisharody, R.; Disalvo, F.J.; Geballe, T.H. Intercalation complexes of lewis bases and layered sulfides: A large class of new superconductors. *Science* **1971**, *174*, 493–497. [CrossRef]
2. Whittingham, M.S. Electrical Energy Storage and Intercalation Chemistry. *Science* **1976**, *192*, 1126–1127. [CrossRef]
3. Mizushima, K.; Jones, P.C.; Wiseman, P.J.; Goodenough, J.B. LixCoO2 (0 < x < −1): A new cathode material for batteries of high energy density. *Mater. Res. Bull.* **1980**, *15*, 783–789. [CrossRef]
4. Delmas, C.; Fouassier, C.; Hagenmuller, P. Structural classification and properties of the layered oxides. *Physica B+C* **1980**, *99*, 81–85. [CrossRef]
5. Ceder, G.; Van der Ven, A.; Marianetti, C.; Morgan, D. First-principles alloy theory in oxides. *Model. Simul. Mater. Sci. Eng.* **2000**, *8*, 311–321. [CrossRef]
6. Wu, E.J.; Tepesch, P.D.; Ceder, G. Size and charge effects on the structural stability of $LiMO_2$ (M = transition metal) compounds. *Philos. Mag. B* **1998**, *77*, 1039–1047. [CrossRef]

7. Hewston, T.A.; Chamberland, B.L. A Survey of first-row ternary oxides LiMO$_2$ (M = Sc-Cu). *J. Phys. Chem. Solids* **1987**, *48*, 97–108. [CrossRef]
8. Kim, S.; Ma, X.; Ong, S.P.; Ceder, G. A comparison of destabilization mechanisms of the layered Na$_x$MO$_2$ and Li$_x$MO$_2$ compounds upon alkali de-intercalation. *Phys. Chem. Chem. Phys.* **2012**, *14*, 15571–15578. [CrossRef] [PubMed]
9. Orman, H.J.; Wiseman, P.J. Cobalt(III) lithium oxide, CoLiO$_2$: Structure refinement by powder neutron diffraction. *Acta Crystallogr. Sect. C* **1984**, *40*, 12–14. [CrossRef]
10. Rüdorff, W.; Becker, H. Notizen: Die Strukturen von LiVO$_2$, NaVO$_2$, LiCrO$_2$ und NaCrO$_2$. *Z. Nat. B* **1954**, *9*, 614–615. [CrossRef]
11. Dyer, L.D.; Borie, B.S.; Smith, G.P. Alkali Metal-Nickel Oxides of the Type MNiO$_2$. *J. Am. Chem. Soc.* **1954**, *76*, 1499–1503. [CrossRef]
12. Manthiram, A.; Goodenough, J.B. Layered lithium cobalt oxide cathodes. *Nat. Energy* **2021**, *6*, 323. [CrossRef]
13. Fu, X.; Beatty, D.N.; Gaustad, G.G.; Ceder, G.; Roth, R.; Kirchain, R.E.; Olivetti, E.A. Perspectives on Cobalt Supply through 2030 in the Face of Changing Demand. *Environ. Sci. Technol.* **2020**, *54*, 2985–2993. [CrossRef] [PubMed]
14. Delmas, C.; Saadoune, I. Electrochemical and physical properties of the Li$_x$Ni$_{1-y}$Co$_y$O$_2$ phases. *Solid State Ion.* **1992**, *53–56*, 370–375. [CrossRef]
15. Ceder, G. The Stability of Orthorhombic and Monoclinic-Layered LiMnO$_2$. *Electrochem. Solid-State Lett.* **1999**, *2*, 550. [CrossRef]
16. Capitaine, F.; Gravereau, P.; Delmas, C. A new variety of LiMnO$_2$ with a layered structure. *Solid State Ion.* **1996**, *89*, 197–202. [CrossRef]
17. Armstrong, A.R.; Bruce, P.G. Synthesis of layered LiMnO$_2$ as an electrode for rechargeable lithium batteries. *Nature* **1996**, *381*, 499–500. [CrossRef]
18. Reed, J.; Ceder, G.; Van Der Ven, A. Layered-to-Spinel Phase Transition in Li$_x$MnO$_2$. *Electrochem. Solid-State Lett.* **2001**, *4*, A78. [CrossRef]
19. Shao-Horn, Y.; Hackney, S.A.; Armstrong, A.R.; Bruce, P.G.; Gitzendanner, R.; Johnson, C.S.; Thackeray, M.M. Structural Characterization of Layered LiMnO$_2$ Electrodes by Electron Diffraction and Lattice Imaging. *J. Electrochem. Soc.* **1999**, *146*, 2404–2412. [CrossRef]
20. Ceder, G.; Van der Ven, A. Phase diagrams of lithium transition metal oxides: Investigations from first principles. *Electrochim. Acta* **1999**, *45*, 131–150. [CrossRef]
21. Jang, Y.; Huang, B.; Wang, H.; Sadoway, D.R.; Ceder, G.; Chiang, Y.; Liu, H.; Tamura, H. LiAl$_y$Co$_{1-y}$O$_2$ (R-3m) Intercalation Cathode for Rechargeable Lithium Batteries. *J. Electrochem. Soc.* **1999**, *146*, 862–868. [CrossRef]
22. Dahn, J.R.; Zheng, T.; Thomas, C.L. Structure and Electrochemistry of Li$_2$Cr$_x$Mn$_{2-x}$O$_4$ for $1.0 \leq x \leq 1.5$. *J. Electrochem. Soc.* **1998**, *145*, 851–859. [CrossRef]
23. Davidson, I.J.; McMillan, R.S.; Murray, J.J.; Greedan, J.E. Lithium-ion cell based on orthorhombic LiMnO$_2$. *J. Power Sources* **1995**, *54*, 232–235. [CrossRef]
24. Reed, J.S. Ab-Initio Study of Cathode Materials for Lithium Batteries. Ph.D. Thesis, Massachusetts Institute of Technology, Cambridge, MA, USA, 2003.
25. Ohzuku, T.; Makimura, Y. Layered Lithium Insertion Material of LiCo$_{1/3}$Ni$_{1/3}$Mn$_{1/3}$O$_2$ for Lithium-Ion Batteries. *Chem. Lett.* **2001**, *30*, 642–643. [CrossRef]
26. Ohzuku, T.; Makimura, Y. Layered Lithium Insertion Material of LiNi$_{1/2}$Mn$_{1/2}$O$_2$: A Possible Alternative to LiCoO$_2$ for Advanced Lithium-Ion Batteries. *Chem. Lett.* **2001**, *30*, 744–745. [CrossRef]
27. Lu, Z.; MacNeil, D.D.; Dahn, J.R. Layered Li[Ni$_x$Co$_{1-2x}$Mn$_x$]O$_2$ Cathode Materials for Lithium-Ion Batteries. *Electrochem. Solid-State Lett.* **2001**, *4*, A200. [CrossRef]
28. Lu, Z.; MacNeil, D.D.; Dahn, J.R. Layered Cathode Materials Li[Ni$_x$Li$_{1/3-2x/3}$Mn$_{2/3-x/3}$]O$_2$ for Lithium-Ion Batteries. *Electrochem. Solid-State Lett.* **2001**, *4*, A191. [CrossRef]
29. Reed, J.; Ceder, G. Charge, Potential, and Phase Stability of Layered Li(Ni$_{0.5}$Mn$_{0.5}$)O$_2$. *Electrochem. Solid-State Lett.* **2002**, *5*, A145. [CrossRef]
30. Manthiram, A.; Song, B.; Li, W. A perspective on nickel-rich layered oxide cathodes for lithium-ion batteries. *Energy Storage Mater.* **2017**, *6*, 125–139. [CrossRef]
31. Kim, J.; Lee, H.; Cha, H.; Yoon, M.; Park, M.; Cho, J. Prospect and Reality of Ni-Rich Cathode for Commercialization. *Adv. Energy Mater.* **2018**, *8*, 1702028. [CrossRef]
32. Zhu, L.; Bao, C.; Xie, L.; Yang, X.; Cao, X. Review of synthesis and structural optimization of LiNi$_{1/3}$Co$_{1/3}$Mn$_{1/3}$O$_2$ cathode materials for lithium-ion batteries applications. *J. Alloys Compd.* **2020**, *831*, 154864. [CrossRef]
33. Aydinol, M.K.; Ceder, G. First-Principles Prediction of Insertion Potentials in Li-Mn Oxides for Secondary Li Batteries. *J. Electrochem. Soc.* **1997**, *144*, 3832–3835. [CrossRef]
34. Li, W.; Reimers, J.N.; Dahn, J.R. Lattice-gas-model approach to understanding the structures of lithium transition-metal oxides LiMO$_2$. *Phys. Rev. B* **1994**, *49*, 826–831. [CrossRef] [PubMed]
35. Van der Ven, A.; Aydinol, M.K.; Ceder, G.; Kresse, G.; Hafner, J. First-principles investigation of phase stability in Li$_x$CoO$_2$. *Phys. Rev. B* **1998**, *58*, 2975–2987. [CrossRef]
36. Reimers, J.N.; Dahn, J.R. Electrochemical and In Situ X-ray Diffraction Studies of Lithium Intercalation in Li$_x$CoO$_2$. *J. Electrochem. Soc.* **1992**, *139*, 2091–2097. [CrossRef]

37. Shao-Horn, Y.; Levasseur, S.; Weill, F.; Delmas, C. Probing Lithium and Vacancy Ordering in O3 Layered Li_xCoO_2 ($x \approx 0.5$). *J. Electrochem. Soc.* **2003**, *150*, A366. [CrossRef]
38. Li, W.; Reimers, J.N.; Dahn, J.R. In situ x-ray diffraction and electrochemical studies of $Li_{1-x}NiO_2$. *Solid State Ion.* **1993**, *67*, 123–130. [CrossRef]
39. Peres, J.P.; Weill, F.; Delmas, C. Lithium/vacancy ordering in the monoclinic Li_xNiO_2 ($0.50 \leq x \leq 0.75$) solid solution. *Solid State Ion.* **1999**, *116*, 19–27. [CrossRef]
40. Arai, H.; Okada, S.; Ohtsuka, H.; Ichimura, M.; Yamaki, J. Characterization and cathode performance of $Li_{1-x}Ni_{1+x}O_2$ prepared with the excess lithium method. *Solid State Ion.* **1995**, *80*, 261–269. [CrossRef]
41. Kim, H.; Ji, H.; Wang, J.; Ceder, G. Next-Generation Cathode Materials for Non-aqueous Potassium-Ion Batteries. *Trends Chem.* **2019**, *1*, 682–692. [CrossRef]
42. Lee, W.; Kim, J.; Yun, S.; Choi, W.; Kim, H.; Yoon, W.-S. Multiscale factors in designing alkali-ion (Li, Na, and K) transition metal inorganic compounds for next-generation rechargeable batteries. *Energy Environ. Sci.* **2020**, *13*, 4406–4449. [CrossRef]
43. Tian, Y.; Zeng, G.; Rutt, A.; Shi, T.; Kim, H.; Wang, J.; Koettgen, J.; Sun, Y.; Ouyang, B.; Chen, T.; et al. Promises and Challenges of Next-Generation "Beyond Li-ion" Batteries for Electric Vehicles and Grid Decarbonization. *Chem. Rev.* **2021**, *121*, 1623–1669. [CrossRef] [PubMed]
44. Berthelot, R.; Carlier, D.; Delmas, C. Electrochemical investigation of the $P2–Na_xCoO_2$ phase diagram. *Nat. Mater.* **2011**, *10*, 74–80. [CrossRef] [PubMed]
45. Xu, J.; Lee, D.A.E.H.O.E.; Meng, Y.S. Recent advances in sodium intercalation positive electrode materials for sodium ion batteries. *Funct. Mater. Lett.* **2013**, *6*, 1330001. [CrossRef]
46. Hironaka, Y.; Kubota, K.; Komaba, S. P2- and P3-K_xCoO_2 as an electrochemical potassium intercalation host. *Chem. Commun.* **2017**, *53*, 3693–3696. [CrossRef] [PubMed]
47. Kim, H.; Kim, J.C.; Bo, S.-H.; Shi, T.; Kwon, D.-H.; Ceder, G. K-Ion Batteries Based on a P2-Type $K_{0.6}CoO_2$ Cathode. *Adv. Energy Mater.* **2017**, *7*, 1700098. [CrossRef]
48. Sun, Y.-K.; Han, J.-M.; Myung, S.-T.; Lee, S.-W.; Amine, K. Significant improvement of high voltage cycling behavior AlF_3-coated $LiCoO_2$ cathode. *Electrochem. Commun.* **2006**, *8*, 821–826. [CrossRef]
49. Radin, M.D.; Van der Ven, A. Stability of Prismatic and Octahedral Coordination in Layered Oxides and Sulfides Intercalated with Alkali and Alkaline-Earth Metals. *Chem. Mater.* **2016**, *28*, 7898–7904. [CrossRef]
50. DeKock, R.L.; Gray, H.B. *Chemical Structure and Bonding*; The Benjamin/Cummings Publishing Company: Menlo Park, CA, USA, 1980.
51. Sun, J.; Ruzsinszky, A.; Perdew, J.P. Strongly Constrained and Appropriately Normed Semilocal Density Functional. *Phys. Rev. Lett.* **2015**, *115*, 36402. [CrossRef]
52. Aydinol, M.K.; Kohan, A.F.; Ceder, G.; Cho, K.; Joannopoulos, J. Ab initio study of lithium intercalation in metal oxides and metal dichalcogenides. *Phys. Rev. B* **1997**, *56*, 1354–1365. [CrossRef]
53. Marianetti, C.A.; Morgan, D.; Ceder, G. First-principles investigation of the cooperative Jahn-Teller effect for octahedrally coordinated transition-metal ions. *Phys. Rev. B* **2001**, *63*, 224304. [CrossRef]
54. Akimoto, J.; Gotoh, Y.; Oosawa, Y. Synthesis and Structure Refinement of $LiCoO_2$ Single Crystals. *J. Solid State Chem.* **1998**, *141*, 298–302. [CrossRef]
55. Wolverton, C.; Zunger, A. First-Principles Prediction of Vacancy Order-Disorder and Intercalation Battery Voltages in $LiCoO_2$. *Phys. Rev. Lett.* **1998**, *81*, 606–609. [CrossRef]
56. Yin, S.-C.; Rho, Y.-H.; Swainson, I.; Nazar, L.F. X-ray/Neutron Diffraction and Electrochemical Studies of Lithium De/Re-Intercalation in $Li_{1-x}Co_{1/3}Ni_{1/3}Mn_{1/3}O_2$ ($x = 0 \rightarrow 1$). *Chem. Mater.* **2006**, *18*, 1901–1910. [CrossRef]
57. Reed, J.; Ceder, G. Role of Electronic Structure in the Susceptibility of Metastable Transition-Metal Oxide Structures to Transformation. *Chem. Rev.* **2004**, *104*, 4513–4534. [CrossRef] [PubMed]
58. Park, H.-S.; Hwang, S.-J.; Choy, J.-H. Relationship between Chemical Bonding Character and Electrochemical Performance in Nickel-Substituted Lithium Manganese Oxides. *J. Phys. Chem. B* **2001**, *105*, 4860–4866. [CrossRef]
59. Chernova, N.A.; Roppolo, M.; Dillon, A.C.; Whittingham, M.S. Layered vanadium and molybdenum oxides: Batteries and electrochromics. *J. Mater. Chem.* **2009**, *19*, 2526–2552. [CrossRef]
60. Radin, M.D.; Vinckeviciute, J.; Seshadri, R.; Van der Ven, A. Manganese oxidation as the origin of the anomalous capacity of Mn-containing Li-excess cathode materials. *Nat. Energy* **2019**, *4*, 639–646. [CrossRef]
61. Vinckeviciute, J.; Kitchaev, D.A.; Van der Ven, A. A Two-Step Oxidation Mechanism Controlled by Mn Migration Explains the First-Cycle Activation Behavior of Li_2MnO_3-Based Li-Excess Materials. *Chem. Mater.* **2021**, *33*, 1625–1636. [CrossRef]
62. Olivetti, E.A.; Ceder, G.; Gaustad, G.G.; Fu, X. Lithium-Ion Battery Supply Chain Considerations: Analysis of Potential Bottlenecks in Critical Metals. *Joule* **2017**, *1*, 229–243. [CrossRef]
63. Van der Ven, A. Lithium Diffusion in Layered Li_xCoO_2. *Electrochem. Solid-State Lett.* **1999**, *3*, 301. [CrossRef]
64. Xia, H.; Lu, L.; Ceder, G. Substrate effect on the microstructure and electrochemical properties of $LiCoO_2$ thin films grown by PLD. *J. Alloys Compd.* **2006**, *417*, 304–310. [CrossRef]
65. Jang, Y.-I.; Neudecker, B.J.; Dudney, N. Lithium diffusion in Li_xCoO_2 ($0.45 < x < 0.7$) Intercalation cathodes. *Electrochem. Solid-State Lett.* **2001**, *4*, A74. [CrossRef]

66. Ohzuku, T.; Ueda, A. Solid-State Redox Reactions of LiCoO$_2$ (R3m) for 4 Volt Secondary Lithium Cells. *J. Electrochem. Soc.* **1994**, *141*, 2972–2977. [CrossRef]
67. Amatucci, G.G.; Tarascon, J.M.; Klein, L.C. CoO$_2$, the End Member of the Li$_x$CoO$_2$ Solid Solution. *J. Electrochem. Soc.* **1996**, *143*, 1114–1123. [CrossRef]
68. Van der Ven, A.; Aydinol, M.K.; Ceder, G. First-Principles Evidence for Stage Ordering in Li$_x$CoO$_2$. *J. Electrochem. Soc.* **1998**, *145*, 2149–2155. [CrossRef]
69. Van der Ven, A.; Ceder, G. Lithium diffusion mechanisms in layered intercalation compounds. *J. Power Sources* **2001**, *97–98*, 529–531. [CrossRef]
70. Urban, A.; Lee, J.; Ceder, G. The Configurational Space of Rocksalt-Type Oxides for High-Capacity Lithium Battery Electrodes. *Adv. Energy Mater.* **2014**, *4*, 1400478. [CrossRef]
71. Van der Ven, A.; Bhattacharya, J.; Belak, A.A. Understanding Li Diffusion in Li-Intercalation Compounds. *Acc. Chem. Res.* **2013**, *46*, 1216–1225. [CrossRef] [PubMed]
72. Kang, K.; Ceder, G. Factors that affect Li mobility in layered lithium transition metal oxides. *Phys. Rev. B* **2006**, *74*, 94105. [CrossRef]
73. Kang, K.; Meng, Y.S.; Bréger, J.; Grey, C.P.; Ceder, G. Electrodes with High Power and High Capacity for Rechargeable Lithium Batteries. *Science* **2006**, *311*, 977–980. [CrossRef]
74. Lee, J.; Urban, A.; Li, X.; Su, D.; Hautier, G.; Ceder, G. Unlocking the Potential of Cation-Disordered Oxides for Rechargeable Lithium Batteries. *Science* **2014**, *343*, 519–522. [CrossRef]
75. Ji, H.; Urban, A.; Kitchaev, D.A.; Kwon, D.-H.; Artrith, N.; Ophus, C.; Huang, W.; Cai, Z.; Shi, T.; Kim, J.C.; et al. Hidden structural and chemical order controls lithium transport in cation-disordered oxides for rechargeable batteries. *Nat. Commun.* **2019**, *10*, 592. [CrossRef]
76. Lun, Z.; Ouyang, B.; Kwon, D.-H.; Ha, Y.; Foley, E.E.; Huang, T.-Y.; Cai, Z.; Kim, H.; Balasubramanian, M.; Sun, Y.; et al. Cation-disordered rocksalt-type high-entropy cathodes for Li-ion batteries. *Nat. Mater.* **2021**, *20*, 214–221. [CrossRef]
77. Clément, R.J.; Lun, Z.; Ceder, G. Cation-disordered rocksalt transition metal oxides and oxyfluorides for high energy lithium-ion cathodes. *Energy Environ. Sci.* **2020**, *13*, 345–373. [CrossRef]
78. Baur, C.; Källquist, I.; Chable, J.; Chang, J.H.; Johnsen, R.E.; Ruiz-Zepeda, F.; Mba, J.M.A.; Naylor, A.J.; Garcia-Lastra, J.M.; Vegge, T.; et al. Improved cycling stability in high-capacity Li-rich vanadium containing disordered rock salt oxyfluoride cathodes. *J. Mater. Chem. A* **2019**, *7*, 21244–21253. [CrossRef]
79. Chen, R.; Ren, S.; Yavuz, M.; Guda, A.A.; Shapovalov, V.; Witter, R.; Fichtner, M.; Hahn, H. Li$^+$ intercalation in isostructural Li$_2$VO$_3$ and Li$_2$VO$_2$F with O^{2-} and mixed O^{2-}/F$^-$ anions. *Phys. Chem. Chem. Phys.* **2015**, *17*, 17288–17295. [CrossRef]
80. Lee, J.; Kitchaev, D.A.; Kwon, D.-H.; Lee, C.-W.; Papp, J.K.; Liu, Y.-S.; Lun, Z.; Clement, R.J.; Shi, T.; McCloskey, B.D.; et al. Reversible Mn^{2+}/Mn^{4+} double redox in lithium-excess cathode materials. *Nature* **2018**, *556*, 185–190. [CrossRef]
81. Takeda, N.; Hoshino, S.; Xie, L.; Chen, S.; Ikeuchi, I.; Natsui, R.; Nakura, K.; Yabuuchi, N. Reversible Li storage for nanosize cation/anion-disordered rocksalt-type oxyfluorides: LiMoO$_{2-x}$ LiF ($0 \leq x \leq 2$) binary system. *J. Power Sources* **2017**, *367*, 122–129. [CrossRef]
82. Glazier, S.L.; Li, J.; Zhou, J.; Bond, T.; Dahn, J.R. Characterization of Disordered Li$_{(1+x)}$Ti$_{2x}$Fe$_{(1-3x)}$O2 as Positive Electrode Materials in Li-Ion Batteries Using Percolation Theory. *Chem. Mater.* **2015**, *27*, 7751–7756. [CrossRef]
83. Cambaz, M.A.; Vinayan, B.P.; Euchner, H.; Pervez, S.A.; Geßwein, H.; Braun, T.; Gross, A.; Fichtner, M. Design and Tuning of the Electrochemical Properties of Vanadium-Based Cation-Disordered Rock-Salt Oxide Positive Electrode Material for Lithium-Ion Batteries. *ACS Appl. Mater. Interfaces* **2019**, *11*, 39848–39858. [CrossRef] [PubMed]
84. Urban, A.; Abdellahi, A.; Dacek, S.; Artrith, N.; Ceder, G. Electronic-Structure Origin of Cation Disorder in Transition-Metal Oxides. *Phys. Rev. Lett.* **2017**, *119*, 176402. [CrossRef] [PubMed]
85. Lun, Z.; Ouyang, B.; Kitchaev, D.A.; Clément, R.J.; Papp, J.K.; Balasubramanian, M.; Ceder, G. Improved Cycling Performance of Li-Excess Cation-Disordered Cathode Materials upon Fluorine Substitution. *Adv. Energy Mater.* **2019**, *9*, 1802959. [CrossRef]

Article

Electrochemical Assessment of Indigo Carmine Dye in Lithium Metal Polymer Technology

Margaud Lécuyer [1,2], Marc Deschamps [1], Dominique Guyomard [2], Joël Gaubicher [2,*] and Philippe Poizot [2,*]

1. Blue*Solutions*, Odet, Ergué Gabéric, CEDEX 9, 29556 Quimper, France; margaud.lecuyer@blue-solutions.fr (M.L.); marc.deschamps@blue-solutions.fr (M.D.)
2. Université de Nantes, CNRS, Institut des Matériaux Jean Rouxel, IMN, 44322 Nantes, France; dominique.guyomard@cnrs-imn.fr
* Correspondence: joel.gaubicher@cnrs-imn.fr (J.G.); philippe.poizot@cnrs-imn.fr (P.P.)

Abstract: Lithium metal batteries are inspiring renewed interest in the battery community because the most advanced designs of Li-ion batteries could be on the verge of reaching their theoretical specific energy density values. Among the investigated alternative technologies for electrochemical storage, the all-solid-state Li battery concept based on the implementation of dry solid polymer electrolytes appears as a mature technology not only to power full electric vehicles but also to provide solutions for stationary storage applications. With an effective marketing started in 2011, Blue*Solutions* keeps developing further the so-called lithium metal polymer batteries based on this technology. The present study reports the electrochemical performance of such Li metal batteries involving indigo carmine, a cheap and renewable electroactive non-soluble organic salt, at the positive electrode. Our results demonstrate that this active material was able to reversibly insert two Li at an average potential of $\approx$2.4 V vs. Li^+/Li with however, a relatively poor stability upon cycling. Post-mortem analyses revealed the poisoning of the Li electrode by Na upon ion exchange reaction between the Na countercations of indigo carmine and the conducting salt. The use of thinner positive electrodes led to much better capacity retention while enabling the identification of two successive one-electron plateaus.

Keywords: indigo carmine; solid polymer electrolyte; solid state battery; LMP® technology; organic battery

Citation: Lécuyer, M.; Deschamps, M.; Guyomard, D.; Gaubicher, J.; Poizot, P. Electrochemical Assessment of Indigo Carmine Dye in Lithium Metal Polymer Technology. *Molecules* 2021, 26, 3079. https://doi.org/10.3390/molecules26113079

Academic Editor: Amor M. Abdelkader

Received: 30 April 2021
Accepted: 17 May 2021
Published: 21 May 2021

Publisher's Note: MDPI stays neutral with regard to jurisdictional claims in published maps and institutional affiliations.

Copyright: © 2021 by the authors. Licensee MDPI, Basel, Switzerland. This article is an open access article distributed under the terms and conditions of the Creative Commons Attribution (CC BY) license (https://creativecommons.org/licenses/by/4.0/).

1. Introduction

The year 2021 marks the 30th anniversary of the first commercialization, by the Sony Corporation, of rechargeable lithium-ion batteries (LIBs), paving the way for higher energy density batteries. Their appearance in the world market heralded first a revolution in consumer electronics giving rise to much more flexibility and comfort while making our everyday life easier and safer. In the early 2000s, in the face of global warming and finite fossil-fuel supplies, it was thought possible to consider their integration in other important applications of our energy engineering thanks to the constant and rapid improvement of this technology coupled with potential cost reductions [1]. Today, LIBs support the deployment of decarbonized transportation systems through the massive use of electric motors (the so-called "electromobility" or "e-mobility") and could enable large-scale storage of electricity produced by the increasingly widespread use of renewable energy sources. Unambiguously, the rise of the Li-ion technology related to the "rocking-chair battery" concept can be perceived as one of the most prominent examples of how chemistry could change our daily life. This fact was highlighted by Zhang et al. [2] in a recent article focused on the history of the LIBs, which honors the pioneer scientists behind it and especially the 2019 Nobel Prize laureates in Chemistry (John B. Goodenough, M. Stanley Whittingham, and Akira Yoshino) as well as M. Armand. Other monographs dealing with this fantastic odyssey are also available in the literature [3–8].

In a very simplified point of view, two main pillars have underpinned the development of high energy Li-based batteries: (*i*) the use of the lithium element, which is characterized by low atomic mass and high reducing properties of the metallic state, and (*ii*) the reversible hosting properties of redox-active materials for Li$^+$ ions. Starting with the latter, the earliest forms of redox-active insertion materials belong to low-dimensional solids, especially 2-D inorganic structures such as graphite or TiS$_2$ (denoted intercalation materials at that time). Seminal contributions on the chemical reduction of such layered materials accompanied with concomitant Li$^+$ insertion came from A. Hérold [9] for graphite and W. von Rüdorff [10] and J. Rouxel for TiS$_2$ [11–13]. Furthermore, in 1972, B. C. H. Steele and M. Armand set the foundations for application of insertion compounds as electrode materials for the electrochemical storage during the *NATO*-sponsored conference held on the shores of Lake Maggiore (Italy) and discussed notably the seminal "rocking-chair battery" concept [14]. Four years later, Steele's group [15] and S. Whittingham [16] (from the Exxon company at that time) reported simultaneously the electrochemical performance of the first Li-battery based on TiS$_2$ as the positive electrode working in a liquid electrolyte medium. On the advice of M. Armand and J. Rouxel, French research and development studies started in 1976 at the Laboratory of the CGE in Marcoussis with A. Le Méhauté, which led to successful collaborative works with the Rouxel's group in Nantes (including R. Brec, G. Ouvrard, A. Louisy, D. M. Schleich) especially on the study of the LiAl | | NiPS$_3$ rechargeable battery [17–19]. However, LiAl electrodes were not stable enough (rapid passivation and volume expansion during the cycling) while safety issues due to dendrite formation (leading to possible internal short-circuits and thermal runaway) were a major drawback for practical application of Li metal negative electrodes. Following the "rocking-chair battery" concept proposed by M. Armand [14] aiming also at using an insertion material for the negative electrode, in 1980, M. Lazzari and B. Scrosati [20] reported the first Li-ion battery based on the Li$_x$WO$_2$ | | Li$_y$TiS$_2$ rocking-chair cell assembly. The electrochemical performance of the next LIBs was then widely improved by pairing the layered oxide LiCoO$_2$ (developed by the Goodenough's group in 1980) as the positive electrode with efficient carbonaceous materials (developed by Yoshino and co-workers) on the negative side [4]. Note that a substantial contribution to the good cycling properties of the next commercialized LIBs was achieved with the electrolyte formulation proposed by D. Guyomard and J.-M. Tarascon [21] based on LiPF$_6$ as the supporting salt dissolved in ethylene carbonate (EC)/dimethyl carbonate (DMC), which still constitutes the standardized electrolyte formulation of today's LIBs. Beyond the "rocking-chair battery" technology, which prevents, in principle, the dangerous formation of lithium dendrites at the negative electrode, M. Armand also proposed, in the late 70s, the fabrication of safer Li-batteries by using a dry solid polymer electrolyte (SPE) membrane [22,23]. This SPE membrane is obtained by dissolving a conducting lithium salt in a poly(ethylene) oxide-based matrix (PEO); P. V. Wright and co-workers [24,25] having previously demonstrated that PEO was sufficiently solvating to dissolve alkali salts. Although the formation of Li dendrites was not fully suppressed upon cycling, this SPE membrane can offer an efficient mechanical resistance that is able to protect the positive electrode against short-circuits. However, the best ionic conductivity values in SPE are typically obtained when the amorphous regions of the polymer are favored, which corresponds to temperatures above 60 °C.

After 30 years of intensive research and development to produce ever more efficient LIBs, the current status is that they have probably reached their threshold limit of $\approx$350 Wh kg^{-1} [26] that falls short of meeting the projected needs in a near future (although it corresponds to a gain of $\approx$400% in gravimetric energy density compared to the first LIBs produced by the Sony corporation in 1991 [5]). Hence, the all-solid-state battery (SSB) technology based on the implementation of an SPE membrane is inspiring renewed interest in the battery community because it allows the direct use of metallic lithium as the negative electrode. This direction therefore paves the way for developing batteries that are not only safer (free of flammable solvents) but also have a higher energy density [26]. Other kinds of solid-state batteries are intensively investigated by both academic and industrial researchers [27], but the maturity of the all-solid-state batteries fabricated by using a PEO-based solid polymer electrolyte is a great advantage. The Bolloré group and its subsidiary Blue*Solutions* have been developing the so-called lithium metal polymer (LMP®) batteries based on SPE since the early 2000s and proved the reliability of this technology in different applications. With the first practical deployment in the world of SSBs in full electric vehicles achieved in 2011, the LMP® technology is now widely used not only in multiple e-mobility scenarios but also in stationary storage applications all over the world (Europe, North America, Asia, and Africa). Like many other electrochemical storage technologies, the search for efficient, cheaper but also earth-abundant insertion electroactive compounds for LMP® batteries remains an ongoing challenge to mitigate their environmental impact [28]. It is now recognized that the use of redox-active organics in batteries, combined with recycling solutions, could decrease the pressure on inorganic compounds and offer valid options to improve the life cycle assessment of cells "from cradle to grave" [29]. Although the volumetric considerations are known to be limiting for organic-based batteries [30], this is not systematically a real issue for some practical LMP® battery applications. Note that the electrochemical performance of *n*-type organic cathode materials in SPE-based SSBs ("*n*-type" electrode reactions involve an ionic compensation with cation release upon oxidation (such as Li$^+$) whereas "*p*-type" electrode reactions imply an anion uptake (such as ClO$_4^-$)) has been seldom reported (to the best of our knowledge, by only two studies [31,32]), compared to the hundreds of publications describing organic batteries operating in liquid electrolyte media [33]. However, M. Armand has reported in 1990 the reversible electrochemical accommodation of anions in the *p*-type poly(decaviologen) [34].

Considering the potential interest of organic electrode materials for the LMP® technology, the company Blue*Solutions* and IMN joined their know-how to electrochemically investigate selected organic lithium insertion materials. In our first study [31], we planned to test small (neutral) electroactive molecules. Although quite promising in terms of specific capacity, small molecules readily dissolved in conventional liquid electrolytes, ruining rapidly the electrochemical performance. Tetramethoxy-*p*-benzoquinone (TMQ) was chosen due to its unusual but necessary high thermal stability (beyond 100 °C) [31]. Interestingly, we obtained better electrochemical results by using the PEO-based solid polymer electrolyte compared to previous data obtained in liquid electrolyte media, although slow diffusion of the TMQ molecule inside the SPE membrane occurred with time, giving rise to some capacity decay upon cycling. For this second investigation, inspired by our recurrent strategy successfully applied in conventional aprotic liquid electrolytes to circumvent solubilization of organics [33,35–37]), we selected an organic salt—namely, disodium 5,5′-indigotin disulfonate, also known as indigo carmine (IC)—as the active material for several reasons. First, it exhibits two permanent and delocalized negative charges in its redox-active organic backbone (Scheme 1), which makes polar interactions with PEO chains difficult.

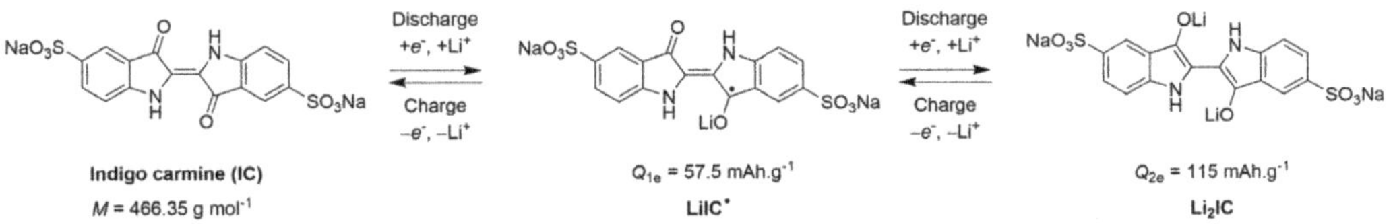

Scheme 1. Expected two-electron electrochemical reaction of IC vs. Li.

Indigo carmine (IC)
M = 466.35 g mol^{-1}

Q_{1e} = 57.5 mAh·g^{-1}
LiIC·

Q_{2e} = 115 mAh·g^{-1}
Li$_2$IC

Second, approved for use as a food colorant, this hydrosoluble synthetic dye is also cheap, renewable, and commercially available at a large scale. Last but not least, IC is already known for its reversible two-lithium insertion process (Q_{2e} = 115 mAh g^{-1}, Scheme 1) thanks to the electroactivity of its two carbonyl groups at an average potential of ≈2.4 V vs. Li$^+$/Li [38,39]. Herein, we report for the first time the electrochemical behavior of

IC measured at 100 °C in the LMP technology using a SPE membrane developed by Blue*Solutions*.

2. Results and Discussion

As previously reported [31], the cell assembly used in this second benchmark study on the electrochemical assessment of positive organic electrode materials derived from that presently integrated in LMP® batteries developed and commercialized by Blue*Solutions* since 2011. In short, the all-solid-state lithium organic battery consisted in a simple adaptation of the commercial Blue*Solutions* battery technology using a PEO-based solid polymer electrolyte by replacing the currently used LiFePO$_4$ positive electrode material with an active organic material (IC), as illustrated in Figure 1. Hence, results presented in this work are representative of realistic and commercial configurations. All the electrochemical measurements were performed at 100 °C, which is a common temperature for the assessment of the LMP® battery prototypes. Note that IC is thermally stable at more than 200 °C [40], as it is commonly observed in the case of alkali salts of organic compounds [33]. Before presenting the as-obtained results in the LMP® battery technology, we have thought useful to recall the typical electrochemical features of IC-based composite electrodes reported in the literature to date [38–41].

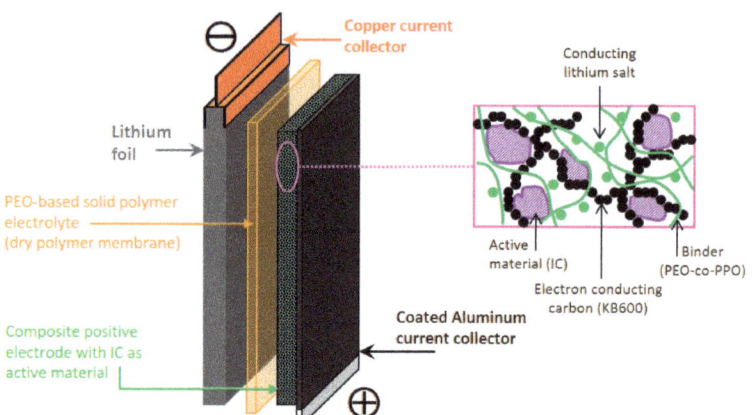

Figure 1. Schematic illustration of an LMP® cell as developed by Blue*Solutions*.

2.1. Recap of the Electrochemical Behavior Concerning Li | | IC Cells Measured in Carbonate-Based Liquid Electrolytes at Room Temperature

Basically, all the reported cycling data were systematically measured in Li- or Na-half cell configuration at 25–30 °C by using carbonate-based liquid electrolytes. Figure 2 shows a representative galvanostatic cycling curve plotted versus both the lithium composition (part a) and the specific capacity (part b) within the 1.6–3.2 V vs. Li$^+$/Li potential range. Note that the corresponding composite electrode was specially formulated with only 10 wt.% of conductive carbon thanks to an original aqueous processing previously developed at IMN [40]. During the first discharge, the potential dropped step by step to 1.6 V, giving rise to an overall discharge capacity of 110 mAh g^{-1}, which matched the expected two-electron electrode reaction (Scheme 1). The corresponding differential capacity vs. potential curve (Figure 2c) showed the occurrence of four successive steps located at 2.45, 2.30, 2.06, and 1.84 V vs. Li$^+$/Li. Upon recharge, most of the inserted lithium ions could be removed of the IC host structure, leading to a reversible capacity above 100 mAh.g^{-1}. However, the electrochemical profile differed from the first discharge. First, a plateau occurred at 2.35 V vs. Li$^+$/Li, involving half of the reversible capacity, that suggested the occurrence of a first order structural phase transition between the fully reduced compound (Li$_2$IC) and the expected radical form (LiIC$^\bullet$) based on the description of the electrochemical process at the molecular level (Scheme 1). Afterwards, the potential continuously increased as the delithiation proceeds through two broad peaks (Figure 2c) centered on 2.55 and 2.77 V vs. Li$^+$/Li. On the following discharge/charge cycles, these electrochemical features were qualitatively and quantitatively reiterated, giving rise to stable capacity retention. Similar data were reported by the Yao's group in their seminal electrochemical investigations of IC in Li/Na batteries [38,39].

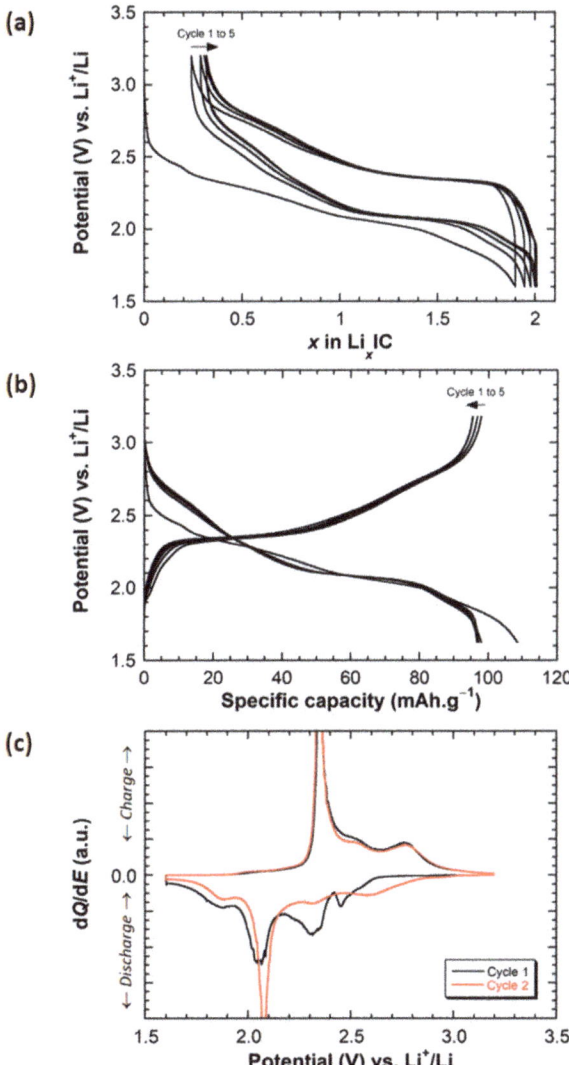

Figure 2. (**a**,**b**) Typical galvanostatic cycling curve of IC measured at 25 °C in Li-half cell configuration for a composite electrode made of IC (85 wt.%) mixed with conductive carbon C-Nergy super C45 (10 wt.%) and carboxymethylcellulose (5 wt.%) as the binder (electrolyte: 1 M LiPF$_6$ in EC/DMC; cycling rate: 1 Li$^+$ exchanged per mole of IC in 10 h (C/20); electrode thickness: 24 μm) [40]. (**c**) Corresponding differential capacity vs. the potential curve.

2.2. Electrochemical Behavior of Li | | IC Cells Using the LMP® Technology

Our first electrochemical assessment of IC in an LMP® battery was then inspired by our former study performed in a conventional liquid electrolyte medium. Therefore, we aimed at obtaining, for the composite electrode, a similar conductive carbon/IC mass ratio ($\approx$0.14) and a comparable thickness ($\approx$25 µm) although the two electrode formulation technologies were quite different. In practice, the composite electrode was prepared by using carbon black (Ketjenblack EC600JD) as the conductive carbon and LiTFSI as the supporting salt. The same carbon additive was also used in our first organic LMP® batteries based on tetramethoxy-p-benzoquinone [31]. The practical details are reported in Section 3.1. The composite electrode was 27 µm thick, its carbon content being 12.5 wt.%.

Unfortunately, as opposed to performance reported in the literature using conventional carbonate-based liquid electrolytes, the as-obtained electrochemical results did not meet our expectations (Figure 3). More precisely, the representative galvanostatic cycling curve of such LMP® cells (Figure 3a) plotted within the 3.2–1.5 V vs. Li$^+$/Li potential window showed an obvious capacity fading coupled with an increase of the cell polarization upon cycling. Figure 3b allows a better visualization of this trend with an irregular but continuous capacity decay of about 40% over the 20 first cycles, when the coulombic efficiency was instable. The corresponding evolution of the apparent cell resistance ($R_{app.}$) provided complementary and useful information (Figure 3c). After a period of relative stability at $\approx$180 Ω.cm^2, a sharp resistance increase could be noticed after 10 cycles, especially in charge reaching more than 440 Ω.cm^2 at the 20th cycle. As the polyanionic nature of IC should have prevented the possible dissolution/diffusion within the SPE, as explained above, such an electrochemical behavior pointed to the occurrence of interfacial issues that worsened with operating time. Let us recall that such poor electrochemical features were not observed when using the neutral TMQ molecule as the positive active material in LMP® cells despite its slow diffusion within the PEO-based network [31]. In this trial, despite the poor cycling stability of the cells, the electrochemical behavior was consistent with previous results, either already published in the literature or obtained in carbonate-based electrolytes. Basically, the shape of the electrochemical curve recorded during the first galvanostatic cycle remained roughly similar to that reported in conventional carbonate-based liquid electrolytes, which confirmed the reversible electroactivity of IC vs. Li at 100 °C in the LMP® cell.

Thicker IC-based electrodes (48 µm) were then prepared and electrochemically tested in similar cycling conditions, but we experienced faster loss in the cycling stability with rapid increase of the cell polarization. This second series of experiments was an occasion to probe more accurately the electrochemical lithium insertion processes into IC by running a potential-controlled mode: potentiodynamic intermittent titration technique (PITT). An advantage of the PITT method is that the electrochemically driven phase transitions in insertion electrodes can be more readily visualized by analyzing both the potential evolution and the corresponding current response [42,43]. Figure 4 shows the as-obtained potential/current profiles recorded during the first discharge. Despite the thickness of the electrode, the potential trace reveals the existence of four main steps (pseudo-plateaus) located at 2.39, 2.30, 2.06, and 1.96 V vs. Li$^+$/Li, which are in very good agreement with the values reported by using conventional carbonate-based liquid electrolytes (Figure 2). Those pseudo-plateaus are correlated with bell-shape-type responses in current (non-Cottrellian-type decay) revealing four successive phase transformations (I to IV). Interestingly, one can notice a larger potential gap (i.e., energy gap) between steps II and III at half of the reversible capacity, which formally corresponds to the formation of the expected radical form (LiIC$^\bullet$), as underlined above.

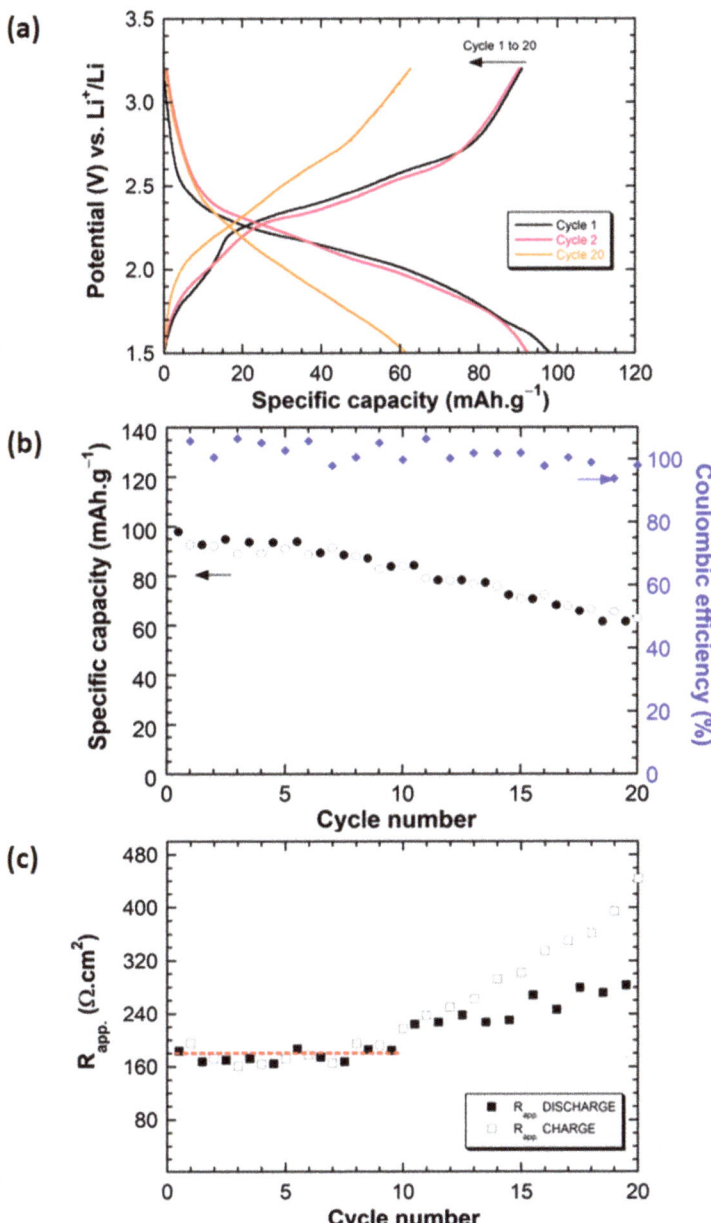

Figure 3. (a) Galvanostatic cycling curve evolution (three selected cycles) measured at 100 °C in LMP® cell for a composite electrode made of IC (70 wt.%) mixed with KB600 (10 wt.%), LiTFSI (4 wt.%), and PEO-PPO (16 wt.%); cycling rate: 1 Li$^+$ exchanged per mole of IC in 2 h (C/4); electrode thickness: 27 µm. (b) Corresponding capacity retention curve together with the coulombic efficiency. (c) Evolution of the apparent cell resistance, both on charge and discharge, upon cycling. The red dashed line highlights the relative stability of $R_{app.}$ during the first ten cycles.

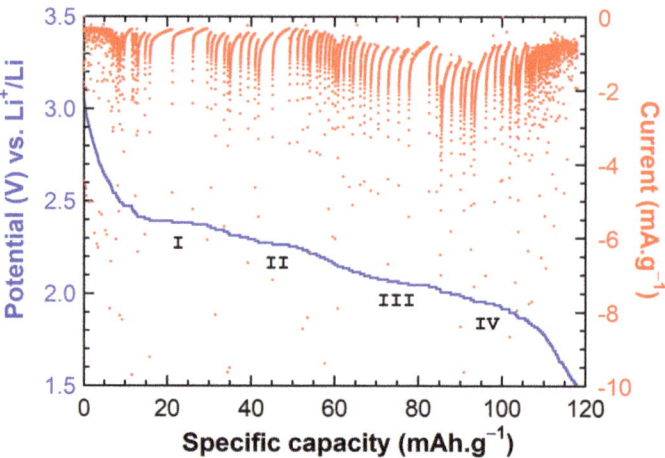

Figure 4. PITT measurements at 100 °C in an LMP® cell for a composite electrode made of IC (70 wt.%) mixed with KB600 (10 wt.%), LiTFSI (4 wt.%), and PEO-PPO (16 wt.%); electrode thickness: 48 µm; ΔE = 10 mV with a current limitation corresponding to C/400 (i.e., i_{min} = 0.29 mA g^{-1}).

Thinner IC-based electrodes (6 µm) were also fabricated and electrochemically tested to complete our screening. To our surprise, the as-obtained electrochemical behavior was very different compared to previous data. An example of a galvanostatic cycling curve recorded at C/2 is shown in Figure 5a. The first striking feature was the drastic simplification of the potential-capacity trace. First, the four pseudo-plateaus conventionally observed in the first discharge merged, giving rise to two successive and well-defined plateaus, fully reversible, and corresponding to the storing of one e^-/Li$^+$ per IC each. The two characteristic average potential values can be determined at 2.43 and 2.11 V vs. Li$^+$/Li. These two successive one-electron electrode reactions are well aligned with the molecular-level description of the electrochemical process (Scheme 1) confirming unambiguously the particular stability of the radical LiIC• phase in the solid state, in agreement with the PITT measurements using a thick electrode (Figure 4) and the electrochemical data obtained in organic liquid electrolyte media [38–40]. This thin electrode was able to sustain its capacity over dozens of cycles, as shown in Figure 5b. Nevertheless, a capacity decay became visible after 60 cycles (≈25% of loss after 100 cycles). The corresponding evolution of the apparent cell resistance (Figure 5c) was certainly much more constant, on average, than that observed with the 27 µm thick electrode, but a certain instability remained observable. In short, these characteristics confirmed again some interfacial limitations, even if the use of a thin electrode enables much better stability on cycling and a markedly improved electrode kinetics.

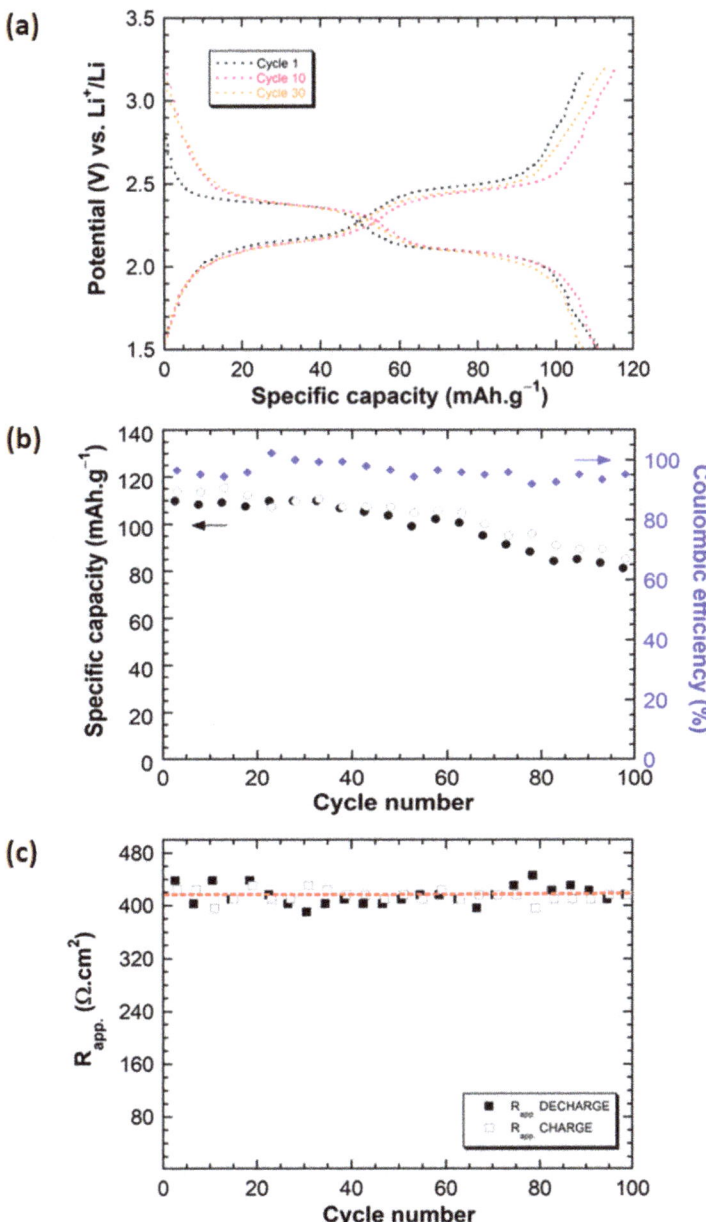

Figure 5. (**a**) Galvanostatic cycling curve evolution (three selected cycles) measured at 100 °C in LMP® cell for a thin composite electrode made of IC (70 wt.%) mixed with KB600 (10 wt.%), LiTFSI (4 wt.%), and PEO-PPO (16 wt.%); cycling rate: 1 Li$^+$ exchanged per mole of IC in 1 h (C/2); electrode thickness: 6 μm. (**b**) Corresponding capacity retention curve together with the coulombic efficiency. (**c**) Evolution of the apparent cell resistance, both on charge and discharge, upon cycling. The red dashed line highlights the relative stability of $R_{app.}$ upon 100 cycles.

2.3. Post-Mortem Analyses and Failure Identification

In order to better understand what could hinder the proper functioning of IC as an active electrode material in LMP® cell, especially when increasing the electrode loading, post-mortem investigations were performed by coupling SEM imaging and EDS analysis (see details in Section 3.2). During the preparation of the sample for SEM/EDS characterizations, we did not notice any change of the pristine colorless electrolyte film (IC being a blue dye), contrary to our former observations with TMQ as the active material; the typical orange color of TMQ was indeed clearly visible to the naked eye in the SPE layer [31]. However, the simultaneous presence of sulfur atoms, both in the IC chemical structure and in the conducting salt (LiTFSI), complicated the interpretations of the EDS measurements. To avoid this interference, a new series of LMP® cells was prepared by using $LiClO_4$ as the conducting salt. Note that the fabrication of relatively thick composite electrodes was preferred for this investigation to make the SEM/EDS characterizations easier and because rapid capacity decay was anticipated. Figure 6a,b shows the corresponding electrochemical cycling data. As expected, the reversible capacity drops rapidly upon cycling, recovering less than 20 mAh g^{-1} after 80 cycles. In addition, the peculiar sharp increase of $R_{app.}$ was also noticed after 10 cycles, in agreement with the data reported in Figure 3c where LiTFSI was used as the conducting salt. However, the galvanostatic trace revealed more clearly the occurrence of two successive plateaus. Figure 6c,d summarizes the main characteristics obtained from such post-mortem studies. Interestingly, no obvious degradation of the components of the LMP® cell was observed by SEM after cycling (Figure 6a). Moreover, the sulfur content being close to zero in the electrolyte compartment, it was well confirmed that no diffusion of the IC dye occurred upon cycling through the cell, in accordance with the macroscopic observation (to the naked eye) of the SPE layer and our initial expectations based on the polyanionic nature of the IC backbone that would impede polar interactions with the PEO chains. However, the EDS line profile pointed out the presence of sodium traces through the full thickness of the cell with an obvious accumulation at the lithium interface. This important result reveals that a Na/Li exchange reaction occurred between the active material and the lithiated conducting salt; Na^+ ions being subsequently transported by the PEO matrix and reduced at the Li negative electrode. The contamination of the reactive lithium interface by Na supports well the interfacial perturbations observed during the electrochemical measurements, especially the rapid increase of the apparent cell resistance. Indeed, the magnification by the SEM, shown in Figure 6c, revealed some damages at the electrolyte/Li interface making it more resistive. Consequently, one could easily understand that such perturbations were found to be more pronounced with the operation time and for higher electrode loadings (i.e., when more Na atoms were present inside the cell). This possibility of ion exchange reaction between these two alkali metal ions could have been anticipated because the Na^+ mobility in PEO matrix had been demonstrated by the pioneer work of P. V. Wright and co-workers [24] but not necessarily this marked instability of the electrochemical interfaces. Therefore, the promising cycling data obtained when using thin IC-based composite electrodes confirmed the stable reversible electrochemical activity of the sodium (*E*)-3,3′-dioxo-[2,2′-biindolinylidene]-5,5′-disulfonate redox center and thus it can be anticipated that replacing IC with the corresponding lithiated salt should allow better electrochemical performance in LMP® technology (note this salt is not, however, commercially available).

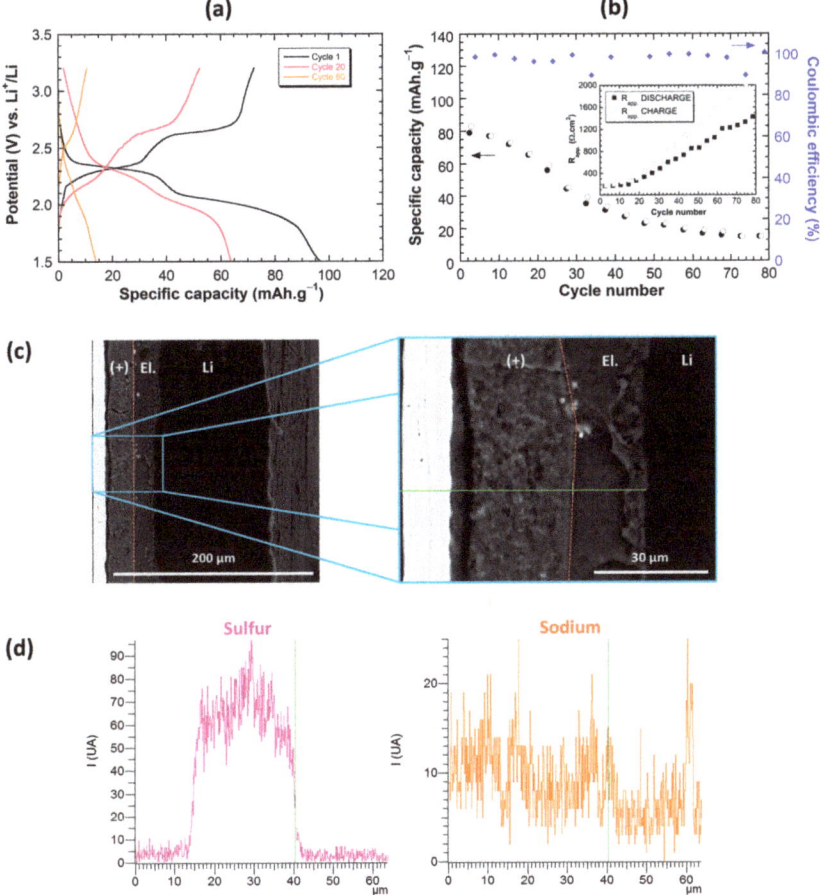

Figure 6. Post-mortem investigations of a LMP® cell stopped after 80 cycles for a composite electrode made of IC (70 wt.%) mixed with KB600 (10 wt.%), LiClO$_4$ (2 wt.%), and PEO-PPO (18%); electrode thickness: 23 μm. (**a**) Galvanostatic cycling curve evolution of the cell (cycling rate: 1 Li$^+$ exchanged per mole of IC in 1 h (C/2)). (**b**) Corresponding capacity retention curve together with the coulombic efficiency. Inset, evolution of the apparent cell resistance upon cycling. (**c**) SEM imaging (edge view) of the full cell together with a magnification of a representative part of the current collector/composite electrode/electrolyte surface area. The red dashed line indicates the boundary between the positive electrode and the electrolyte; (+): IC-based composite electrode; **El.**: solid polymer electrolyte; **Li**: metallic lithium for the negative electrode. (**d**) EDS spectroscopy results (line profile along the green line shown in (c)) highlighting the presence of sodium traces through the full thickness of the cell and more especially at the Li interface. The vertical green line corresponds to the boundary between the composite electrode and the SPE layer.

3. Materials and Methods

3.1. Reagents, Electrode Preparation, and LMP® Cell Assembly

Indigo carmine (Sigma Aldrich, St. Louis, MO, USA, for microscopy), lithium bis (trifluoromethanesulfonyl)imide (LiTFSI, 3M), tri-hydrated lithium perchlorate LiClO$_4$·3H$_2$O (Aldrich), carbon black Ketjenblack® EC600JD (denoted KB600, AkzoNobel), poly(ethylene oxide)/poly(propylene oxide) copolymer (denoted PEO-co-PBO or ICPSEB, Nippon Shokubaï) were used as received. The composite electrode was prepared as follows: An amount of 4 g of carbon black KB600, 1.6 g of LiTFSI, 6.4 g of ICPSEB (binder), and 28 g of IC were mixed together in water and introduced in a plastograph® Brabender® and maintained at 80 °C at a rotation speed of 80 rpm for 20 min. The as-obtained paste was spread between two heating-rolls (95 °C) over an aluminum current-collector coated with carbon. Then, water was evaporated in an oven for 20 min at 105 °C in a dry room with controlled moisture (dew point of −55 °C). Two representative electrode thicknesses were more specifically selected: 27 µm (0.30 mAh cm^{-2}) and 6 µm (0.07 mAh cm^{-2}). Note that 0.81 g of LiClO$_4$·3H$_2$O and 7.55 g of ICPSEB were used in the case of composite electrodes containing LiClO$_4$ as the conducting salt (LiTFSI-free composite electrodes, thickness: 23 µm). Lithium foils for the negative electrode as well as PEO-based electrolyte membranes (O/Li ratio = 25) were provided by the Blue*Solutions* company. The polymer cells were assembled in a dry room by stacking lithium, solid polymer electrolyte membrane, and the composite IC-based positive electrode onto its current collector at 80 °C. On the negative side, the current collector was a copper foil hand-welded to lithium. Each current-collector was assembled to copper-connectors by spot welding. Cells were air tight sealed in coffee-bags and vacuumed in order to avoid air pockets (geometrical surface area: ≈20 cm^2).

3.2. Electrochemical Measurements and Characterization Techniques

The galvanostatic cycling data of the cells were measured by using a MCV 64-1/0.1/0.01/-5 CE battery testing equipment (Bitrode). The apparent cell resistance (R_{app}), also known as internal cell resistance, was determined by the simple current interrupt technique by using the open circuit voltage (OCV) periods integrated in the cycling program at each half-cycle. A VMP3 system (Bio-Logic SAS, Seyssinet-Pariset, France) was used for the PITT measurements. The reported capacity values referred to the mass of the active material (IC) in the positive electrode. For the post-mortem study, the cell was opened in a dry room. A 1 cm^2 piece of this cell was cut with a blade, frozen in liquid nitrogen, and finally sliced with a microtome to get proper cross-sections. Morphology of the as-prepared samples was then investigated by scanning electron microscopy (SEM) using a Hitachi TM3000 equipped with a Bruker Quantax 70 for energy-dispersive X-ray spectroscopy (EDS) analysis.

4. Conclusions

The electrochemical behavior of IC was described for the first time in an SPE-based SSB using the LMP® technology commercialized by Blue*Solutions* at an operating temperature of 100 °C and for a low conductive carbon/IC mass ratio (≈0.14). First, ≈25 µm thick electrodes were evaluated to be compared with the cycling data, reported in the literature, obtained at 25–30 °C in Li-half cell configuration by using carbonate-based liquid electrolytes. The as-obtained galvanostatic cycling curves were well aligned with the expected electrochemical features associated with the occurrence of a reversible two-electron process. However, the cycling performances were altered as a result of interfacial issues at the lithium negative electrode. The latter stemmed from the presence of sodium upon the Na$^+$/Li$^+$ ion exchange between the SPE electrolyte and IC. Much better capacity retention curves were obtained when we decreased the Na contamination thanks to the use of thinner electrodes (6 µm). For such electrodes, the galvanostatic cycling profile was, in addition, significantly modified and exhibited two fully reversible and well-defined plateaus, corresponding to the storing of one e^-/Li$^+$ per IC in line with the description of the electrochemical process at the molecular level. This investigation therefore reinforces the interest of using organic salts (rather than neutral molecules) as active materials to suppress the possible solubilization/diffusion into the SPE membrane and highlights that caution should be taken to avoid alkaline ions mixing within a Li cell.

Author Contributions: Conceptualization, M.D., J.G. and P.P.; Methodology, M.L., M.D., J.G. and P.P.; Validation, M.D., D.G., J.G. and P.P.; Formal Analysis, M.D., J.G. and P.P.; Investigation, M.D.; Resources, M.D., D.G., J.G. and P.P.; Writing—Original Draft Preparation, M.L. and P.P.; Writing—Review & Editing, M.L., M.D., D.G., J.G. and P.P.; Supervision, M.D., D.G., J.G. and P.P.; Funding Acquisition, M.D. All authors have read and agreed to the published version of the manuscript.

Funding: This research was funded by the Blue*Solutions* company.

Institutional Review Board Statement: Not applicable.

Informed Consent Statement: Not applicable.

Data Availability Statement: Data are the property of Blue*Solutions*.

Acknowledgments: The authors would like to dedicate this article to the memory of Jean Rouxel (1935–1998), founder of the 'Institut des Matériaux de Nantes' (IMN), recipient of the CNRS Gold Medal in 1997, member of the French Academy of Sciences, a chemist known worldwide especially for pioneer developments in the chemistry and physics of intercalation compounds (low-dimensional materials), including their first applications for the electrochemical energy storage. The authors would like also to thank P. Deniard for helpful discussions about the manuscript.

Conflicts of Interest: The authors declare no conflict of interest.

Sample Availability: Samples of the compounds are not available from the authors.

References

1. Ziegler, M.S.; Trancik, J.E. Re-Examining Rates of Lithium-Ion Battery Technology Improvement and Cost Decline. *Energy Environ. Sci.* **2021**, *14*, 1635–1651. [CrossRef]
2. Zhang, H.; Li, C.; Eshetu, G.G.; Laruelle, S.; Grugeon, S.; Zaghib, K.; Julien, C.; Mauger, A.; Guyomard, D.; Rojo, T.; et al. From Solid-Solution Electrodes and the Rocking-Chair Concept to Today's Batteries. *Angew. Chem. Int. Ed.* **2020**, *59*, 534–538. [CrossRef] [PubMed]
3. Brandt, K. Historical Development of Secondary Lithium Batteries. *Solid State Ion.* **1994**, *69*, 173–183. [CrossRef]
4. Yoshino, A. The Birth of the Lithium-Ion Battery. *Angew. Chem. Int. Ed.* **2012**, *51*, 5798–5800. [CrossRef]
5. Blomgren, G.E. The Development and Future of Lithium Ion Batteries. *J. Electrochem. Soc.* **2017**, *164*, A5019–A5025. [CrossRef]
6. Goodenough, J.B. How We Made the Li-Ion Rechargeable Battery. *Nat. Electron.* **2018**, *1*, 204. [CrossRef]
7. Mauger, A.; Julien, C.M.; Goodenough, J.B.; Zaghib, K. Tribute to Michel Armand: From Rocking Chair—Li-Ion to Solid-State Lithium Batteries. *J. Electrochem. Soc.* **2020**, *167*, 070507. [CrossRef]
8. Shanmukaraj, D.; Ranque, P.; Ben Youcef, H.; Rojo, T.; Poizot, P.; Grugeon, S.; Laruelle, S.; Guyomard, D. Review—Towards Efficient Energy Storage Materials: Lithium Intercalation/Organic Electrodes to Polymer Electrolytes—A Road Map (Tribute to Michel Armand). *J. Electrochem. Soc.* **2020**, *167*, 070530. [CrossRef]
9. Hérold, A. Insertion Compounds of Graphite with Bromine and the Alkali Metals. *Bull. Soc. Chim. Fr.* **1955**, *187*, 999–1012.
10. Rüdorff, W. Über Die Einlagerung von Unedlen Metallen in Graphit Sowie in Metallchalkogenide Vom Typ MeX$_2$. *Chimia* **1965**, *19*, 489–499.
11. Danot, M.; Le Blanc, A.; Rouxel, J. Les Composés Intercalaires K$_x$TiS$_2$. *Bull. Soc. Chim. Fr.* **1969**, *8*, 2670–2675.
12. Rouxel, J.; Danot, M.; Bichon, J. Les Composés Intercalaires Na$_x$TiS$_2$. Etude Structurale Générale des Phases Na$_x$TiS$_2$ et K$_x$TiS$_2$. *Bull. Soc. Chim. Fr.* **1971**, *11*, 3930–3935.
13. Bichon, J.; Danot, M.; Rouxel, J. Systématique Structurale pour les Séries d'Intercalaires M$_x$TiS$_2$ (M = Li, Na, K, Rb, Cs). *C. R. Acad. Sci.* **1973**, *276*, 1283–1286.
14. van Gool, W. Fast Ion Transport in Solids: Solid State Balteries and Devices. In *Proceedings of the NATO Sponsored Advanced Study Institute on Fast Ion Transport in Solids, Solid State Balteries and Devices*, Belgirate, Italy, 5–16 September 1972.
15. Winn, D.A.; Shemilt, J.M.; Steele, B.C.H. Titanium Disulphide: A Solid Solution Electrode for Sodium and Lithium. *Mater. Res. Bull.* **1976**, *11*, 559–566. [CrossRef]
16. Whittingham, M.S. Electrical Energy Storage and Intercalation Chemistry. *Science* **1976**, *192*, 1126–1127. [CrossRef]
17. Rouxel, J. Recent Progress in Intercalation Chemistry: Alkali Metals in Chacogenide Host Structures. *Rev. Inorg. Chem.* **1979**, *1*, 245–279.
18. Brec, R.; Schleich, D.M.; Ouvrard, G.; Louisy, A.; Rouxel, J. Physical Properties of Lithium Intercalation Compounds of the Layered Transition-Metal Chalcogenophosphites. *Inorg. Chem.* **1979**, *18*, 1814–1818. [CrossRef]
19. Brec, R.; Ouvrard, G.; Louisy, A.; Rouxel, J.; Lemehaute, A. The Influence, on Lithium Electrochemical Intercalation, of Bond Ionicity in Layered Chalcogenophosphates of Transition Metals. *Solid State Ion.* **1982**, *6*, 185–190. [CrossRef]
20. Lazzari, M.; Scrosati, B. A Cyclable Lithium Organic Electrolyte Cell Based on Two Intercalation Electrodes. *J. Electrochem. Soc.* **1980**, *127*, 773–774. [CrossRef]
21. Guyomard, D.; Tarascon, J.M. Rechargeable Li$_{1+x}$Mn$_2$O$_4$/Carbon Cells with a New Electrolyte Composition: Potentiostatic Studies and Application to Practical Cells. *J. Electrochem. Soc.* **1993**, *140*, 3071–3081. [CrossRef]
22. Armand, M.B.; Chabagno, J.M.; Duclot, M. Extended Abstracts. In *Proceedings of the Second International Meeting on Solid Electrolytes*, St Andrews, Scotland, 20–22 September 1978.
23. Armand, M.B.; Chabagno, J.M.; Duclot, M. *Fast Ion Transport in Solids*; Vashishta, P., Mundy, J.N., Shenoy, J.K., Eds.; North Holland Publishers: Amsterdam, The Netherlands, 1979; p. 131.

24. Fenton, D.E.; Parker, J.M.; Wright, P.V. Complexes of Alkali Metal Ions with Poly(Ethylene Oxide). *Polymer* **1973**, *14*, 589. [CrossRef]
25. Wright, P.V. Electrical Conductivity in Ionic Complexes of Poly(Ethylene Oxide). *Brit. Poly. J.* **1975**, *7*, 319–327. [CrossRef]
26. Liu, B.; Zhang, J.-G.; Xu, W. Advancing Lithium Metal Batteries. *Joule* **2018**, *2*, 833–845. [CrossRef]
27. Albertus, P.; Anandan, V.; Ban, C.; Balsara, N.; Belharouak, I.; Buettner-Garrett, J.; Chen, Z.; Daniel, C.; Doeff, M.; Dudney, N.J.; et al. Challenges for and Pathways toward Li-Metal-Based All-Solid-State Batteries. *ACS Energy Lett.* **2021**, 1399–1404. [CrossRef]
28. Vandepaer, L.; Cloutier, J.; Amor, B. Environmental Impacts of Lithium Metal Polymer and Lithium-Ion Stationary Batteries. *Renew. Sustain. Energy Rev.* **2017**, *78*, 46–60. [CrossRef]
29. Esser, B.; Dolhem, F.; Becuwe, M.; Poizot, P.; Vlad, A.; Brandell, D. A Perspective on Organic Electrode Materials and Technologies for next Generation Batteries. *J. Power Sources* **2021**, *482*, 228814. [CrossRef]
30. Judez, X.; Qiao, L.; Armand, M.; Zhang, H. Energy Density Assessment of Organic Batteries. *ACS Appl. Energy Mater.* **2019**, *2*, 4008–4015. [CrossRef]
31. Lécuyer, M.; Gaubicher, J.; Barrès, A.-L.; Dolhem, F.; Deschamps, M.; Guyomard, D.; Poizot, P. A Rechargeable Lithium/Quinone Battery Using a Commercial Polymer Electrolyte. *Electrochem. Commun.* **2015**, *55*, 22–25. [CrossRef]
32. Li, W.; Chen, L.; Sun, Y.; Wang, C.; Wang, Y.; Xia, Y. All-Solid-State Secondary Lithium Battery Using Solid Polymer Electrolyte and Anthraquinone Cathode. *Solid State Ion.* **2017**, *300*, 114–119. [CrossRef]
33. Poizot, P.; Gaubicher, J.; Renault, S.; Dubois, L.; Liang, Y.; Yao, Y. Opportunities and Challenges for Organic Electrodes in Electrochemical Energy Storage. *Chem. Rev.* **2020**, *120*, 6490–6557. [CrossRef]
34. Bouridah, A.; Dalard, F.; Armand, M.B. Electrochemical Properties of Poly(Decaviologen) in Polymer Media. *J. Appl. Electrochem.* **1990**, *20*, 1040–1044. [CrossRef]
35. Chen, H.; Armand, M.; Demailly, G.; Dolhem, F.; Poizot, P.; Tarascon, J.-M. From Biomass to a Renewable $Li_xC_6O_6$ Organic Electrode for Sustainable Li-Ion Batteries. *ChemSusChem* **2008**, *1*, 348–355. [CrossRef] [PubMed]
36. Chen, H.; Armand, M.; Courty, M.; Jiang, M.; Grey, C.P.; Dolhem, F.; Tarascon, J.-M.; Poizot, P. Lithium Salt of Tetrahydroxybenzoquinone: Toward the Development of a Sustainable Li-Ion Battery. *J. Am. Chem. Soc.* **2009**, *131*, 8984–8988. [CrossRef] [PubMed]
37. Renault, S.; Geng, J.; Dolhem, F.; Poizot, P. Evaluation of Polyketones with N-Cyclic Structure as Electrode Material for Electrochemical Energy Storage: Case of Pyromellitic Diimide Dilithium Salt. *Chem. Commun.* **2011**, *47*, 2414–2416. [CrossRef]
38. Yao, M.; Araki, M.; Senoh, H.; Yamazaki, S.; Sakai, T.; Yasuda, K. Indigo Dye as a Positive-Electrode Material for Rechargeable Lithium Batteries. *Chem. Lett.* **2010**, *39*, 950–952. [CrossRef]
39. Yao, M.; Kuratani, K.; Kojima, T.; Takeichi, N.; Senoh, H.; Kiyobayashi, T. Indigo Carmine: An Organic Crystal as a Positive-Electrode Material for Rechargeable Sodium Batteries. *Sci. Rep.* **2015**, *4*. [CrossRef]
40. Deunf, E.; Poizot, P.; Lestriez, B. Aqueous Processing and Formulation of Indigo Carmine Positive Electrode for Lithium Organic Battery. *J. Electrochem. Soc.* **2019**, *166*, A747–A753. [CrossRef]
41. Kato, M.; Sano, H.; Kiyobayashi, T.; Takeichi, N.; Yao, M. Improvement of the Battery Performance of Indigo, an Organic Electrode Material, Using PEDOT/PSS with D-Sorbitol. *ACS Omega* **2020**, *5*, 18565–18572. [CrossRef]
42. Poizot, P.; Laruelle, S.; Grugeon, S.; Dupont, L.; Tarascon, J.-M. From the Vanadates to 3d-Metal Oxides Negative Electrodes. *Ionics* **2000**, *6*, 321–330. [CrossRef]
43. Delacourt, C.; Poizot, P.; Morcrette, M.; Tarascon, J.-M.; Masquelier, C. One-Step Low-Temperature Route for the Preparation of Electrochemically Active $LiMnPO_4$ Powders. *Chem. Mater.* **2004**, *16*, 93–99. [CrossRef]

Article

Fast Li-Ion Conduction in Spinel-Structured Solids

Jan L. Allen [1,*], Bria A. Crear [2], Rishav Choudhury [3], Michael J. Wang [3], Dat T. Tran [1], Lin Ma [1], Philip M. Piccoli [4], Jeff Sakamoto [3] and Jeff Wolfenstine [5]

[1] Energy Sciences Division, Sensors & Electron Devices Directorate, US Army Research Laboratory, Adelphi, MD 20783, USA; dat.t.tran4.civ@mail.mil (D.T.T.); liam.l.ma.civ@outlook.com (L.M.)
[2] Department of Chemistry, Howard University, Washington, DC 20059, USA; bria.a.crear@gmail.com
[3] Department of Materials Science and Engineering, University of Michigan, Ann Arbor, MI 48109, USA; rishavc@umich.edu (R.C.); micwan@umich.edu (M.J.W.); jeffsaka@umich.edu (J.S.)
[4] Department of Geology, University of Maryland, College Park, MD 20742, USA; piccoli@umd.edu
[5] Solid Ionic Consulting, 9223 Matthews Ave, Seattle, WA 98115, USA; jeffyspeak@outlook.com
* Correspondence: jan.l.allen8.civ@mail.mil

Citation: Allen, J.L.; Crear, B.A.; Choudhury, R.; Wang, M.J.; Tran, D.T.; Ma, L.; Piccoli, P.M.; Sakamoto, J.; Wolfenstine, J. Fast Li-Ion Conduction in Spinel-Structured Solids. *Molecules* **2021**, *26*, 2625. https://doi.org/10.3390/molecules26092625

Academic Editors: Stephane Jobic, Claude Delmas and Myung-Hwan Whangbo

Received: 18 March 2021
Accepted: 29 April 2021
Published: 30 April 2021

Publisher's Note: MDPI stays neutral with regard to jurisdictional claims in published maps and institutional affiliations.

Copyright: © 2021 by the authors. Licensee MDPI, Basel, Switzerland. This article is an open access article distributed under the terms and conditions of the Creative Commons Attribution (CC BY) license (https://creativecommons.org/licenses/by/4.0/).

Abstract: Spinel-structured solids were studied to understand if fast Li$^+$ ion conduction can be achieved with Li occupying multiple crystallographic sites of the structure to form a "Li-stuffed" spinel, and if the concept is applicable to prepare a high mixed electronic-ionic conductive, electrochemically active solid solution of the Li$^+$ stuffed spinel with spinel-structured Li-ion battery electrodes. This could enable a single-phase fully solid electrode eliminating multi-phase interface incompatibility and impedance commonly observed in multi-phase solid electrolyte–cathode composites. Materials of composition $Li_{1.25}M(III)_{0.25}TiO_4$, M(III) = Cr or Al were prepared through solid-state methods. The room-temperature bulk Li$^+$-ion conductivity is 1.63×10^{-4} S cm^{-1} for the composition $Li_{1.25}Cr_{0.25}Ti_{1.5}O_4$. Addition of Li_3BO_3 (LBO) increases ionic and electronic conductivity reaching a bulk Li$^+$ ion conductivity averaging 6.8×10^{-4} S cm^{-1}, a total Li-ion conductivity averaging 4.2×10^{-4} S cm^{-1}, and electronic conductivity averaging 3.8×10^{-4} S cm^{-1} for the composition $Li_{1.25}Cr_{0.25}Ti_{1.5}O_4$ with 1 wt. % LBO. An electrochemically active solid solution of $Li_{1.25}Cr_{0.25}Mn_{1.5}O_4$ and $LiNi_{0.5}Mn_{1.5}O_4$ was prepared. This work proves that Li-stuffed spinels can achieve fast Li-ion conduction and that the concept is potentially useful to enable a single-phase fully solid electrode without interphase impedance.

Keywords: solid electrolyte; fast Li$^+$ ion conductor; Li-ion battery; spinel; solid-state battery; cathode-electrolyte interface

1. Introduction

Interest in solid-state electrolytes has intensified owing to the discovery of fast Li-ion conduction in the cubic garnet structure [1], the continued push for higher energy density batteries and the allure of the safety of an inorganic all solid-state battery. The spinel is a suitable cubic structure to search for fast Li-ion conduction owing to its network of empty edge-shared MO_6 octahedra bridged by face-shared LiO_4 tetrahedra, which connect in three dimensions thereby providing a path for 3D Li$^+$-ion conduction [2]. In fact, $LiMn_2O_4$ spinel's favorable mixed electronic-ionic conductivity has enabled its use as a positive electrode [2]. $LiMn_2O_4$ ideally crystallizes in the normal spinel structure in the $Fd\bar{3}m$ space group: Mn occupies the 16d octahedral site, O occupies the 32e position and Li occupies the 8a tetrahedral site which share faces with an empty 16c octahedral site within the spinel's pseudo-cubic closed packed oxygen framework, thus forming a three-dimensional 8a → 16c → 8a Li ion conduction pathway. Furthermore, the spinel's cubic unit cell is desirable for solid-state battery application since differences in thermal expansion coefficients of different crystallographic directions in large-grained non-cubic ceramic materials, may lead to micro-cracking [3–5] during cooling after densification which is unfavorable for mechanical properties and ionic conductivity [6–9]. Additionally, the use of a spinel-structured solid electrolyte as a separator in an all solid-state battery and pursuit of high conductivity in the spinel structure can lead to insights that may improve rate capability of spinel structured electrodes for use with liquid based electrolytes or as a catholyte or anolyte in a fully solid-state configuration.

Limited work has been done on oxide spinel structured solid electrolytes. Kawai et al. discovered a conductivity of about 10^{-7} S cm^{-1} for the ordered (P_432 space group) spinel LiNi$_{0.5}$Ge$_{1.5}$O$_4$ at 63 °C [10]. The Ni and Ge are ordered on the octahedral sites of this compound. In 1985, Thackeray and Goodenough proposed the all-solid all-spinel battery as a means to reduce interfacial impedance at the interface of the solid-state cathode, electrolyte and anode but did not identify a suitable solid-state electrolyte [11]. Rosciano et al. suggested the Li doped MgAl$_2$O$_4$ spinel as a potential solid-state electrolyte based on high Li diffusivity as measured by nuclear magnetic resonance (NMR) as a means to enable a full spinel concept [12]. However, Djenadic et al. reported that the Li motion is localized in Li doped MgAl$_2$O$_4$ and therefore the long-range Li conductivity is insufficient for realization of an all-solid all-spinel battery [13].

Spinel is similar to garnet in that in both structures, the occupied tetrahedral and empty octahedral sites form an interconnected 3-D array for Li$^+$-ion transport. Conventional garnets are described by the formula A$_3$B$_3$C$_2$O$_{12}$ where A, B, and C have 8, 4, and 6 oxygen coordination, respectively. However, in garnet structured Li$_3$Nd$_3$Te$_2$O$_{12}$, where Li only occupies tetrahedral sites, the Li-ion is practically immobile at room temperature and ionic conductivity can only be measured at elevated temperature, only achieving 10^{-5} S cm^{-1} at 600 °C [14]. In contrast, when additional Li is added to occupy both tetrahedral and octahedral sites in a "Li-stuffed" garnet such as Li$_5$La$_3$Ta$_2$O$_{12}$ [15] room temperature conductivity rises to 1.2×10^{-6} S cm^{-1}. Further Li leads to even higher conductivity in Li$_{7-3x}$La$_3$Zr$_2$Al$_x$O$_{12}$ which has room temperature ionic conductivity greater than 10^{-4} S cm^{-1} [1]. In this work, we studied a substitutional strategy to form a "Li-stuffed" spinel, with Li occupying both the tetrahedral 8a and octahedral 16d sites, in a somewhat analogous fashion to Li-stuffed cubic garnet. Since we have different energy (tetrahedral and octahedral) sites, the presence of Li on the octahedral sites tends to reduce the energy potential between the two sites making Li motion easier [16]. A room-temperature Li$^+$ ionic conductivity greater than 10^{-4} S cm^{-1} is observed for the Li-stuffed spinel. In the composition of highest conductivity, Li$_{1.25}$Cr$_{0.25}$Ti$_{1.5}$O$_4$, the 8a tetrahedral site is fully occupied by Li and the 16d octahedral site is 75% occupied by Ti and the remaining 25% split evenly by Li and Cr. The remaining tetrahedral, 8b, and octahedral, 16c, are unoccupied. Li$_{1.25}$Cr$_{0.25}$Ti$_{1.5}$O$_4$ contains Ti(IV), which is unstable to Li reduction; however, the use of an interfacial layer such as Li$_3$N or a conductive polymer can fix this problem [17] to enable its use with Li or LiC$_6$ anodes or more likely the concept can be used in solid solution formation with spinel-structured cathodes, where mixed electronic ionic conductivity is desirable. The integration of solid electrolytes with lithium has progressed rapidly since the report of high conductivity in garnet; however, there has been a lack of progress with cathode–solid electrolyte integration to achieve high ionic and electronic transport through the cathode [18,19]. Unlike conventional Li-ion cells where liquid electrolyte fills the cathode pores providing ion transport and carbon black provides electron transport, cathodes for all solid-state batteries require both ion-conducting as well as electron-conducting additives to enable mixed ionic/electronic transport. However, these solid–solid interfaces have high resistance compared to solid–liquid interfaces, and transport through composite cathodes is a major challenge to enable all solid-state batteries [20]. In a first small step towards application of the concept as solid-state catholyte, we show a solid solution of a Li-stuffed, spinel-structured electrolyte and the LiNi$_{0.5}$Mn$_{1.5}$O$_4$ high voltage, spinel-structured positive electrode material to be an electrochemically active cathode material in a liquid-containing cell thus showing that no high resistance solid–solid interfaces are formed.

2. Results

2.1. X-ray Diffraction (XRD) and Structural Refinement

The XRD patterns of the $Li_{1.25}Cr_{0.25}Ti_{1.5}O_4$ (LCTO, bottom) and $Li_{1.25}AlTi_{1.5}O_4$ (LATO, top) powders are shown in Figure 1. The patterns are indexed to the cubic spinel structure, space group, $Fd\bar{3}m$. The LATO pattern has two small, unidentified peaks at 39.64 and 46.18° 2Θ which disappear after hot-pressing (Figure 2). The lattice parameters determined from Rietveld refinement and using Si as internal peak position standard are 8.3440 Å and 8.3574 Å for $Li_{1.25}Cr_{0.25}Ti_{1.5}O_4$ (LCTO) and, $Li_{1.25}AlTi_{1.5}O_4$ (LATO), respectively. Since, Cr^{3+} is larger than Al^{3+}, 0.615 Å vs. 0.535 Å [21,22], the fact that the unit cell of LATO is larger than LCTO might suggest a small amount of Al^{3+} mixing onto the tetrahedral 8a spinel site in exchange for Li^+ (0.76 Å) on the 16d octahedral site in the $Fd\bar{3}m$ space group. Site mixing is highly unlikely for d^3 Cr^{3+} owing to its well-known high crystal field stabilization energy in octahedral coordination [23]. Site mixing of Al onto the 8a tetrahedral sites is common in spinels and is, for example, observed in $MgAl_2O_4$ [23]. Partial occupation of the 8a tetrahedral site by the heavier Al atom relative to Li would be evidenced in the XRD pattern by an increase in the intensity of the 220 (~30° 2Θ, Cu K α radiation) and the 422 (~54° 2Θ, Cu K α radiation) peaks [24].

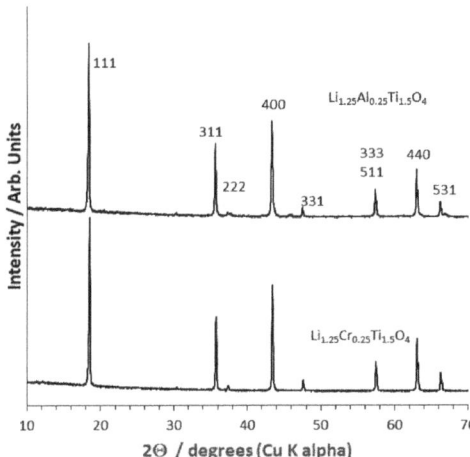

Figure 1. XRD of $Li_{1.25}Cr_{0.25}Ti_{1.5}O_4$ solid electrolyte powder (**bottom**) and $Li_{1.25}Al_{0.25}Ti_{1.5}O_4$ solid electrolyte powder (**top**). XRD peaks are indexed to the $Fd\bar{3}m$ spinel structure.

However, comparison of the LCTO XRD pattern versus the LATO XRD pattern (Figure 1) does not indicate any significant difference in the intensity of the peaks for the two samples, suggesting that any site mixing between Al and Li in LATO is negligible. The lack of significant site mixing is further evidenced from Rietveld structure refinement using power XRD data. Structural analysis by Rietveld refinement of XRD data was done with the Fullprof program [25].

The Rietveld refinement results are plotted with WINPLOTR program [26] in the supplementary information (Figures S1 and S2) and the atomic positions and final refinement information is contained in the supplementary information (Tables S1 and S2), for LCTO and LATO, respectively. In the starting Rietveld structural model for both LCTO and LATO, Li was placed on the 8a tetrahedral site, Li, Ti and Cr or Al were randomly distributed on the 16d octahedral site and oxygen on the 32e site of the $Fd\bar{3}m$ space group. During the refinement, site mixing of Ti, Al, or Cr on the 8a tetrahedral site was explored but led to a lower goodness of fit. That is, in the fully converged refinements, the 8a octahedral site is occupied by Li and the 16d octahedral site is randomly occupied by Li, Ti, and (Cr or Al). The oxygen positional parameter (u), the atomic position coordinate of oxygen in the 32e site, was refined to a value of 0.26321 and 0.26340 for LCTO and LATO, respectively. The u values should be taken as an approximation since it is relatively difficult to locate oxygen positions through X-ray diffraction. Neutron diffraction will be needed to definitively define the oxygen position. The relative value of the positional parameter is in accordance with what is expected based on the relative sizes of Cr^{3+} and Al^{3+} [27]. If the origin of the unit cell is taken as the center of symmetry, in an ideal, cubic closed packed oxygen lattice, the oxygen parameter, u, has a value of 0.250. In the case of $u = 0.250$, the octahedral cation–oxygen bond length is 1.155 times longer than the tetrahedral–oxygen bond length [27]. However, the spinel oxygen parameter changes depend on the size and charge of the cations occupying the tetrahedral and octahedral sites, distorting the cubic closed packed oxygen lattice to accommodate different ions. In fact, more than 30 different ions of varying size can be accommodated in the spinel structure [27]. To accommodate large cations on the tetrahedral sites, oxygens are displaced along the [11] direction increasing the tetrahedral cation–oxygen bond length and concurrently decreasing the octahedral cation–oxygen bond length, leading to an increase in u and vice versa. Manipulation of the u parameter through changes in site occupation of octahedral sites might be a tool to optimize Li-ion conductivity in spinels.

In summary, both LCTO and LATO are single phase spinel structured materials with Li occupying 8a and 16d sites, Cr, Al and Ti randomly occupying the 16d sites in a nearly cubic close packed oxygen framework. The oxygen framework adjusts the positions of the oxygens within the unit cell to accommodate the relative difference in size of Cr versus Al.

XRD patterns of hot-pressed and Li_3BO_3 (LBO)-containing samples are shown in Figure 2, indicating no new peaks, i.e., no new phases, and retention of the single-phase spinel structure. LBO, a low-melting sintering aid, was added to increase density and thereby improve conductivity and mechanical properties.

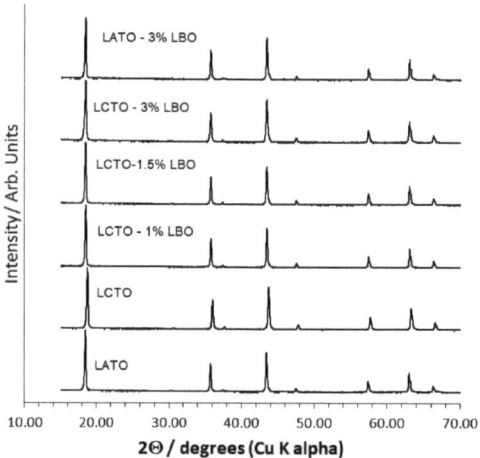

Figure 2. XRD of hot-pressed pellets of $Li_{1.25}Cr_{0.25}Ti_{1.5}O_4$ (LCTO) and $Li_{1.25}Al_{0.25}Ti_{1.5}O_4$ (LATO) with variable amounts of Li_3BO_3 (LBO). Pellets were ground to a powder prior to XRD data collection.

LBO is not detected in the XRD pattern owing to the light elements present and the low concentration. Generally, phases at or below a few weight% are not detected and some site substitution on spinel is also a possibility. The lattice constant as a function of Li_3BO_3 content for hot-pressed pellets was determined by grinding the pellets to a powder, the addition of NIST-traceable Si as an internal peak position standard and Rietveld refinement of the XRD pattern. The obtained lattice constants are tabulated and plotted in the supplementary information, Table S3 and Figure S3, respectively. In the case of LATO, there is an increase in the lattice constant of hot-pressed pellets with the addition of 3% LBO to LATO from 8.3459(1) to 8.3474(1) Å. The LCTO lattice constants change in a more complicated manner. There is an initial increase in the lattice constant from 0% to 1% LBO, 8.3444(1) to 8.3456(1) Å and then a roughly linear decrease in lattice constant from 1, 1.5 and 3% LBO, 8.3456(1), 8.3450(1), 8.3449(1) Å, respectively. An increase might be attributed to substitution of additional relatively large Li onto the lattice and the decrease to substitution of the relatively small B. The evidence for site substitution by B is in agreement with the WDS analysis, which showed B distributed throughout the samples and not only at grain boundaries.

2.2. Microstructure

Representative micrographs of the fracture surfaces of the hot-pressed LCTO sample without LBO (left) and with 1% LBO (right) are shown in Figure 3. From the SEM analysis, a couple of important points are noted. First, the LBO-containing sample is very dense in agreement with the high relative density ~98%, determined from the physical dimensions, weight and the theoretical density in contrast to the ~94% density of the sample without LBO. Almost no porosity is observable in the LBO-containing sample. A high relative density is extremely important for device applications because it leads to increased mechanical strength and higher ionic conductivity.

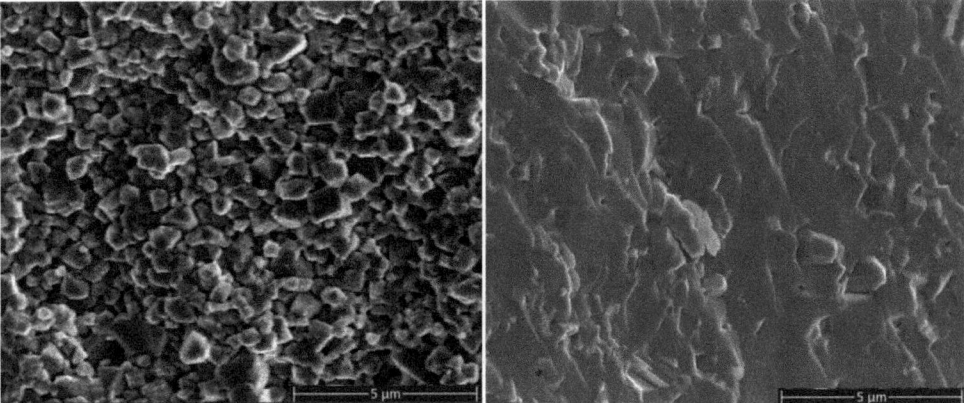

Figure 3. Representative SEM image of fracture surface images of hot-pressed $Li_{1.25}Cr_{0.25}Ti_{1.5}O_4$ without LBO, **left**, and with 1% LBO, **right**.

Second, the fracture surface is very flat indicating transgranular fracture revealing high grain boundary strength, which should lead to low inter-granular ionic resistance whereas for the LBO free sample the fracture node is primarily intergranular leading to higher grain boundary resistance. Third, the average grain size observable for the LBO free sample is about 1 µm. In the LBO-containing sample, the grain size is roughly estimated to be about 2–5 µm, though the grains are difficult to distinguish.

2.3. Conductivity

The room temperature impedance plot for hot-pressed $Li_{1.25}Cr_{0.25}Ti_{1.5}O_4$ (~94% relative density) and the equivalent circuit (inset) which models the data are shown in Figure 4. In the equivalent circuit, R refers to resistance, and CPE to constant phase element. The impedance spectra shows a single semi-circle at higher frequency and starts to level off at lower frequency, which we interpret as the precursor to an upward sloping line. A fit of the data, based on this interpretation using the indicated equivalent circuit, is included in the supplementary information (Figure S4, Table S4). The equivalent circuit for this system where ionic conduction is predominant includes R_b, bulk or intra-grain impedance,

R_{gb}, the grain boundary or inter-grain impedance, CPE_{gb}, the grain boundary constant phase element and CPE_{int}, the sample electrode interface or dual layer constant phase element which is physically attributed to charge build-up at the electrode [28–30]. Since we used Li-ion blocking electrodes, the shape of the curve represents a material, which is predominantly a Li-ion conductor with low electronic conductivity [28–30]. From Figure 4, several important points are noted.

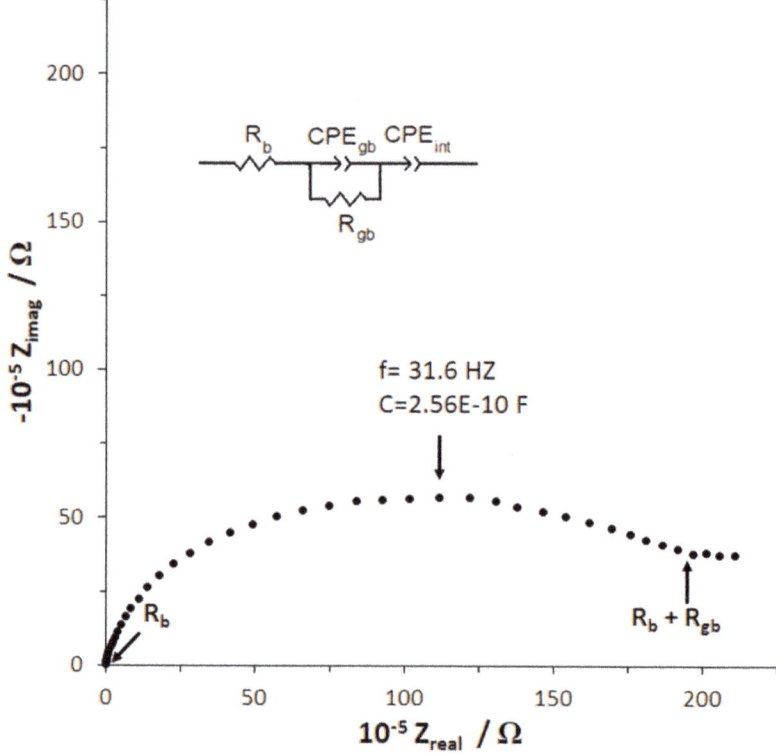

Figure 4. Room temperature impedance plot of hot-pressed $Li_{1.25}Cr_{0.25}Ti_{1.5}O_4$ and the equivalent circuit used to interpret the data.

First, the calculated value of the capacitance using the frequency at the maximum point of the semi-circle is shown on Figure 4. This capacitance, 2.56×10^{-10} F, was calculated from $C_{gb} = (2\pi f R)^{-1}$, using $f = 31$ Hz and R (diameter of the semi-circle) = 1.97×10^5 Ω [31]. Second, this capacitance value is characteristic of a grain boundary [31] confirming the assignment of this semi-circle to a grain boundary phenomenon. The bulk impedance value, R_b can be taken from the Z_{real} intercept at the high frequency of the semi-circle and the total impedance, $R_{total} = R_b + R_{gb}$, is taken from the Z_{real} low frequency intercept. Third, the values of R_b and R_{gb} and the physical dimensions of the sample are then used to determine the Li-ion conductivity. The bulk ionic conductivity of the $Li_{1.25}Cr_{0.25}Ti_{1.5}O_4$ pellet at room temperature is 1.63×10^{-4} S cm^{-1} and the total ionic conductivity of $Li_{1.25}Cr_{0.25}Ti_{1.5}O_4$ is 2.84×10^{-8} S cm^{-1}. This bulk ionic conductivity is in the range of Al substituted $Li_7La_3Zr_2O_{12}$ cubic garnet solid-state electrolyte when first reported by Murugan et al. [1], however, the total ionic conductivity is three to four orders of magnitude lower than cubic garnet indicating relatively high grain boundary impedance. In fact, the ratio of grain boundary impedance to the total ionic impedance is 99.93%. We compare the ratio of grain boundary impedance to the total ionic impedance since we cannot calculate a grain boundary conductivity as the volume of the grain boundaries is unknown.

The electronic conductivity of $Li_{1.25}Cr_{0.25}Ti_{1.5}O_4$ at room temperature obtained from the steady state current found through DC polarization [32,33] is about 1.84×10^{-8} S cm^{-1}. Thus, the ionic transport number, t_{ionic}, for Li-ions in $Li_{1.25}Cr_{0.25}Ti_{1.5}O_4$ ($t_{ionic} = \sigma_{ionic}/\sigma_{total}$, where $\sigma_{total} = \sigma_{ionic} + \sigma_{electronic}$; σ_{ionic} is the total ionic conductivity) is about 0.6, confirming that $Li_{1.25}Cr_{0.25}Ti_{1.5}O_4$ is an ionic conductor, yet having a significant electronic conductivity. Similarly, $Li_{1.25}Al_{0.25}Ti_{1.5}O_4$ was prepared, densified through hot-pressing and analyzed. The bulk ionic, total ionic and electronic conductivities obtained from the 97% relative density $Li_{1.25}Al_{0.25}Ti_{1.5}O_4$ pellet are 5.11×10^{-5}, 4.08×10^{-7}, and 9.79×10^{-8} S cm^{-1}, respectively, a slightly lower bulk conductivity but a slightly higher bulk ionic and electronic conductivity than that of the 94% relative density LCTO pellet. The ionic transport number, t_{ionic}, for Li-ions in $Li_{1.25}Al_{0.25}Ti_{1.5}O_4$ is 0.81, evidencing a higher ionic component compared to LCTO as might be expected owing to the substitution of more easily reducible Cr^{3+} compared to Al^{3+}. The lower bulk conductivity in LATO relative to LCTO most likely results from the difference in lattice constant, since the Li content is similar. Similarly to LCTO, almost all ionic impedance in the LATO pellet, 99.20%, originates at grain boundaries.

In order to overcome the high grain boundary impedance and to increase the density of the samples, Li_3BO_3 (LBO) was used as a sintering and hot-pressing aid. The room temperature impedance plot of $Li_{1.25}Cr_{0.25}Ti_{1.5}O_4$ hot-pressed with 3 wt% LBO to form a pellet of 98% relative density and the equivalent circuit (inset) which models the data are shown in Figure 5. This figure will be used to illustrate the interpretation of the EIS data for all of the LBO-containing samples. From Figure 5, several points can be made. First, the capacitance is calculated as previously described in the discussion of Figure 4 for the first two semi-circles from higher (right) to lower frequency and the values are noted on Figure 5. The calculated capacitances are characteristic of grain boundary and bulk phenomena for the higher frequency and lower frequency semicircles, respectively.

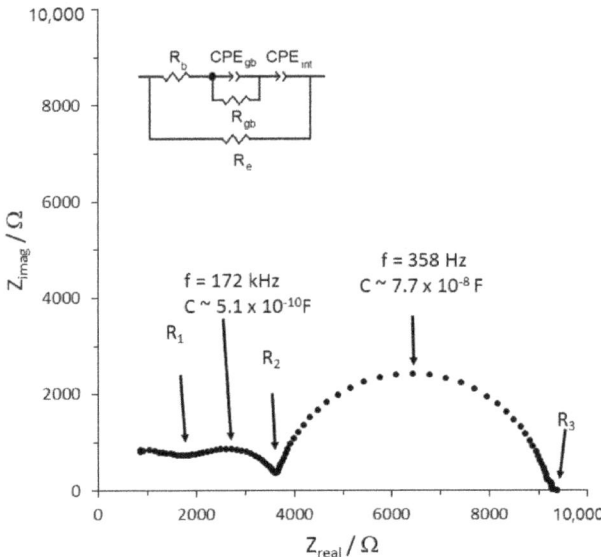

Figure 5. Room temperature impedance plot of hot-pressed $Li_{1.25}Cr_{0.25}Ti_{1.5}O_4$/3% LBO and the equivalent circuit used to interpret the data where $R_1 = R_eR_b/(R_e + R_b)$, $R_2 = R_e(R_b + R_{gb})/(R_e + R_b + R_{gb})$, and $R_3 = R_e$.

Second, the shape of the impedance plot is characteristic of an ionic conductor with electronic conductance [28], in agreement with the observation that hot-pressing with the LBO sintering aid changed the color of the samples from green (LCTO) and white (LATO) to black. Modelling the transport, therefore, requires the addition of a parallel electronic resistance, R_e, to the ionic-conduction circuit [28,29]. The values of the resistances in the equivalent circuit, bulk ionic resistance (R_b), grain boundary ionic resistance (R_{gb}), and electronic resistance (R_e) can be determined from the intercepts, R_1, R_2 and R_3, respectively, based on the following relationships: $R_1 = R_e R_b / (R_e + R_b)$, $R_2 = R_e(R_b + R_{gb})/(R_e + R_b + R_{gb})$, and $R_3 = R_e$ [28,29]. Finally, the values of R_b, R_{ion}, and R_e and the physical dimensions of the sample are then used to determine the Li-ion and electronic conductivities.

For the LBO-containing samples, two pellets were analyzed for each composition. The highest total ionic conductivity was found for the $Li_{1.25}Cr_{0.25}Ti_{1.5}O_4$/1 wt. % LBO composition and is 4.17×10^{-4} S cm^{-1}. The room temperature total Li$^+$ ion conductivity measured for a pellet of $Li_{1.25}Cr_{0.25}Ti_{1.5}O_4$—1 wt.% LBO is near the range of the highest ever reported for an oxide [34]. By comparison, substituted cubic garnet $Li_7La_3Zr_2O_{12}$ has reported total Li-ion conductivity ranging from 5×10^{-4} to 1×10^{-3} S cm^{-1} [34]. The percent of ionic impedance from the grain boundary dropped to 19.31% in this sample. The DC electronic conductivity of $Li_{1.25}Cr_{0.25}Ti_{1.5}O_4$/1 wt.% LBO averaged 3.8×10^{-4} S cm^{-1} which means this composition may have applicability as an anolyte or catholyte where mixed electronic and ionic conductivity is important. The AC electronic conductivity value was in agreement with the DC measurement.

With the addition of 1 wt. % LBO, density increased, bulk ionic conductivity was slightly increased, total ionic conductivity increased by two to three orders of magnitude, and the electronic conductivity increased by four orders of magnitude. Thus, use of an optimal amount (~1 wt.%) of LBO might be particularly attractive to increase electronic conductivity and total ionic conductivity as a catholyte or anolyte. Data for all samples are tabulated in Table 1.

Table 1. Average room temperature (298 K) bulk ionic, σ_{bulk}, grain boundary ionic impedance as percentage of total impedance, Z_{gb}/Z_{tot}, total ionic conductivity, σ_{ion}, electronic conductivity, σ_{elec} and relative density, D, of $Li_{1.25}CrTi_{1.5}O_4$ (LCTO) and $Li_{1.25}Al_{0.25}Ti_{1.5}O_4$ (LATO) solid electrolytes with and without Li_3BO_3 (LBO) sintering aid.

Sample	σ_{bulk} (S cm^{-1})	Z_{gb}/Z_{tot} (%)	σ_{ion} (S cm^{-1})	σ_{elec} (S cm^{-1})	D (%)
LCTO	1.63×10^{-4}	99.9	1.19×10^{-7}	1.84×10^{-8}	94
LATO	5.11×10^{-5}	99.2	4.08×10^{-7}	9.79×10^{-8}	97
LCTO/1% LBO	6.77×10^{-4}	19.3	4.17×10^{-4}	3.76×10^{-4}	98
LCTO/1.5% LBO	4.01×10^{-4}	94.0	1.96×10^{-4}	1.78×10^{-4}	99
LCTO/3% LBO	8.75×10^{-5}	51.9	5.32×10^{-5}	4.06×10^{-5}	97
LATO/3% LBO	5.00×10^{-5}	27.4	1.78×10^{-5}	1.20×10^{-7}	99

Temperature-dependent conductivity data collection focused on the higher conductivity LBO-containing samples. Bulk ionic, grain boundary ionic, total ionic and electronic conductivities of $Li_{1.25}Cr_{0.25}Ti_{1.5}O_4$ (LCTO) and $Li_{1.25}Al_{0.25}Ti_{1.5}O_4$ (LATO) with varied weight percent LBO are shown in Figure 6 as a function of temperature. Log (1/Rgb) versus 1/T was plotted for the grain boundary data since one cannot calculate σ_{gb} since the grain boundary volume is unknown and it only differs from plotting log(σ_{gb}) versus 1/T by a constant [35]. In Figure 6, log σ is plotted as a function of 1/T in order to ease the reading of the conductivity values; however, all values of the activation energies E_A were calculated based on log (σT) plotted as a function of 1/T, where σ is the conductivity (S cm^{-1}) and T is the temperature (K). For the 1/T vs. log (σT) plots, see the supplementary information, Figure S5. From Figure 6, several points can be made. First, all samples show fast room temperature Li-ion conductivity ranging from ~10^{-4} to ~10^{-3} S cm^{-1} and comparable electronic conductivity, suggesting applicability as mixed ionic electronic conductors and the LCTO-1% LBO composition stands out for both high ionic and electronic conductivity. Use of these materials as solid electrolytes will require the discovery of an alternate sintering aid or an alternate densification process to increase density and reduce grain boundary impedance perhaps under oxygen in order to maintain low electronic conductivity or the use of an interfacial layer such as Li$_3$N [17]. However, it may find greater applicability as a catholyte or anolyte where mixed electronic ionic conductivity is desirable.

Turning attention to the bulk ionic conductivity activation energies, the values range from 0.18 to 0.28, 0.32 and 0.32, respectively for LCTO-1%LBO, LCTO-1.5%LBO, LCTO-3%LBO and LATO-3%LBO, respectively. The bulk activation energies are close to what is reported for other fast Li-ion conductors indicating fast Li-ion mobility [34,36–39].

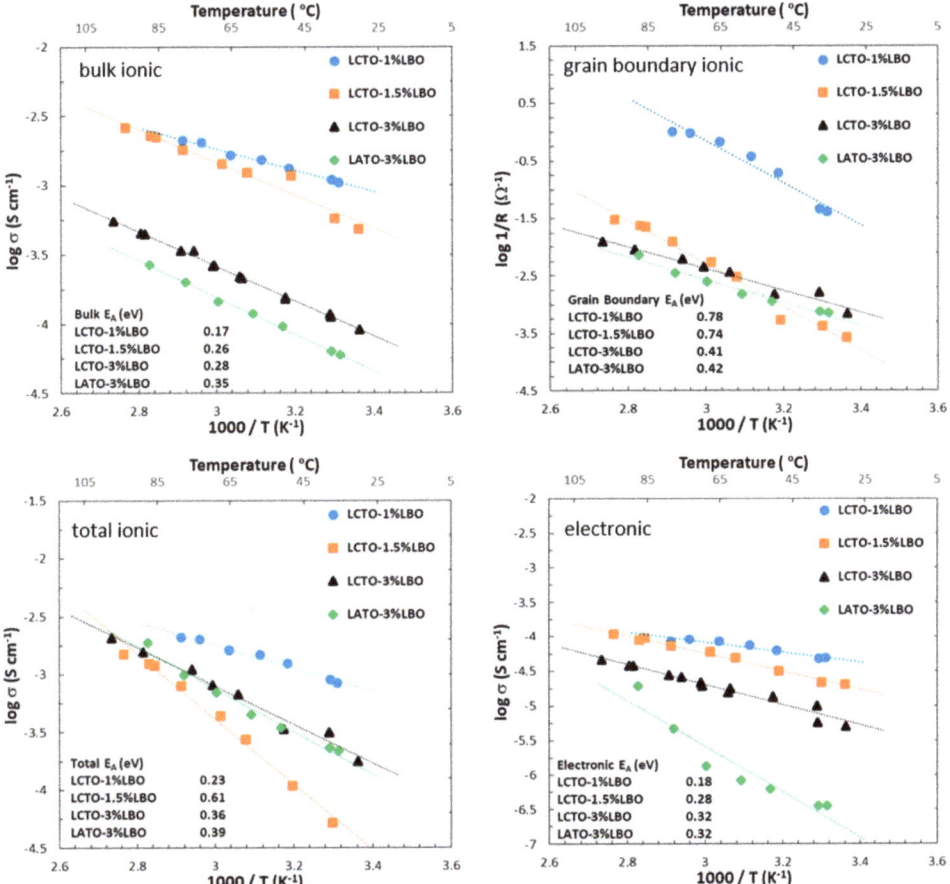

Figure 6. Bulk ionic, total ionic and electronic conductivities and (1/R_{gb}) grain boundary resistance plots of Li$_{1.25}$CrTi$_{1.5}$O$_4$ (LCTO) and Li$_{1.25}$Al$_{0.25}$Ti$_{1.5}$O$_4$ (LATO) with varied weight percent Li$_3$BO$_3$ (LBO) as a function of temperature. E_A is the activation energy. Plotting (1/R_{gb}) was done since the grain boundary conductivity cannot be calculated because the grain boundary volume is unknown [35].

The especially low bulk activation energy of LCTO-1% helps to explain the very high Li-ion conductivity of this composition. These bulk activation energies are lower than the activation energy of 0.35 eV reported from Li NMR line broadening experiments on spinel-structured, Li-doped MgAl$_2$O$_4$ [12]. The grain boundary activation energies range from 0.78 to 0.74, 0.41 and 0.42 eV for LCTO-1%LBO, LCTO-1.5%LBO, LCTO-3%LBO and LATO-3%LBO, respectively, suggesting that the addition of a higher concentration of LBO has a strong effect to lower the activation energy for ionic conductivity at the grain boundary.

However, looking at the total ionic conductivities shows a clear superior performance for the LCTO-1%LBO sample, its total conductivity predominantly controlled by its higher bulk ionic mobility despite higher activation energy at its grain boundaries. The total ionic conductivity activation energies range from 0.23 to 0.61, 0.36, 0.39 eV for LCTO-1%LBO, LCTO-1.5%LBO, LCTO-3%LBO and LATO-3%LBO, respectively. The addition of excess LBO, >1%, negatively affects the total Li-ion conductivity although at 3% LBO the grain boundary ionic activation energy is considerably lower. The LCTO-1.5%LBO sample appears to be an outlier, as one would expect it to fall between the 1% and the 3% LBO samples. The electronic conductivities activation energies range from 0.18 to 0.28, 0.32, and 0.32 eV for LCTO-1%LBO, LCTO-1.5%LBO, LCTO-3%LBO and LATO-3%LBO, respectively. It is observed that at a low level of LBO the electronic conductivity is highest and as more LBO is added the electronic conductivity decreases. Overall, it appears that the 1%LBO sample has the maximum electronic and ionic conductivity. It might be that the higher electronic conductivity improves the ionic conductivity owing to an enhancement effect of the transport of two species [40–42]. That is, as a Li-ion hops, in order to preserve charge neutrality both species must move at the same speed which means the faster moving electron is slowed down while the speed of the slower moving Li$^+$ ion is increased.

2.4. Electrochemical Properties of Solid Solutions of $LiNi_{0.5}Mn_{1.5}O_4$ and "$Li_{1.25}Cr_{0.25}Mn_{1.5}O_4$"

Demonstration of the practical applicability of $Li_{1.25}(Cr,Al)_{0.25}Ti_{1.5}O_4$ spinels as solid electrolyte to function as a separator will require finding a new densification aid that does not lead to high electronic conductivity. Subsequently, if such a densification aid is found, we will study the decomposition window of the electrolyte and build lithium symmetric cells. However, the use of the LBO densification aid leads to significant electronic conductivity, therefore this suggests that the most likely application of LBO-spinel composites is as catholyte or anolyte. Furthermore, the unique feature of demonstrating fast Li-ion conduction in spinel structured solids is the possibility to demonstrate that an electrochemically active solid solution can be formed from a solid solution with a spinel structured electrode material. A liquid based cell was used as a simple way to demonstrate this principle. Demonstration in a solid-state cell will require optimization of multiple properties and will be part of a future study.

Since $LiNi_{0.5}Mn_{1.5}O_4$ is known to form high-impedance phases when formed into a composite with $Li_7La_3Zr_2O_{12}$ garnet and other well-studied solid electrolytes, it was decided that it would be a good test case with practical applicability. We found that LCTO and LATO form single-phase, spinel-structured solid solutions with $LiNi_{0.5}Mn_{1.5}O_4$ and also $Li_4Ti_5O_{12}$. However, as a first test-case, Mn was substituted for Ti in the solid electrolyte component owing to the known deleterious effect of significant Ti substitution for Mn in $LiNi_{0.5}Mn_{1.5}O_4$ [43]. This substitution can be done to produce a Li-stuffed spinel of composition $Li_{1.075}Cr_{0.075}Ni_{0.35}Mn_{1.5}O_4$, though the $Li_{1.25}Cr_{0.25}Mn_{1.5}O_4$ end component could not be successfully synthesized owing to the formation of Li_2MnO_3 impurity. Follow-up studies will look in detail at the electrochemical properties, the mixed electronic-ionic conductivities, spinel phase stability, and electrochemical stability windows of the vast solid-solution range of materials in the families $Li_{1.25}(Cr,Al)_{0.25}(Mn,Ti)_{1.5})O_4$: $LiNi_{0.5}Mn_{1.5}O_4$ and $Li_{1.25}(Cr,Al)_{0.25}(Mn,Ti)_{1.5})O_4$: $Li_4Ti_5O_{12}$). The range of solid solution formation will be part of this future study. For the current study, we chose a 3:7 ratio to compare to a typical amount of porosity (30%) in a standard cell. In other words, the "electrolyte" component of the solid solution was set at 30%. This was a first estimate and is not yet optimized. The XRD pattern of nominal composition $0.3[Li_{1.25}Cr_{0.25}Mn_{1.5}O_4]$ $0.7[LiNi_{0.5}Mn_{1.5}O_4]$, i.e., $Li_{1.075}Cr_{0.075}Ni_{0.35}Mn_{1.5}O_4$ composition is shown in Figure 7. The lattice constant was refined to 8.1704(1) Å, which is comparable to $LiNi_{0.5}Mn_{1.5}O_4$ (8.1785 Å) [44] the predominant component of the solid solution. The slight decrease in the lattice constant is to be expected based on the slightly smaller average size of Li$^+$, Cr^{3+} (0.76 Å, 0.615 Å, averaging 0.6875 Å) compared to Ni^{2+} (0.69Å) in octahedral coordination [21,22]. The pattern is indexed to the cubic spinel structure, space group, $Fd\bar{3}m$, indicating a single-phase composition of spinel structure. Structural analysis by Rietveld refinement of XRD data was done with the Fullprof program. The Rietveld refinement result is plotted in the supplementary information, Figure S6, and the atomic positions and final refinement information is contained in the supplementary information, Table S5. In the Rietveld structural model for $Li_{1.075}Cr_{0.075}Ni_{0.35}Mn_{1.5}O_4$, the 8a tetrahedral site is occupied by Li, the 16d octahedral site contains Li, Mn, Cr and Ni and oxygen occupies the 32e site of the $Fd\bar{3}m$ space group. During the refinement, site mixing of Cr, Ni or Mn on the 8a tetrahedral site was explored but did not improve the fit. That is, in the fully converged refinements, Li fully occupies the 8a tetrahedral and the 16d octahedral is randomly occupied by Li, Mn, Ni and Cr.

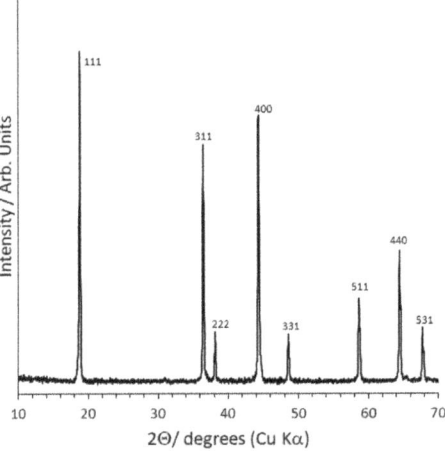

Figure 7. XRD pattern of 0.3 $[Li_{1.25}Cr_{0.25}Mn_{1.5}O_4]$: 0.7 $[LiNi_{0.5}Mn_{1.5}O_4]$ solid solution formed at 850 °C (nominal composition: $Li_{1.025}Cr_{0.025}Ni_{0.45}Mn_{1.5}O_4$). XRD peaks are indexed to the $Fd\bar{3}m$ spinel structure.

The oxygen positional parameter (u), the atomic position coordinates of oxygen in the 32e site refined to a value of 0.26266, in accordance with expectation from the ionic radii of the tetrahedrally and octahedrally coordinated cations [27]. In order to demonstrate expediently that the $Li_{1.025}Cr_{0.025}Ni_{0.45}Mn_{1.5}O_4$ composition is electrochemically active, a liquid electrolyte-based cell was built. In the future, as composition, LBO content, microstructure and the ability to fabricate a thin, dense catholyte is optimized and developed, the concept will be tested in a solid-state configuration. The electrochemical discharge curve at 0.2C charge and discharge rate of the nominal composition $0.3[Li_{1.25}Cr_{0.25}Mn_{1.5}O_4]$ $0.7[LiNi_{0.5}Mn_{1.5}O_4]$, i.e., $Li_{1.075}Cr_{0.075}Ni_{0.35}Mn_{1.5}O_4$ composition is shown in Figure 8a. A discharge capacity ~120 mAh g^{-1} is observed. Assuming electrochemical activity based on Ni^{2+}/Ni^{3+}, Ni^{3+}/Ni^{4+} and Cr^{3+}/Cr^{4+} couples yields a theoretical capacity of about 116 mAh g^{-1} for $Li_{1.075}Cr_{0.075}Ni_{0.35}Mn_{1.5}O_4$. The theoretical discharge capacity is observed at >4.6 V. Additional capacity, beyond the theoretical, can be attributed to the Mn^{3+}/Mn^{4+} couple observed as a shoulder around 4 V. The good rate performance resulting from high Li^+ ionic conductivity and electronic conductivity is shown in Figure 8b. The charge and discharge rates are varied from 0.2 to 10 C, with the charge and discharge rates remaining equal for each individual cycle. Additionally, the material shows good cycle life and no damage from the high rate of charge and discharge is evidenced as the capacity returns to 120 mAh g^{-1} for 1 C after charging and discharging at 10 C rate. Further measurements and optimization of the mixed ionic electronic conductivity through composition and electrode design will be needed to fully realize the maximum rate performance. Further studies will be needed to characterize the conductivity of the solid electrolyte–solid cathode solid solutions as a function of composition and with the addition of LBO to improve grain boundary conductivity. This preliminary result confirms the formation of an electrochemically active solid solution.

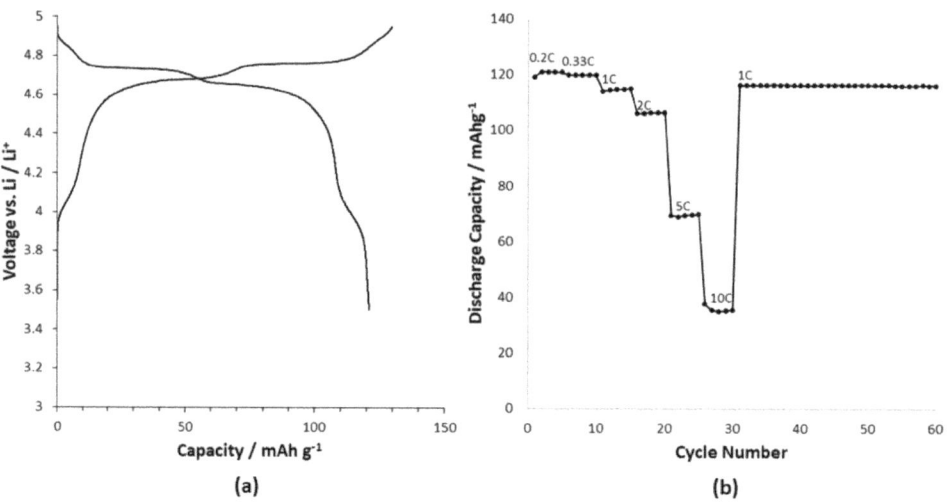

Figure 8. (a) Electrochemical charge and discharge curve of 30% $Li_{1.25}Cr_{0.25}Mn_{1.5}O_4$ and 70% $LiNi_{0.5}Mn_{1.5}O_4$ solid solution formed at 850 °C (nominal composition: $Li_{1.025}Cr_{0.025}Ni_{0.45}Mn_{1.5}O_4$) (b) Discharge capacity of $Li_{1.075}Cr_{0.075}Ni_{0.35}Mn_{1.5}O_4$ as function of charge and discharge rate at a loading of about 6 mg cm^{-1} (~ 0.7 mAh cm^{-1}). Symmetrical charge and discharge rate indicated on the figure were varied each five cycles (cycles 1–30) and fixed at 1C for cycles 31–60.

2.5. Composition

The elemental composition of the $Li_{1.25}Cr_{0.25}Ti_{1.5}O_4$ and $Li_{1.25}Al_{0.25}Ti_{1.5}O_4$ powders as obtained from ICP is tabulated in the supplementary information (Table S6) and is in excellent agreement with the nominal composition. In the LBO-containing samples, the semi-quantitative distribution of boron (B) was probed by wavelength dispersive spectroscopy (WDS) analysis of the cross-section of two representative pellets with 1 and 3% LBO. Over multiple spots spanning the length thickness of the pellet, B was spread through the whole sample, suggesting that the B is not concentrated at the grain boundaries. Since it was not possible to differentiate the grain boundary from the bulk during the data collection, we cannot definitively locate B since there is some chance that the measurement were done only on grain boundary or only bulk spots, the spot size is rather large (5 μ) and the error in measurement (±2% based on counting statistics) is on the order of the concentration (1–3% LBO). However, the analysis of XRD data and the changes in both bulk and grain boundary conductivity described later all support the conclusion that B is distributed throughout the sample. The B distribution from WDS is tabulated in the supplementary information (Table S7) and the images of the pellets used for determination of B distribution are shown in Figure S7.

3. Discussion

The introduction of Li on multiple sites has been previously demonstrated to be a successful strategy to attain higher Li$^+$ ion conductivity [1,15,16]. As an example, in garnet structured $Li_7La_3Zr_2O_{12}$, one of the most well-known oxide-based Li$^+$ conductors (10^{-3} to 10^{-4} S cm^{-1} at room temperature) Li$^+$ is sited on both tetrahedral (24d) and octahedral (96h) sites and the occupancy ratio between the sites is critical to optimization of the conductivity [45]. Similarly, the bulk ionic conductivity of $LiTi_2(PO_4)_3$ jumps by 3 orders of magnitude when additional Li occupies another crystallographic site as a result of partial substitution of Ti^{4+} by Al^{3+} to form $Li_{1.3}Ti_{0.7}Al_{0.3}(PO_4)_3$ [46,47]. By contrast, $LiNi_{0.5}Ge_{1.5}O_4$ of modest Li^+ conductivity (10^{-8} S cm^{-1} at 63 °C) has Li only on the tetrahedral 8c site and Ni and Ge ordered 1:3 over the 4b and 12d octahedral sites of the ordered P_432 spinel.

The high Li-ion conductivity observed in LCTO and LATO can be compared to the $Fd\bar{3}m$ spinel structured anode $Li_4Ti_5O_{12}$ which has been reported to be a rather poor Li ion conductor [48–50] based on Li-NMR studies. This Li-NMR study is in conflict with the fact that $Li_4Ti_5O_{12}$ is a high-rate anode material [51,52], and that discrepancy was attributed by the authors to a fast ion conducting interphase which forms immediately upon Li insertion. Another study based on muon spin spectroscopy of $Li_4Ti_5O_{12}$ and $LiTi_2O_4$ spinels come to a different conclusion. Their muon spectroscopy results point to very mobile Li ions in $Li_4Ti_5O_{12}$ and lower activation energy for Li motion relative to $LiTi_2O_4$ [53]. In $Li_4Ti_5O_{12}$, the 8a tetrahedral site is occupied by Li as in LCTO and LATO, and the 16d octahedral site is occupied by Li and Ti in ratio of 1:5. In $LiTi_2O_4$, Li only occupies the 8a tetrahedral site and Ti completely occupies the 16d octahedral site. This is therefore further evidence of the positive effect of multiple site occupation on Li mobility.

Integration of solid electrolytes into electrodes has been problematic thus far in published studies of solid-state batteries. Density functional theory (DFT) computational studies have shown reactivity of cubic garnet structured $Li_7La_3Zr_2O_7$ and common Li-ion cathode materials [54,55] and experimental studies have shown the reactivity of $LiNi_{0.5}Mn_{1.5}O_4$ cathode with LLZO during electrochemical cycling [56]. Furthermore, $LiCoO_2$ forms unfavorable interfaces during densification and requires a $LiNbO_3$ coating to reduce reactivity [57]. Solid solutions of $Li_{1.25}(Al,Cr)_{0.25}(Ti,Mn)_{1.5}O_4$ with known electrode materials such as $LiNi_{0.5}Mn_{1.5}O_4$, e.g., x[$LiNi_{0.5}Mn_{1.5}O_4$] 1-x[$Li_{1.25}(Al,Cr)_{0.25}(Ti,Mn)_{1.5}O_4$] (0 < x < 1) with grain boundary engineering through use of LBO or other sintering aids offer an alternative, simpler route since the solid electrolyte-solid electrode interface is eliminated and the LBO increases the electronic conductivity, which is needed for use as an electrode. This should lead to increased power owing to improved Li-ion and electronic conductivity within the electrode and the solid solutions could be used as part of an all solid-state battery with a garnet based separator and Li metal anode. If a new sintering aid is found that does not lead to electronic conductivity, these spinel electrolytes might be used in a fully spinel structured all-solid battery as envisioned by Thackeray and Goodenough [11] or with an interfacial layer such as Li_3N [17] separating it from a Li metal or carbon anode. The $LiNi_{0.5}Mn_{1.5}O_4$ spinel is particularly attractive for solid-state application owing to its high energy storage density, high voltage (~4.7 V), use of abundant chemicals, small lattice change during charge and discharge and high Li diffusivity throughout the range of Li composition [58].

4. Materials and Methods

4.1. Powder Preparation

The compound $Li_{1.25}Cr_{0.25}Ti_{1.5}O_4$ was prepared by solid-state reaction from a stoichiometric ratio of TiO_2 (Sigma-Aldrich, Saint Louis, MO, USA) and Cr_2O_3 (Alfa Aesar, Ward Hill, MA, USA) and a 3% stoichiometric excess of Li_2CO_3 (Alfa Aesar, Ward Hill, MA, USA) to counteract volatilization of Li. The precursors were ground by hand using a mortar and pestle, then the fine, mixed powder was heated in an uncovered alumina crucible at 10 °C per minute to 600 °C and held at this temperature for 10 h in air and allowed to furnace cool. The resulting powder was reground and pelletized using a SPEX Sample Prep 13 mm diameter pellet die (Spex Sampleprep LLC, Metuchen, NJ, USA) and Carver laboratory press (Fred S. Carver Company, Wabash, IN, USA) at a load of about 2300 kg. The pellet was placed in a covered alumina crucible and heated at 10 to 850 °C and held at this temperature for 24 h in air and then allowed to furnace cool. $Li_{1.25}Al_{0.25}Ti_{1.5}O_4$ and $Li_{1.25}Cr_{0.25}Mn_{1.5}O_4$ was prepared similarly substituting Al_2O_3 (Alfa Aesar, Ward Hill, MA, USA) for Cr_2O_3 and $MnCO_3$ for Al_2O_3 or Cr_2O_3, respectively. Li_3BO_3 (LBO) was prepared from a stoichiometric mixture of Li_2CO_3 (Alfa Aesar, Ward Hill, MA, USA) and H_3BO_3 (Alfa Aesar, Ward Hill, MA, USA). The starting mixture was ground in a mortar and pestle and heated at 600 °C in air for 4 h. Solid solutions of the solid electrolyte and the LNMO cathode were prepared through an aqueous solution based route from a mixture Li$_2$CO$_3$ (Alfa Aesar, Ward Hill, MA, USA), MnCO$_3$ (Alfa Aesar, Ward Hill, MA, USA), Ni(OH)$_2$ (Alfa Aesar, Ward Hill, MA, USA), Cr(NO$_3$)$_3$·9H$_2$O (Alfa Aesar, Ward Hill, MA, USA) precursors dissolved in a citric acid (Sigma-Aldrich, Saint Louis, MO, USA)/nitric acid (Sigma-Aldrich, Saint Louis, MO, USA) solution. As an example, Li$_{1.075}$Cr$_{0.075}$Ni$_{0.35}$Mn$_{1.5}$O$_4$ was prepared from 0.2659 g Li$_2$CO$_3$ (3% excess) 1.1208 g MnCO$_3$, 0.2109 g Ni(OH)$_2$, and 0.2401 g Cr(NO$_3$)$_3$·9H$_2$O, 1 g citric acid and 6 g concentrated HNO$_3$ diluted to 30 mL with H$_2$O. The clear green solution obtained from heating the mixture was evaporated to dryness and then heated under air in a Lindbergh furnace at 10 °C per minute to 450 °C, held for 3 h, heated at 10 °C per minute to 850 °C, held for 6 h, then furnace cooled (Lindbergh/MPH, Riverside, MI, USA).

4.2. Consolidation of Samples for Conductivity Measurements

Sintering to obtain dense pellets was attempted in air at 850 °C. Sintering pure Li$_{1.25}$Cr$_{0.25}$Ti$_{1.5}$O$_4$ led to pellets of low density (~60–70% relative density) and the temperature could not be increased owing to the formation of a ramsdellite-structured phase at higher temperature. The addition of LBO led to much higher density pellets (~80–85%). LBO was chosen owing to the fact that its melting point of ~700 °C is below the 850 °C consolidation temperature, so that it can act as a liquid-phase sintering aid and also because LBO itself has moderate ionic conductivity which might enhance conduction at the grain boundaries unlike other typical sintering additives such as LiF which are poor Li-ion conductors and have melting point above the temperature at which the samples are converted from the spinel to the ramsdellite structure. A similar spinel to ramsdellite phase transformation upon heating has been previously documented for LiTi$_2$O$_4$ [59]. Dense discs (>90%) were prepared by rapid induction hot-pressing (custom built by University of Michigan, Ann Arbor, MI, USA) without (93–97%) and with LBO (near 100%). For the higher conducting Li$_{1.25}$Cr$_{0.25}$Ti$_{1.5}$O$_4$, three different amounts of LBO were tested (1, 1.5 and 3 wt. %) in an attempt to optimize conductivity. For the Li$_{1.25}$Al$_{0.25}$Ti$_{1.5}$O$_4$, a 3 wt. % LBO-containing hot-pressed sample was prepared based on sintering studies to increase density. The powders were densified at 850 °C at 40 MPa for 40 min under Ar using a rapid induction hot-pressing technique. The spinel powders were hot-pressed in a graphite die. During the hot-pressing, the die is contained in a stream of Ar, creating a reducing atmosphere. After, densification in the presence of LBO, the pellets changed to a black color. In the absence of LBO, no color change was observed. Attempts to oxidize reduced LBO-containing samples at 850 °C in air, in the presence of mother power to reduce Li loss by heating under air, were unsuccessful. The bulk density of the sample was determined from the weight and physical dimensions. The relative density values were estimated by dividing the measured density by the theoretical crystal density based on the spinel structure and the measured lattice constants. The presence of 1–3 wt.% lower density LBO (2.16 g cm^{-3} versus 3.48 and 3.61 g cm^{-3} for LATO and LCTO, respectively) was used to calculate a theoretical density based on the rule of mixtures and the relative density of each sample was then calculated. This can vary from actual by 1–2% based on whether LBO is incorporated into spinel or present as a distinct separate phase.

4.3. X-ray Diffraction

X-ray diffraction (Cu Kα radiation, Rigaku Miniflex 600, D/teX Ultra silicon strip detector, Rigaku Americas Inc., The Woodlands, TX, USA) was used to characterize the phase purity of the powders and the material after hot-pressing. To determine phase purity and for Rietveld structural analysis [60], data were collected from 10–90° 2θ at 0.02° increments at 4° per minute. Lattice constants were calculated from Rietveld Refinement of an X-ray diffraction pattern collected for the sample mixed with a NIST traceable Si internal peak position standard. Data were collected from 10–90° 2θ at 0.02° increments at 4° per minute. Rietveld refinements of XRD data were carried out in the $Fd\bar{3}m$ space group with Li in 8a sites and Li, Cr or Al and Ti or Mn randomly distributed in 16d sites, occupancies fixed to the starting composition.

4.4. Elemental Analysis

Elemental analysis of the powders was performed at Galbraith Laboratories (Galbraith Laboratories, Inc., Knoxville, TN, USA) using Inductively Coupled Plasma-Optical Emission Spectroscopy (ICP-OES). Manually fractured, hot-pressed pellet cross-sections containing Li_3BO_3 were analyzed for boron concentrations via WDS using a JEOL JXA-8900 Electron Probe Microanalyzer (EPMA, Jeol USA, Inc., Peabody, MA, USA) in the Advanced Imaging and Microscopy Laboratory (AIMLab) at the University of Maryland. Analyses were conducted with a beam current of 50 nA and accelerating voltage of 15 kV. The beam diameter was 5 microns. Boron K-alpha x-rays were observed using an LDEB (Mo/B_4C layered synthetic microstructure) analytical crystal. Raw counts were corrected using a ZAF (Z, atomic number, A, absorption, F, fluorescence) algorithm.

4.5. Microstructure

Hot-pressed pellets were manually fractured for cross-sectional microstructural analysis. Cross-sectional analysis was conducted using Thermo Fisher Helios (Thermo Fisher Scientific, Waltham, MA, USA) and an FEI Quanta 200F scanning electron microscopes under a 5kV accelerating voltage (FEI Company, Hillsboro, OR, USA).

4.6. Conductivity

The temperature-dependent ionic conductivity was determined from AC impedance measurements with a Bio-logic VMP300 (Bio-logic USA, Knoxville, TN, USA) (applied frequency range 0.1 Hz to 7 MHz) and/or a Solartron Modulab (Ametek Scientific Instruments, Oak Ridge, TN, USA) (0.1 Hz–300 kHz) with an amplitude of 10–100 mV. Ni was sputtered on the top and bottom of the hot-pressed discs to serve as Li—blocking electrode. The equivalent circuit was modelled and each data set was normalized to the geometric dimensions of the disc to determine the Li-ion conductivity. The Li-ion conduction activation energies were determined from the Arrhenius plot of the relationship of the ionic conductivities to temperature in the range of ~298K to 373K.

The electronic conductivity at room temperature was measured using DC polarization measurements at a voltage of 2 V (Solartron Modulab, Ametek Scientific Instruments, Oak Ridge, TN, USA). The steady-state current and applied voltage were used to determine the resistance, which was converted to the electronic conductivity using the specimen dimensions. Electronic conductivity was also estimated from the AC impedance data.

4.7. Electrochemical Measurements

Solid solution electrodes of composition $Li_{1.25}Cr_{0.25}Mn_{1.5}O_4$:$LiNi_{0.5}Mn_{1.5}O_4$ of 3:7 mole ratio ($Li_{1.075}Cr_{0.075}Ni_{0.35}Mn_{1.5}O_4$) were mixed with carbon and PVDF in an NMP slurry to produce an 80:15:5 composite coating of the active: carbon black: PVDF on an Al foil current collector. The active loading was ~6 mg/cm^3. The C rate was based on a capacity of 147 mAh g^{-1} for $LiNi_{0.5}Mn_{1.5}O_4$. Coin cells (Hohsen, Al clad, Pred Materials, New York, NY, USA) were fabricated using an electrolyte 1 M $LiPF_6$ dissolved in EC:EMC 1:1 (weight ratio, Sigma-Aldrich, Saint Louis, MO, USA) and 2% tris (trimethylsilyl) phosphate (TCI Americas, Portland, OR, USA) an electrolyte stabilizing additive for use at high voltage [61] and Li foil (Johnson Matthey, Alpharetta, GA, USA) as anode. The electrochemical data was collected on a Maccor 4000 Battery cycler (Maccor Inc., Tulsa, OK, USA) The charge and discharge rates were equal for each charge/discharge cycle and charge and discharge rates were varied each five cycles (cycles 1–30) in order from 0.2C, 0.33C, 1C, 2C, 5C to 10C and then fixed at 1C for cycles 31–60.

5. Conclusions

Herein, we report the synthesis and the fast Li-ion conductivity of the spinel structured $Li_{1.25}(Al\ or\ Cr)_{0.25}(Ti\ or\ Mn)_{1.5}O_4$ and a solid solution with the $LiNi_{0.5}Mn_{1.5}O_4$ high voltage positive electrode as examples of a large class of fast Li-ion conducting potential electrolytes and cathodes based on the spinel structure. Bulk and total ionic conductivities for 1% LBO LCTO of 6.8×10^{-4} and 4.2×10^{-4} S cm^{-1}, respectively, is comparable to that of the first reported bulk and total conductivities of garnet structured Al substituted $Li_7La_3Zr_2O_{12}$ [1], 4.9×10^{-4} and 5.1×10^{-4} S cm^{-1}, respectively. Li is located on both octahedral and tetrahedral sites to form a fast 3D Li$^+$ ion conduction pathway in $Li_{1.25}(Al,Cr)_{0.25}(Ti,Mn)_{1.5}O_4$, potentially enabling the all-solid all-spinel-structured battery concept with $Li_4Ti_5O_{12}$ spinel structured anode and $LiMn_2O_4$ or $LiNi_{0.5}Mn_{1.5}O_4$ spinel structured cathode. Significant electronic conductivity of Cr-containing samples points towards application as a catholyte or anolyte in a solid solution with known spinel structured electrode materials. Sintering with LBO leads to a highly dense mixed ionic, electronic conductor which may have application as a catholyte or a coating layer to form an artificial solid electrolyte interface to reduce reactivity with electrolytes. Electrochemical activity in liquid electrolyte-containing cells has been demonstrated for solid solutions of $Li_{1.25}Cr_{0.25}Mn_{1.5}O_4$ and $LiNi_{0.5}Mn_{1.5}O_4$ with discharge capacity of near or greater than the theoretical capacity of $LiNi_{0.5}Mn_{1.5}O_4$ demonstrating the concept of a spinel catholyte and a spinel cathode reacted to form a single-phase solid solution of spinel structure. This is a small step towards demonstrating their potential applications as catholyte or anolyte in a fully solid-state electrode.

Supplementary Materials: The following are available online; Figure S1: Rietveld fit of $Li_{1.25}Cr_{0.25}Ti_{1.5}O_4$; Figure S2: Rietveld fit of $Li_{1.25}Al_{0.25}Ti_{1.5}O_4$; Figure S3: Lattice constants of LCTO and LATO pellets as a function of LBO content; Figure S4: Fit of LCTO EIS data to equivalent circuit; Figure S5: Log (σT) or Log(T/R_{gb}) plotted as a function of 1/T used for determination of bulk ionic, grain boundary ionic, total ionic and electronic activation energies of LCTO and (LATO) with varied weight percent of LBO as a function of temperature; Figure S6: Rietveld fit of $Li_{1.075}Cr_{0.075}Ni_{0.35}Mn_{1.5}O_4$; Figure S7: SEM image of pellets used for analysis of B distribution; Table S1: Refined structural model of LCTO; Table S2: Refined structural model of LATO; Table S3: Lattice constants of LCTO and LATO as a function of LBO content; Table S4: Equivalent circuit elements and value for fit of LCTO EIS data; Table S5: Refined structural model of $Li_{1.075}Cr_{0.075}Ni_{0.35}Mn_{1.5}O_4$; Table S6: Elemental analysis of LCTO and LATO; Table S7: Boron distribution in LCTO pellets.

Author Contributions: J.L.A. conceptualization, methodology, validation, formal analysis, investigation, data curation, writing—original draft preparation, writing—review and editing, visualization, supervision; B.A.C., investigation, data curation; R.C., methodology, investigation, validation, data curation; M.J.W., methodology, investigation, validation, data curation, writing—review and editing; D.T.T., investigation, methodology; L.M., investigation, methodology; P.M.P., investigation, methodology, resources; J.S., resources, methodology, writing—review and editing, supervision; J.W., conceptualization, methodology, validation, writing—review and editing. All authors have read and agreed to the published version of the manuscript.

Funding: J.L.A. and D.T.T. acknowledge the U.S. Army Combat Capabilities Development Command Army Research Laboratory (ARL) for funding. B.A.C. acknowledges funding from the ARL's Historically Black Colleges and Universities/Minority Institutions (HBCU/MI) program and the Army Educational Outreach College Qualified Leader program. L.M. acknowledges ARL for providing financial support under the Dr. Brad. E. Forch Distinguished Postdoctoral Fellowship administered by the National Research Council.

Institutional Review Board Statement: Not applicable.

Informed Consent Statement: Not applicable.

Conflicts of Interest: The authors declare no conflict of interest.

Sample Availability: Samples of the compounds are available from the authors through collaborative agreement.

References

1. Murugan, R.; Thangadurai, V.; Weppner, W. Fast Lithium Ion Conduction in Garnet-type $Li_7La_3Zr_2O_{12}$. *Angew. Chem. Int. Ed.* **2007**, *46*, 7778–7781. [CrossRef] [PubMed]
2. Thackeray, M.M.; David, W.I.F.; Bruce, P.G.; Goodenough, J.B. Lithium Insertion into Manganese Spinels. *Mater. Res. Bull.* **1983**, *11*, 461–472. [CrossRef]
3. Ota, T.; Yamai, I. Thermal Expansion Behavior of $NaZr_2(PO_4)_3$ Type Compounds. *J. Am. Ceram. Soc.* **1986**, *69*, 1–6. [CrossRef]
4. Yamai, I.; Ota, T. Grain Size-Microcracking Relation for $NaZr_2(PO_4)_3$ Family Ceramics. *J. Am. Ceram. Soc.* **1993**, *76*, 487–491. [CrossRef]
5. Smith, S.; Thompson, T.; Sakamoto, J.; Allen, J.L.; Baker, D.R.; Wolfenstine, J. Electrical, mechanical and chemical behavior of $Li_{1.2}Zr_{1.9}Sr_{0.1}(PO_4)_3$. *Solid-State Ion.* **2017**, *300*, 38–45. [CrossRef]
6. Case, E.D. The effect of microcracking upon the Poisson's ratio for brittle materials. *J. Mater. Sci. Lett.* **1984**, *19*, 3702–3712.
7. Kim, Y.; Case, E.D.; Gaynor, S. The effect of surface-limited microcracks on the effective Young's modulus of ceramics. *J. Mater. Sci.* **1993**, *28*, 1910–1918. [CrossRef]
8. Jackman, S.D.; Cutler, R.A. The effect of microcracking on ionic conductivity in LATP. *J. Power Sources* **2012**, *218*, 65–72. [CrossRef]
9. Hupfer, T.; Bucharsky, E.C.; Schell, K.G.; Senyshyn, A.; Monchak, M.; Hoffmann, M.J.; Ehrenberg, H. Evolution of microstructure and its relation to ionic conductivity in $Li_{1+x}Al_xTi_{2-x}(PO_4)_3$. *Solid-State Ion.* **2016**, *288*, 235–239. [CrossRef]
10. Kawai, H.; Tabuchi, M.; Nagata, M.; Tukamoto, H.; West, A.R. Crystal chemistry and physical properties of complex lithium spinels $Li_2MM'_3O_8$ (M = Mg, Co, Ni, Zn; M' = Ti, Ge). *J. Mater. Chem.* **1998**, *8*, 1273–1280. [CrossRef]
11. Thackeray, M.M.; Goodenough, J.B. Solid State Cell Wherein an Anode, Solid Electrolyte and Cathode Each Comprise a Cu-Bic-Close-Packed Framework Structure. U.S. Patent No. 4,507,371, 26 March 1985.
12. Rosciano, F.; Pescarmona, P.P.; Houthoofd, K.; Persoons, A.; Bottke, P.; Wilkening, M. Towards a Lattice-Matching Solid-State Battery: Synthesis of a New Class of Lithium-Ion Conductors with the Spinel Structure. *Phys. Chem. Chem. Phys.* **2013**, *15*, 6107–6112. [CrossRef] [PubMed]
13. Djenadic, R.; Botros, M.; Hahn, H. Is Li-Doped $MgAl_2O_4$ a Potential Solid Electrolyte for an All-Spinel Li-Ion Battery? *Solid State Ion.* **2016**, *287*, 71–76. [CrossRef]
14. O'Callaghan, M.P.; Lynham, D.R.; Cussen, E.J.; Chen, G.Z. Structure and Ionic-Transport Properties of Lithium-Containing Garnets $Li_3Ln_3Te_2O_{12}$ (Ln = Y, Pr, Nd, Sm—Lu). *Chem. Mater.* **2006**, *18*, 4681–4689. [CrossRef]
15. Thangadurai, V.; Weppner, W. $Li_6ALa_2Ta_2O_{12}$ (A = Sr, Ba): Novel Garnet-like Oxides for Fast Lithium Ion Conduction. *Adv. Funct. Mater.* **2005**, *15*, 107–112. [CrossRef]
16. Goodenough, J.B. Fast Ionic Conduction in Solids. *Proc. R. Soc. A* **1984**, *393*, 215–234. [CrossRef]
17. Xie, H.; Li, Y.; Goodenough, J.B. NASICON-Type $Li_{1+2x}Zr_{2-x}Ca_x(PO_4)_3$ with High Ionic Conductivity at Room Temperature. *RSC Adv.* **2011**, *1*, 1728–1731. [CrossRef]
18. Kerman, K.; Luntz, A.; Viswanathan, V.; Chiang, Y.-M.; Chen, Z. Review—Practical Challenges Hindering the Development of Solid-state Li Ion Batteries. *J. Electrochem. Soc.* **2017**, *161*, A1731–A1744. [CrossRef]
19. Bielefeld, A.; Weber, D.A.; Janek, J. Microstructural Modeling of Composite Cathodes for All-Solid-State Batteries. *J. Phys. Chem. C Nanomater. Interfaces* **2019**, *121*, 1626–1634. [CrossRef]
20. Kotobuki, M.; Munakata, H.; Kanamura, K.; Sato, Y.; Yoshida, T. Compatibility of $Li_7La_3Zr_2O_{12}$ Solid Electrolyte to All-Solid-State Battery Using Li Metal Anode. *J. Electrochem. Soc.* **2010**, *151*, A1076–A1079. [CrossRef]
21. Shannon, R.D.; Prewitt, C.T. Effective Ionic Radii in Oxides and Fluorides. *Acta Crystallogr. B* **1969**, *21*, 925–946. [CrossRef]
22. Shannon, R.D. Revised Effective Ionic Radii and Systematic Studies of Interatomic Distances in Halides and Chalcogenides. *Acta Crystallogr. A* **1976**, *31*, 751–767. [CrossRef]
23. Burdett, J.K.; Price, G.D.; Price, S.L. Role of the Crystal-Field Theory in Determining the Structures of Spinels. *J. Am. Chem. Soc.* **1982**, *101*, 92–95. [CrossRef]
24. Le, M.-L.-P.; Strobel, P.; Colin, C.V.; Pagnier, T.; Alloin, F. Spinel-Type Solid Solutions Involving Mn^{4+} and Ti^{4+}: Crystal Chemistry, Magnetic and Electrochemical Properties. *J. Phys. Chem. Solids* **2011**, *71*, 124–135. [CrossRef]
25. Rodríguez-Carvajal, J. Recent Advances in Magnetic Structure Determination by Neutron Powder Diffraction. *Phys. B Condens. Matter* **1993**, *192*, 55–69. [CrossRef]
26. Roisnel, T.; Rodríquez-Carvajal, J. WinPLOTR: A Windows Tool for Powder Diffraction Pattern Analysis. Available online: https://www.scientific.net/MSF.378-381.118 (accessed on 28 April 2021).
27. Hill, R.J.; Craig, J.R.; Gibbs, G.V. Systematics of the Spinel Structure Type. *Phys. Chem. Miner.* **1979**, *1*, 317–339.
28. Huggins, R.A. Simple Method to Determine Electronic and Ionic Components of the Conductivity in Mixed Conductors a Review. *Ionics (Kiel)* **2002**, *8*, 300–313. [CrossRef]
29. Thangadurai, V.; Huggins, R.A.; Weppner, W. Use of Simple Ac Technique to Determine the Ionic and Electronic Conductivities in Pure and Fe-Substituted $SrSnO_3$ Perovskites. *J. Power Sources* **2002**, *108*, 64–69. [CrossRef]
30. Bauerle, J.E. Study of Solid Electrolyte Polarization by a Complex Admittance Method. *J. Phys. Chem. Solids* **1969**, *31*, 2657–2670. [CrossRef]
31. Irvine, J.T.S.; Sinclair, D.C.; West, A.R. Electroceramics: Characterization by Impedance Spectroscopy. *Adv. Mater.* **1990**, *1*, 132–138. [CrossRef]

32. Kennedy, J.H.; Kimura, N.; Stuber, S.M. Measurement of electronic conductivity in Fe-doped β-alumina. *J. Electrochem. Soc.* **1982**, *129*, 1968–1973. [CrossRef]
33. Hema, M.; Selvasekerapandian, S.; Hirankumar, G.; Sakunthala, A.; Arunkumar, D.; Nithya, H. Structural and Thermal Studies of PVA:NH4I. *J. Phys. Chem. Solids* **2009**, *71*, 1098–1103. [CrossRef]
34. Wu, Z.; Xie, Z.; Yoshida, A.; Wang, Z.; Hao, X.; Abudula, A.; Guan, G. Utmost Limits of Various Solid Electrolytes in All-Solid-State Lithium Batteries: A Critical Review. *Renew. Sustain. Energy Rev.* **2019**, *109*, 367–385. [CrossRef]
35. Bayard, M.A. Complex Impedance Analysis of the Ionic Conductivity of $Na_{1+x}Zr_2Si_xP_{3-x}O_{12}$ Ceramics. *J. Electroanal. Chem.* **1978**, *91*, 201–209. [CrossRef]
36. Allen, J.L.; Wolfenstine, J.; Rangasamy, E.; Sakamoto, J. Effect of Substitution (Ta, Al, Ga) on the Conductivity of $Li_7La_3Zr_2O_{12}$. *J. Power Sources* **2012**, *206*, 315–319. [CrossRef]
37. Knauth, P. Inorganic Solid Li Ion Conductors: An Overview. *Solid-State Ion.* **2009**, *18*, 911–916. [CrossRef]
38. Harada, Y.; Ishigaki, T.; Kawai, H.; Kuwano, J. Lithium ion conductivity of polycrystalline perovskite $La_{0.67-x}Li_{3x}TiO_3$ with ordered and disordered arrangements of the A-site ions. *Solid-State Ion.* **1998**, *108*, 407–413. [CrossRef]
39. Adachi, G.-Y.; Imanaka, N.; Aono, H. Fast Li⊕ Conducting Ceramic Electrolytes. *Adv. Mater.* **1996**, *1*, 127–135. [CrossRef]
40. Wagner, C. The theory of the warm-up process. *Z. Phys. Chem.* **1933**, *21*, 25–41.
41. Weppner, W.; Huggins, R.A. Determination of the Kinetic Parameters of Mixed-conducting Electrodes and Application to the System Li_3Sb. *J. Electrochem. Soc.* **1977**, *121*, 1569–1578. [CrossRef]
42. West, A.R. *Solid-State Chemistry and Its Applications*, 2nd ed.; John Wiley & Sons: West Sussex, UK, 2014; p. 421.
43. Alcántara, R.; Jaraba, M.; Lavela, P.; Tirado, J.L.; Biensan, P.; de Guibert, A.; Jordy, C.; Peres, J.P. Structural and Electrochemical Study of New $LiNi_{0.5}Ti_xMn_{1.5-x}O_4$ Spinel Oxides for 5-V Cathode Materials. *Chem. Mater.* **2003**, *11*, 2376–2382. [CrossRef]
44. Liu, D.; Lu, Y.; Goodenough, J.B. Rate Properties and Elevated-Temperature Performances of $LiNi_{0.5-x}Cr_{2x}Mn_{1.5-x}O_4$ ($0 \leq 2x \leq 0.8$) as 5 V Cathode Materials for Lithium-Ion Batteries. *J. Electrochem. Soc.* **2010**, *151*, A1269. [CrossRef]
45. Thompson, T.; Sharafi, A.; Johannes, M.D.; Huq, A.; Allen, J.L.; Wolfenstine, J.; Sakamoto, J. A Tale of Two Sites: On Defining the Carrier Concentration in Garnet-Based Ionic Conductors for Advanced Li Batteries. *Adv. Energy Mater.* **2015**, *1*, 1500096. [CrossRef]
46. Aono, H.; Sugimoto, E.; Sadaoka, Y.; Imanaka, N.; Adachi, G. Ionic Conductivity of Solid Electrolytes Based on Lithium Titanium Phosphate. *J. Electrochem. Soc.* **1990**, *131*, 1023–1027. [CrossRef]
47. Monchak, M.; Hupfer, T.; Senyshyn, A.; Boysen, H.; Chernyshov, D.; Hansen, T.; Schell, K.G.; Bucharsky, E.C.; Hoffmann, M.J.; Ehrenberg, H. Lithium Diffusion Pathway in $Li_{1.3}Al_{0.3}Ti_{1.7}(PO_4)_3$ (LATP) Superionic Conductor. *Inorg. Chem.* **2016**, *51*, 2941–2945. [CrossRef] [PubMed]
48. Wilkening, M.; Amade, R.; Iwaniak, W.; Heitjans, P. Ultraslow Li Diffusion in Spinel-Type Structured $Li_4Ti_5O_{12}$—a Comparison of Results from Solid-state NMR and Impedance Spectroscopy. *Phys. Chem. Chem. Phys.* **2007**, *1*, 1239–1246. [CrossRef] [PubMed]
49. Wagemaker, M.; van Eck, E.R.H.; Kentgens, A.P.M.; Mulder, F.M. Li-Ion Diffusion in the Equilibrium Nanomorphology of Spinel $Li_{4+x}Ti_5O_{12}$. *J. Phys. Chem. B* **2009**, *111*, 224–230. [CrossRef]
50. Ganapathy, S.; Vasileiadis, A.; Heringa, J.R.; Wagemaker, M. The Fine Line between a Two-Phase and Solid-Solution Phase Transformation and Highly Mobile Phase Interfaces in Spinel $Li_{4+x}Ti_5O_{12}$. *Adv. Energy Mater.* **2017**, *1*, 1601781. [CrossRef]
51. Wang, G.X.; Bradhurst, D.H.; Dou, S.X.; Liu, H.K. Spinel $Li[Li1/3Ti5/3]O_4$ as an Anode Material for Lithium Ion Batteries. *J. Power Sources* **1999**, *83*, 156–161. [CrossRef]
52. Panero, S.; Reale, P.; Ronci, F.; Albertini, V.R.; Scrosati, B. Structural and Electrochemical Study on $Li(Li_{1/3}Ti_{5/3})O_4$ Anode Material for Lithium Ion Batteries. *Ionics (Kiel)* **2000**, *6*, 461–465. [CrossRef]
53. Sugiyama, J.; Nozaki, H.; Umegaki, I.; Mukai, K.; Miwa, K.; Shiraki, S.; Hitosugi, T.; Suter, A.; Prokscha, T.; Salman, Z.; et al. Li-Ion Diffusion In $Li_4Ti_5O_{12}$ and $LiTi_2O_4$ battery Materials Detected by Muon Spin Spectroscopy. *Phys. Rev. B Condens. Matter Mater. Phys.* **2015**, *91*. [CrossRef]
54. Miara, L.J.; Richards, W.D.; Wang, Y.E.; Ceder, G. First-Principles Studies on Cation Dopants and Electrolyte|cathode Interphases for Lithium Garnets. *Chem. Mater.* **2015**, *21*, 4040–4047. [CrossRef]
55. Zhu, Y.; He, X.; Mo, Y. First Principles Study on Electrochemical and Chemical Stability of Solid Electrolyte–Electrode Interfaces in All-Solid-State Li-Ion Batteries. *J. Mater. Chem. A Mater. Energy Sustain.* **2016**, *4*, 3253–3266. [CrossRef]
56. Hänsel, C.; Afyon, S.; Rupp, J.L.M. Investigating the All-Solid-State Batteries Based on Lithium Garnets and a High Potential Cathode—$LiMn_{1.5}Ni_{0.5}O_4$. *Nanoscale* **2016**, *1*, 18412–18420. [CrossRef]
57. Ohta, N.; Takada, K.; Sakaguchi, I.; Zhang, L.; Ma, R.; Fukuda, K.; Osada, M.; Sasaki, T. $LiNbO_3$-Coated $LiCoO_2$ as Cathode Material for All Solid-State Lithium Secondary Batteries. *Electrochem. Commun.* **2007**, *9*, 1486–1490. [CrossRef]
58. Xia, H.; Meng, Y.S.; Lu, L.; Ceder, G. Electrochemical Properties of Nonstoichiometric $LiNi_{0.5}Mn_{1.5}O_{4-\delta}$ Thin-Film Electrodes Prepared by Pulsed Laser Deposition. *J. Electrochem. Soc.* **2007**, *151*, A737. [CrossRef]
59. Gover, R.K.B.; Irvine, J.T.S.; Finch, A.A. Transformation of $LiTi_2O_4$ from Spinel to Ramsdellite on Heating. *J. Solid-State Chem.* **1997**, *131*, 382–388. [CrossRef]
60. Rietveld, H.M. A Profile Refinement Method for Nuclear and Magnetic Structures. *J. Appl. Crystallogr.* **1969**, *2*, 65–71. [CrossRef]
61. Allen, J.L.; Allen, J.L.; Thompson, T.; Delp, S.A.; Wolfenstine, J.; Jow, T.R. Cr and Si Substituted-$LiCo_{0.9}Fe_{0.1}PO_4$: Structure, Full and Half Li-Ion Cell Performance. *J. Power Sources* **2016**, *327*, 229–234. [CrossRef]

Review

The Fascinating World of Low-Dimensional Quantum Spin Systems: Ab Initio Modeling

Tanusri Saha-Dasgupta

S.N. Bose National Centre for Basic Sciences, JD Block, Sector III, Salt Lake, Kolkata 700106, India; tanusri@bose.res.in; Tel.: +91-33-2335-5707

Abstract: In recent times, ab initio density functional theory has emerged as a powerful tool for making the connection between models and materials. Insulating transition metal oxides with a small spin forms a fascinating class of strongly correlated systems that exhibit spin-gap states, spin–charge separation, quantum criticality, superconductivity, etc. The coupling between spin, charge, and orbital degrees of freedom makes the chemical insights equally important to the strong correlation effects. In this review, we establish the usefulness of ab initio tools within the framework of the N-th order muffin orbital (NMTO)-downfolding technique in the identification of a spin model of insulating oxides with small spins. The applicability of the method has been demonstrated by drawing on examples from a large number of cases from the cuprate, vanadate, and nickelate families. The method was found to be efficient in terms of the characterization of underlying spin models that account for the measured magnetic data and provide predictions for future experiments.

Keywords: magnetism; density functional theory; spin Hamiltonian; quantum Monte Carlo

Citation: Saha-Dasgupta, T. The Fascinating World of Low-Dimensional Quantum Spin Systems: Ab Initio Modeling. *Molecules* **2021**, 26, 1522. https://doi.org/10.3390/molecules26061522

Academic Editor: Stephane Jobic

Received: 11 February 2021
Accepted: 3 March 2021
Published: 10 March 2021

Publisher's Note: MDPI stays neutral with regard to jurisdictional claims in published maps and institutional affiliations.

Copyright: © 2021 by the author. Licensee MDPI, Basel, Switzerland. This article is an open access article distributed under the terms and conditions of the Creative Commons Attribution (CC BY) license (https://creativecommons.org/licenses/by/4.0/).

1. Introduction

Compounds that have a dimensionality of less than three dimensions have long caught the attention of researchers due to their unconventional properties. Reductions in dimensionality can be structural, as in the case of two-dimensional compounds, such as graphene [1] and metal dichalcogenides [2], or as in nanoclusters [3] and nanowires [4]. Reductions in dimensionality can be also electronic, which may occur in otherwise structurally three-dimensional compounds due to interplay between their geometry and the directional nature of the chemical bonding. Magnetic systems of low dimensionality arise when the anisotropic electronic interaction translates into anisotropic magnetic interaction, thereby reducing the effective dimensionality of the magnetic system.

The history of low-dimensional magnetism begins with the Ising model [5], which considers an infinite chain of spins with nearest-neighbor interactions between preferred components of the spin S.

$$H_{Ising} = J \sum_n S_n^z S_{n+1}^z$$

The other limit of the Ising model is the isotropic Heisenberg model [6],

$$H_{Heisen} = J \sum_n (S_n^x S_{n+1}^x + S_n^y S_{n+1}^y + S_n^z S_{n+1}^z).$$

The ground states of one-dimensional (1-d) uniform chains of S = 1/2 spins are strikingly different in these two models. While the Ising model leads to an ordered ground state, the ground state is disordered even at T = 0 K in the Heisenberg model. The Onsager's famous solution [7] of the Ising model extended to a two-dimensional (2-d) square lattice showed the existence of a long-range order at a finite temperature with a magnetic transition temperature comparable to the value of the exchange interaction *J*. On the other hand, the two-dimensional Heisenberg system remains disordered at T $\neq$ 0, but is ordered at T = 0 K. The low-dimensional magnetism in the isotropic Heisenberg model is given by the Mermin–Wagner theorem [8], which states that a 1-d or 2-d array of spins that is described by an underlying isotropic Heisenberg Hamiltonian cannot show a transition to a magnetically ordered state above absolute zero temperature, with the long-range order being destroyed by any finite thermal fluctuation.

A model that is distinct from those of Ising and Heisenberg arises when the magnetic moments lie perpendicularly to the chosen axis, giving rise to what is known as the XY model [9],

$$H_{XY} = J \sum_n (S_n^x S_{n+1}^x + S_n^y S_{n+1}^y).$$

This model shows a unique form of a long-range topological order formed by the bound pairs of vortices below a certain temperature, which is known as the Berezinskii–Kosterlitz–Thouless (BKT) transition temperature [10].

An important distinction between half-integer and integer spin was put forward by Haldane [11]. S = 1 Heisenberg antiferromagnetic (AF) chains, also known as Haldane chains, are conjectured to have spin singlet ground states with an energy gap between the singlet and triplet excited states, which is in marked contrast with the S = 1/2 Heisenberg antiferromagnetic chains, which show a gapless continuum of spinon with the algebraic decay of spin–spin correlation.

Thus, the magnetic behavior crucially depends on the symmetry (discrete for the Ising model, continuous abelian for the XY model, and continuous non-abelian for the Heisenberg model) of the spin Hamiltonian and dimensionality. It came as a further surprise that the crossover between one and two dimensions of S = 1/2 magnets was not found to be at all smooth. In a sense, spin ladders [12] that are formed by spin chains that are put next to each other show unconventional behavior. While ladders with odd numbers of legs display properties similar to those of spin chains and have a power-law decay of their spin–spin correlations, ladders with even numbers of legs have a spin-liquid ground state with an exponential decay of their spin–spin correlations.

Interestingly, the above-described exotic phenomena happen in magnets with small spins—either S = 1/2 or S = 1—and not in large classical moment systems, as they are governed by the quantum nature of the spins. The importance of quantum fluctuation can be appreciated by considering a two-site problem connected by an antiferromagnetic Heisenberg interaction. The bond energy is minimized if the two spins form a singlet in which $|S_i + S_j| = 0$. The classical antiparallel alignment gains energy only from the z–z part of the Heisenberg interaction, while the fluctuation of the z-component also allows for energy to be gained from the spin–flip part, with the quantum correction to classical z–z energy given by $(-JS/(-JS^2) = 1/S$ [13]. The quantum fluctuation is therefore expected to be strongest for small spins, particularly for S = 1/2 or S = 1.

The realization of abstract low-dimensional quantum spin models for real systems started in the period of the 1970s and 1980s, when real materials whose magnetic behavior resembled that predicted in models were synthesized [14]. The presence of strong quantum fluctuations in the low-dimensional magnetic subsystems of high-T_C superconductors [15] triggered renewed interest in both theoretical and experimental studies of low-dimensional quantum spin systems. The original motivation was guided by the possible connection between the spin gap and superconductivity, an issue that is yet to be settled. However, with the extensive investigation of spin ladders, it became clear that even purely insulating low-dimensional quantum magnets can exhibit very rich phenomena that deserve attention in their own right. With the aid of advanced computational methods, theorists were able to solve a variety of more complex low-dimensional quantum spin lattices—examples include the Shastry–Shutherland model [16], the Kagomé lattice model [17], the Kitaev [18] honeycomb model, etc., and many of them have yet to be found in the real world.

An understanding of this complex world requires close collaboration between chemists and experimental and theoretical physicists. Recently, the huge importance of quantum-mechanical calculations based on ab initio electronic structure calculations has been realized; it can become crucial for the identification of the underlying spin model corresponding to a studied material. The measured magnetic susceptibility is often fitted with assumed magnetic models. This procedure may give rise to non-unique answers due to the rather insensitive nature of the magnetic susceptibility with respect to the details of magnetic models, which is complicated by the effect of inter-chain/inter-layer coupling, crystal fields, spin anisotropy, dilution, and other effects that are present in real compounds. Microscopic understanding is thus required for the sake of uniqueness.

One prominent candidate for the realization of low-dimensional quantum spin models is inorganic transition-metal compounds. Although the exchange integrals for these systems are typically several hundred degrees Kelvin, their specific geometry can reduce them substantially in the spirit of the Goodenough–Kanamori–Anderson rules [19,20], making the low-temperature properties easily accessible. They are better choices than organic systems, for which, in most cases, it is impossible to grow large single crystals, and the exchange paths involving organic ligands can be very complex.

Within the limited scope of the present review, we discuss the theoretical attempts within the framework of ab initio density functional theory coupled with the solution of a model Hamiltonian derived based on ab initio inputs, which is applied to understanding and predictions of low-dimensional quantum spin compounds. Specifically, we will discuss the applicability of an ab initio tool of a muffin-tin orbital (MTO)-based method—namely, N-th order muffin orbital (NMTO)-downfolding—for this purpose. The review will draw examples from cuprates, vanadates, and nickelates, which are listed in Table 1.

It is worth mentioning at this point that other attempts at ab initio modeling of low-dimensional quantum spin systems also exist, which use a variety of methods, such as the extended Huckel tight-binding method (EHTB) [21], tight-binding fitting of the density function theory (DFT) band structure in terms of Slater–Koster parametrization, and total energy calculations. For representative references, see [22–26]. In addition to quantum Monte Carlo (QMC) and exact diagonalization, methods like the density matrix renormalization group (DMRG) [27], bond-operator theory, variational Ansätze, etc. have also been used for solving the spin Hamiltonian.

Table 1. Examples of the quantum spin systems studied in this review.

Compound Name	Magnetic Ion	Occupancy	Spin
$SrCu_2O_3$	Cu^{2+}	d^9	$S = 1/2$
$CaCuGe_2O_6$	Cu^{2+}	d^9	$S = 1/2$
$Cu_2Te_2O_5X_2$ (X=Cl/Br)	Cu^{2+}	d^9	$S = 1/2$
$Na_3Cu_2Te(Sb)O_6$	Cu^{2+}	d^9	$S = 1/2$
$CuTe_2O_5$	Cu^{2+}	d^9	$S = 1/2$
$Cs_2CuAl_4O_8$	Cu^{2+}	d^9	$S = 1/2$
γ-LiV_2O_5	V^{4+}	d^1	$S = 1/2$
α-NaV_2O_5	V^{4+}	d^1	$S = 1/2$
CsV_2O_5	V^{4+}	d^1	$S = 1/2$
$VOSeO_3$	V^{4+}	d^1	$S = 1/2$
$Na_2V_3O_7$	V^{4+}	d^1	$S = 1/2$
$Zn_2VO(PO_4)_2$	V^{4+}	d^1	$S = 1/2$
Ti-$Zn_2VO(PO_4)_2$	Ti^{4+}/V^{4+}	d^0/d^1	$S = 1/2$
$NiAs_2O_6$	Ni^{2+}	d^8	$S = 1$
$NiRh_2O_4$	Ni^{2+}	d^8	$S = 1$
$Sr_3NiPt(Ir)O_6$	Ni^{2+}	d^8	$S = 1$
$SrNi_2V_2O_8$	Ni^{2+}	d^8	$S = 1$

2. Theoretical Framework

The presence of transition metal ions in quantum spin systems makes the electron–electron correlations in their unfilled d-shell a dominant effect. Together with the strong correlation effect, the true nature of a magnetic exchange network is often found to not be what is expected from the crystal structure. The theoretical framework must thus include the structural and chemical details. Microscopic investigations demand the involvement of both ab initio methods and many-body effects.

Starting from a Hubbard model [28] description, which describes the competition between the kinetic energy—governed by the hopping interaction, t_{ij}—and electron–electron correlation—governed by the Hubbard onsite interaction, U—and integrating out the double occupancy in the strong correlation limit ($U/t \gg 1$), the t-J model [29] is obtained (where J is the magnetic exchange). For half-filling, the t-J model gives rise to the relevant spin Hamiltonian for studying the quantum spin system. To add chemical reality to such physicists' models, density function theory (DFT) [30] calculations are carried out with the choice of an exchange-correlation functional of the local density approximation (LDA) [31] or generalized gradient approximation (GGA) [32]. In order to represent only the degrees of freedom associated with magnetic ions, in terms of construction of an effective low-energy Hamiltonian, a highly successful approach has been the downfolding technique within the framework of the n-th-order muffin-tin orbital (NMTO) method [33], which relies on the self-consistent DFT potential borrowed from linear muffin-tin orbital (LMTO) [34] calculations.

Within the NMTO method [33], the basis sets may be chosen to span selected energy bands with as few basis orbitals as there are bands by using the downfolding method of integrating degrees of freedom that are not of interest. The method can be used for direct generation of Wannier and Wannier-like functions. This leads to a deterministic scheme for deriving a low-energy model Hamiltonian starting from a complicated DFT band structure. As it is free from fitting parameters, this scheme takes into account the proper renormalization effect from degrees of freedom that are integrated out, and thus retains the information of wave-functions and captures the correct material dependence.

Figure 1 shows the application of NMTO-downfolding to construct the low-energy Hamiltonian for V_2O_3 [35] and high-T_c cuprate, $HgBa_2CuO_4$ [36]. In the case of the former, the bands around the Fermi level (set as zero energy in the figures) are spanned by V t_{2g} states, while for the latter, they are spanned by the antibonding Cu $x^2 - y^2$ state. The blue bands in the top panels are the downfolded bands, which show an almost perfect agreement with the DFT band structure in red within ± 1 eV around the Fermi level. The Wannier or Wannier-like functions describing the downfolded bands are shown in the bottom panels, which highlight the $pd\pi$ and $pd\sigma$ antibonding nature of the renormalized V t_{2g} and Cu $x^2 - y^2$ functions, respectively, with the head part of the functions shaped as V t_{2g} and Cu $x^2 - y^2$ and the tail parts shaped as integrated-out degrees of freedom—predominantly O p.

The real space representation of this few-band Hamiltonian facilitates the identification of dominant effective hopping interactions that connect the magnetic centers, which bear information on important exchange pathways. Following this, the magnetic exchanges for the identified exchange paths can be obtained either through use of the super-exchange formula [37] or through calculation of the total energy of the different spin configurations within the LDA + U calculations [38] and mapping them onto Heisenberg model.

To calculate the thermodynamic properties of the DFT-derived spin Hamiltonian H, in the present review, the stochastic series expansion (SSE) implementation of the quantum Monte Carlo (QMC) [39–41] method was primarily used, though in some cases, exact diagonalization was also used.

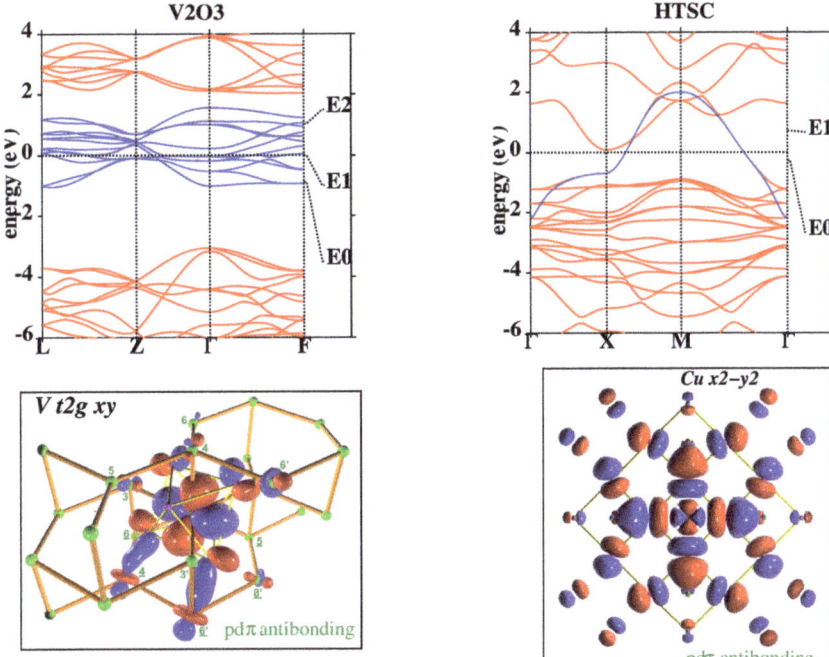

Figure 1. (**Top**) Band structure. of the *n*-th-order muffin-tin orbital (NMTO)-downfolded V t_{2g} bands of V_2O_3 (**left**) and Cu x^2-y^2 bands of Hg cuprate (**right**) in blue lines in comparison to the full density functional theory (DFT) band structure in red lines. E_0, E_1, etc. denote the energy points used in the energy-selective NMTO-downfolding procedure. (**Bottom**) The NMTO-downfolding-generated Wannier-like functions of V xy in V_2O_3 (**left**) and Cu $x^2 - y^2$ in Hg cuprate (**right**). The central parts of the Wannier functions are shaped according to the active degrees of freedom—namely, V xy or Cu $x^2 - y^2$—while the tails of the Wannier functions are shaped according to integrated-out orbitals with significant mixing with active degrees of freedom. The oppositely signed lobes of the Wannier functions are colored in red and blue. This figure is adapted from [35,36].

In the following, a brief description of SSE-QMC is given. For details, see [39]. The thermal expectation value of a quantity A is given by

$$\langle A \rangle = \frac{1}{Z} Tr\{A e^{-\beta H}\}, Z = Tr\{e^{-\beta H}\}.$$

Within the stochastic series expansion implementation of the quantum Monte Carlo method, one chooses a basis and performs a Taylor expansion of the exponential operator:

$$Z = \sum_{\alpha} \sum_{n=0}^{\infty} \frac{\beta^n}{n!} \langle \alpha | (-H)^n | \alpha \rangle,$$

where $H = J \sum_{i,j} S_i \cdot S_j$. One chooses a standard z-component basis, $|\alpha\rangle = |S_1^z, S_2^z, \ldots, S_N^z\rangle$, and performs the summation with Monte Carlo technique. We note that, for the quantum case, H consists of non-commuting operators.

For a practical implementation, the DFT-derived magnetic exchanges are used as a starting guess, following which the optimal values of the dominant magnetic exchange, J, and the effective g factor are obtained by fitting the QMC results for the susceptibility:

$$\chi^{th} = \langle (S^z - \langle S^z \rangle)^2 \rangle,$$

where μ_B and k_B denote the Bohr magneton and the Boltzmann constant, respectively, and the experimental susceptibility (in [emu/mol]) at intermediate to high temperatures is given via $\chi = 0.375\,(g^2/J)\chi^{th}$. To simulate the low-temperature region of the susceptibility data, the respective Curie contribution from impurities, such as $\chi^{CW} = C_{imp}/T$, is included.

With the stochastic series expansion implementation of the quantum Monte Carlo method, it is possible to simulate quantum spin models in an external field, examples of which will be given in the following.

3. Selected Inorganic Quantum Spin Systems

3.1. Cuprates

The cuprate family, with Cu in its 2+ valence state of the d^9 electronic configuration, which amounts to one hole in the highest occupied d state, is perhaps the most studied S = 1/2 quantum spin system family. The discovery of high T_c superconductivity in layered cuprate compounds has raised interest in the role of low dimensionality and the quantum nature of the Cu spins. The synthesis of various cuprates with different possible realizations of coordinations of magnetic sublattices has made this family one of the most popular families in terms of the study of low-dimensional quantum spin systems.

$SrCu_2O_3$—With the parent compounds of cuprate superconductors considered as the example of 2-d lattices of spin 1/2 antiferromagnets, efforts were put forward to understand the crossover from chains to square lattices. However, as mentioned above, the transition from 1-d to 2-d quantum spin systems was found to be highly non-monotonic. Even leg-spin ladders were predicted to possess a spin-liquid ground state, while ladders with odd numbers of legs were predicted to possess properties similar to those of single chains [12].

In this ladder family, the compound $Sr_{14-x}Ca_xCu_{24}O_{41}$ was experimentally synthesized [42,43] and compared to theoretical predictions of the spin gap and superconductivity. The transport properties were found to be dominated by holes in the ladder planes. The normal state of $x = 11$ was found [44] to show a strong anisotropy between the DC resistivity along and across the ladder direction with $\rho_\perp/\rho_\parallel \sim 30$ at T = 100 K. The microscopic insight into this observation was obtained in terms of DFT calculations performed for $SrCu_2O_3$ [45], a compound that possesses the same kind of Cu_2O_3 ladder planes as $Sr_{14-x}Ca_xCu_{24}O_{41}$, as shown in the left panel of Figure 2.

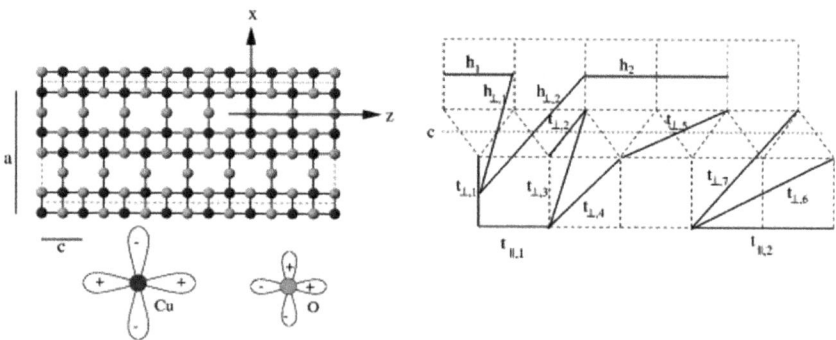

Figure 2. (**Left**) The structure of the ladder compound containing CuO_2 planes, with Cu and O atoms marked as black and gray balls, which host Cu $x^2 - y^2$ and O p_x/p_y orbitals. (**Right**) Various effective Cu–Cu intra-ladder hoppings along the rungs and legs, as well as inter-ladder hoppings. Adapted from [45].

In the first ever application of NMTO-downfolding [33], starting from a full DFT calculation, the low-energy Hamiltonian of $SrCu_2O_3$ was constructed in terms of renormalized Cu $x^2 - y^2$ orbitals, which were obtained by integrating out all other degrees of freedom except Cu $x^2 - y^2$. The effective Cu–Cu hopping interactions between Cu sites along the rungs and legs and between ladders were found to be long ranged, as shown in the right panel of Figure 2. This analysis showed effective inter-ladder hoppings to be much smaller than intra-ladder hoppings [45]. Furthermore, intra-ladder hoppings between nearest-neighbor Cu pairs were found to be anisotropic with $t_{||} \neq t_\perp$ [45]. This was explained [45] as a consequence of anisotropic t_{pd} in the chemical Hamiltonian model involving both Cu $x^2 - y^2$ and O p degrees of freedom due to effective hopping through paths involving Cu 4 s states. Estimates of the conductivity in the model where holes were unbound and confined in the ladder [46] were found to give good agreement with the experiments at temperatures of T >100 K [44].

$CaCuGe_2O_6$—Although the crystal structure of $CaCuGe_2O_6$ [47] consists of zig-zagged 1-d chains running along the c-axis and alternating between two neighboring bc planes, as shown in the top panel of Figure 3, an experimental study involving magnetization and susceptibility measurements was found to be in disagreement with the magnetic properties of the S = 1/2 Heisenberg chain. Instead, the compound was found to show a spin-singlet ground state with an energy gap of 6 meV [48]. However, unlike the well-known related compound $CuGeO_3$ [49], the spin gap is intrinsic, as no spin-Peierls phase transitions were reported between 4.2 and 300 K. This strongly suggests spin dimer characteristics [50]. A question that the experimental measurements could not answer was that of which Cu pairs constitute antiferromagnetic dimers. NMTO-based downfolding of the ab initio band structure together with solution of the effective spin Hamiltonian for computing the thermodynamic properties was carried out to answer the above question [51]. The DFT-derived low-energy Hamiltonian in an effective Cu $x^2 - y^2$ basis showed [51] that longer-ranged magnetic interactions dominated over the short-ranged interactions, and the third-neighbor Cu–Cu pair was the strongest, followed by the nearest-neighbor (NN) Cu–Cu interaction. This led to a description of systems of interacting dimers, given by: $H_{eff} = J_3 \sum_{(i,j)} S_i S_j + J_1 \sum_{(i,j)} S_i S_j$, where J_3 and J_1 are intra- and inter-dimer interactions, respectively. This spin Hamiltonian was solved using SSE-QMC for field-dependent magnetization and magnetic susceptibility, as shown in the bottom panels of Figure 3. The optimal values of $J_1/J_3 = -0.2$ and $J_3 = 67$ K = 5.8 meV were found to provide a good description of both magnetization and susceptibility [48]. The underlying spin model is interesting in its own right, in the sense that in the limit $J_3 = 0$, it consists of decoupled gapless J_1 chains (for both positive and negative J_1), while it shows a gap in the limit $J_1 = 0$. Thus, there should be two quantum-critical points, which were found to be $J_1 \sim 0.55 J_3$ and $J_1 \sim -0.9 J_3$, although the parameters for $CaCuGe_2O_6$ were far from both the critical points. Microscopic analysis thus established that $CaCuGe_2O_6$ can be described as a system of dimers formed by the third NN, s = 1/2 Cu^{2+}, with ferromagnetic one-NN inter-dimer coupling. The edge-shared CuO_6 octahedra at NN positions with Cu-O-Cu angles of 92° and 98° justify the ferromagnetic nature of J_1. In contrast to $CuGeO_3$, which is a frustrated J_1–J_2 system showing a spin-Peierls phase transition, in the present case, the primary role is played by 3rd NN, with possible frustration arising from the second NN being secondary.

$Cu_2Te_2O_5X_2$ (X = Cl/Br)—The $Cu_2Te_2O_5X_2$ compounds were introduced [52] as spin-cluster compounds, where, structurally, the magnetic ions form well-defined clusters (see the left panel of Figure 4), and the crystal is made by periodic repetition of the clusters. At first sight, it appears that the magnetic behavior of the compounds should be dominated by spin clusters with little interaction between them. It thus came as a surprise that these compounds were reported to exhibit long-range magnetic orders with T^N (Br) = 11.4 K and T^N (Cl) = 18.2 K [53]. Modeling of the DFT band structure [54] in terms of effective Cu $x^2 - y^2$ orbitals obtained via application of the NMTO-downfolding procedure showed that the effective hoppings were rather long ranged and involved dominant hopping interactions within the Cu_4 cluster as well as between the Cu_4 clusters (see the right panel of Figure 4). The authors of [54] highlighted the important role of the halogen X_4 ring formed by (X-p)-(X-p) covalent bonding and coupled to respective Cu_4 tetrahedrons to mediate the long-ranged inter-cluster interactions, thus establishing the long-range order.

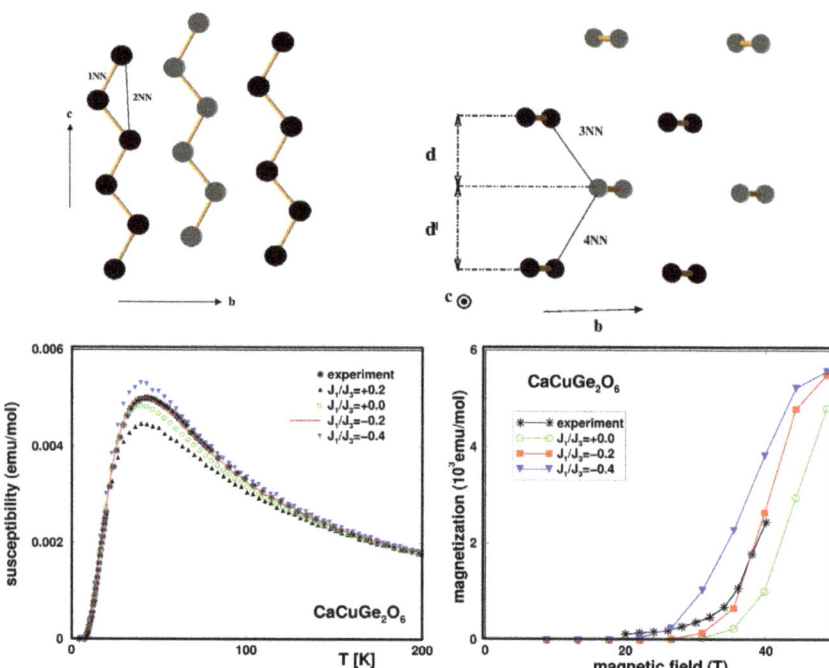

Figure 3. (**Top panels**): Cu-only lattice of CaCuGe$_2$O$_6$ showing zig-zagged chains in the *bc* planes (**left**) and long-ranged dimers formed in the *ab* plane (**right**). Atoms belonging to the same and different planes are marked in black and gray colors. (**Bottom panels**): Calculated magnetic susceptibility (**left**) and magnetization (**right**) of the interacting J_3–J_1 dimer model in comparison to the experiment [48] for different choices of J_1/J_3. Adapted from [51].

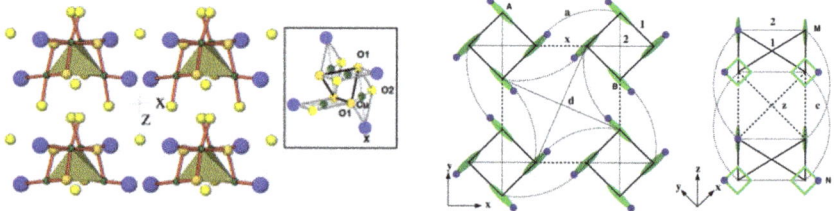

Figure 4. (**Left**) Crystal structure of Cu$_2$Te$_2$O$_5$X$_2$ (X=Cl/Br) with a Cu tetrahedron formed by four Cu atoms in a square environment of oxygen and halogen atoms, which is shown in the inset. Oxygen atoms sitting in the intra-tetrahedral and inter-tetrahedral positions are marked as yellow balls, and halogen atoms, as part of the square environment surrounding Cu^{2+}, are marked as blue balls. (**Right**) Various intra-tetrahedral and inter-tetrahedral Cu–Cu interactions, with filled circles denoting the directions of halogen sites in the square plane surrounding the Cu sites. Adapted from [54].

Na$_3$Cu$_2$Te(Sb)O$_6$—The low-dimensional magnetic behavior of ordered rocksalt oxides [55,56] Na$_3$Cu$_2$TeO$_6$ and Na$_2$Cu$_2$SbO$_6$, which contain layers of edge-sharing Cu–O and Sb/Te–O octahedra (see the left panel of Figure 5) separated by Na$^+$ ions, has raised some debate. The three most significant in-plane Cu–Cu interactions are the NN interactions, J_3, edge-shared Cu–Cu interactions, J_2, and Cu–Cu interactions through intervening Sb/Te octahedra, J_1 (see the middle panel of Figure 5). The long-ranged interactions J_4 and J_5 are expected to be negligible. Interestingly, depending on the signs and relative magnitudes of J_1, J_2, and J_3, the magnetic dimensionality of the compounds can be 0 (J_1

$\gg J_2 \sim J_3$), 1 ($J_1 = J_2 \gg J_3$ or $J_1 > J_2 \gg J_3$), or 2 ($J_2 \sim J_3 \sim J_1$). The measured magnetic susceptibility data (see the right panel of Figure 5) suggested an alternating chain model.

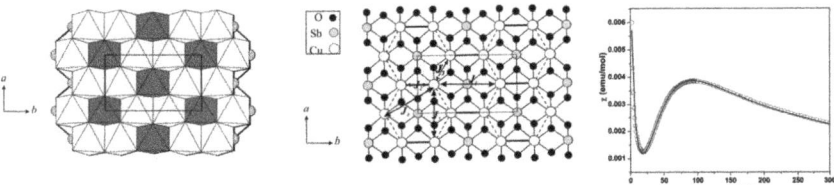

Figure 5. (**Left**) Crystal structure of $Na_3Cu_2SbO_6$, showing edge-shared Cu octahedra separated by SbO_6 octahedra. $Na_3Cu_2TeO_6$ has a similar structure. (**Middle**) Various Cu–Cu interactions, with the thick line denoting nearest neighbor (NN) Cu–Cu interactions (J_2), the thin arrowed line marking interactions through Sb(Te)O_6 octahedra (J_1), and the dashed line denoting inter-chain interactions (J_3. (**Right**) Measured magnetic susceptibility data (circles) fitted with the antiferromagnetic–antiferromagnetic (AF-AF) model (solid line). Adapted from [57].

Fitting the susceptibility data with antiferromagnetic–antiferromagnetic (AF-AF) and antiferromagnetic–ferromagnetic (AF-F) models has been tried, and it has been concluded that, based on the fitting criterion, it is very difficult to distinguish between the AF-AF and AF-F models [56]. NMTO-downfolding calculations, as well as total energy calculations, established [57] that J_1 mediated by the Cu-O-Sb/Te-O-Cu pathway is overwhelmingly the strongest exchange pathway for both materials, and while J_3 is small, the NMTO calculations predicted [57] both J_1 and J_2 to be antiferromagnetic (AF) with J_2(Sb) / J_1(Sb) > J_2(Te) / J_1(Te). Comparison of calculated and observed Curie–Weiss temperatures considering the AF-AF and AF-F models showed [57] that the AF-AF model gives significantly better agreement (−75 K/ −56 K calculated value vs. −87 K/−55 K observed value for Sb/Te compounds) compared to the AF-F model (−14 K/−11 K calculated value vs. −87 K/−55 K observed value for Sb/Te compounds). It was thus concluded that the AF-AF alternating chain is the appropriate model for both compounds [57].

$CuTe_2O_5$—In an attempt to analyze the effect of lone-pair cations, such as Te^{4+}, on the magnetic behavior of Cu^{2+} systems, the $CuTe_2O_5$ compound was synthesized and investigated [58]. The crystal structure of the compound [59] consists of edge-shared Cu octahedra forming Cu–Cu dimers, whose corners are shared with TeO_4 to form a three-dimensional lattice of $CuTe_2O_5$. The measured magnetic susceptibility of $CuTe_2O_5$ shows a maximum at T_{max} = 56.5 K and an exponential drop below T $\sim$ 10 K, signaling the opening of a spin gap [58]. Electron spin resonance (ESR) data [58] suggested that structural dimers did not coincide with the magnetic dimers. Fitting to magnetic susceptibility data gave rise to a number of possibilities, including a dimer model, an alternating chain model, and an interacting dimer model, while the extended Hückel analysis suggested [58] an alternating chain model. The constructed Wannier-like function of the effective Cu $x^2 - y^2$ via NMTO-downfolding (see the top left panel of Figure 6) showed [60] that, in addition to the formation of a strong $pd\sigma$ antibond between Cu- $x^2 - y^2$ and O-p_x/p_y, the O-p_x/p_y tails of the Wannier function bend towards the Te atom, which is responsible for enhancing the Cu–Cu interactions between different structural dimers. The strongest Cu–Cu interaction, J_4, was found to be given by Cu pairs belonging to different structural dimers that were connected by two O-Te-O bridges. Two additional in-plane interactions, J_6 and J_1, one of which is the intra-structural dimer interaction (J_1), were found to be appreciable, giving rise to a 2-d coupled dimer model, as shown in the top right panel of Figure 6. An SSE-QMC calculation [41] of the susceptibility data for the proposed 2-d model was found to be in good agreement with the experimental data (see the bottom left panel of Figure 6). Predictions were made for the temperature dependence of magnetization at different values of an external magnetic field (cf. the bottom right panel of Figure 6), which would help to differentiate between the alternating chain model and the 2-d coupled dimer model. Further experimental investigation is needed to settle this issue.

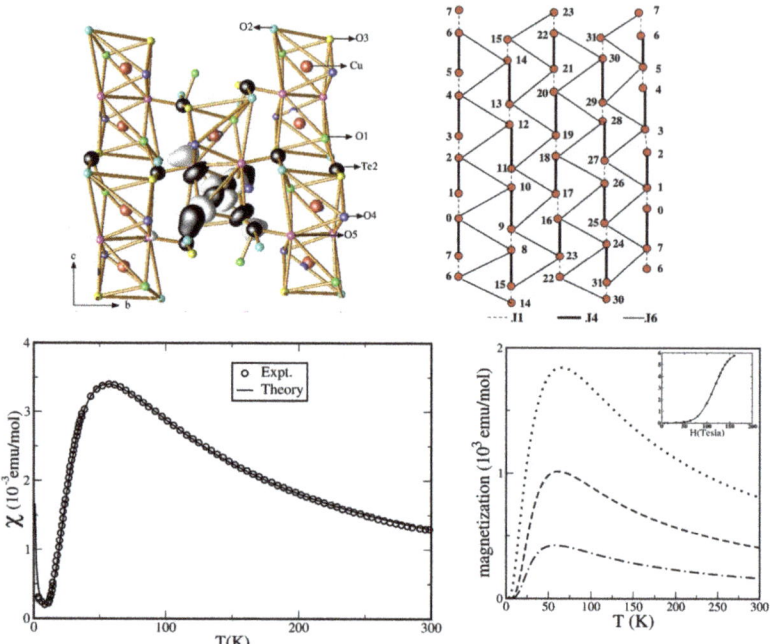

Figure 6. (**Top, left**) Effective Cu $x^2 - y^2$ Wannier function of CuTe$_2$O$_5$, as obtained in an NMTO-downfolding calculation, with lobes of opposite signs colored as black and white. Noticeable is the bending of the tail of the Wannier function at the O site to the Te atoms, thus reflecting strong Te-O hybridization. (**Top, right**) The interacting dimer model of CuTe$_2$O$_5$ defined on a 32-site lattice, with dimer interactions marked as thick, solid lines and two inter-dimer interactions as thick, solid, and dashed lines. (**Bottom, left**) Computed magnetic susceptibility in comparison to experimental data [58]. (**Bottom, right**) Computed temperature-dependent magnetization in external magnetic fields of 12.7 T (dash-dotted), 31.7 T (dashed), and 63.4 T (dotted). The inset shows the plot of magnetization as a function of the magnetic field at 10 K. Adapted from [60].

Cs$_2$CuAl$_4$O$_8$—The introduction of alternation of nearest-neighbor magnetic interactions into a uniform-chain S = 1/2 AFM Heisenberg model causes its gap-less spectrum to be gapped [61]. The excitation spectrum of a uniform spin chain with both nearest-neighbor and next-nearest-neighbor interactions also becomes gapped if the next-nearest-neighbor (NNN) interactions exceed a certain fraction of the nearest neighbor interactions [62]. It is curious to ask what happens in the presence of both alternation and NNN interactions. For this, one must identify a compound that shows (i) heavily suppressed inter-chain interaction, excluding the formation of a 3-d order and (ii) significant NNN interactions together with alternation. Cs$_2$CuAl$_4$O$_8$, a recently synthesized Cu^{2+}-based compound with a novel zeolite-like structure [63], appears to be a perfect candidate for this. From the fit of the susceptibility data [63], it appears that this compound belongs of the category of non-uniform spin-chain compounds, and it is hard to infer anything further. NMTO-downfolding-based first-principle calculations, as well as total energy calculations, were carried out to provide a microscopic understanding [64]. This gave rise to two different NN interactions between crystallographically inequivalent Cu sites, Cu1 and Cu2, Cu1–Cu2, and Cu2–Cu2, as well as two NNN interactions between Cu1 and Cu2 and Cu2 and Cu2, as shown in left panel of Figure 7. This gave rise to a magnetic J-J-J' / J_{nnn}-J'_{nnn} model, giving rise to the first-ever example of a 1-d spin chain with both alternation and competing NNN interactions. Interestingly, the edge-shared NN interactions with near cancellation of Wannier tails at neighboring Cu sites, as shown in the left panel of Figure 7, turned out to be much smaller than NNN interactions for which the Wannier tails at neighboring Cu sites pointed towards each other. The sign of the alternation parameter turned out to be negative, giving rise to the presence of both ferromagnetic and antiferromagnetic nearest-neighbor exchanges, thereby suggesting a rather rich physical system. The solution of the first-principle-derived spin model through the quantum Monte Carlo technique, as shown in the middle panel of Figure 7, provided a reasonable description of the experimentally measured magnetic susceptibility, which shows the presence of a spin gap of ~3–4 K. The curious nature of the derived spin model prompted further investigation of the model parameter space through exact diagonalization, which showed the possibility of a quantum phase transition from a gap-full to a gap-less situation by tuning the value of J_{nnn} in the presence of competing J and J', as shown in the right panel of Figure 7 [64].

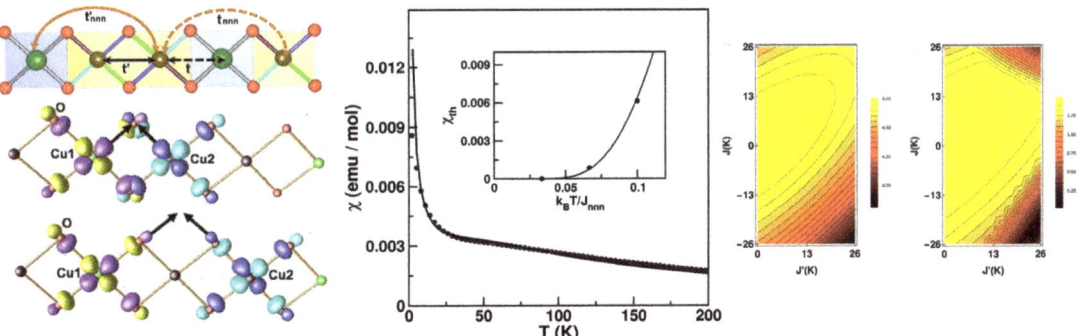

Figure 7. (**Left**) The effective Cu–Cu interactions in the chain compound $Cs_2CuAl_4O_8$ (**top**), and the overlap of effective Cu Wannier functions placed at NN Cu sites (**middle**) and next-nearest-neighbor (NNN) Cu sites (**bottom**). Positive and negative lobes of wave-functions are colored in yellow/blue and magenta/cyan at site 1 and site 2. (**Middle**) The calculated magnetic susceptibility of the derived spin model, which is shown as a solid line, in comparison with the experimental data, which are shown as symbols, in the presence of a magnetic field of 5 T [63]. The inset shows the susceptibility at H = 0 T in the absence of impurity contributions, which shows the presence of a spin gap. (**Right**) The spin gap in the parameter space of J and J' for two choices of J_{nnn} = 78 K (**left**) and = 26 K (**right**). Adapted from [64].

3.2. Vanadates

As opposed to cuprates with the Cu ion primarily in the 2+ valence, which is Jahn–Teller active, with an unfilled occupancy in the e_g manifold, vanadate compounds pose the other limit with an unfilled occupancy of the t_{2g} manifold of V ions. Note that the Jahn–Teller activity of t_{2g} ions is expected to be much less than that of e_g ions. Furthermore, possible oxidation of V as 4+ and 5+ leads to interesting phenomena, such as charge disproportion, charge ordering, and their influence on magnetic properties.

α'-NaV_2O_5 and γ-LiV_2O_5—Layered vanadates (AV_2O_5) [65] form an important family of low-dimensional magnets. While CaV_2O_5 and MgV_2O_5, with their divalent A sites, contain V in its 4+ state and behave as spin-1/2 ladders that exhibit spin gaps [66,67], monovalent A cation compounds, such as α'-NaV_2O_5 [68] γ-LiV_2O_5 [69], and CsV_2O_5 [70], have, on average, $V^{4.5+}$, with important charge and spin fluctuations. The monovalent A cation gives rise to a quarter-filled V d_{xy} band rather than half-filled. Both NaV_2O_5 and LiV_2O_5 crystallize in an orthorhombic space group [71,72], with a layered structure of square VO_5 pyramids separated by Li^+/Na^+ ions between the layers. Within the layers, the VO_5 pyramids form zig-zagged chains running along the y-axis, with two successive zig-zagged chains linked by corner sharing via a bridging O, as shown in the top left panel of Figure 8. The in-plane lattice structure of V ions with possible interactions can be schematically represented as shown in the bottom left panel of Figure 8. Appropriate choices of parameters can present a zig-zagged Heisenberg chain ($J(1) \gg J(b)$), a Heisenberg double chain ($J(b) \gg J(1)$), or a ladder with one electron per V(1)-O-V(2) rung. The nature and degree of charge disproportionation have a big role in deciding on these three possibilities.

For NaV$_2$O$_5$, it has been established that below a critical temperature, T_c = 34 K, a charge disproportionation appears, 2 V$^{4.5+}$ → V^{5+} + V^{4+}, while above T_c, all vanadium ions are equivalent (V$^{4.5+}$) [73]. The magnetic field effect on T_c establishes a zig-zagged ordering of charge-disporportionated V ions [68]. Interestingly, both below and above the charge-ordering transition, the compound has been reported to be insulating [74]. While the insulating nature of charge-ordered phase has been described properly in terms of LDA+U calculations, the description of the insulating state of the charge-disordered phase is challenging. The application of the NMTO-downfolding procedure for the construction of a low-energy Hamiltonian in terms of effective V d_{xy} Wannier functions resulted [75] in strongest rung hopping (t_a) for 0.38 eV, followed by leg hopping (t_b) for 0.08 eV, and inter-ladder hopping for t_1 = 0.03 eV and t_2 = 0.02 eV. Counter-intuitively, this procedure also resulted in a large diagonal hopping (t_d) for 0.08 eV. This strong diagonal hopping had important implications in the description of the underlying Hubbard model corresponding to a two-leg ladder system, which had to include not only local, onsite Coulomb interaction (U), but also inter-site Coulomb interaction (V), a combination of in-rung and diagonal inter-rung V parameters [75]. The resulting density of states obtained with a cluster dynamical mean field theory (DMFT) [76] solution of the extended Hubbard model (see the top right panel of Figure 8) of the two-leg ladder system highlighted the crucial importance of the inter-site charge fluctuation captured through the V parameter in describing the insulating state of charge-disordered NaV$_2$O$_5$.

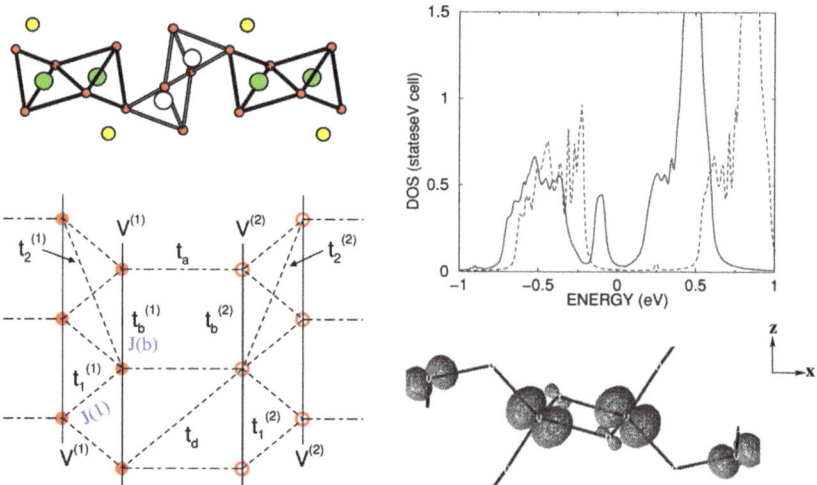

Figure 8. (**Top, left**) The edge-shared and corner-shared VO$_5$ structure of α'-NaV$_2$O$_5$ and γ-LiV$_2$O$_5$ viewed along the direction of the ladder, with the corner-shared VO$_5$ pyramids forming the rung and edge-shared VO$_5$ pyramids forming the inter-ladder neighbors. The V atoms showing possible charge disproportionation in γ-LiV$_2$O$_5$ are shown as filled green and open white balls, and oxygen atoms are marked as small balls. The Na/Li atoms are shown as yellow balls. (**Top, right**) The density of states of α'-NaV$_2$O$_5$ with choices of two inter-site Coulomb interactions, V = 0.17 eV (solid) and V = 0.5 eV, computed with the cluster dynamical mean field theory (DMFT) (**right**). The onsite Coulomb interaction U was fixed at 2.8 eV. (**Bottom, left**) Various chain, inter-chain, and rung interactions in the ladder geometry of α'-NaV$_2$O$_5$ and γ-LiV$_2$O$_5$. (**Bottom, right**) Electronic charge density of low-energy V xy bands of γ-LiV$_2$O$_5$. The bigger and smaller lobe orbitals correspond to partially charge-disproportionated V(1) and V(2) sites. Adapted from [75,77].

Unlike α'-NaV$_2$O$_5$, γ-LiV$_2$O$_5$ does not show any signatures of phase transition, although the crystal structure contains two inequivalent V ions [71], V(1) and V(2), along two legs of the ladder. Modeling in terms of a low-energy Hamiltonian on the basis of electronically active effective V(1)-d_{xy} and V(2)-d_{xy} orbitals (cf. the bottom right panel of Figure 8) showed [77] the onsite energy of V(1) and V(2) to be $\pm \epsilon_0$ with $\epsilon_0 = 0.15$ eV, and showed the rung hopping, t_a, connecting V(1) and V(2) to be 0.35 eV [77], which is close to that estimated for NaV$_2$O$_5$.

Diagonalization of the simple two-site V(1)-V(2) model gave

$$\frac{\epsilon_0}{t_a} = \frac{p(2) - p(1)}{2\sqrt{p(1)p(2)}},$$

which yielded $p(1)/p(2) = 2.3$ for $\epsilon_0/t_a = 0.15/0.35$, where $p(1)/p(2)$ are the occupancy of V(1) and V(2). The non-negligible value of $p(2) = 0.3$ suggests a non-negligible contribution of V(2) to the magnetic moment on the V(1)-O-V(2) rung, which has an important contribution to the microscopic model. For negligible values of $p(2)$, the microscopic model would be that of a zig-zagged chain with $J \sim (t_1^{(1)})^2$, since the estimated $t_b^{(1)}$ (-0.06 eV) is much smaller than $t_1^{(1)}$ (-0.18 eV). Consideration of a non-negligible contribution of V(2) changes the scenario quite a bit, with an effective hopping matrix element in the asymmetric rung state:

$$t_b^{eff} = p(1)t_b^{(1)} + p(2)t_b^{(2)} - 2\sqrt{p(1)p(2)}t_d.$$

The large value of diagonal hopping (-0.1 eV), which was already pointed out for α'-NaV$_2$O$_5$, makes the effective t_b^{eff} large, highlighting the influence of charge ordering on the nature of magnetic coupling [77].

CsV$_2$O$_5$–CsV$_2$O$_5$—Belonging to same group as α'-NaV$_2$O$_5$ and γ-LiV$_2$O$_5$, provides an opportunity to study the influence of A-site cations on the electronic and magnetic properties. The crystal structure of CsV$_2$O$_5$ is somewhat different from those of α-NaV$_2$O$_5$ and γ-LiV$_2$O$_5$. While α-NaV$_2$O$_5$ and γ-LiV$_2$O$_5$ form a double-chained structure of square VO$_5$ pyramids in an orthorhombic space group, the chains are linked together via common corners to form layers, and the layers in the monoclinic structured CsV$_2$O$_5$ are formed by edge-shared V(1)O$_5$ pyramid pairs connected through V(2)O$_4$ tetrahedra, as shown in the top left panel of Figure 9 [78]. The V ions in the tetrahedral coordination are in the 5+ or d^0 state, while those in pyramidal environment are in the 4+ or d^1 state. CsV$_2$O$_5$ thus shares the same monoclinic crystal structure as (VO)$_2$P$_2$O$_7$ (VOPO) [79], and KCuCl$_3$ and TlCuCl$_3$ show alternating spin chain behaviors [80]. The measured susceptibility data have been interpreted in terms of the underlying spin dimer model [70].

The right panel of Figure 9 shows the DFT densities of states (DOSs) [81] projected to V 3d states for CsV$_2$O$_5$ (top), γ-LiV$_2$O$_5$ (middle), and α-NaV$_2$O$_5$ (bottom). Noticeably, while α'-NaV$_2$O$_5$ and γ-LiV$_2$O$_5$ show characteristic quasi-1-d van Hove singularities, the DOSs for CsV$_2$O$_5$ show a more 2-d nature, hinting at appreciable inter-dimer interactions. NMTO-downfolding was used to calculate effective V-V hopping, which revealed [81] that, in addition to intra-dimer interactions, t_1, there are several non-negligible inter-dimer interactions, t_2, t_3, and t_5 (see the bottom left panel of Figure 9) that are mediated by paths through V^{5+}O$_4$ tetrahedra with $t_1 = 0.117$ eV, $t_2 = 0.015$ eV, $t_3 = 0.097$ eV, and $t_5 = 0.050$ eV. The strongest hopping t_1 is, however, significantly smaller than the strongest hopping for α'-NaV$_2$O$_5$ (0.38 eV) and γ-LiV$_2$O$_5$ (0.35 eV), which is rationalized by the edge-shared path for CsV$_2$O$_5$ as opposed to the corner-shared path for α'-NaV$_2$O$_5$ and γ-LiV$_2$O$_5$. The proposed 2-d model showed equally good fit to the measured susceptibility compared to the dimer model, stressing the insensitivity of magnetic susceptibility to the details of the spin model. This indicates the need for further experimental studies, such as ESR, inelastic neutron scattering, and Raman scattering, to settle on an underlying spin model.

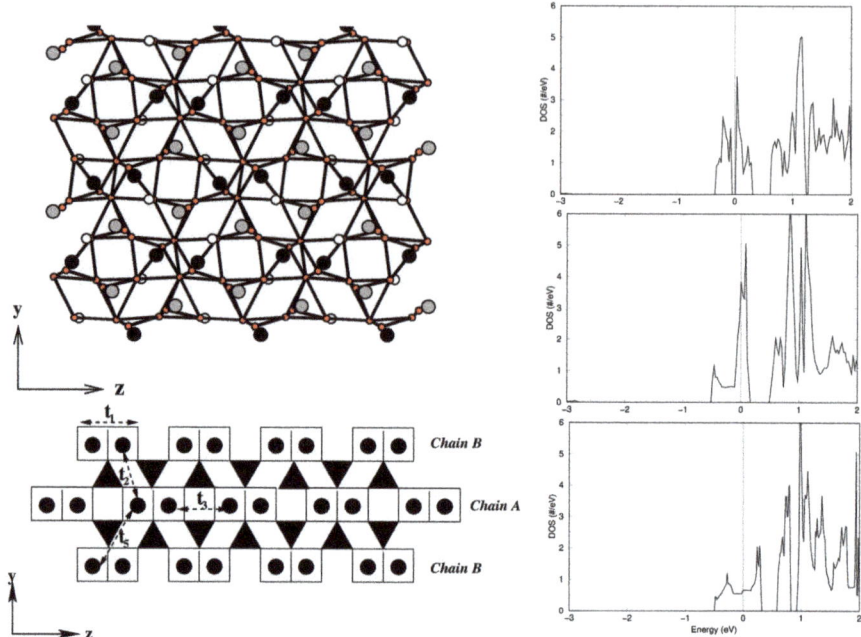

Figure 9. (**Left, top**) The in-plane crystal structure of CsV_2O_5, with V d^1 atoms in a pyramidal coordination marked as black balls and V d^0 atoms in a tetrahedral coordination marked as gray balls. Oxygen atoms are shown as small, red balls, and Cs atoms are not shown. (**Left, bottom**) Various V–V hoppings between edge-shared d^1 V pairs connected by d^0 V in O_4 tetrahedra (marked as solid triangles). (**Right**) Comparison of densities of states of V $3d$ states for CsV_2O_5 (**top**), γ-LiV_2O_5 (**middle**), and α'-NaV_2O_5 (**bottom**). Adapted from [81].

$VOSeO_3$—Spin-gap systems with moderate values of spin gaps are of general interest, as a gap may be closed by a strong enough field, driving a quantum phase transition [82]. Spin dimer systems with weak inter-dimer interactions appear to be attractive systems for realizing this possibility. The $VOSeO_3$ compound, consisting of edge-shared VO_5 pairs, forms a probable candidate belonging to this class [83]. NMTO-downfolding calculation for deriving a V d_{xy}-only Hamiltonian showed [84] the intra-dimer hopping (t_d, see the left panel of Figure 10) to be the strongest, followed by the inter-dimer coupling along the z-direction (t_2), with comparable magnitudes ($t_d = -0.083$ eV, $t_2 = -0.079$ eV). Other hopping parameters in the yz plane—t_4, t_1, and t_3—are smaller but non-negligible. Two more parameters in the xz plane, t_v and t_s (see the right panel of Figure 10) also turned out to be non-negligible. Thus, as opposed to the initial suggestion for a spin dimer system, $VOSeO_3$ turned out to be an alternating spin-chain compound with moderate inter-chain interactions.

$Na_2V_3O_7$—$Na_2V_3O_7$ is the first reported transition-metal-based nanotubular system [85]. The basic structural units are distorted square VO_5 pyramids that share corners and edges to give rise to a 2-d sheet-like structure, which, in turn, folds to provide a tube-like geometry with connected rings of nine V atoms, as shown in the top left panel of Figure 11. The Na atoms sitting inside and outside the tubes provide cohesion to the structure. Since the synthesis of this curious structure, several suggestions have been made for the description of the underlying low-energy spin model, which include nine-leg spin tubes [85], mutually intersecting helical spin chains [86], effective three-leg spin tubes with inter-ring frustration, dimerized vanadium moments [87], etc.

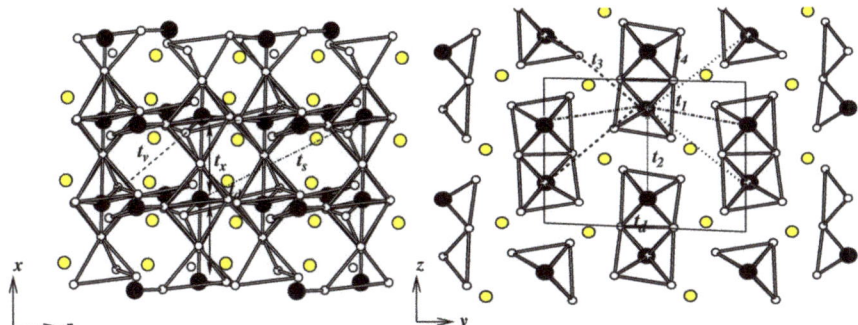

Figure 10. The various out-of-plane (**left**) and in-plane (**right**) inter-dimer V-V interactions in VOSeO$_3$, with V atoms shown as black balls, O atoms as small, white balls, and S atoms as yellow balls. Adapted from [84].

Considering the antiferromagnetic spin-1/2 ladder systems, even leg ladders give rise to a spin-singlet ground state with a spin gap, while odd-leg ladders with open boundary conditions result in free spin along one of the legs, resulting in a gap-less situation. Spin tubes, as applicable for Na$_2$V$_3$O$_7$, can be considered as odd-leg ladder with periodic boundary conditions, which, in addition to the spin degrees of freedom, also possess chirality, as shown schematically in the top right panel of Figure 11, and should exhibit a spin gap [88]. The measured susceptibility shows Curie–Weiss behavior at high and low temperatures, with a reduction in the effective magnetic moment from high to low temperatures, and, importantly, no spin gap [87]. The constructed low-energy model keeping the d_{xy} orbital active at three inequivalent V sites showed [89] that due to the complex geometry, the edge-shared V-V couplings were equally as strong as the corner-shared V-V couplings, which is demonstrated in terms of the overlap of Wannier orbitals at different V pairs in the bottom left panel of Figure 11. Neglecting the inter-ring coupling for a first approximation, which is found to be an order of magnitude smaller than the intra-ring couplings, leads to nine-site rings with partial frustration. The partial frustration arises due to the presence of both NN and NNN interactions, though not all NNN interactions appear due to the complex geometry. Exact diagonalization of the nine-site-ring spin model provides a good description of the experimental susceptibility data down to a temperature of a few K [89]. Importantly, the partially frustrated model, as opposed to the fully frustrated model, was crucial for a proper description of the data, as shown in the bottom right panel of Figure 11.

Zn$_2$VO(PO$_4$)$_2$—In an attempt to modulate the nature of the magnetic ground state, spin dilution has been attempted through the depletion of magnetic centers. A prominent example is CaV$_4$O$_9$, which is formed by 1/5 depletion of the two-dimensional antiferromagnetic lattice [90]. With a similar motivation, Zn$_2$VO(PO$_4$)$_2$ was studied; 1/4 of the V sites were replaced by Ti [91]. The crystal structure of the pristine compound, as shown in the top left panel of Figure 12, consists of VO$_5$ pyramids. In the ab plane, NN VO$_5$ pyramids are connected via PO$_4$ tetrahedra, while NNN VO$_5$ pyramids are connected via two ZnO$_5$ units. The ab layers are connected via corner-shared PO$_4$ and ZnO$_5$ units to form the tetragonal symmetry [92] of the 3-d structure. Starting from the pristine crystal structure of Zn$_2$VO(PO$_4$)$_2$, every fourth V atom was substituted by nonmagnetic d^0 Ti^{4+} ions to achieve dilution of the S = 1/2 lattice formed by d^1 V^{4+} ions, as shown in the top right panel of Figure 12. Ti substitution resulted in three inequivalent V sites; the V atoms had no Ti neighbors, thus retaining four in-plane and two out-of-plane V neighbors. The V1 atoms had in-plane Ti neighbors, with two in-plane and two out-of-plane V atoms. The V2 atoms had both in-plane and out-of-plane Ti neighbors, giving rise to only two V neighbors. The NMTO-downfolding calculation for a low-energy model on the basis of the effective V d_{xy} Wannier function for the pristine compound showed [91] that the compound was best described by a weakly coupled two-dimensional S = 1/2 antiferromagnetic square lattice (cf. Figure 12). The NNN and AF V-V magnetic interaction in the *ab* plane was found to be 2% of the strongest, and the NN and AF V-V magnetic interaction in the *ab* plane with ferromagnetic interlayer magnetic interaction had a strength of 3% of the NN interaction, which was in good agreement with the conclusions drawn from a neutron scattering experiment [93]. The NMTO-derived spin model for the Ti-substituted $Zn_2VO(PO_4)_2$ compound turned out to be a coupled S = 1/2 AFM chain. The missing V sites made the in-plane NN V-V interactions unequal along the *a* and *b* directions; the interlayer coupling was found to be of the dimer type. The computed magnetic susceptibility for the pristine compound showed good agreement with experimental data at H = 10,000 Oe [94]. The calculated susceptibility and magnetization (cf. Figure 12) of the substituted compound and their comparison to those of the pristine compound confirmed [91] the change in the magnetic ground state from a long-ranged ordered phase in the pristine compound to a spin-gapped phase in the Ti-substituted compound. This was corroborated by calculated spin wave spectra [95] (cf. Figure 12).

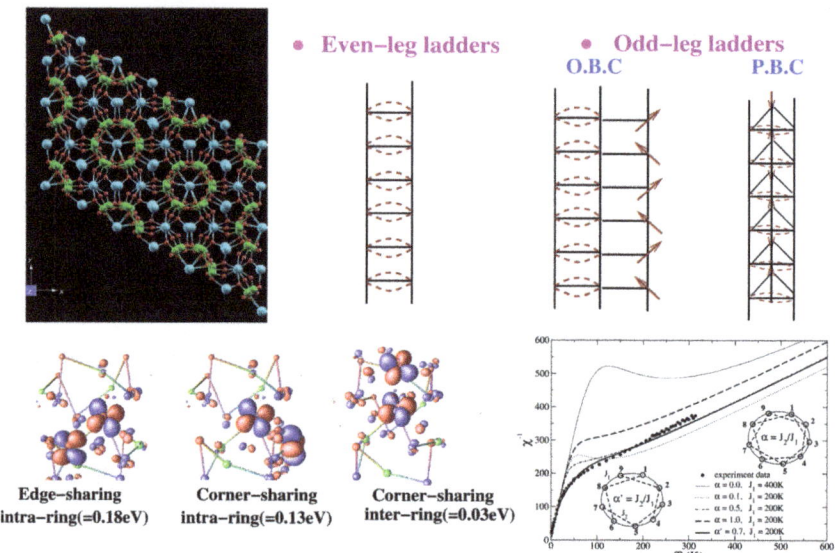

Figure 11. (**Top, left**) The tubular crystal structure of $Na_2V_3O_7$ with edge- and corner-sharing VO_5 pyramids (V atoms in green and O atoms in red), and Na (cyan) atoms occupying the insides and outsides of the tubes. (**Top, right**) Schematic representation of even-leg and odd-leg ladders with open boundary conditions (O.B.C.) and periodic boundary conditions (P.B.C.). (**Bottom, left**) Overlap of effective V *xy* Wannier functions at various edge-shared and corner-shared positions of V sites in the tube. Shown are the corresponding hopping integrals. (**Bottom, right**) Temperature dependence of the inverse magnetic susceptibility of different spin models compared with the experimental data [87] of the compound. Adapted from [89].

The theoretical prediction needs to be verified experimentally, as the ordering between the Ti and V atoms assumed in the calculations needs to be ensured, which may be challenging.

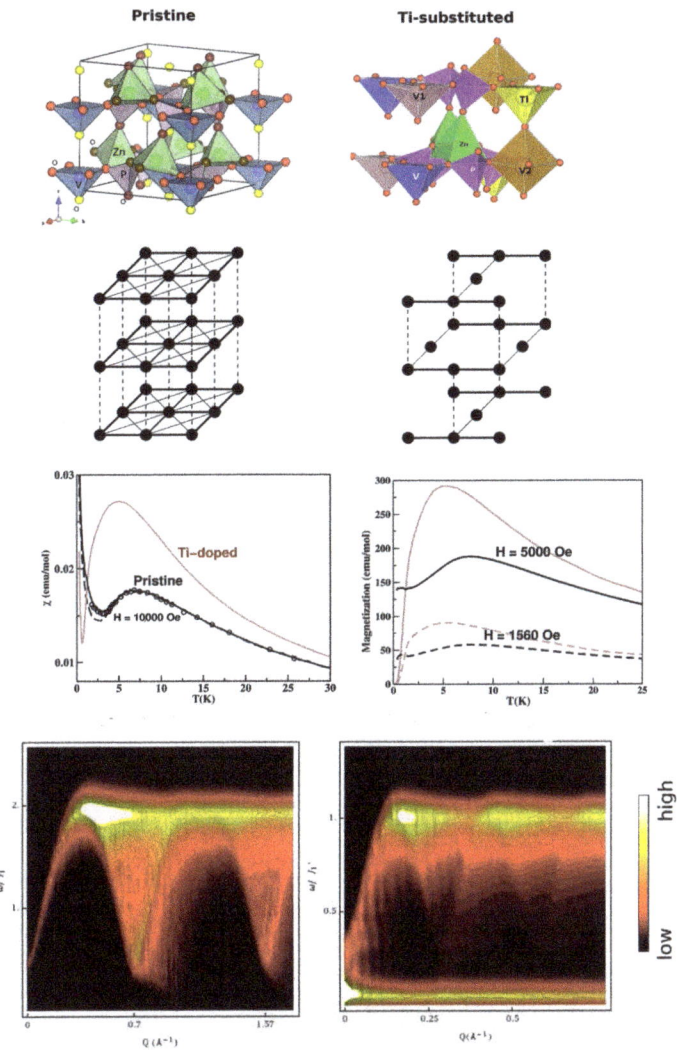

Figure 12. (**Top**) Crystal structures of pristine (**left**) and 1/4 Ti-substituted (**right**) $Zn_2VO(PO_4)_2$ in a shaded polyhedral representation, with oxygen atoms marked by small, red balls. Second from top: Spin models of pristine (**left**) and 1/4 Ti-substituted (**right**) $Zn_2VO(PO_4)_2$, with the strongest interactions shown by thick, solid lines, and the two weaker interactions shown by thin, solid, and dashed lines. See the text for details. Third from top: The computed magnetic susceptibility (**left**) and magnetization (**right**) of the pristine (black) and 1/4 Ti-substituted (brown) $Zn_2VO(PO_4)_2$. For the pristine compound, also shown are the experimental data points for the susceptibility measured at H = 10,000 Oe. (**Bottom**) The computed spin wave spectra for the pristine (**left**) and 1/4 Ti-substituted (**right**) $Zn_2VO(PO_4)_2$. Figure adapted from [91,95].

3.3. Nickelates

As opposed to cuprates and vanadates, which have one hole or one electron in the magnetically active Cu^{2+} or V^{4+} ions, nickelates, with Ni in their 2+ valence state or d^8, serve as examples of S = 1 spins with half-filled Ni e_g states. The presence of active e_g electrons makes the metal–ligand hybridization stronger compared to t_{2g}-based systems, such as vanadates. On other other hand, as Ni is a neighbor to Cu, this serves as an excellent opportunity to look for an alternative to cuprates, resulting in the recent study of nickelate superconductivity [96]. This makes the study of the low-dimensional quantum spin systems of nickelates an important topic.

$NiAs_2O_6$—$NiAs_2O_6$, a member of the $3d$ homologous series AAs_2O_6 (A = Mn, Co, Ni) [97], shows antiferromagnetic ordering with $T_N \sim 30$ K. The situation became curious with the synthesis of $PdAs_2O_6$ with a measured Curie temperature that was five times greater than that of $NiAs_2O_6$, $\sim$150 K [98]. Note that Ni^{2+} and Pd^{2+} are examples of d^8 ions with S = 1 that belong to the $3d$ and $4d$ transition metal series, respectively. The reported magnetically ordered ground states with reasonably high values for the Néel temperature are surprising, since the Ni/PdO_6 octahedra do not share corners, edges, or faces in the AAs_2O_6 structure, which consists of edge-shared As^{5+} ions in an octahedral oxygen coordination that forms hexagonal layers; different layers are connected by the Ni/PdO_6 octahedra [98] (cf. the top left panel of Figure 13). The magnetically active orbitals of d^8 ions in the octahedral coordination are two e_g orbitals, x^2-y^2 and $3z^2 - r^2$. The Wannier orbitals for $x^2 - y^2$ and $3z^2 - r^2$ constructed with the NMTO-downfolding procedure of integrating out everything else other than the A e_g orbitals (cf. the top right panel of Figure 13) highlighted the bending of O p tails of the functions to bond with the sp characters of the nearest As pairs [99]. This enables long-ranged interactions between A-A pairs, although there is no short-ranged interaction mediated by connected O atoms. The dominant hopping paths, as given by the tight-binding Hamiltonian on the downfolded A-e_g basis [99], are shown in the bottom left panel of Figure 13. The third NN interactions were found to dominate over second NN and NN interactions. Accounting for the larger e_g band width of the Pd compound, the hopping interactions for $PdAs_2O_6$ were found to be about 1.4 times larger than those for $PdAs_2O_6$ [99]. With Hubbard U parameters [99] estimated as $U_{Ni}/U_{Pd} \sim 2.4$, this led to ratios of magnetic exchanges in two compounds of 4.7, which was in good agreement with the ratios of the experimentally measured Néel temperatures of the two compounds.

Finally, the magnetic susceptibility computed with the SSE-QMC of the derived spin model of the two compounds showed exceedingly good agreement with measured data [99], as shown in the bottom right panel of Figure 13.

$NiRh_2O_4$—With the goal of understanding symmetry-protected topological spin systems, S = 1 spin models on diamond lattices were proposed as potential candidates [100]. This will be a 3-d analogue of the Haldane chain with a gapless 2-d surface state [101]. The A sublattice of spinel compounds offers the possibility of studying diamond lattice magnetism if a magnetic ion can be put at an A site. Magnetic measurements on A sublattice magnetic ion spinels, such as $MnSc_2S_4$ (S = 5/2), $CoAl_2O_4$ and $CoRh_2O_4$ (S = 3/2), and $CuRh_2O_4$ (S = 1/2), revealed an ordered magnetic ground state [102,103]. The report of $NiRh_2O_4$ with S = 1 ion on the A site therefore created a lot of excitement [104], as the ground state was reported to be non-magnetic.

To provide a microscopic understanding of the nature of the non-magnetic ground state of $NiRh_2O_4$, first-principle calculations together with a model study were carried out [105]. We note that Ni at the A site of a spinel is in a tetrahedral coordination, which results in crystal field splitting of Ni d states into high-lying t_2 and low-lying e states. The d^8 occupation thus leads to partially filled t_2 bands, allowing for spin–orbit coupling to be active among orbitally active degrees of freedom. DFT calculations within a GGA+U formulation gave rise to a half-metallic solution, while the inclusion of SOC was necessary to correctly describe [105] the insulating state of $NiRh_2O_4$ as it was observed experimentally (cf. Figure 14). This highlights the importance of SOC in the description of SOC in d^8 systems in tetragonal coordinations, which, as opposed to octahedral d^8 systems, have active orbital degrees of freedom. The orbital moment at Ni site calculated with the DFT turned out to be [105] large, ~ 1 μ_B, supporting the formation of an S = 1, L_{eff} =1 state. The other important input from DFT [105] was the large Ni–Rh hybridization, triggered by the near degeneracy of Ni t_2 and Rh t_{2g} states in the down-spin channel, as shown in Figure 14. The substantial mixing between Ni–Rh states, in addition to NN Ni–Ni interactions, gave rise to NNN Ni–Ni interactions, as evidenced in the overlap of tails of

Ni t_2 Wannier functions placed at NNN Ni sites and at intervening Rh sites (see Figure 14). The calculated values of magnetic exchanges gave $J_1 \sim 1.2$ meV and J_2 (J'_2, J''_2) $\sim 0.4\, J_1$, suggestive of strong magnetic frustration.

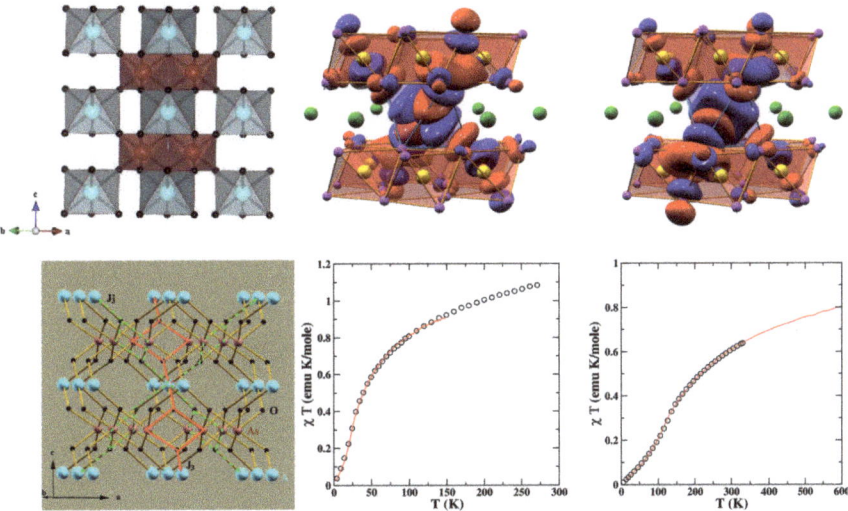

Figure 13. (**Top, left**) Crystal structure of NiAs$_2$O$_6$ with Ni, As, and O atoms shown as blue, brown, and black balls, respectively. PdAs$_2$O$_6$ is isostructural to NiAs$_2$O$_6$. (**Top, right**) The magnetically active Ni/Pd x^2-y^2 and 3^3-r^2 Wannier functions on the downfolded e_g-only basis. The AsO$_6$ octahedra are marked with two oppositely signed lobes of the wave-functions, which are colored in red and blue. (**Bottom, left**) The spin model involving long-ranged third NN exchange paths, J_3 and J'_3. (**Bottom, right**) The calculated magnetic susceptibility of NiAs$_2$O$_6$ (left) and PdAs$_2$O$_6$ (right) in comparison to experimental data, which are shown as circles. Adapted from [99].

Following the DFT results of the L$_{eff}$ = 1 and S = 1 state, one can write the single-site Hamiltonian as

$$H = -\delta L_z^2 + \lambda \vec{L}.\vec{S}$$

The DFT results gave $\delta \gg \lambda$, with δ, the spin-averaged tetrahedral crystal field splitting between Ni d_{xy} and Ni $d_{yz}/d_{xz} \sim 100$ meV, and the spin–orbit coupling $\lambda \sim 10$ meV. The solution of the spin-site Hamiltonian showed the ground state to be a nonmagnetic singlet, as observed experimentally. However, contrary to expectation of a topological quantum paramagnet [106], it turned out to be a spin–orbit entangled singlet state. Incorporating the inter-site interaction via a J_1–J_2 Heisenberg exchange model provided a description of the measured inelastic neutron-scattering results [104].

Sr$_3$NiPt(Ir)O$_6$—These compounds belong to K$_4$CdCl$_6$-type structures [107] consisting of (BB$'$O$_6$)$^{-6}$ chains formed by alternating BO$_6$ trigonal biprisms and B$'$O$_6$ octahedra, where the chains are separated by intervening A^{2+} cations, as shown in the top left panel of Figure 15. The construction of magnetically active Ni $d_{xz/yz}$ Wannier functions [108] highlighted the strong hybridization between Ni and Ir states, which is responsible for the Ni–Ni intra-chain interactions through Ir. For Sr$_3$NiPtO$_6$, the intra-chain Ni–Ni interactions occur between magnetically active half-filled Ni $d_{xz/yz}$ levels through oxygen-mediated superexchange paths, which, in accordance with the Kugel–Khomskii picture, gave rise to antiferromagnetic interactions. On the other hand, the interactions between half-filled Ni $d_{xz/yz}$ and one of the Ir t_{2g} states turned out to be antiferromagnetic. Additionally, there exists a direct exchange between the two, which turned out to be of a ferromagnetic nature. A large antiferromagnetic inter-chain interaction for Sr$_3$NiIrO$_6$ was noticed [108] (cf. Figure 15), presumably explaining the antiferromagnetic couplings observed in experiments [109]. The inclusion of spin–orbit coupling showed [108] magnetocrystallic anisotropy to be an easy axis (chain direction) for Sr_3NiPtO_6, while it to was perpendicular to the chain direction for Sr_3NiIrO_6.

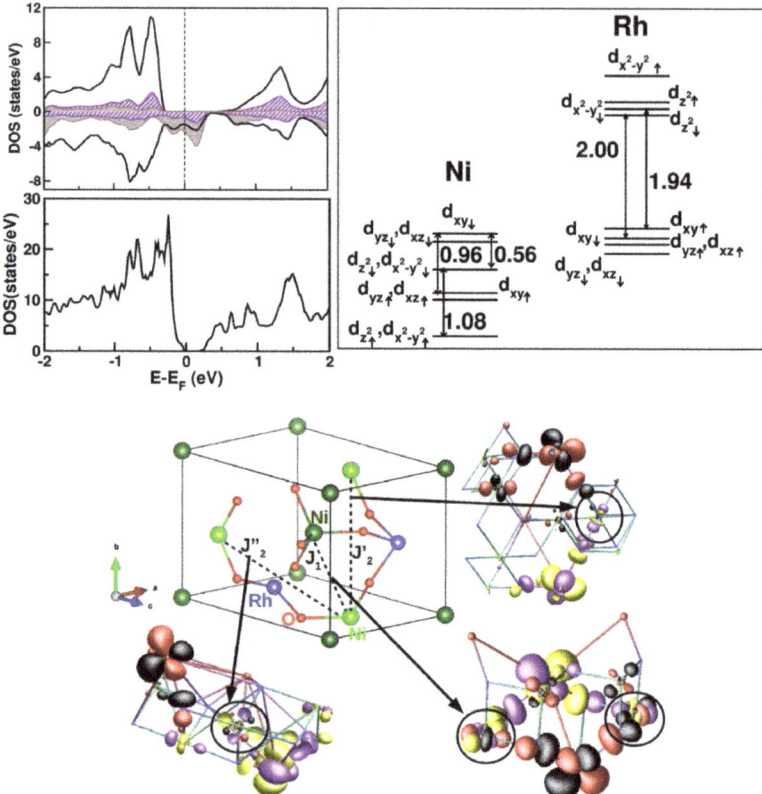

Figure 14. (**Top, left**) The calculated density of states of $NiRh_2O_4$ in the GGA + U (**top**) and GGA + U + SOC schemes of calculations. The zero of energy is set at the respective Fermi energies. For ease of visualization, in the GGA + U density of states plot, the states corresponding to the up- and down-spin channels are shown as positive and negative, respectively. Additionally shown are the states projected to Ni d (solid, black), Rh d (brown shaded), and O p (hatched). (**Top, right**) The crystal field splitting at the Ni and Rh sites, as estimated using NMTO calculations on the tight-binding Wannier basis of Ni d and Rh d. (**Bottom**) The dominant magnetic interactions and the overlap of downfolded Ni Wannier functions at NN and NNN sites. Lobes of opposite signs are colored differently, with the color convention of positive/negative represented by red (yellow)/black (magenta) in sites 1 and 2. Adapted from [105].

$SrNi_2V_2O_8$—$SrNi_2V_2O_8$ serves as a candidate material for studying the closing of the spin gap by an external magnetic field in an S = 1 Haldane chain compound [110]. While the original conjecture of Haldane is applicable to a strictly one-dimensional chain, the behavior of the real compound is complicated by the presence of inter-chain interaction, single-ion anisotropy, etc. [111], resulting in the need for microscopic investigation. In $SrNi_2V_2O_8$, edge-sharing NiO_6 octahedra form zig-zagged chains that are connected to each other by VO_4 tetrahedra, giving rise to three-dimensional connectivity [112]. Low-energy modeling in terms of constructed Ni e_g Wannier functions demonstrated the effectiveness of hybridization with V degrees of freedom and generated [113] well-defined exchange paths beyond nearest-neighbor intra-chain Ni–Ni interactions. This gave rise to both longer-ranged intra-chain (NNN) and inter-chain interactions, as shown in the top left and right panels of Figure 16. The derived spin model thus consisted of J_1, J_2, and J_3, and J_4, J_1, J_2 were the nearest- and next-nearest-neighbor intra-chain interactions, respectively, while the latter two were the inter-chain interactions. Overlap plots of the Ni e_g Wannier functions confirmed [113] (cf. the middle panels in Figure 16) the exchange paths mediated by the V atoms. The magnetic-field-dependent magnetization (cf. the bottom left panel of Figure 16) calculated through the application of SSE-QMC to the derived spin model showed [113] that the critical magnetic field necessary for closing of the spin gap was markedly different from the estimated value of 0.4 J_1 when considering a strictly 1-d spin-chain model, and further established the effectiveness of J_2 in tuning the spin gap. Studies carried out with bi-axial strain [113] showed a monotonic decrease in the spin gap value upon increase in the in-plane lattice constant (cf. the bottom right panel of Figure 16). This was found to be caused by the modulation of the J_2/J_2 ratio due to compressive strain and the change in the J_1 value due to tensile strain. This study predicts that bi-axial strain is an effective tool for tuning the spin gap of this compound, which should be verified experimentally. This also opens up the possible strain-induced closing of the spin gap by driving a quantum phase transition from a gap-full to a gap-less situation.

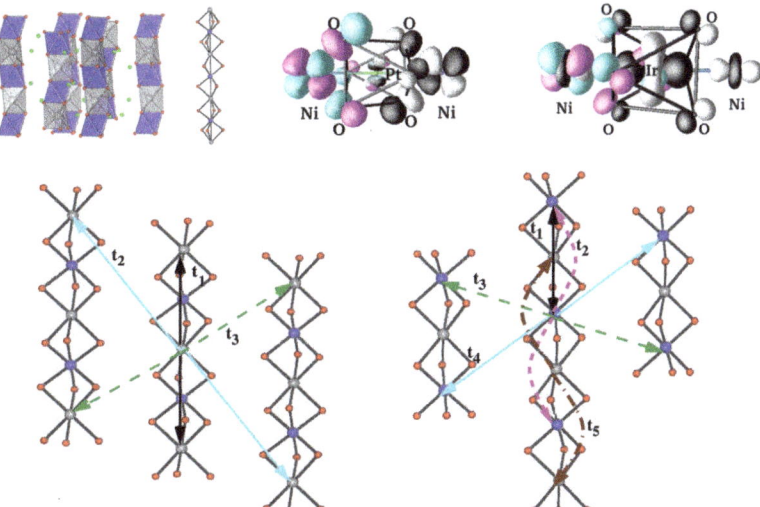

Figure 15. (**Top, left**) The chain structure of $Sr_3NiPt(Ir)O_6$ with face-shared NiO_6 trigonal biprisms and Ir/PtO$_6$ octahedra that alternate along the chain direction. (**Top, right**) The magnetically active Ni d Wannier function for Ni–Pt and Ni–Ir compounds. Note the tail with a weight at neighboring Pt/Ir sites, which is appreciable for the Ni–Ir compound, signaling strong hybridization between the Ni and It d states. (**Bottom**) Effective Ni–Ni hopping interactions for Ni–Pt (**left**) and Ni–Ir (**right**) compounds. Adapted from [108].

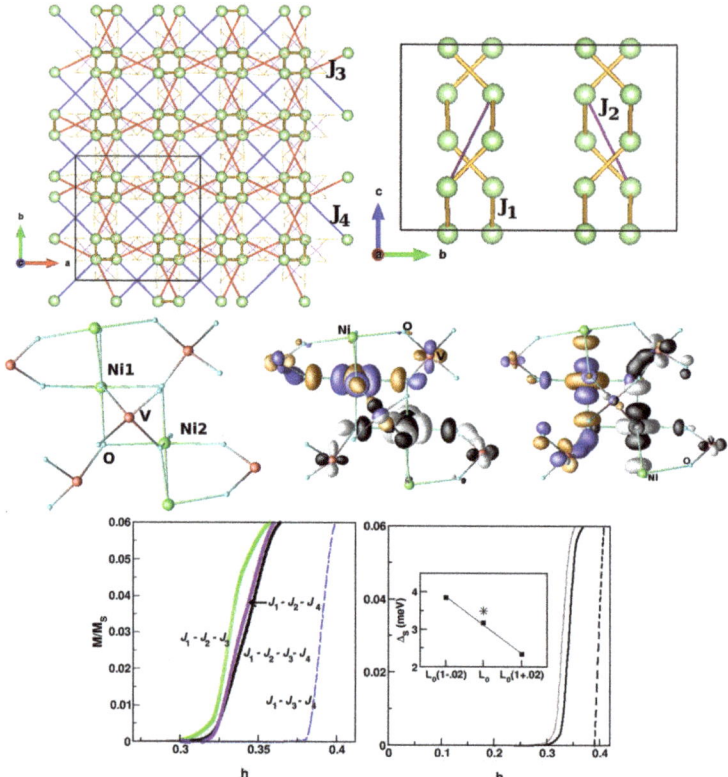

Figure 16. (**Top**) Ni–Ni magnetic interactions in SrNi$_2$V$_2$O$_8$, shown in the plane perpendicular to the chain direction (**left**) and in the plane containing the chain (**right**). Marked are the two dominant intra-chain (J_1, J_2) interactions and two inter-chain (J_3, J_4) interactions. Middle: Overlap of Ni $x^2 - y^2$ and Ni $3z^2 - r^2$ Wannier functions at NN positions, connected through edge-sharing oxygens and V atoms. The tails of the Wannier functions at the O positions bend towards the neighboring V atoms due to appreciable V–O hybridization. Bottom: Magnetization calculated as a function of the applied magnetic field for different spin-chain models (**left**) and for unstrained (black, solid line), 2% tensile strained (brown, solid line), and 2% compressive strained (dashed) compounds (**right**). The inset shows the magnitude of the spin gap plotted as a function of the in-plane lattice constant. The experimental estimate of the spin gap for the unstrained compound is marked with an asterisk. Adapted from [113].

4. Summary and Outlook

Low-dimensional quantum magnetism offers a playground for envisaging highly nontrivial and versatile phenomena. One of the key requirements for studying these systems is the identification of the appropriate spin model for describing a given material. Given the success of density functional theory in describing complex materials, it is a natural choice to apply ab initio DFT methods to the problem of low-dimensional quantum magnetism. The strong correlation effect that dictates the properties of quantum spin systems, however, prohibits the direct usage of the DFT for this purpose. Instead, it is much more pragmatic to filter out the DFT's output to arrive at a low-energy Hamiltonian that contains only magnetic degrees of freedom. In this review, we advocate for the NMTO-downfolding technique as an intelligible, fast, and accurate DFT tool to be used for the filtering. This procedure, which takes the renormalization using the degrees of freedom associated with non-magnetic ions into account, provides information on the relevant exchange paths that connect two magnetic ions. Armed with the knowledge of relevant exchange paths, the corresponding magnetic exchanges can be obtained by employing a superexchange formula from information on the real-space representation of a low-energy Hamiltonian, or in terms of the LDA + U total energy calculations of different spin arrangements. The applicability of the proposed method has been illustrated by considering a variety of low-dimensional quantum spin compounds belonging to the cuprate, vanadate, and nickelate families. Table 2 lists a description of the proposed spin model in each case. Most often, the description of the underlying model turned to be different from what may be anticipated based of the crystal structure. The validity of the models was checked by comparing the computed magnetic susceptibility and magnetization with measured magnetic data. Predictions were also made for future experiments in terms of designing spin gaps or closing spin gaps through the application of an external magnetic field, biaxial strain, etc. The theoretical predictions should motivate future experiments in this exciting area of quantum materials.

Finally, we would like to mention that the above discussion is pertinent for 3-d TM-based spin systems only. For 4-d or 5-d TM-based spin systems, in addition to isotropic Heisenberg terms in the Hamiltonian, the presence of non-negligible spin–orbit coupling may give rise to Kitaev, Dzyaloshinskii–Moriya, and off-diagonal anisotropic terms, as discussed, for example, for α-RuCl3, Na_2IrO_3, and α-Li_2IrO_3 [114].

Table 2. The theoretically predicted spin models of the cuprate, vanadate, and nickelate compounds covered in this review.

Compound Name	Proposed Spin Model	Ref.
$SrCu_2O_3$	Two-leg ladder	Ref. [45]
$CaCuGe_2O_6$	Coupled dimer	Ref. [51]
$Cu_2Te_2O_5X_2$ (X=Cl/Br)	Interacting spin cluster	Ref. [54]
$Na_3Cu_2Te(Sb)O_6$	AF-AF chain	Ref. [57]
$CuTe_2O_5$	2-d coupled dimer	Ref. [60]
$Cs_2CuAl_4O_8$	1-d AF-F chain with NN–NNN interaction	Ref. [64]
γ-LiV_2O_5	Asymmetric spin ladder	Ref. [77]
α-NaV_2O_5	Spin ladder	Ref. [75]
CsV_2O_5	2-d coupled spin dimer	Ref. [81]
$VOSeO_3$	Alternating spin chain	Ref. [84]
$Na_2V_3O_7$	Partially frustrated spin tube	Ref. [89]
$Zn_2VO(PO_4)_2$	2-d AF spin model	Ref. [91]
Ti-$Zn_2VO(PO_4)_2$	Coupled AF chain	Ref. [91]
$NiAs_2O_6$	2-d spin lattice formed by third-NN interactions	Ref. [99]
$NiRh_2O_4$	Spin–orbit entangled singlet	Ref. [105]
$Sr_3NiPt(Ir)O_6$	Interacting spin chain	Ref. [108]
$SrNi_2V_2O_8$	Coupled Haldane chain	Ref. [113]

Funding: The research was funded by J.C. Bose National Fellowship (grant no. JCB/2020/000004).

Institutional Review Board Statement: Not applicable.

Informed Consent Statement: Not applicable.

Data Availability Statement: Not applicable.

Acknowledgments: The author acknowledges the J.C. Bose Fellowship (grant no. JCB/2020/000004) for the research support. The author gratefully acknowledges the collaboration with O. K. Andersen, I. Dasgupta, R. Valentí, H. Das, S. Kanungo, K. Samanta, S. Kar, C. Gross, T. M. Rice, F. Mila, H. Rosner, A. Vasiliev, V. Anisimov, M. Jansen, B. Keimer, A. Paramekanti, S. Das, and D. Nafday for some of the work presented in this review article.

Conflicts of Interest: The author declares no conflict of interest.

Sample Availability: Not applicable.

References

1. Geim, A.K.; Novoselov, K.S. The rise of graphene. *Nat. Mater.* **2007**, *6*, 183. [CrossRef]
2. Eftekhari, A. Tungsten dichalcogenides (WS_2, WSe_2, and WTe_2): Materials chemistry and applications. *J. Mater. Chem. A* **2017**, *5*, 18299. [CrossRef]
3. Chakraborty, I.; Pradeep, T. Atomically Precise Clusters of Noble Metals: Emerging Link between Atoms and Nanoparticles. *Chem. Rev.* **2017**, *117*, 8208. [CrossRef]
4. Boston, R.; Schnepp, Z.; Nemoto, Y.; Sakka, Y.; Hall, S.R. In Situ TEM Observation of a Microcrucible Mechanism of Nanowire Growth. *Science* **2014**, *344*, 623. [CrossRef]
5. Ising, E. Beitrag zur Theorie des Ferromagnetismus. *Z. Phys.* **1925**, *31*, 253. [CrossRef]
6. Heisenberg, W. Zur Theorie des Ferromagnetismus. *Z. Phys.* **1928**, *49*, 619. [CrossRef]
7. Onsager, L. Crystal Statistics. I. A Two-Dimensional Model with an Order-Disorder Transition. *Phys. Rev.* **1944**, *65*, 117. [CrossRef]
8. Mermin, N.D.; Wagner, H. Absence of Ferromagnetism or Antiferromagnetism in One- or Two-Dimensional Isotropic Heisenberg Models. *Phys. Rev. Lett.* **1966**, *17*, 1133. [CrossRef]
9. Berezinskii, V.L. Destruction of Long-range Order in One-dimensional and Two-dimensional Systems Possessing a Continuous Symmetry Group. II. Quantum Systems. *Sov. Phys. JETP* **1972**, *34*, 610.
10. Kosterlitz, J.M.; Thouless, D.J. Long range order and metastability in two dimensional solids and superfluids. (Application of dislocation theory). *J. Phys. C Solid State Phys.* **1972**, *5*, L124. [CrossRef]
11. Haldane, F.D.M. Continuum dynamics of the 1-D Heisenberg antiferromagnet: Identification with the O(3) nonlinear sigma model. *Phys. Lett. A* **1983**, *93*, 464. [CrossRef]
12. Dagotto, E.; Rice, T.M. Surprises on the Way from 1D to 2D Quantum Magnets: The Novel Ladder Materials. *Science* **1996**, *271*, 618. [CrossRef]
13. Fazekas P. *Lecture Notes on Electron Correlation and Magnetism*; World Scientific: Singapore, 1999.
14. Lemmens, P.; Güntherodtb, G.; Gros, C. Magnetic light scattering in low-dimensional quantum spin systems. *Phys. Rep.* **2003**, *375*, 1. [CrossRef]
15. Bednorz, J.G.; Mueller, K.A. Possible high Tc superconductivity in the Ba-La-Cu-O system. *Z. Phys. B* **1986**, *64*, 189. [CrossRef]
16. Shastry, B.S.; Sutherland, B. Exact ground state of a quantum mechanical antiferromagnet. *Phys. B* **1981**, *108*, 1069. [CrossRef]
17. Shores, M.P.; Nytko, E.A.; Bartlett, B.M.; Nocera, D.G. A structurally perfect S = 1/2 kagome antiferromagnet. *J. Am. Chem. Soc.* **2005**, *127*, 13462. [CrossRef]
18. Kitaev, A. Anyons in an exactly solved model and beyond. *Ann. Phys.* **2006**, *321*, 2. [CrossRef]
19. Goodenough, J.B. Theory of the Role of Covalence in the Perovskite-Type Manganites [La, M(II)]MnO_3. *Phys. Rev.* **1955**, *100*, 564. [CrossRef]
20. Kanamori, J. Superexchange interaction and symmetry properties of electron orbitals. *J. Phys. Chem. Solids* **1959**, *10*, 87. [CrossRef]
21. Hoffmann, R.; Lipscomb, W.N. Theory of Polyhedral Molecules. I. Physical Factorizations of the Secular Equation. *J. Chem. Phys.* **1962**, *36*, 2179. [CrossRef]
22. Whangbo, M.-H.; Koo, H.-J.; Kremer, R.K. Spin Exchanges Between Transition Metal Ions Governed by the Ligand p-Orbitals in Their Magnetic Orbitals. *Molecules* **2021**, *26*, 531. [CrossRef] [PubMed]
23. Kaul, E.E.; Rosner, H.; Yushankhai, V.; Sichelschmidt, J.; Shpanchenko, R.V.; Geibel, C. $Sr_2V_3O_9$ and $Ba_2V_3O_9$: Quasi-one-dimensional spin-systems with an anomalous low temperature susceptibility. *Phys. Rev. B* **2003**, *67*, 174417. [CrossRef]
24. Tocchio, L.; Feldner, H.; Becca, F.; Valentí, R.; Gros, C. Spin-liquid versus spiral-order phases in the anisotropic triangular lattice. *Phys. Rev. B* **2013**, *87*, 035143. [CrossRef]
25. Danilovich, I.L.; Karpova, E.V.; Morozov, I.V.; Ushakov, A.V.; Streltsov, S.V.; Shakin, A.A.; Volkova, O.S.; Zvereva, E.A.; Vasiliev, A.N. Spin-singlet Quantum Ground State in Zigzag Spin Ladder $Cu(CF_3COO)_2$. *Chem. Phys. Chem.* **2017**, *18*, 2482. [CrossRef] [PubMed]
26. Weickert, F.; Aczel, A.A.; Stone, M.B.; Garlea, V.O.; Dong, C.; Kohama, Y.; Movshovich, R.; Demuer, A.; Harrison, N.; Gamża, M.N.; et al. Field-induced double dome and Bose-Einstein condensation in the crossing quantum spin chain system $AgVOAsO_4$. *Phys. Rev. B* **2019**, *100*, 104422. [CrossRef]
27. Hallberg, K. New trends in density matrix renormalization. *Adv. Phys.* **2006**, *55*, 477. [CrossRef]
28. Hubbard, J. Electron Correlations in Narrow Energy Bands. *Proc. R. Soc. Lond.* **1963**, *276*, 238.
29. Spałek, J. t-J Model Then and Now: A Personal Perspective from the Pioneering Times. *Acta Phys. Pol. A* **2007**, *111*, 409. [CrossRef]
30. Kohn, W.; Sham, L.J. Self-Consistent Equations Including Exchange and Correlation Effects. *Phys. Rev.* **1965**, *140*, A1133. [CrossRef]
31. Perdew, J.P.; Wang, Y. Accurate and simple analytic representation of the electron-gas correlation energy. *Phys. Rev. B.* **1992**, *45*, 13244. [CrossRef]
32. Perdew, J.P.; Burke, K.; Ernzerhof, M. Generalized Gradient Approximation Made Simple. *Phys. Rev. Lett.* **1996**, *77*, 3865. [CrossRef]
33. Andersen, O.K.; Saha-Dasgupta, T. Muffin-tin orbitals of arbitrary order. *Phys. Rev. B* **2000**, *62*, R16219. [CrossRef]
34. Andersen, O.K.; Jepsen, O. Explicit, First-Principles Tight-Binding Theory. *Phys. Rev. Lett.* **1984**, *53*, 2571. [CrossRef]
35. Saha-Dasgupta, T.; Andersen, O.K.; Nuss, J.; Poteryaev, A.I.; Georges, A.; Lichtenstein, A.I. Electronic structure of V_2O_3: Wannier orbitals from LDA-NMTO calculations. *arXiv* **2009**, arXiv:0907.2841.

36. Andersen, O.K.; Saha-Dasgupta, T.; Ezhov, S.; Tsetseris, L.; Jepsen, O.; Tank, R.W.; Arcangeli, C.; Krier, G. Third-generation MTOs. *Psi-k Newsl.* **2001**, *45*, 86.
37. Anderson, P.W. Antiferromagnetism. Theory of Superexchange Interaction. *Phys. Rev.* **1950**, *79*, 350. [CrossRef]
38. Lichtenstein, A.I.; Zaanen, J.; Anisimov, V.I. Density-functional theory and strong interactions: Orbital ordering in Mott-Hubbard insulators. *Phys. Rev. B* **1995**, *52*, R5467. [CrossRef] [PubMed]
39. Sandvik, A.W. Stochastic series expansion method with operator-loop update. *Phys. Rev. B* **1999**, *59*, R14157. [CrossRef]
40. Dorneich, A.; Troyer, M. Accessing the dynamics of large many-particle systems using the stochastic series expansion. *Phys. Rev. E* **2001**, *64*, 066701. [CrossRef]
41. Louis, K.; Gros, C. Stochastic cluster series expansion for quantum spin systems. *Phys. Rev. B* **2004**, *70*, 100410. [CrossRef]
42. McCarron, E.M.; Subramanian, M.A.; Calabrese, J.C.; Harlow, R.L. The incommensurate structure of $(Sr_{14-x}Ca_x)Cu_{24}O_{41}$ ($0 < x \sim 8$) a superconductor byproduct. *Mater. Res. Bull.* **1988**, *23*, 1355.
43. Siegrist, T.; Schneemeyer, L.F.; Sunshine, S.A.; Waszczak, J.V. A new layered cuprate structure-type, $(A_{1-x}A'_x)_{14}Cu_{24}O_{41}$. *Mater. Res. Bull.* **1988**, *23*, 1429. [CrossRef]
44. Motoyama, N.; Osafune, T.; Kakeshita, T.; Eisaki, H.; Uchida, S. Effect of Ca substitution and pressure on the transport and magnetic properties of $Sr_{14}Cu_{24}O_{41}$ with doped two-leg Cu-O ladders. *Phys. Rev. B* **1997**, *55*, R3386. [CrossRef]
45. Müller, T.A.; Anisimov, V.; Rice, T.M.; Dasgupta, I.; Saha-Dasgupta, T. Electronic structure of ladder cuprates. *Phys. Rev. B* **1998**, *57*, R12655. [CrossRef]
46. Osafune, T.; Motoyama, N.; Eisaki, H.; Uchida, S. Optical Study of the $Sr_{14-x}Ca_xCu_{24}O_{41}$ System: Evidence for Hole-Doped Cu_2O_3 Ladders. *Phys. Rev. Lett.* **1997**, *78*, 1980. [CrossRef]
47. Behruzi, M.; Breuer, K.-H.; Eysel, W. Copper (II) silicates and germanates with chain structures. *Z. Kristallogr.* **1986**, *176*, 205. [CrossRef]
48. Sasago, Y.; Hase, M.; Uchinokura, K.; Tokunaga, M.; Miura, N. Discovery of a spin-singlet ground state with an energy gap in CaCuGe2O6. *Phys. Rev. B* **1995**, *52*, 3533. [CrossRef]
49. Hase, M.; Terasaki, I.; Uchinokura, K. Observation of the spin-Peierls transition in linear Cu^{2+} (spin-1/2) chains in an inorganic compound CuGeO3. *Phys. Rev. Lett.* **1993**, *70*, 3651. [CrossRef] [PubMed]
50. Zheludev, A.; Shirane, G.; Sasago, Y.; Hase, M.; Uchinokura, K. Dimerized ground state and magnetic excitations in CaCuGe2O. *Phys. Rev. B* **1996**, *53*, 11642. [CrossRef] [PubMed]
51. Valentí, R.; Saha-Dasgupta, T.; Gros, C. Nature of the spin-singlet ground state in CaCuGe2O. *Phys. Rev. B* **2002**, *66*, 054426. [CrossRef]
52. Johnsson, M.; Törnroos, K.W.; Mila, F.; Millet, P. Tetrahedral Clusters of Copper(II): Crystal Structures and Magnetic Properties of Cu2Te2O5X2. (X = Cl, Br) *Chem. Mater.* **2000**, *12*, 2853. [CrossRef]
53. Lemmens, P.; Choi, K.-Y.; Kaul, E.E.; Geibel, C.; Becker, K.; Brenig, W.; Valenti, R.; Gros, C.; Johnsson, M.; Millet, P.; et al. Evidence for an Unconventional Magnetic Instability in the Spin-Tetrahedra System Cu2Te2O5Br2. *Phys. Rev. Lett.* **2001**, *87*, 227201. [CrossRef] [PubMed]
54. Valenti, R.; Saha-Dasgupta, T.; Gros, C.; Rosner, H. Halogen-mediated exchange in the coupled-tetrahedra quantum spin systems Cu2Te2O5X2(X=Br,Cl). *Phys. Rev. B* **2003**, *67*, 245110. [CrossRef]
55. Xu, J.; Assoud, A.; Soheinia, N.; Derakhshan, S.; Cuthbert, H.L.; Greedan, J.; Whangbo, M.H.; Kleinke, H. Synthesis, Structure, and Magnetic Properties of the Layered Copper(II) Oxide Na2Cu2TeO6. *Inorg. Chem.* **2005**, *44*, 5042. [CrossRef]
56. Miura, Y.; Hirai, R.; Kobayashi, Y.; Sato, M. Spin-Gap Behavior of Na3Cu2SbO6 with Distorted Honeycomb Structure. *J. Phys. Soc. Jpn.* **2006**, *75*, 847071. [CrossRef]
57. Derakhshan, S.; Cuthbert, L.H.; Greedan, J.E.; Rahaman, B.; Saha-Dasgupta, T. Electronic structures and low-dimensional magnetic properties of the ordered rocksalt oxides Na3Cu2SbO6 and Na2Cu2TeO6. *Phys. Rev. B* **2007**, *76*, 104403. [CrossRef]
58. Deisenhofer, J.; Eremina, R.M.; Pimenov, A.; Gavrilova, T.; Berger, H.; Johnsson, M.; Lemmens, P.; Krug von Nidda, H.-A.; Loidl, A.; Lee, K.-S.; et al. Structural and magnetic dimers in the spin-gapped system CuTe2O5. *Phys. Rev. B* **2006**, *74*, 174421. [CrossRef]
59. Hanke, K.; Kupcik, V.; Lindqvist, O. The crystal structure of CuTe2O5. *Acta Crystallogr. Sect. B Struct. Crystallogr. Cryst. Chem.* **1973**, *29*, 963. [CrossRef]
60. Das, H.; Saha-Dasgupta, T.; Gros, C.; Valentí, R. Proposed low-energy model Hamiltonian for the spin-gapped system CuTe2O5. *Phys. Rev. B* **2008**, *77*, 224437. [CrossRef]
61. Diederix, K.M.; Groen, J.P.; Klaassen, T.O.; Poulis, N.J. Spin Dynamics in an Alternating Linear-Chain Heisenberg Antiferromagnet. *Phys. Rev. Lett.* **1978**, *41*, 1520. [CrossRef]
62. Majumdar, C.K.; Ghosh, D.K. On Next-Nearest-Neighbor Interaction in Linear Chain. II. *J. Math. Phys.* **1969**, *10*, 1388. [CrossRef]
63. Shvanskaya, L.; Yakubovich, O.; Massa, W.; Vasiliev, A. Two-dimensional zeolite-like network in the new caesium copper aluminate Cs2CuAl4O8. *Acta Crystallogr. Sect. B* **2015**, *71*, 498. [CrossRef]
64. Rahaman, B.; Kar, S.; Vasiliev, A.; Saha-Dasgupta, T. Interplay of alternation and further neighbor interaction in S=12 spin chains: A case study of Cs2CuAl4O8. *Phys. Rev. B* **2018**, *98*, 144412. [CrossRef]
65. Ueda, Y. Vanadate Family as Spin-Gap Systems. *Chem. Mater.* **1998**, *10*, 2653. [CrossRef]
66. Onoda, M.; Nishiguchi, N. Crystal Structure and Spin Gap State of CaV2O5. *J. Solid State Chem.* **1996**, *127*, 359. [CrossRef]
67. Millet, P.; Satto, C.; Bonvoisin, J.; Normand, B.; Penc, K.; Albrecht, M.; Mila F. Magnetic properties of the coupled ladder system MgV2O5. *Phys. Rev. B* **1998**, *57*, 5005. [CrossRef]

68. Smolinski, H.; Gros, C.; Weber, W.; Peuchert, U.; Roth, G.; Weiden, M.; Geibel, C. NaV2O5 as a Quarter-Filled Ladder Compound. *Phys. Rev. Lett.* **1998**, *80*, 5164. [CrossRef]
69. Takeo, Y.; Yosihama, T.; Nishi, M.; Nakajima, K.; Kakurai, K.; Isobe, M.; Ueda, Y. Inelastic neutron scattering study of LiV2O5. *J. Phys. Chem. Solids* **1999**, *60*, 1145. [CrossRef]
70. Isobe, M.; Ueda, Y. Magnetic Susceptibilities of AV2O5 (A=Li and Cs) with Square Pyramidal V(IV)O5. *J. Phys. Soc. Jpn.* **1996**, *65*, 3142. [CrossRef]
71. Galy, J.; Hardy, A. Structure cristalline du bronze de vanadium–lithium LiV2O5. *Acta Crystallogr.* **1955**, *19*, 432. [CrossRef]
72. Anderson, D.N.; Willett, R.D. Refinement of the structure of LiV2O5. *Acta Crystallogr. B* **1971**, *27*, 1476. [CrossRef]
73. Nakao, H.; Ohwada, K.; Takesue, N.; Fujii, Y.; Isobe, M.; Ueda, Y.; Zimmermann, M.V.; Hill, J.P.; Gibbs, D.; Woicik, J.C.; et al. X-ray Anomalous Scattering Study of a Charge-Ordered State in NaV2O5. *Phys. Rev. Lett.* **2000**, *85*, 4349. [CrossRef]
74. Presura, C.; van der Marel, D.; Damascelli, A.; Kremer, R. K. Low-temperature ellipsometry of α'-NaV2O5. *Phys. Rev. B* **2000**, *61*, 15762. [CrossRef]
75. Mazurenko, V.V.; Lichtenstein, A.I.; Katsnelson, M.I.; Dasgupta, I.; Saha-Dasgupta, T.; Anisimov, V.I. Nature of insulating state in NaV2O5 above charge-ordering transition: A cluster dynamical mean-field study. *Phys. Rev. B* **2002**, *66*, 081104. [CrossRef]
76. Lichtenstein, A.I.; Katsnelson, M.I. Antiferromagnetism and d-wave superconductivity in cuprates: A cluster dynamical mean-field theory. *Phys. Rev. B* **2000**, *62*, R9283. [CrossRef]
77. Valentí, R.; Saha-Dasgupta, T.; Alvarez, J.V.; Pozgajcic, K.; Gros, C. Modeling the Electronic Behavior of γ-LiV2O5: A Microscopic Study. *Phys. Rev. Lett.* **2001**, *86*, 5381. [CrossRef]
78. Waltersson, K.; Forslund, B. A refinement of the crystal structure of CsV2O5. *Acta Crystallogr. Sect. B Struct. Crystallogr. Cryst. Chem.* **1977**, *33*, 789. [CrossRef]
79. Garrett, A.W.; Nagler, S.E.; Tennant, D.A.; Sales, B.C.; Barnes, T. Magnetic Excitations in the S=1/2 Alternating Chain Compound (VO)2P2O7. *Phys. Rev. Lett.* **1997**, *79*, 745. [CrossRef]
80. Oosawa, A.; Kato, T.; Tanaka, H.; Kakurai, K.; Müller, M.; Mikeska, H.-J. Magnetic excitations in the spin-gap system TlCuCl3. *Phys. Rev. B* **2002**, *65*, 094426. [CrossRef]
81. Valentí, R.; Saha-Dasgupta, T. Electronic and magnetic structure of CsV2O5. *Phys. Rev. B* **2002**, *65*, 144445. [CrossRef]
82. Mila, F. Ladders in a magnetic field: A strong coupling approach. *Eur. Phys. J. B* **1998**, *6*, 201. [CrossRef]
83. Trombe, J.-C.; Gleizes, A.; Galy, J.; Renard, J.-P.; Journaux, Y.; Verdaguer, M. Structure and magnetic properties of vanadyl chains: Crystal structure of VOSeO3 and comparative study of VOSeO3 and VOSe2O5. *New J. Chem.* **1987**, *11*, 321.
84. Valentí, R.; Saha-Dasgupta, T.; Mila, F. Ab initio investigation of VOSeO3: A spin gap system with coupled spin dimers. *Phys. Rev. B* **2003**, *68*, 024411. [CrossRef]
85. Millet, P.; Henry, J.Y.; Mila, F.; Galy, J. Vanadium(IV)–Oxide Nanotubes: Crystal Structure of the Low-Dimensional Quantum Magnet Na2V3O7. *J. Solid State Chem.* **1999**, *147*, 676. [CrossRef]
86. Whangbo, M.-H.; Koo, H.-J. Investigation of the spin exchange interactions in the nanotube system Na2V3O7 by spin dimer analysis. *Solid State Commun.* **2000**, *115*, 675. [CrossRef]
87. Gavilano, J.; Rau, D.; Mushkolaj, S.; Ott, H.R.; Millet, P.; Mila, F. Low-Dimensional Spin S=1/2 System at the Quantum Critical Limit: Na2V3O7. *Phys. Rev. Lett.* **2003**, *90*, 167202. [CrossRef] [PubMed]
88. Subrahmanyam, V. Effective chiral-spin Hamiltonian for odd-numbered coupled Heisenberg chains. *Phys. Rev. B* **1994**, *50*, 16109. [CrossRef]
89. Saha-Dasgupta, T.; Valentí, R.; Capraro, F.; Gros, C. Na2V3O7: A Frustrated Nanotubular System with Spin-1/2 Diamond Ring Geometry. *Phys. Rev. Lett.* **2005**, *95*, 107201. [CrossRef] [PubMed]
90. Taniguchi, S.; Nishikawa, Y.; Yasui, Y.; Kobayashi, Y.; Sato, M.; Nishikawa, T.; Kontani, M.; Sano, K. Spin Gap Behavior of S=1/2 Quasi-Two-Dimensional System CaV4O9. *J. Phys. Soc. Jpn.* **1995**, *64*, 2758. [CrossRef]
91. Kanungo, S.; Kar, S.; Saha-Dasgupta, T. Tuning of magnetic ground state of the spin-12 square-lattice compound Zn2VO(PO4)2 through chemical substitution. *Phys. Rev. B* **2013**, *87*, 054431. [CrossRef]
92. Lii, K.H.; Tsai, H.J. Synthesis and crystal structure of Zn2VO(PO4)2, a vanadyl(IV) orthophosphate containing a dimer of edge-sharing ZnO5 square pyramids. *J. Solid State Chem.* **1991**, *90*, 291. [CrossRef]
93. Yusuf, S.M.; Bera, A.K.; Kini, N.S.; Mirebeau, I.; Petit, S. Two- and three-dimensional magnetic correlations in the spin-12 square-lattice system Zn2VO(PO4)2. *Phys. Rev. B* **2010**, *82*, 094412. [CrossRef]
94. Kini, N.S.; Kaul, E.E.; Geibel, C. Zn2VO(PO4)2: An S = 1/2 Heisenberg antiferromagnetic square lattice system. *J. Phys. Condens. Matter* **2006**, *18*, 1303. [CrossRef]
95. Kar, S.; Saha-Dasgupta, T. Quasi-2D J1–J2 antiferromagnet Zn2VO(PO4)2 and its Ti-substituted derivative: A spin-wave analysis. *Physica B* **2014**, *432*, 71. [CrossRef]
96. Adhikary, P.; Bandyopadhyay, S.; Das, T.; Dasgupta, I.; Saha-Dasgupta, T. Orbital-selective. *Phys. Rev. B* **2020**, *102*, 100501. [CrossRef]
97. Nakua, A.M.; Greedan, J.E. Structural and Magnetic Properties of Transition Metal Arsenates, AAs2O6, A = Mn, Co, and Ni. *J. Solid State Chem.* **1995**, *118*, 402. [CrossRef]
98. Orosel, D.; Jansen, M. PdAs2O6, das erste paramagnetische Palladiumoxid. *Z. Anorg. Allg. Chem.* **2006**, *632*, 1131. [CrossRef]
99. Reehuis, M.; Saha-Dasgupta, T.; Orosel, D.; Nuss, J.; Rahaman, B.; Keimer, B.; Andersen, O.K.; Jansen, M. Magnetic properties of PdAs2O6: A dilute spin system with an unusually high Néel temperature. *Phys. Rev. B* **2012**, *85*, 115118. [CrossRef]

100. Wang, C.; Nahum, A.; Senthil, T. Topological paramagnetism in frustrated spin-1 Mott insulators. *Phys. Rev. B* **2015**, *91*, 195131. [CrossRef]
101. Affleck, I.; Kennedy, T.; Lieb, E.H.; Tasaki, H. Rigorous results on valence-bond ground states in antiferromagnets. *Phys. Rev. Lett.* **1987**, *59*, 799. [CrossRef]
102. Fritsch, V.; Hemberger, J.; Büttgen, N.; Scheidt, E.-W.; Krug von Nidda, H.-A.; Loidl, A.; Tsurkan, V. Spin and Orbital Frustration in $MnSc_2S_4$ and FeSc2S4. *Phys. Rev. Lett.* **2004**, *92*, 116401. [CrossRef]
103. Tristan, N.; Hemberger, J.; Krimmel, A.; Krug von Nidda, H.-A.; Tsurkan, V.; Loidl, A. Geometric frustration in the cubic spinels MAl_2O_4 (M=Co, Fe, and Mn). *Phys. Rev. B* **2005**, *72*, 174404. [CrossRef]
104. Chamorro, J.R.; Ge, L.; Flynn, J.; Subramanian, M.A.; Mourigal, M.; McQueen, T.M. Frustrated spin one on a diamond lattice in $NiRh_2O_4$. *Phys. Rev. Mater.* **2018**, *2*, 034404. [CrossRef]
105. Das, S.; Nafday, D.; Saha-Dasgupta, T.; Paramekanti, A. $NiRh_2O_4$: A spin-orbit entangled diamond-lattice paramagnet. *Phys. Rev. B* **2019**, *100*, 140408. [CrossRef]
106. Vishwanath, A.; Senthil, T. Physics of Three-Dimensional Bosonic Topological Insulators: Surface-Deconfined Criticality and Quantized Magnetoelectric Effect. *Phys. Rev. X* **2013**, *3*, 011016. [CrossRef]
107. Wu, H.; Haverkort, M.W.; Hu, Z.; Khomskii, D.I.; Tjeng, L.H. Nature of Magnetism in $Ca_3Co_2O_6$. *Phys. Rev. Lett.* **2005**, *95*, 186401. [CrossRef]
108. Sarkar, S.; Kanungo, S.; Saha-Dasgupta, T. Ab initio study of low-dimensional quantum spin systems Sr_3NiPtO_6, Sr_3CuPtO_6, and Sr_3NiIrO_6. *Phys. Rev. B* **2010**, *82*, 235122. [CrossRef]
109. Flahaut, D.; Hebert, S.; Amignan, A.; Hardy, V.; Martin, C.; Hervieu, M.; Costes, M.; Raquet, B.; Broto, J.M. A magnetic study of the one dimensional $Sr_3 NiIrO_6$ compound. *Eur. Phys. J. B* **2003**, *35*, 317–323. [CrossRef]
110. Rice, T.M. To Condense or Not to Condense. *Science* **2002**, *298*, 760. [CrossRef] [PubMed]
111. Sakai, T.; Takahashi, M. Effect of the Haldane gap on quasi-one-dimensional systems. *Phys. Rev. B* **1990**, *42*, 4537. [CrossRef] [PubMed]
112. Wichmann, R.; Müller-Buschbaum, H. $SrNi_2V_2O_8$: Ein neuer Strukturtyp der Erdalkali-Oxometallate. *Rev. Chim. Miner.* **1986**, *23*, 1.
113. Samanta, K.; Kar, S.; Saha-Dasgupta, T. Magnetic modeling and effect of biaxial strain on the Haldane chain compound $SrNi_2V_2O_8$. *Phys. Rev. B* **2016**, *93*, 224404. [CrossRef]
114. Winter, S.M.; Li, Y.; Jeschke, H.O.; Valentí, R. Challenges in design of Kitaev materials: Magnetic interactions from competing energy scales. *Phys. Rev. B* **2016**, *93*, 214431. [CrossRef]

Review

Towards Reversible High-Voltage Multi-Electron Reactions in Alkali-Ion Batteries Using Vanadium Phosphate Positive Electrode Materials

Edouard Boivin [1,2,3], Jean-Noël Chotard [1,3,4], Christian Masquelier [1,3,4] and Laurence Croguennec [2,3,4,*]

[1] Laboratoire de Réactivité et de Chimie des Solides, CNRS-UMR 7314, Université de Picardie Jules Verne, CEDEX 1, F-80039 Amiens, France; edouard.boivin@univ-lille.fr (E.B.); jean-noel.chotard@u-picardie.fr (J.-N.C.); christian.masquelier@u-picardie.fr (C.M.)
[2] CNRS, Université Bordeaux, Bordeaux INP, ICMCB UMR 5026, F-33600 Pessac, France
[3] RS2E, Réseau Français sur le Stockage Electrochimique de l'Energie, FR CNRS 3459, CEDEX 1, F-80039 Amiens, France
[4] ALISTORE-ERI European Research Institute, FR CNRS 3104, CEDEX 1, F-80039 Amiens, France
* Correspondence: laurence.croguennec@icmcb.cnrs.fr; Tel.: +33-(0)5-4000-2647

Citation: Boivin, E.; Chotard, J.-N.; Masquelier, C.; Croguennec, L. Towards Reversible High-Voltage Multi-Electron Reactions in Alkali-Ion Batteries Using Vanadium Phosphate Positive Electrode Materials. *Molecules* **2021**, *26*, 1428. https://doi.org/10.3390/molecules 26051428

Academic Editor: Myung-Hwan Whangbo

Received: 30 January 2021
Accepted: 1 March 2021
Published: 6 March 2021

Publisher's Note: MDPI stays neutral with regard to jurisdictional claims in published maps and institutional affiliations.

Copyright: © 2021 by the authors. Licensee MDPI, Basel, Switzerland. This article is an open access article distributed under the terms and conditions of the Creative Commons Attribution (CC BY) license (https:// creativecommons.org/licenses/by/ 4.0/).

Abstract: Vanadium phosphate positive electrode materials attract great interest in the field of Alkali-ion (Li, Na and K-ion) batteries due to their ability to store several electrons per transition metal. These multi-electron reactions (from V^{2+} to V^{5+}) combined with the high voltage of corresponding redox couples (e.g., 4.0 V vs. for V^{3+}/V^{4+} in $Na_3V_2(PO_4)_2F_3$) could allow the achievement the 1 kWh/kg milestone at the positive electrode level in Alkali-ion batteries. However, a massive divergence in the voltage reported for the V^{3+}/V^{4+} and V^{4+}/V^{5+} redox couples as a function of crystal structure is noticed. Moreover, vanadium phosphates that operate at high V^{3+}/V^{4+} voltages are usually unable to reversibly exchange several electrons in a narrow enough voltage range. Here, through the review of redox mechanisms and structural evolutions upon electrochemical operation of selected widely studied materials, we identify the crystallographic origin of this trend: the distribution of PO_4 groups around vanadium octahedra, that allows or prevents the formation of the vanadyl distortion ($O\cdots V^{4+}=O$ or $O\cdots V^{5+}=O$). While the vanadyl entity massively lowers the voltage of the V^{3+}/V^{4+} and V^{4+}/V^{5+} couples, it considerably improves the reversibility of these redox reactions. Therefore, anionic substitutions, mainly O^{2-} by F^-, have been identified as a strategy allowing for combining the beneficial effect of the vanadyl distortion on the reversibility with the high voltage of vanadium redox couples in fluorine rich environments.

Keywords: batteries; positive electrode; vanadium phosphates; covalent vanadyl bond; mixed anion

1. Introduction

In the 2000s, the research on polyanion compounds as positive electrode materials was mainly motivated by the interesting properties of the low cost triphylite $LiFePO_4$ [1–5] (olivine-type structure) providing long-term structural stability, essential for extensive electrochemical cycling and safety issues. In this material, the high voltage for the Fe^{2+}/Fe^{3+} redox couple delivered in $LiFePO_4$ (i.e., 3.45 V vs. Li^+/Li vs. ca. 2.2 V in oxides) is due to the inductive effect of the phosphate group. Further exploitation of the inductive effect with fluorine and/or sulfate has led to materials such as $LiFeSO_4F$ (Tavorite or Triplite structure) delivering an even higher voltage than $LiFePO_4$ (i.e., 3.6 V and 3.9 V vs. Li^+/Li for Tavorite and Triplite structures, respectively) [6,7]. However, these materials suffer from a deficit of capacity compared to the current best commercially available Li-ion positive electrode materials. Li_2FeSiO_4 has been proposed to overcome this issue by triggering both Fe^{2+}/Fe^{3+} and Fe^{3+}/Fe^{4+} redox couples but the strong structural changes involved seem to be detrimental to long-term performances [8]. To the best of our knowledge, this material is the only one providing a multi-electron reaction (i.e., exchange of more than one electron per transition metal) in iron-based polyanion systems while vanadium phosphate materials offer numerous such examples. Indeed, the ability of vanadium to be stabilized in the large range of oxidation states, from V^{2+} to V^{5+} (e.g., from V^{3+} in Li_2VPO_4O to V^{5+} in VPO_4O) [9–13] combined with the rather high voltage of the corresponding redox couples (e.g., 4.25 V vs. Li^+/Li for V^{3+}/V^{4+} in $LiVPO_4F$) [14] could allow the achievement of high energy density thanks to reversible high-voltage multi-electron redox in Alkali-ion batteries (Figure 1).

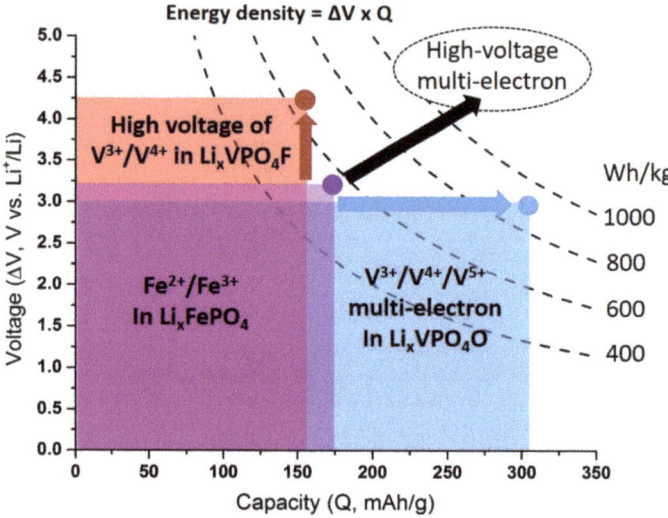

Figure 1. Voltage vs. capacity plot for $LiFePO_4$ (**purple**), $LiVPO_4F$ (**red**), $LiVPO_4O$ (**blue**). Combining the high voltage of $LiVPO_4F$ with the high capacity of the multi-electron redox in $LiVPO_4O$ could allow the achievement of higher energy density through high voltage multi-electron redox. The dash lines represent constant energy densities in Wh/kg of active positive electrode material.

However, depending on the geometry of the VO_n polyhedra, the positions of the V^{3+}/V^{4+} and V^{4+}/V^{5+} redox couples massively change. For instance, Tavorite $LiVPO_4F$ operates at 4.25 V vs. Li^+/Li while in the homeotype $LiVPO_4O$, the apparent same V^{3+}/V^{4+} redox couple is activated at 2.3 V vs. Li^+/Li. This large difference cannot be attributed only to the inductive effect, Li site energy or even cation-cation repulsion (i.e., main mechanisms reported to govern the voltage of a given redox couple): it is actually due to the vanadyl distortion observed in $LiVPO_4O$ and not in $LiVPO_4F$. The first order Jahn Teller (FOJT) distortion is known to be weak in d^1 (V^{4+}) and d^2 (V^{3+}) and even inexistent for d^3 (V^{2+}) and d^0 (V^{5+}) cations (Figure 2). Therefore, the second order Jahn Teller (SOJT) effect drives the distortion of the V^{4+} and V^{5+} polyhedra while this distortion can be prevented for V^{4+} in a mixed O^{2-}/F^- environment. This wide range of oxidation states for vanadium cations (V^{2+}, V^{3+}, V^{4+} and V^{5+}) and the extended panel of environments that they can adopt (regular octahedra, distorted octahedra, square pyramids and tetrahedra) with very different electronic configurations depending on the ligand distribution (JT active or JT inactive) confer to the vanadium phosphate a very rich crystal chemistry.

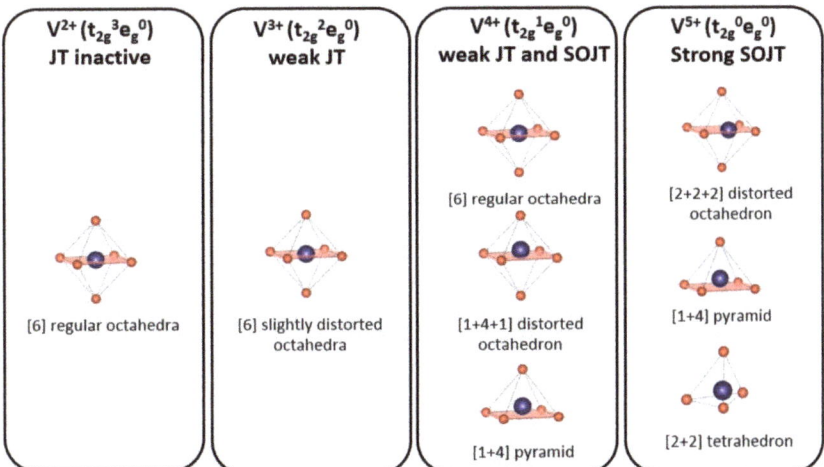

Figure 2. Stable environments of vanadium according to its oxidation state [15]. The number in square brackets correspond to the number of "equivalent bonds".

Beyond their fascinating electrochemical properties, vanadium phosphate materials possess very interesting catalytic and magnetic properties. The relation between structures and these properties was already reviewed by Raveau's group 20 years ago [15] and despite the existence of several reviews on polyanionic structures in Li and Na-ion batteries [16–19], or even specific to vanadyl phosphates (i.e., $A_x(VO)PO_4$) [20,21], none of them focused on the relation between electrochemical properties and crystallographic structure in vanadium phosphates. Therefore, this article aims at clarifying this relation in order to unveil the structural features that dictate the redox voltage in such compounds. Through the fine description of the redox mechanism and the structural evolution observed during cycling of some widely studied materials (NASICON $Na_3V_2(PO_4)_3$, anti-NASICON $Li_3V_2(PO_4)_3$, $Na_3V_2(PO_4)_2F_{3x}O_x$, and Tavorite-like $LiVPO_4F_{1-x}O_x$) we propose to sort all the vanadium phosphates reported as positive electrode materials for Li and Na-ion batteries according to the distribution of phosphate groups around the vanadium polyhedra. This classification gives a holistic picture of such systems and allows for identifying the strategies available to tend towards reversible high-voltage multi-electron reactions in Alkali-ion batteries.

2. Irreversible Multi-Electron Reactions in NASICON and Anti-NASICON $A_xV_2(PO_4)_3$ (A = Li, Na) Structures

The NASICON (Na-super ionic conductor) and anti-NASICON structures have the general formula $A_xM_2(XO_4)_3$ (with M = Fe, Ti, Sc, Hf, V, Ti, Zr, etc. or mixtures of them and X = W, P, S, Si, Mo or mixtures of them) [22,23]. These versatile structures have provided a great playground for solid state chemists. Manthiram and Goodenough demonstrated experimentally the inductive effect which modulates the voltage of the Fe^{3+}/Fe^{2+} redox couple in NASICON [24] which is at the origin of all advances on polyanion materials as positive electrode materials for Alkali-ion batteries. The crystallographic arrangements of NASICON and anti-NASICON are closely related. Indeed, they are built on a three-dimensional framework of VO_6 octahedra sharing all their corners with PO_4 tetrahedra and conversely forming basic $V_2(PO_4)_3$ repeating units commonly named "lantern units" (Figure 3). The connectivity of the lantern units generates different ion conduction paths, vanadium environments and hence different electrochemical properties.

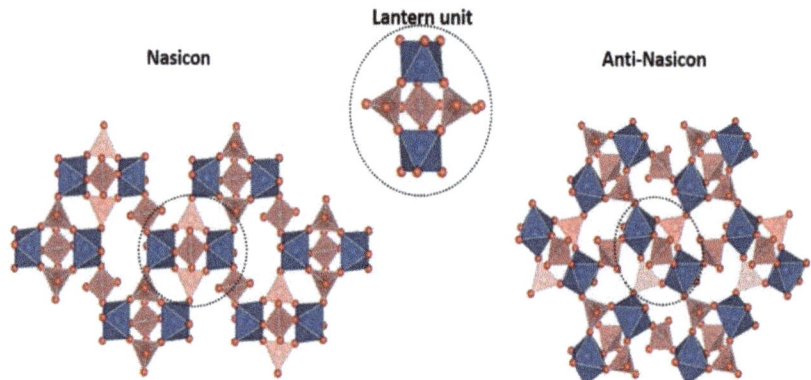

Figure 3. Structural relationship between Nasicon (**left**) and anti-Nasicon (**right**) structures adapted from ref. [18].

In the structure of the anti-NASICON polymorph of $Li_3V_2(PO_4)_3$, the lithium ions fully occupy three crystallographic sites (one tetrahedral $Li(1)O_4$ and two pseudo tetrahedral $Li(2)O_4O$ and $Li(3)O_4O$ sites) [25–27]. The electrochemical extraction of lithium from $Li_3V_2(PO_4)_3$ occurs according to several biphasic reactions involving the V^{3+}/V^{4+} redox couple at 3.6, 3.7 and 4.1 V vs. Li^+/Li and then the V^{4+}/V^{5+} one at 4.5 V vs. Li^+/Li (Figure 4).

Nazar and coworkers [27] studied the complex phase diagram involved during lithium extraction from $Li_3V_2(PO_4)_3$ through X-ray and neutron diffraction (XRD and ND) and solid state nuclear magnetic resonance spectroscopy (NMR), as summarized in Figure 5. The first delithiation step leads to the formation of $Li_{2.5}V_2(PO_4)_3$ with partial depopulation of the pseudo tetrahedral $(Li(3)O_4O)$ site and to a complex short range ordering of V^{3+}/V^{4+} cations [28]. The following delithiation stage affects only the remaining Li(3) ions to yield to $Li_2V_2(PO_4)_3$ characterized by lithium/vacancies and V^{3+}/V^{4+} orderings as suggested by diffraction. The further oxidation of vanadium allows reaching the V^{4+}-rich $LiV_2(PO_4)_3$ phase in which only one Li site remains (Li(2)) as fully occupied. In this phase, there are two very similar crystallographic sites for vanadium ((V(1)-O = V(2)-O = 1.91 Å in average).

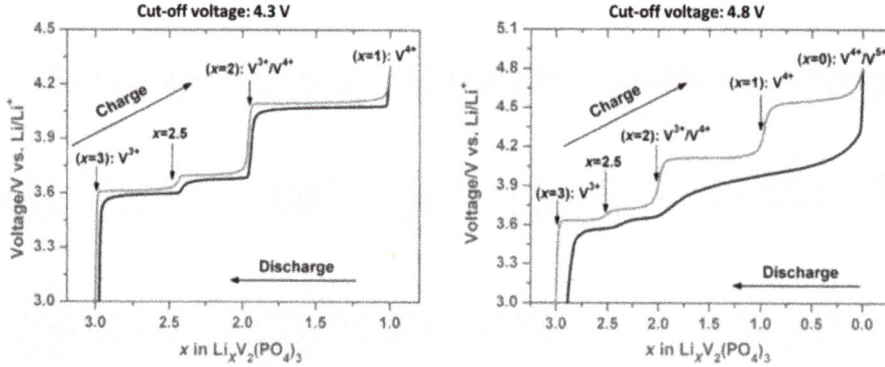

Figure 4. Electrochemical signature of $Li_3V_2(PO_4)_3$ cycled between (left) 3.0 and 4.3 V vs. Li^+/Li or between (right) 3.0 and 4.8 V vs. Li^+/Li adapted from ref. [29]. Reproduced with permission from Rui et al., Journal of Power Sources; published by Elsevier, 2014.

The last process leading to the $V_2(PO_4)_3$ composition is kinetically more limited with a large over-potential (around 500 mV) [27]. At this state of charge, the environments of vanadium (with a mixed valence V^{4+}/V^{5+}) become more distorted, although without significant modification compared to the average V-O distances observed in the V^{IV}-rich $LiV_2(PO_4)_3$ phase. This extraction/insertion process is asymmetrical as the lithium ordering observed for $LiV_2(PO_4)_3$ during charge is not observed during discharge. A disordered Lithium re-intercalation is observed until the $Li_2V_2(PO_4)_3$ composition is reached [30]. This asymmetrical mechanism is not observed for a lower cut-off voltage (i.e., 4.3 V when the V^{4+}/V^{5+} redox couple is not activated, see Figure 4). Under this electrochemical cycling conditions the charge and discharge superimposes [31]. That was tentatively explained, in ref. [27], by the occurrence of Lithium/vacancies ordering observed in $LiV_2(PO_4)_3$ which involves an ordered depopulation of Lithium, whereas from the disordered fully delithiated phase the lithium is free to be inserted randomly until the $Li_2V_2(PO_4)_3$ composition is recovered. More recently, operando XAS at V K-edge investigation of this irreversible mechanism suggested the formation of anti-site Li/V defects at high voltage (Figure 5) providing V^{5+} with a much more stable tetrahedral environment than its initial distorted octahedral one [32].

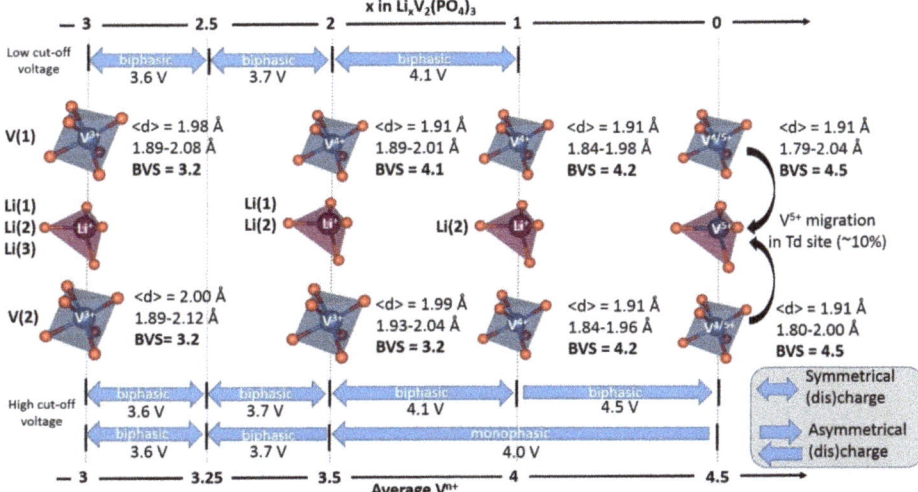

Figure 5. Structural evolution during Lithium extraction/insertion from/into $Li_3V_2(PO_4)_3$ [27,32].

In $Li_3V_2(PO_4)_3$, the electronically insulating phosphate groups isolate the valence electrons of transition metals within the lattices resulting in low intrinsic electronic conductivities—a trend common to all polyanion compounds. Therefore, the use of carbon coating or/and doping elements are required to improve the electrochemical performances: all the works applying these strategies are reviewed in ref [29]. The majority of these studies reports good performances only in the small voltage range (i.e., 3.0–4.3 V vs. Li^+/Li, in which only the V^{3+}/V^{4+} is activated). Indeed, due to the strong distortion of vanadium environments and the Li/V anti-site defects generated, the kinetic limitations of the V^{4+}/V^{5+} process is difficult to overcome.

The lithium insertion into $Li_3V_2(PO_4)_3$ reveals a complex series of reactions as well, to reach $Li_5V_2(PO_4)_3$ by activating the V^{2+}/V^{3+} redox couple [33]. The whole lithium insertion process into $Li_{3+x}V_2(PO_4)_3$ involves four consecutive two-phase regions to reach $Li_5V_2(PO_4)_3$. Approximately 0.5 Li^+ is inserted at every potential plateau around 1.95, 1.86, 1.74 and 1.66 V vs. Li^+/Li [26]. To the best of our knowledge, the crystallographic details of this complex mechanism have never been fully studied yet.

The anti-NASICON polymorph of $Li_3V_2(PO_4)_3$ is most thermodynamically stable but Gaubicher et al. [34] obtained the NASICON form by Na^+/Li^+ ionic exchange from $Na_3V_2(PO_4)_3$. This material reveals a similar electrochemical signature compared to the one of $Na_3V_2(PO_4)_3$ with a single plateau until the $LiV_2(PO_4)_3$ composition at 3.7 V vs. Li^+/Li. The crystal structure of $Na_3V_2(PO_4)_3$ was originally reported by Delmas et al. [35] 40 years ago using the standard rhombohedral unit cell, S.G. $R\text{-}3c$. Since then, $Na_3V_2(PO_4)_3$ has almost always been reported to adopt the rhombohedral symmetry with a partial occupancy of both Na(1) (6b Wyckoff position) and Na(2) (18e Wyckoff position) sodium sites. However, a recent article reveals that a $C2/c$ space group is more appropriate to describe this structure at room temperature and below due to Sodium-vacancies ordering [36] within five sites (one 4a and four others 8f) fully occupied. Several transitions between 10 and 230 °C involving four distinct phases (α ordered, β and β' with incommensurate modulations and γ disordered) were also reported. The transition between the α and β forms occurring close to the ambient temperature (i.e., 27 °C) may impact the sodium diffusion and a fortiori the electrochemical performances while the vanadium environment is hardly impacted by this phase transition. In both cases the VO_6 entities are slightly distorted with distances ranging between 1.97 and 2.03 Å for the rhombohedral description (2.00 Å in average on a single vanadium site) or 1.94 and 2.06 Å for the monoclinic one (2.00 Å in average on the three vanadium sites).

The electrochemical sodium extraction from $Na_3V_2(PO_4)_3$ occurs at a 3.4 V vs. Na^+/Na according to a biphasic reaction until the $NaV_2(PO_4)_3$ composition is reached (Figure 6). The structure of this V^{4+} phase, reported by Jian et al. [37], keeps a NASICON framework (Rhombohedral, $R\bar{3}c$) with only one fully occupied sodium site (6b Wyckoff site). During the sodium extraction, the V-O distances in VO_6 octahedra undergo an inequivalent shortening leading to distorted VO_6 octahedra (with 3 V-O distances at 1.86 Å and three others at 1.95 Å). The electrochemical extraction of the third sodium has never been reported despite the apparent successful chemical extraction realized by Gopalakrishnan et al. [38]. However, they did not report the detailed structure of this mixed-valence $V_2(PO_4)_3$.

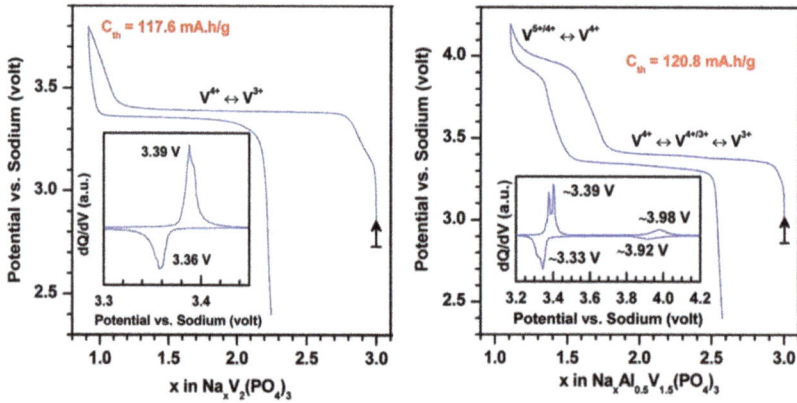

Figure 6. High voltage signature of Nasicon $Na_3V_2(PO_4)_3$ and $Na_3V_{1.5}Al_{0.5}(PO_4)_3$, adapted from ref. [39]. Reproduced from Ref. [39] with permission from the Centre National de la Recherche Scientifique (CNRS) and The Royal Society of Chemistry.

Even though the third Na^+ of $Na_3V_2(PO_4)_3$ doesn't seem electrochemically removable, the V^{4+}/V^{5+} redox couple in the NASICON was reported to lie at around 4 V vs. Na^+/Na thanks to the partial substitution of a part of Vanadium by Aluminum [39], Iron [40] or Chromium [41]. The Aluminum substituted material presents two advantages as it allows an increase in the capacity due to the lower weight of aluminum compared to vanadium (and also iron and chromium) as well as to reach the mixed valence V^{4+}/V^{5+} state at rather high voltage (i.e., 4.0 V vs. Na^+/Na, see Figure 6). However, in the Al^{3+} substituted compound, the V^{4+}/V^{5+} capacity is limited contrarily to that observed in $Na_3VCr(PO_4)_3$ where nearly 1.5 electrons/vanadium are exchanged [42]. At room temperature, this redox process induces a rapid degradation of the performance due to the migration of vanadium into the vacant Na site, while at lower temperature (i.e., −15 °C), vanadium is pinned in its original position leading to a rather reversible process is observed [43]. The V^{4+}/V^{5+} redox couple has also been reported in Na-rich NASICON such as $Na_4MnV(PO_4)_3$ [44–47]. From this compound, ca. 3 Na^+ are exchanged based on the V^{3+}/V^{4+}, Mn^{2+}/Mn^{3+} and then V^{4+}/V^{5+} redox achieving a capacity of 155 mAh/g. However, this latter appears to be poorly reversible inducing a higher irreversible capacity upon discharge during which the highly polarized S-shape voltage profile contrasts with staircase charge curve [45–47].

The replacement of a $(PO_4)^{3-}$ group in $Na_3V_2(PO_4)_3$ by 3 F^- leads to the $Na_3V_2(PO_4)_2F_3$ composition, often named as a "NASICON composition" but its crystal structure is fundamentally different.

3. Irreversible Multi-Electron Reactions in $Na_3V_2(PO_4)_2F_3$

The first physico-chemical investigation of the $Na_3M_2(PO_4)_2F_3$ system was conducted 20 years ago by Le Meins et al. [48]. They demonstrated a great compositional tunability of this framework which can accommodate many trivalent cations in octahedral sites (M = Al, V, Cr, Fe and Ga) and proposed the description of the structure of the vanadium phase in the $P4_2/mnm$ space group. Later, a combined synchrotron X-ray and neutron diffraction investigation revealed a tiny orthorhombic distortion at room temperature [49].

The *Amam* space group (i.e., S.G. #63, *Cmcm*) used leads to a different sodium distribution in the cell with three Na sites, one *4c* fully occupied and two *8f* partially occupied (approximatively distributed as 1/3:2/3) (Figure 7). The host structure is composed of $V_2O_8F_3$ bi-octahedra sharing a fluorine aligned along the [001] direction and connected to each other through PO_4 tetrahedra aligned in parallel with the (001) plane (Figure 7). The VO_4F_2 octahedra in this structure are non-centrosymmetric and hence vanadium does not occupy the inversion center. Indeed, a displacement along the c direction leads to two slightly different V-F bonds (V-F(1) = 1.968(6) Å and V-F(2) = 1.981(2) Å).

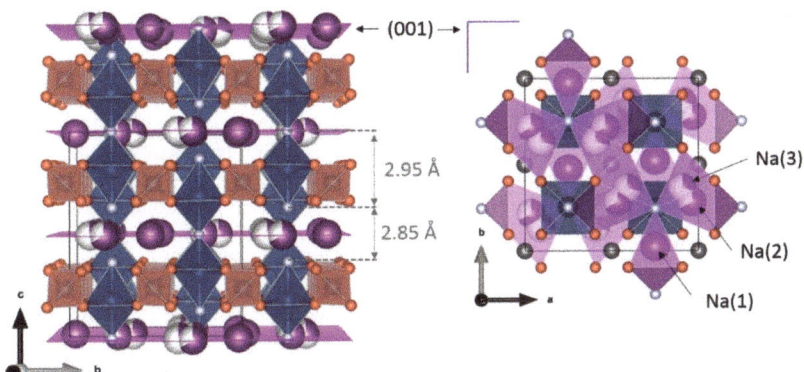

Figure 7. Structure of $Na_3V_2(PO_4)_2F_3$ (**left**) and sodium distribution (**right**) [49].

Slow electrochemical galvanostatic cycling shows the presence of four distinct reversible voltage-composition features at 3.70, 3.73, 4.18 and 4.20 V vs. Na^+/Na (Figure 8) suggesting a complex phase diagram upon sodium extraction/reinsertion [50].

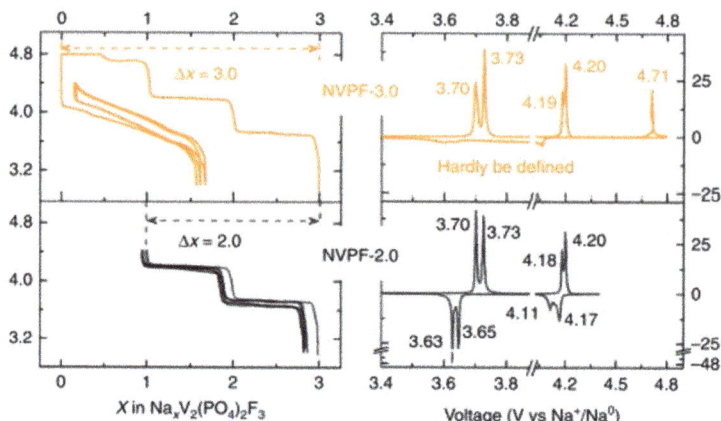

Figure 8. Galvanostatic electrochemical voltage-composition data of $Na_3V_2(PO_4)_2F_3$ at C/10 per exchanged ion and the corresponding derivative curve in the 3.0–4.4 V or 3.0–4.8 V voltage range adapted from ref. [51]. Reproduced with permission from Guochan Yan et al., Nature Communications; published by Springer Nature, 2019.

The operando synchrotron XRD investigation conducted by Bianchini et al. [52] is summarized in Figure 9. The phase diagram involves several intermediate phases of compositions $Na_xV_2(PO_4)_2F_3$ with x = 2.4, 2.2, 2, 1.8 and 1.3 before the $NaV_2(PO_4)_2F_3$ is reached. During extraction of the first sodium, an alternation between ordered and disordered phases (Na^+/vacancy and/or V^{3+}/V^{4+} ordering and disordering) is observed. The superstructure peaks observed for the $Na_{2.4}V_2(PO_4)_2F_3$ disappear for $Na_{2.2}V_2(PO_4)_2F_3$ and the diffraction pattern of $Na_2V_2(PO_4)_2F_3$ reveals the reappearance of a series of additional contributions non-indexed in the tetragonal cell. In the V^{3+}-rich phase, the two symmetrically inequivalent V-F bonds are very similar and as the oxidation of vanadium is increased, two kinds of bonds gradually appear as a short one at 1.88 Å and a longer one at 1.94 Å, whereas the equatorial V-O bonds decrease uniformly (from 1.99 to 1.95 Å). The extraction of the second sodium also involves intermediate phases at x = 1.8 and x = 1.3 accompanied by the disappearance of the superstructure peaks and finally leads to the formation of $NaV_2(PO_4)_2F_3$. This phase contains a single Na site and two vanadium sites conferring to vanadium cations two very different environments despite an average oxidation state of V^{4+}. Indeed, the BVS calculation suggests the formation of a V^{3+}-V^{5+} pair in bi-octahedra at this composition (Figure 9). The investigation of the redox mechanism involved during sodium extraction was conducted by Broux et al. [53] through operando XANES at V K-edge. They evidenced that V^{4+} starts to disproportionate from $Na_2V_2(PO_4)_2F_3$ and hence the formation of V^{3+}-V^{5+} pairs are confirmed for $Na_1V_2(PO_4)_2F_3$.

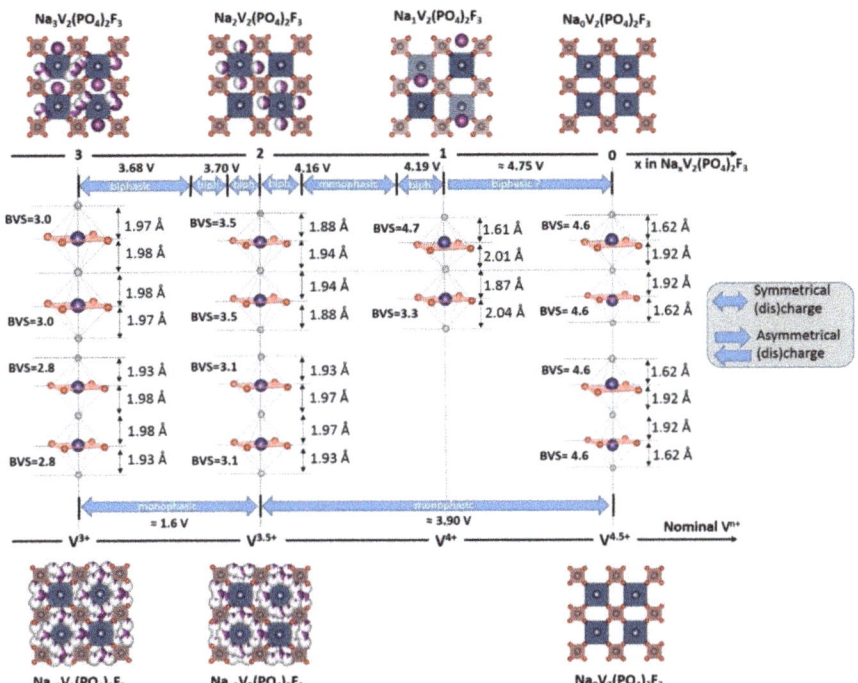

Figure 9. Evolution of vanadium environments and Na/vacancies ordering upon cycling of Na$_3$V$_2$(PO$_4$)$_2$F$_3$ [51,52,54].

Kang and coworkers predicted that the extraction of the third Na$^+$ towards the mixed valence V^{4+}/V^{5+} V$_2$(PO$_4$)$_2$F$_3$ composition would occur only at very high voltage (ca. 4.9 V vs. Na$^+$/Na) [55]. This was confirmed experimentally by Tarascon's group, under sever oxidative conditions (i.e., potentiostatic step at 4.8 V vs. Na$^+$/Na see Figure 8) in an optimized electrolyte [51]. In this structure, vanadium is displaced from the inversion center of the VO$_4$F$_2$ octahedra in such a way as to generate a short 1.62 Å and a longer 1.92 Å V-F bond lengths within the bi-octahedra. However, these extreme cycling conditions imply an irreversible reaction and only 2 Na$^+$ could be reinserted upon discharge down to 3.0 V, according to a solid solution mechanism, the third Na$^+$ being reinserted at much lower voltage (i.e., 1.6 V vs. 3.7 V for the same composition range upon charge). The subsequent charge/discharge allows for the reversible extraction/insertion of 3 Na$^+$ in a wide voltage range (1.0–4.4 V vs. Na$^+$/Na). This new β-Na$_3$V$_2$(PO$_4$)$_2$F$_3$ polymorph exhibits a different symmetry, different V-X bond lengths and a disordered Na distribution (see Figure 9) which bears strong resemblance with the one of the high temperature phase (T > 400 K) [49,54]. Due to the low voltage associated to the reinsertion of third Na$^+$, only 2 Na$^+$ can be exchanged in a real battery system where the third one acts as an alkali reservoir to compensate for the solid electrolyte interface (SEI) formation at negative electrode [51], which allows offering up to 460 Wh/kg in full cell vs. hard carbon (+18% compared with a conventional α-Na$_3$V$_2$(PO$_4$)$_2$F$_3$), corresponding to the highest energy density reported so far in Na-ion battery [56].

Most of the Na$_3$V$_2$(PO$_4$)$_2$F$_3$ materials reported as stoichiometric in the literature actually present various amounts of vanadyl-type defects (i.e., partial substitution of F$^-$ by O^{2-} with a charge compensation by partial oxidation of V^{3+}-F into vanadyl V^{4+}=O) impacting on the electrochemical performance. Several authors studied in detail the crystallographic changes generated by this substitution in Na$_3$V$_2$(PO$_4$)$_2$F$_{3-x}$O$_x$ (with $0 \leq x \leq 0.5$ [54], $0 \leq x \leq 2$ [55] and x = 1.6 [57,58]). This oxidation has strong effects on the local environments of vanadium and on the sodium distribution and appears to be beneficial for enhancing the charge rates of the battery. Kang's group [57] was the first to investigate the performance of $Na_3V_2(PO_4)_2F_{1.4}O_{1.6}$ (i.e., $V^{3.8+}$) as a positive electrode material and reported high charge and discharge rate capabilities, assigned to a low activation energy for Na^+ diffusion (~350 meV) inside the framework, despite the poor electronic conductivity (~2.4×10^{-12} S.cm^{-1}) and its great cycling stability was assigned to the small volume change during sodium extraction/insertion (~3%). The same group later published a promising result about the computed voltage for the extraction of the third sodium around 4.7 V for $Na_3V_2(PO_4)_2F_{1.5}O_{1.5}$ [55] (lower than the one computed up to 4.9 V in $Na_3V_2(PO_4)_2F_3$) and experimentally realized the reversible exchange of more than one electron per vanadium at high voltage (ca. 3.8V vs. Na^+/Na in average) with a symmetrical charge/discharge profile and an improved capacity retention.

4. Low Voltage Multi-Electron Reactions in Tavorites Li_xVPO_4Y (Y = O, F and/or OH)

Tavorite-type compositions of general formula A_xMXO_4Y are a third class of very interesting polyanion structures in which A is an alkali cation (i.e., Li, Na and $0 \leq x \leq 2$) and M a metal (i.e., Mg, Al, Ti, V, Fe, Mn, Zn or mixture of them). The polyanionic group, XO_4, is either PO_4 or SO_4 and the bridging anion, Y, is a halide, hydroxide, oxygen, H_2O group or a mixture of them [18]. The multiple redox center combined with this double inductive effect bring a strong interest at both practical and fundamental levels as it allows scanning a wide range of working voltages, from 1.5 V for Ti^{3+}/Ti^{4+} in $LiTiPO_4O$ [59] to 4.26 V for the V^{3+}/V^{4+} redox couple in $LiVPO_4F$ [60]. The high voltages provided by the V^{3+}/V^{4+} and V^{4+}/V^{5+} redox couples confer high theoretical energy densities to the vanadium-based Tavorite compositions.

The crystal structure of Tavorite-like materials can be described in either triclinic (*P-1*, with Z = 2 or Z = 4) or monoclinic (*C2/c* or *P2$_1$/c*) systems [14,61,62]. Tavorite-like therefore gathers Tavorite (*P-1*, Z = 2, $LiVPO_4F$ and $LiVPO_4OH$), Montebrasite (*P-1*, Z = 4, $LiVPO_4O$), Maxwellite (*C2/c*, $NaVPO_4F$ and $HVPO_4.OH$) and even Talisite (*P2$_1$/c*, $NaVPO_4O$) structures. Their crystallographic arrangements present common features which can be broadly described as vanadium octahedra (VO_4Y_2) sharing a bridging anion Y in order to form infinite chains $[-Y - VO_4 - Y-]_\infty$. These chains are connected to each other through PO_4 tetrahedra sharing their four oxygen atoms with four vanadium octahedra belonging to three different chains. This 3D framework accommodates Li^+ or Na^+ in hexagonal channels. The symmetry of vanadium octahedra is dictated by the nature of the V^{n+}-Y bond. Indeed, in V^{3+}-rich $LiVPO_4F$, $NaVPO_4F$ and $VPO_4.H_2O$, the vanadium is located on an inversion center of the VO_4Y_2 octahedra, whereas in V^{4+}-rich $NaVPO_4O$ and $LiVPO_4O$ a loss of the centrosymmetry of the vanadium environment is observed. Indeed, in the Tavorite-like structure, for an oxidation state of vanadium superior to +3, vanadium likely forms the vanadyl bond resulting from the Jahn Teller activity of V^{4+} (d^1 $t2g^1eg^0$). This strongly covalent V=O bond can be formed only with oxygen atoms which are not already involved in a covalent PO_4 group. Only the bridging oxygen, Y, fulfils these requirements and, therefore, in the V^{4+} compounds, an ordering between short and long bonds takes place along the chains (Figure 10). This ordering generates a change of space group (from *C2/c* for $NaVPO_4F$ to *P2$_1$/c* for $NaVPO_4O$) or a doubling of the cell size (Z = 2 for $LiVPO_4F$ to Z = 4 for $LiVPO_4O$).

The lithium content in Li_xVPO_4O can vary from 0 to 2, leading to a capacity of 300 mAh/g at an average voltage of ca. 3.1 V allowing the achievement a stable energy density > 900 Wh/kg using surface engineering and nanosizing strategies [9–13]. However, the large difference between the voltage for oxidation of V^{4+} into V^{5+} (i.e., 3.95 V vs. Li^+/Li) and that for the reduction in V^{4+} to V^{3+} (around 2.3 V vs. Li^+/Li) makes this multi-electron reaction unsuitable for a real battery system (Figure 11).

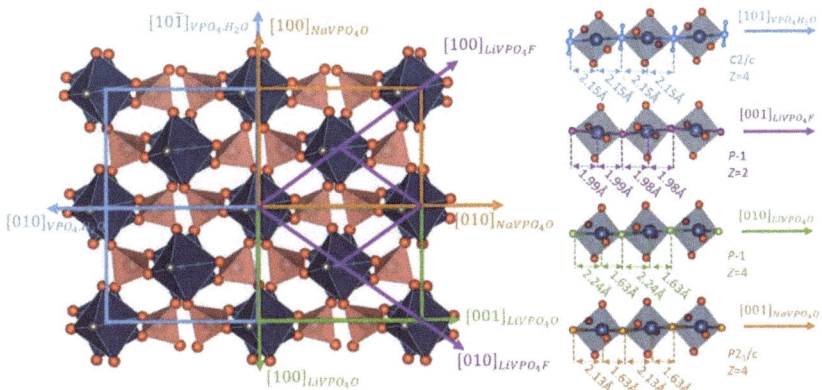

Figure 10. Structural relationships between different Tavorite-type materials.

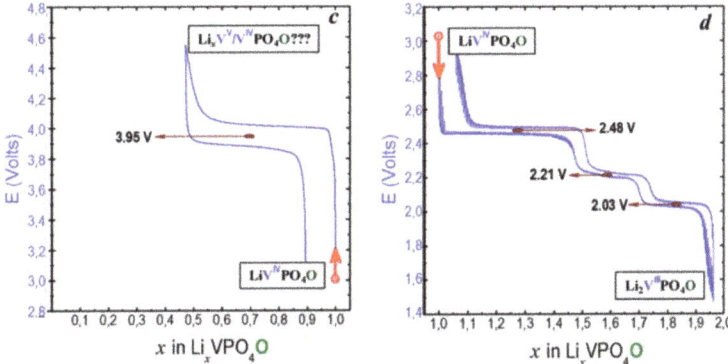

Figure 11. Voltage profile of Li_xVPO_4O cycled between 3.0–4.6 V vs. Li^+/Li (**left**) and between 3.0 and 1.5 V vs. Li^+/Li in GITT mode (**right**) adapted from ref. [14]. Reproduced with permission from Ateba Mba et al., Chemistry of Materials; published by American Chemical Society, 2012.

In the high voltage region (i.e., 3.0–4.6 V vs. Li^+/Li involving the V^{4+}/V^{5+} redox couple), the oxidation process occurs via a biphasic mechanism between $LiVPO_4O$ and VPO_4O [14]. The crystal structure of this V^{5+} phase (ε-VPO_4O) is described in a Cc space group allowing the formation of vanadyl bonds appearing as shorter than the ones observed in $LiVPO_4O$ (i.e., 1.59 vs. 1.67 Å, Figure 10). Conversely, the antagonist $V^{5+}\cdots O$ bond along the chains elongates from 2.2 Å in $LiVPO_4O$ to 2.5 Å in VPO_4O leading to an unconventional increase in the cell volume during lithium extraction ($\Delta V/V$ = 4.1%) [63]. This VPO_4O polymorph can also be obtained while deintercalating the homeotype $LiVPO_4OH$ (and also $VPO_4 \cdot H_2O$), according to an original mechanism [64,65]. Indeed, VPO_4OH appears instable vs. $LiVPO_4OH$ and VPO_4O as this V^{4+}-rich phase is not formed upon Li^+ deintercalation from $LiVPO_4OH$. On the contrary, VPO_4O is formed showing that the V^{3+}-O/V^{5+}=O redox couple is activated at a constant equilibrium voltage of 3.95 V vs. Li^+/Li [65]. Indeed, in the VPO_4OH hypothetical phase the competition between the two highly covalent bonds, V^{4+}=O on one side and O-H bond on the other side, would destabilize the V^{IV}-O-H sequence, leading to the concomitant extraction of Li^+ and H^+ and to the atypical two-electron V^{3+}/V^{5+}=O redox reaction at a constant voltage. Unfortunately, on the contrary to the two-electron reaction observed in Li_xVPO_4O over 3.2 V, which is reversible but not practical, this one observed at a constant high voltage leads to an irreversible phase transformation.

The Li⁺ insertion within LiVPO₄O involves two intermediate phases, Li$_{1.5}$VPO$_4$O and Li$_{1.75}$VPO$_4$O, before reaching the Li$_2$VPO$_4$O [66]. Although this V^{3+}-rich composition is described in a triclinic (P-1, Z = 4) system allowing the formation of a vanadyl-type distortion along the chains, the refined V-O distances do not reveal significant differences between them [63], in agreement with the weak Jahn Teller activity of V^{3+} (d^2 t$_{2g}^2$e$_g^0$). Lin et al. [67] studied in detail the structural evolutions at the local scale during the lithium insertion in Li$_{1+x}$VPO$_4$O, and V K-edge EXAFS shows the disappearance of vanadyl bond for Li$_{1.5}$VPO$_4$O and the persistence of the longer antagonist until Li$_{1.75}$VPO$_4$O in good agreement with the phase transitions observed (Figure 11).

The investigation of LiVPO$_4$F started in 2003 with a series of studies conducted by Barker and co-workers [60,68,69] who highlighted the promising performance of this material. Indeed, the high voltage delivered for the Lithium extraction (4.25 V vs. Li⁺/Li for the V^{3+}/V^{4+} redox voltage, Figure 12) and a capacity very close to the theoretical one even at high C-rate confer to this material a higher practical energy density compared to the ones of commercially available LiFePO$_4$ and LiCoO$_2$ (655 vs. 585 and 525 Wh/kg respectively).

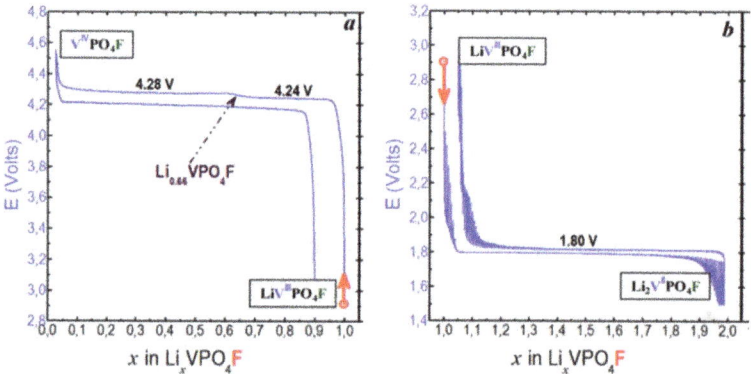

Figure 12. Voltage profile of Li$_x$VPO$_4$F cycled between 3.0–4.6 V vs. Li⁺/Li (**left**) and between 3.0 and 1.5 V vs Li⁺/Li in GITT mode (**right**) adapted from ref. [14]. Reproduced with permission from Ateba Mba et al., Chemistry of Materials; published by American Chemical Society, 2012.

The Lithium extraction from LiVPO$_4$F involves an intermediate phase, Li$_{2/3}$VPO$_4$F, and then VPO$_4$F through two biphasic reactions. The crystal structure of VPO$_4$F was reported by Ellis et al. [70], its C2/c space group involving centrosymmetric vanadium octahedra with V-F distances of 1.95 Å (Figure 13) whereas the actual nature of the Li$_{2/3}$VPO$_4$F phase is still unclear, although superstructure peaks have been identified and indexed by doubling the b parameter [71]. This intermediate phase is not formed during discharge where a biphasic reaction between the end-member compositions VPO$_4$F and LiVPO$_4$F takes place [72]. This asymmetric charge/discharge mechanism is not understood at the moment even though it was first attributed by Ellis et al. to the presence of two lithium sites partially occupied (0.8/0.2) in the starting LiVPO$_4$F. Nevertheless, this hypothesis was ruled out later by Ateba Mba et al. [67] who localized Lithium in a single fully occupied site. Piao et al. [73] conducted operando V-K edge XANES in order to probe the redox mechanism during delithiation of LiVPO$_4$F. By a principal component analysis, three components were required to fit the series of spectra recorded upon charge. This might suggest at least a V^{3+}/V^{4+} ordering for Li$_{2/3}$VPO$_4$F. The lithium insertion into LiVPO$_4$F occurs at low voltage, typical for the V^{3+}/V^{2+} redox couple (i.e., 1.8 V vs. Li⁺/Li) through a biphasic reaction leading to the formation of Li$_2$VPO$_4$F (Figure 12). The structure of Li$_2$VPO$_4$F is described in a C2/c space group with V^{2+} sitting in a centrosymmetric VO$_4$F$_2$ octahedra with V-F distances at 2.10 Å and equatorial V-O ones at 2.13 Å in average [70]

(Figure 13) while Li$^+$ ions are distributed between two 8f Wyckoff sites half occupied in LiO$_3$F$_2$ environments.

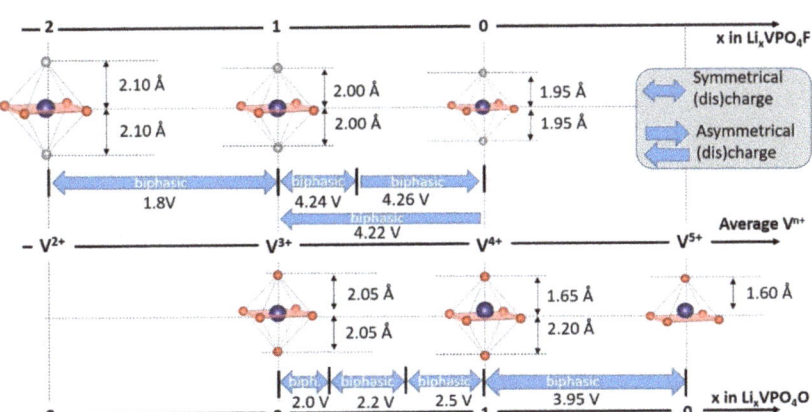

Figure 13. Structural evolution during Lithium extraction/insertion from/into LiVPO$_4$F [72] and LiVPO$_4$O [63,67].

Various chemical routes to obtain polycrystalline powders of Tavorite LiVPO$_4$F were reported: sol-gel-assisted carbo thermal reduction (CTR) [74], ionothermal [75]. The majority of these reports highlight the difficulty to obtain pure powders (i.e., without anti-NASICON Li$_3$V$_2$(PO$_4$)$_3$ impurity) or vanadyl-free compounds. Indeed, a series of recent papers demonstrated, by ^{7}Li NMR (and its 2D analogue) and DFT calculations, the presence of various amounts of vanadyl-type defects in crystallographically pure "LiVPO$_4$F" [76,77]. Recently, B. Kang and co-workers [78] reported on an ingenious strategy to avoid the fluorine loss during synthesis, using PTFE as an additional fluorine source. The material thus obtained reveals high electrochemical performance with a stable discharge capacity of 120 mAh/g at 10C over 500 cycles. The same group also published for the first time the electrochemical properties of the mixed valence V^{3+}/V^{4+} LiVPO$_4$F$_{0.25}$O$_{0.75}$ [79]. This strategy aimed at decreasing the difference in voltage between Li insertion and extraction reactions, conferring to the material a high energy density (i.e., 820 Wh/kg) in a reduced voltage range (i.e., 2.0–4.5 V vs. Li$^+$/Li) with the activation of the V^{3+}/V^{4+} and V^{4+}/V^{5+} redox couples, respectively. Further investigation of the LiVPO$_4$F-LiVPO$_4$O tie-line has allowed several compositions to stabilize in which the competition between ionicity of the V^{3+}-F bond and covalency of the V^{4+}=O bond distorts the structure, freezes the framework upon Li extraction and hence allows for improved rate capabilities compared with the end-member phases [80,81]. Interestingly, upon Li deintercalation from these materials, the V^{4+}=O/V^{5+}=O redox couple is triggered first before the V^{3+}/V^{4+} is activated in fluorine rich environments leading to the formation of a mixed valence V^{3+}-V^{5+} phase at half charge [81]. Although surprising, this redox mechanism is in full agreement with the operating voltage of the end-member phases, the V^{4+}/V^{5+} redox couple being activated at 3.95 V in LiVPO$_4$O and the V^{3+}/V^{4+} redox couple at 4.25 V in LiVPO$_4$F due to the absence of vanadyl distortion in LiVPO$_4$F and VPO$_4$F.

Most of the vanadium phosphates discussed above operate at a rather high V^{3+}/V^{4+} redox voltage, suggesting a massive improvement of the energy density delivered while triggering the V^{4+}/V^{5+} redox couple. However, materials that operate at such a high V^{3+}/V^{4+} voltage are usually unable to reversibly exchange several electrons in a narrow enough voltage range. In the following section we will clarify the crystallographic origin of this trend and identify the strategies able to overcome this issue.

5. Towards Reversible High-Voltage Multi-Electron Reactions

Many other vanadium phosphates (as well as pyrophosphates and phosphites, see Table 1) have been stabilized and studied as positive electrode materials for Li(Na)-ion batteries. This article does not aim at providing an exhaustive review of all of them, however careful descriptions of selected systems, provided above, now allow us to generalize and predict part of their properties (especially working voltages, redox mechanisms and structural evolutions) from the only consideration of their crystal structures in their pristine state.

Table 1. List of the vanadium phosphate, pyro-phosphate and phosphite materials with their redox voltage and corresponding practical capacity based on vanadium redox. More details about the classification of these materials (Type I, II or III) are provided in the text and at the Figure 14.

As Synthesized Compositions	Initial V^{n+}	M/P Ratio	V^{2+}/V^{3+}		V^{3+}/V^{4+}		V^{4+}/V^{5+}		Ref.
			E (V vs. Li^+/Li)	Capacity (mAh/g)	E (V vs. Li^+/Li)	Capacity (mAh/g)	E (V vs. Li^+/Li)	Capacity (mAh/g)	
Type I Materials									
$Na_3V_2(PO_4)_3$	V^{3+}	0.67	1.9 *	59	3.7 *	118	/	/	[39]
$Na_3V_{1.5}Al_{0.5}(PO_4)_3$	V^{3+}	0.67	1.9 *	60	3.7 *	85	4.3 *	28	[39]
$Na_3VCr(PO_4)_3$	V^{3+}	0.67	/	/	3.7 *	60	4.4 *	50	[42]
$Na_4VMn(PO_4)_3$	V^{3+}	0.67	/	/	3.7 *	60	4.2 *	50	[47]
$r-Li_3V_2(PO_4)_3$	V^{3+}	0.67	/	/	3.7	131	/	/	[34]
$m-Li_3V_2(PO_4)_3$	V^{3+}	0.67	1.8	131	3.9	131	4.5	33	[27]
$LiVP_2O_7$	V^{3+}	0.5	2	116	4.3	95	/	/	[82,83]
$Na_7V_4(P_2O_7)_3(PO_4)_2$	V^{3+}	0.5	/	/	4.2 *	90	/	/	[84]
$Na_7V_3Al_1(P_2O_7)_3(PO_4)_2$	V^{3+}	0.5	/	/	4.2 *	77	4.5	46	[85]
$Na_3V(PO_4)_2$	V^{3+}	0.5	/	/	3.8 *	90	4.4 *	20	[86,87]
$LiV(HPO_3)_2$	V^{3+}	0.5	/	/	4.1	75	/	/	[88]
$Li_9V_3(P_2O_7)_3(PO_4)_2$	V^{3+}	0.375	/	/	3.7	55	4.5	55	[89]
$Na_7V_3(P_2O_7)_4$	V^{3+}	0.375	/	/	4.3 *	80	/	/	[90]
$Na_3V(PO_3)_3N$	V^{3+}	0.33	/	/	4.3 *	74	/	/	[91]
Type II Materials									
$LiVPO_4F$	V^{3+}	1	1.8	156	4.2	156	/	/	[14]
$NaVPO_4F$	V^{3+}	1	/	/	≈4.2 *	20	/	/	[92]
$LiVPO_4OH$	V^{3+}	1	1.4	155	/	/	/	/	[65]
$(Li,K)VPO_4F$	V^{3+}	1	/	/	4.0	110	/	/	[93]
$Na_3V_2(PO_4)_2F_3$	V^{3+}	0.67	1.5 *	64	4.0 *	64	≈4.8 *	64	[51,53,94]
$Li_5V(PO_4)_2F_2$	V^{3+}	0.5	/	/	4.15	85	4.7	85	[95]
$t-Na_5V(PO_4)_2F_2$	V^{3+}	0.5	/	/	3.7 *	62	/	/	[96]
$o-Na_5V(PO_4)_2F_2$	V^{3+}	0.5	/	/	3.9 *	65	/	/	[96]
Type III Materials									
$\alpha-LiVPO_4O$	V^{4+}	1	/	/	2.4	155	3.95	150	[14]
$\beta-LiVPO_4O$	V^{4+}	1	/	/	2.2	155	4	130	[97]
$\beta-NaVPO_4O$	V^{4+}	1	/	/	/	/	3.6 *	58	[98]
$\gamma-LiVPO_4O$	V^{4+}	1	/	/	2	80	4	150	[99]
$Li_4VO(PO_4)_2$	V^{4+}	0.5	/	/	2	94	4.1	94	[100]
$Na_4VO(PO_4)_2$	V^{4+}	0.5	/	/	/	/	3.8 *	77	[101]
$Li_2VOP_2O_7$	V^{4+}	0.5	/	/	/	/	4.1	64	[102]

The voltage values are reported vs. Li^+/Li even for those obtained in Na-cell (according to $E(Na^+/Na) = 0.3$ V vs. Li^+/Li), in that case the voltage is marked by *

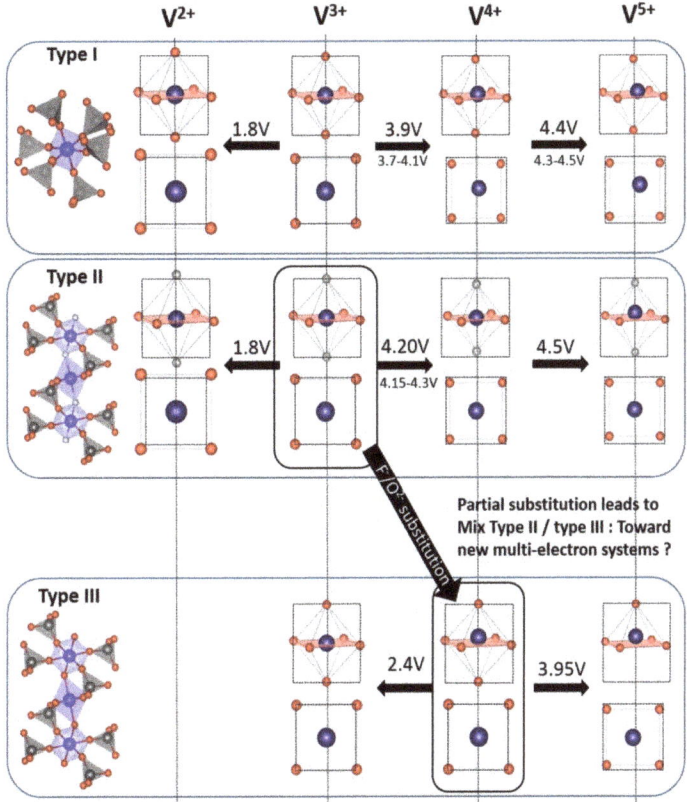

Figure 14. Typical evolution of vanadium environments according to the oxidation state of vanadium for type I, type II and type III materials.

Table 1 highlights the significant divergence in the V^{3+}/V^{4+} redox voltages which cannot be attributed to the inductive effect, the cation-cation repulsions or even Li site energy, which are the main reported features impacting the voltage in polyanions [103,104]. Indeed, the voltage for the V^{3+}/V^{4+} redox couple in the Tavorite system Li_xVPO_4Y (with Y = O or F) varies from 2.4 V for $Li_{1+x}VPO_4O$ ($0 \leq x \leq 1$) to 4.26 V in $Li_{1-x}VPO_4F$ ($0 \leq x \leq 1$). This is attributed to the effect of the highly covalent vanadyl bond which is observed for oxidation states of vanadium strictly superior to 3 in $Li_{1-x}VPO_4O$, $Na_4VO(PO_4)_2$ ... These structures present a common crystallographic feature: at least one oxygen around vanadium is not involved in a covalent P-O bond and hence could be engaged in a vanadyl bond. In the compounds where the VO_6 octahedra share all their oxygen atoms with PO_4 (or P_2O_7) groups, the structure of the de-alkalinated V^{4+} phases are vanadyl free with VO_6 octahedra slightly distorted. The corresponding vanadyl free V^{3+}/V^{4+} redox couple is located at 3.9 V in monoclinic $Li_{3-x}V_2(PO_4)_3$ and 4.2 V in $Li_{1-x}VP_2O_7$, a much higher voltage than the V^{3+}/V^{4+} couple involved in $Li_{1+x}VPO_4O$ polymorphs (around 2.3 V vs. Li^+/Li).

Boudin et al. [15] proposed a classification of the vanadium phosphates into three groups according to size of the "clusters" of vanadium polyhedra ($[VO_x]_n$ with $1 < n < \infty$). Although this classification is pertinent to the discussion of the catalytic or magnetic properties of vanadium phosphates, it does not really make sense for a discussion of electrochemical properties. Therefore, we chose to sort these materials considering vanadylforbidden (type I and type II) and vanadyl-allowed (type III) structures (summarized in Table 1 and Figure 14):

- For type I materials (e.g., $Li_3V_2(PO_4)_3$), in which the vanadyl bond cannot appear due to the involvement of each oxygen atom of VO_6 octahedra in a PO_4-type entity, the typical evolution of the vanadium environment upon oxidation (from V^{2+} to V^{5+}) follows a quasi-homogeneous shortening of V-O bonds from V^{2+} to V^{4+} and a strong increase in VO_6 distortion to reach the V^{5+} state with corresponding voltages of 1.8 V vs. Li^+/Li for V^{2+}/V^{3+}, 3.9 V vs. Li^+/Li for V^{3+}/V^{4+} and 4.4 V vs. Li^+/Li for V^{4+}/V^{5+} redox couples (on average for all the type I materials reported in Table 1).
- In type II materials (e.g., $LiVPO_4F$), at least one of the ligands around vanadium is unshared with a phosphate group and hence would be available to form the vanadyl bond. However, in that case, the nature of this ligand (F^- instead of O^{2-}) inhibits its formation. From V^{2+} to V^{4+}, the evolution of the vanadium environment follows a similar trend with slightly higher voltages than for type I due to the higher ionicity of V-F versus V-O. For V^{5+}, for instance in deintercalated $Na_3V_2(PO_4)_2F_3$, a "vanadyl-like" distortion appears with V-F bond length of 1.6 Å and 1.9 Å. Such an F$\cdots$V-F sequence has never been reported elsewhere and the precise nature of the V-F bonds formed is still to be clarified.
- Type III group (e.g., $LiVPO_4O$) gathers the structures having at least one oxygen belonging to VO_6 octahedra available to form the covalent vanadyl bond for vanadium oxidation states higher than 3. In this class of materials, the V^{3+} environments are quasi undistorted. As the oxidation state of vanadium is increased, vanadium leaves the inversion center of the VO_6 octahedra in order to form the vanadyl bond. The formation of a distorted $V^{IV}O_6$ octahedra (with typical distances ranging between 1.6 and 2.4 Å along dz^2 and quasi equivalent equatorial distances around 2 Å) and V^VO_5 pyramids (in which the short V=O bond is about 1.6 Å and a shortening of the equatorial distances is observed around 1.8–1.9 Å) are observed. The corresponding voltages appear completely different to those of type I and type II materials: 2.4 V vs. Li^+/Li for the V^{3+}-O/V^{4+}=O and 3.95 V vs. Li^+/Li for the V^{4+}=O/V^{5+}=O redox couples.

Note that type II materials are crystallographically pseudo type III ones in which the oxygen involved in the vanadyl bond is replaced by Fluorine. Therefore, partial substitution of this fluorine by oxygen leads to mixed type II/III materials—which is actually the case for most of the type III materials, difficult to obtain as vanadyl-free. Extended oxy-fluorine solid solutions were investigated for $Na_3V_2(PO_4)_2F_{3-y}O_y$ [54,55,57,58] and for $LiVPO_4F_{1-y}O_y$ [79–81,105,106]. The particularity of these compounds resides in the redox paradox of vanadium where the V^{4+}=O/V^{5+}=O is activated at lower voltage than the V^{3+}-F/V^{4+}-F [55,81,107]. Depending on the distribution of ligands around V, it behaves as type II ($V^{3+}O_4F_2$), type III ($V^{4+}O_4O_2$) or mixed type II/III ($V^{3/4+}O_4OF$) [81]. For this latter environment, the V^{5+}=O vanadyl-like distortion is allowed upon cycling, promoting the reversibility of the process, but is observed at higher voltage than type III materials thanks to the antagonist fluorine. This highlights the importance of the heteroleptic units formed in statistically distributed or in peculiar O/F ordered compounds which are somewhat difficult to obtain due to the different nature of the V^{4+}=O and V^{3+}-F bonds promoting their clusterization [108].

This classification makes further sense regarding the ability of each type of vanadium phosphates to reversibly exchange several electrons per transition metal at high voltage and in a narrow enough voltage range. In type III materials multi-electron redox through V^{3+}-O/V^{4+}=O and V^{4+}=O/V^{5+}=O couples have often been reported [9–12]. The oxygen atoms unshared with PO_4 facilitate the formation of the vanadyl bond allowing for two rather reversible electron processes and thus allow the achievement of cycling stability with energy density higher than 900 Wh/kg [9]. However, this multi-electron reaction cannot be used in a real battery system due to the large voltage difference between the V^{3+}-O/V^{4+}=O and V^{4+}=O/V^{5+}=O (ca. 2.5 V) redox couples. Substituting oxygen by fluorine in such a way to obtain $LiVPO_4F_{0.75}O_{0.25}$ allows raising the voltage of the V^{3+}/V^{4+} redox and hence reversibly intercalating 1.6 electrons per vanadium in a reduced voltage range [79]. However, this material suffers from rapid capacity fading under such conditions. Since then, the possibility to stabilize multiple compositions along the LiVPO$_4$F-LiVPO$_4$O tie-line has been demonstrated and a systemic investigation of substitution ratio (i.e., y) vs. the voltage range could allow fixing this issue by controlling the Δx in Li$_{1\pm x}$VPO$_4$F$_{1-y}$O$_y$.

In type I and type II materials, while the low voltage V^{2+}/V^{3+} ($\approx$1.8 V vs. Li$^+$/Li) and high voltage V^{3+}/V^{4+} (3.9–4.2 V vs. Li$^+$/Li) redox are easily triggered, the V^{4+}/V^{5+} redox is rarely reported (see Table 1). Moreover, this latter is often kinetically limited and/or irreversible, most probably due to the structural rearrangements required to provide to V^{5+} cations a satisfying environment. Indeed, the V^{5+} cations are stable either in a pyramidal ([1+4] coordination) or in a tetrahedral ([2+2] coordination) or even in a very distorted octahedral ([2+2+2] coordination) [15] environments. In each case, at least one covalent vanadyl bond must be formed, but this formation is not privileged by the crystallographic arrangements adopted by type I and II materials. In order to provide to the V^{5+} cations with a more stable environment than this distorted octahedral one, migration of V^{5+} in tetrahedral site has been proposed [32]. Therefore, kinetic limitations and/or an irreversible capacity, which can be compensated only at very low voltage (as seen in Na$_3$V$_2$(PO$_4$)$_2$F$_3$, Li$_3$V$_2$(PO$_4$)$_3$, Li$_5$V(PO$_4$)$_2$F$_2$ and Li$_9$V$_3$(P$_2$O$_7$)$_3$(PO$_4$)$_2$), appear. Although V^{5+} migration in tetrahedral sites has been reported only in Li$_3$V$_2$(PO$_4$)$_3$ so far, analyzing the electrochemical response upon subsequent discharge for other compounds gives insight about the nature of the irreversible reaction taking place. For instance, in Na$_3$V$_2$(PO$_4$)$_2$F$_3$, the Na re-insertion into V$_2$(PO$_4$)$_2$F$_3$ (i.e., V$^{4.5+}$) occurs at 3.9 V vs. Na$^+$/Na in average, until the composition Na$_2$V$_2$(PO$_4$)$_2$F$_3$ (V$^{4.5+}$ to V$^{3.5+}$): this voltage range is associated to the Na$_3$(V^{3+})-Na$_1$(V^{4+}) composition range during the previous charge. The further Na insertion occurs at 1.6 V vs. Na$^+$/Na until the composition Na$_3$V$_2$(PO$_4$)$_2$F$_3$ is recovered. Moreover, the length of this low voltage plateau is proportional to the amount of vanadium oxidized above V^{4+} during the previous charge. Therefore, this low voltage feature is more likely to correspond to the reduction in V^{3+} (i.e., $\approx$1.5 V vs. Na$^+$/Na for V^{3+}/V^{2+} redox in type I and type II materials) rather than to the reduction in the V^{4+} into V^{3+} (i.e., 3.6–3.9 V vs. Na$^+$/Na). This behavior could agree with the presence of V^{5+} in Td sites. Indeed, as seen in transition metal vanadates used as anode in alkali-ion batteries [109], V^{5+}$_{Td}$ is not reduced above 1.5 V vs. Li$^+$/Li without migrating back in an octahedral site. Therefore, the V^{3+} reduction would occur at a higher voltage than the V^{5+}$_{Td}$ reduction. The presence of oxygen in the fluorine site would help in accommodating V^{5+} cation in distorted octahedral site in the charged state. Indeed, it has been shown by theoretical calculations that the partial substitution of fluorine by oxygen in such a way to obtain Na$_3$V$_2$(PO$_4$)$_2$F$_{3-x}$O$_x$ composition tends to decrease the voltage of extraction of the third Na$^+$ cations (from 4.9 V to 4.7 V vs. Na$^+$/Na from pure fluoride to oxy-fluoride) leading to the reversible exchange of more than one electron per vanadium with an excellent rate capability [55].

Finally, this review reveals that the versatility of the vanadium chemistry with a large number of stable oxidation states stabilized in very different environments opens the road towards the formation of new materials whose strains imposed by the crystal field give attractive electrochemical properties. While, in the battery field, the search for new polyanion positive electrode materials slows down for few years, maintaining the efforts towards the stabilization of new phases is crucial. LiVPO$_4$F$_{1-x}$O$_x$ and Na$_3$V$_2$(PO$_4$)$_2$F$_{3-x}$O$_x$ are the only vanadium phosphate oxy-fluorides studied as positive electrode materials and have shown very promising properties. Further playing with anionic substitution, not only with vanadium phosphate oxy-fluorides but also oxy-nitrides (as recently reported with Na$_3$V$_2$(PO$_3$)$_3$N [91]) and even oxy-sulfides, would offer new degrees of flexibility for such versatile polyanion systems and could allow the achievement of high energy density (ca. 1 kWh/kg of active positive electrode material corresponding to ca. 400 Wh/kg at the cell level) through reversible high-voltage multi-electron redox.

6. Conclusions

This review has identified the vanadyl distortion as the main feature governing the operating voltage in vanadium phosphates and their ability to reversibly store several electrons per transition metal. The classification of such materials in three groups, according to the nature of the ligands in the vanadium octahedra and to the distribution of PO_4 around them, has allowed to unveil the strategies to increase their energy density. Indeed, anionic substitutions have led to vanadium phosphate oxy-fluorides which allow to combine the beneficial effect of the vanadyl distortion on the reversibility with the high voltage of vanadium redox couples in fluorine rich environments. Further investigation of these anionic substitutions could allow to tend towards reversible high-voltage multi-electron reactions in Alkali-ion batteries.

Funding: The authors acknowledge FEDER, the Reégion Hauts-de-France and the RS2E Network for the funding of EB's PhD thesis, as well as the financial support of Région Nouvelle Aquitaine, of the French National Research Agency (STORE-EX Labex Project ANR-10-LABX-76-01) and of the European Union's Horizon 2020 research and innovation program under grant agreement No 875629 (NAIMA project).

Institutional Review Board Statement: Not applicable.

Informed Consent Statement: Not applicable.

Data Availability Statement: Not applicable.

Conflicts of Interest: The authors declare no conflict of interest.

References

1. Padhi, A.K.; Nanjundaswamy, K.S.; Goodenough, J.B. Phospho-Olivines as Positive-Electrode Materials for Rechargeable Lithium Batteries. *J. Electrochem. Soc.* **1997**, *144*, 1188–1194. [CrossRef]
2. Andersson, A.S.; Thomas, J.O.; Kalska, B.; Häggström, L. Thermal Stability of LiFePO4-Based Cathodes. *Electrochem. Solid State Lett.* **2000**, *3*, 66–68. [CrossRef]
3. Andersson, A.S.; Kalska, B.; Haggstrom, L.; Thomas, J.O. Lithium Extraction/Insertion in LiFePO4: An X-Ray Diffraction and Mössbauer Spectroscopy Study. *Solid State Ion.* **2000**, *130*, 41–52. [CrossRef]
4. Yamada, A.; Chung, S.C.; Hinokuma, K. Optimized LiFePO4 for Lithium Battery Cathodes. *J. Electrochem. Soc.* **2001**, *148*, 224–229. [CrossRef]
5. Huang, H.; Yin, S.; Nazar, L.F. Approaching Theoretical Capacity of LiFePO4 at Room Temperature at High Rates. *Electrochem. Solid State Lett.* **2001**, *4*, 170–172. [CrossRef]
6. Recham, N.; Chotard, J.-N.; Jumas, J.-C.; Laffont, L.; Armand, M.; Tarascon, J.-M. Ionothermal Synthesis of Li-Based Fluorophosphates Electrodes. *Chem. Mater.* **2010**, *22*, 1142–1148. [CrossRef]
7. Barpanda, P.; Ati, M.; Melot, B.C.; Rousse, G.; Chotard, J.-N.; Doublet, M.-L.; Sougrati, M.T.; Corr, S.A.; Jumas, J.-C.; Tarascon, J.-M. A 3.90 V Iron-Based Fluorosulphate Material for Lithium-Ion Batteries Crystallizing in the Triplite Structure. *Nat. Mater.* **2011**, *10*, 772–779. [CrossRef]
8. Lv, D.; Bai, J.; Zhang, P.; Wu, S.; Li, Y.; Wen, W.; Jiang, Z.; Mi, J.; Zhu, Z.; Yang, Y. Understanding the High Capacity of Li 2 FeSiO 4: In Situ XRD/XANES Study Combined with First-Principles Calculations. *Chem. Mater.* **2013**, *25*, 2014–2020. [CrossRef]
9. Siu, C.; Seymour, I.D.; Britto, S.; Zhang, H.; Rana, J.; Feng, J.; Omenya, F.O.; Zhou, H.; Chernova, N.A.; Zhou, G.; et al. Enabling Multi-Electron Reaction of ε-VOPO4 to Reach Theoretical Capacity for Lithium-Ion Batteries. *Chem. Commun.* **2018**, *No. 54*, 7802–7805. [CrossRef]
10. Shi, Y.; Zhou, H.; Seymour, I.D.; Britto, S.; Rana, J.; Wangoh, L.W.; Huang, Y.; Yin, Q.; Reeves, P.J.; Zuba, M.; et al. Electrochemical Performance of Nanosized Disordered LiVOPO4. *ACS Omega* **2018**, *3*, 7310–7323. [CrossRef] [PubMed]
11. Shi, Y.; Zhou, H.; Britto, S.; Seymour, I.D.; Wiaderek, K.M.; Omenya, F.; Chernova, N.A.; Chapman, K.W.; Grey, C.P.; Whittingham, M.S. A High-Performance Solid-State Synthesized LiVOPO4 for Lithium-Ion Batteries. *Electrochem. Commun.* **2019**, *105*, 106491. [CrossRef]
12. Chung, Y.; Cassidy, E.; Lee, K.; Siu, C.; Huang, Y.; Omenya, F.; Rana, J.; Wiaderek, K.M.; Chernova, N.A.; Chapman, K.W.; et al. Nonstoichiometry and Defects in Hydrothermally Synthesized ε-LiVOPO4. *ACS Appl. Energy Mater.* **2019**, *2*, 4792–4800. [CrossRef]
13. Rana, J.; Shi, Y.; Zuba, M.J.; Wiaderek, K.M.; Feng, J.; Zhou, H.; Ding, J.; Tianpin, W.; Cibin, G.; Balasubramanian, M.; et al. Role of Disorder in Limiting the True Multi-Electron Redox in ε-LiVOPO4. *J. Mater. Chem. A* **2018**, *42*, 1–10. [CrossRef]
14. Ateba Mba, J.; Masquelier, C.; Suard, E.; Croguennec, L. Synthesis and Crystallographic Study of Homeotypic LiVPO4F and LiVPO4O. *Chem. Mater.* **2012**, *24*, 1223–1234. [CrossRef]

15. Boudin, S.; Guesdon, A.; Leclaire, A.; Borel, M.M. Review on Vanadium Phosphates with Mono and Divalent Metallic Cations: Syntheses, Structural Relationships and Classification, Properties. *Int. J. Inorg. Mater.* **2000**, *2*, 561–579. [CrossRef]
16. Whittingham, M.S. Ultimate Limits to Intercalation Reactions for Lithium Batteries. *Chem. Rev.* **2014**, *114*, 11414–11443. [CrossRef] [PubMed]
17. Croguennec, L.; Palacin, M.R. Recent Achievements on Inorganic Electrode Materials for Lithium-Ion Batteries. *J. Am. Chem. Soc.* **2015**, *137*, 3140–3156. [CrossRef]
18. Masquelier, C.; Croguennec, L. Polyanionic (Phosphates, Silicates, Sulfates) Frameworks as Electrode Materials for Rechargeable Li (or Na) Batteries. *Chem. Rev.* **2013**, *113*, 6552–6591. [CrossRef] [PubMed]
19. Yabuuchi, N.; Kubota, K.; Dahbi, M.; Komaba, S. Research Development on Sodium-Ion Batteries. *Chem. Rev.* **2014**, *114*, 11636–11682. [CrossRef] [PubMed]
20. Shi, H.-Y.; Jia, Z.; Wu, W.; Zhang, X.; Liu, X.-X.; Sun, X. The Development of Vanadyl Phosphate Cathode Materials for Energy Storage Systems: A Review. *Chem. A Eur. J.* **2020**, *26*, 8190–8204. [CrossRef] [PubMed]
21. Beneš, L.; Melánová, K.; Svoboda, J.; Zima, V. Intercalation Chemistry of Layered Vanadyl Phosphate: A Review. *J. Incl. Phenom. Macrocycl. Chem.* **2012**, *73*, 33–53. [CrossRef]
22. Jian, Z.; Hu, Y.-S.; Ji, X.; Chen, W. NASICON-Structured Materials for Energy Storage. *Adv. Mater.* **2017**, *29*, 1601925. [CrossRef]
23. Anantharamulu, N.; Koteswara Rao, K.; Rambabu, G.; Vijaya Kumar, B.; Radha, V.; Vithal, M. A Wide-Ranging Review on Nasicon Type Materials. *J. Mater. Sci.* **2011**, *46*, 2821–2837. [CrossRef]
24. Manthiram, A.; Goodenough, J.B. Lithium Insertion into Fe2(SO4)3 Frameworks. *J. Power Sources* **1989**, *26*, 403–408. [CrossRef]
25. Huang, H.; Yin, S.C.; Kerr, T.; Taylor, N.; Nazar, L.F. Nanostructured Composites: A High Capacity, Fast Rate Li3V2(PO4)3/Carbon Cathode for Rechargeable Lithium Batteries. *Adv. Mater.* **2002**, *14*, 1525–1528. [CrossRef]
26. Patoux, S.; Wurm, C.; Morcrette, M.; Rousse, G.; Masquelier, C. A Comparative Structural and Electrochemical Study of Monoclinic Li3Fe2(PO4)3 and Li3V2(PO4)3. *J. Power Sources* **2003**, *119*, 278–284. [CrossRef]
27. Yin, S.-C.; Grondey, H.; Strobel, P.; Anne, M.; Nazar, L.F. Electrochemical Property: Structure Relationships in Monoclinic Li3-YV2(PO4)3. *J. Am. Chem. Soc.* **2003**, *125*, 10402–10411. [CrossRef] [PubMed]
28. Yin, S.C.; Strobel, P.S.; Grondey, H.; Nazar, L.F. Li2.5V2(PO4)3: A Room-Temperature Analogue to the Fast-Ion Conducting High-Temperature γ-Phase of Li3V2(PO4)3. *Chem. Mater.* **2004**, *16*, 1456–1465. [CrossRef]
29. Rui, X.; Yan, Q.; Skyllas-Kazacos, M.; Lim, T.M. Li3V2(PO4)3 Cathode Materials for Lithium-Ion Batteries: A Review. *J. Power Sources* **2014**, *258*, 19–38. [CrossRef]
30. Kang, J.; Mathew, V.; Gim, J.; Kim, S.; Song, J.; Im, W.B.; Han, J.; Lee, J.Y.; Kim, J. Pyro-Synthesis of a High Rate Nano-Li3V2(PO4)3/C Cathode with Mixed Morphology for Advanced Li-Ion Batteries. *Sci. Rep.* **2014**, *4*, 1–9. [CrossRef]
31. Saïdi, M.Y.; Barker, J.; Huang, H.; Swoyer, J.L.; Adamson, G. Performance Characteristics of Lithium Vanadium Phosphate as a Cathode Material for Lithium-Ion Batteries. *J. Power Sources* **2003**, *119*, 266–272. [CrossRef]
32. Kim, S.; Zhang, Z.; Wang, S.; Yang, L.; Cairns, E.J.; Penner-Hahn, J.E.; Deb, A. Electrochemical and Structural Investigation of the Mechanism of Irreversibility in Li3V2(PO4)3 Cathodes. *J. Phys. Chem. C* **2016**, *120*, 7005–7012. [CrossRef]
33. Rui, X.H.; Yesibolati, N.; Chen, C.H. Li3V2(PO4)3/C Composite as an Intercalation-Type Anode Material for Lithium-Ion Batteries. *J. Power Sources* **2011**, *196*, 2279–2282. [CrossRef]
34. Gaubicher, J.; Wurm, C.; Goward, G.; Masquelier, C.; Nazar, L. Rhombohedral Form of Li3V2(PO4)3 as a Cathode in Li-Ion Batteries. *Chem. Mater.* **2000**, *12*, 3240–3242. [CrossRef]
35. Delmas, C.; Olazcuaga, R.; Cherkaoui, F.; Brochu, R.; Leflem, G. New Family of Phosphates with Formula Na3M2(PO4)3 (M= Ti,V,Cr,Fe). *C. R. Seances Acad. Sci.* **1978**, *287*, 169–171.
36. Chotard, J.-N.; Rousse, G.; David, R.; Mentré, O.; Courty, M.; Masquelier, C. Discovery of a Sodium-Ordered Form of Na3V2(PO4)3 below Ambient Temperature. *Chem. Mater.* **2015**, *27*, 5982–5987. [CrossRef]
37. Jian, Z.; Yuan, C.; Han, W.; Lu, X.; Gu, L.; Xi, X.; Hu, Y.; Li, H.; Chen, W.; Chen, D.; et al. Atomic Structure and Kinetics of NASICON NaxV2(PO4)3 Cathode for Sodium-Ion Batteries. *Adv. Funct. Mater.* **2014**, *24*, 4265–4272. [CrossRef]
38. Golalakrishnan, J.; Kasthuri Rangan, K. NASICON-Type Vanadium Phosphate Synthesized by Oxidative Deintercalation of Sodium from Sodium Vanadium Phosphate (Na3V2(PO4)3). *Chem. Mater.* **1992**, *4*, 745–747. [CrossRef]
39. Lalère, F.; Seznec, V.; Courty, M.; David, R.; Chotard, J.N.; Masquelier, C. Improving the Energy Density of Na3V2(PO4)3-Based Positive Electrodes through V/Al Substitution. *J. Mater. Chem. A* **2015**, *3*, 16198–16205. [CrossRef]
40. Mason, C.W.; Gocheva, I.; Hoster, H.E.; Dennis, Y.W. Activating Vanadium's Highest Oxidation State in the NASICON Structure. *ECS Trans.* **2014**, *58*, 41–46. [CrossRef]
41. Aragón, M.J.; Lavela, P.; Ortiz, G.F.; Tirado, J.L. Benefits of Chromium Substitution in Na3V2(PO4)3 as a Potential Candidate for Sodium-Ion Batteries. *ChemElectroChem* **2015**, *2*, 995–1002. [CrossRef]
42. Liu, R.; Xu, G.; Li, Q.; Zheng, S.; Zheng, G.; Gong, Z.; Li, Y.; Kruskop, E.; Fu, R.; Chen, Z.; et al. Exploring Highly Reversible 1.5-Electron Reactions (V3+/V4+/V5+) in Na3VCr(PO4)3 Cathode for Sodium-Ion Batteries. *ACS Appl. Mater. Interfaces* **2017**, *9*, 43632–43639. [CrossRef]
43. Liu, R.; Zheng, S.; Yuan, Y.; Yu, P.; Liang, Z.; Zhao, W.; Shahbazian-Yassar, R.; Ding, J.; Lu, J.; Yang, Y. Counter-Intuitive Structural Instability Aroused by Transition Metal Migration in Polyanionic Sodium Ion Host. *Adv. Energy Mater.* **2020**, *2003256*, 1–9. [CrossRef]

44. Zhou, W.; Xue, L.; Gao, H.; Li, Y.; Xin, S.; Fu, G.; Cui, Z.; Zhu, Y.; Goodenough, J.B. MV(PO4)3 (M= Mn, Fe, Ni) Structure and Properties for Sodium Extraction. *Nano Lett.* **2016**, *16*, 7836–7841. [CrossRef]
45. Zakharkin, M.V.; Drozhzhin, O.A.; Tereshchenko, I.V.; Chernyshov, D.; Abakumov, A.M.; Antipov, E.V.; Stevenson, K.J. Enhancing Na+ Extraction Limit through High Voltage Activation of the NASICON-Type Na4MnV(PO4)3 Cathode. *ACS Appl. Energy Mater.* **2018**, *1*, 5842–5846. [CrossRef]
46. Zakharkin, M.V.; Drozhzhin, O.A.; Ryazantsev, S.V.; Chernyshov, D.; Kirsanova, M.A.; Mikheev, I.V.; Pazhetnov, E.M.; Antipov, E.V.; Stevenson, K.J. Electrochemical Properties and Evolution of the Phase Transformation Behavior in the NASICON-Type Na3+xMnxV2-x(PO4)3 ($0 \leq x \leq 1$) Cathodes for Na-Ion Batteries. *J. Power Sources* **2020**, *470*, 1–8. [CrossRef]
47. Chen, F.; Kovrugin, V.M.; David, R.; Mentré, O.; Fauth, F.; Chotard, J.N.; Masquelier, C. A NASICON-Type Positive Electrode for Na Batteries with High Energy Density: Na4MnV(PO4)3. *Small Methods* **2019**, *3*, 1–9. [CrossRef]
48. Le Meins, J.; Crosnier-Lopez, M.-P.; Hemon-Ribaud, A.; Courbion, G. Phase Transitions in the Na3M2(PO4)2F3 Family (M= Al3+,V3+,Cr3+,Fe3+,Ga3+): Synthesis, Thermal, Structural, and Magnetic Studies. *J. Solid State Chem.* **1999**, *148*, 260–277. [CrossRef]
49. Bianchini, M.; Brisset, N.; Fauth, F.; Weill, F.; Elkaim, E.; Suard, E.; Masquelier, C.; Croguennec, L. Na3V2(PO4)2F3 Revisited: A High-Resolution Diffraction Study. *Chem. Mater.* **2014**, *26*, 4238–4247. [CrossRef]
50. Shakoor, R.A.; Seo, D.-H.; Kim, H.; Park, Y.-U.; Kim, J.; Kim, S.-W.; Gwon, H.; Lee, S.; Kang, K. A Combined First Principles and Experimental Study on Na3V2(PO4)2F3 for Rechargeable Na Batteries. *J. Mater. Chem.* **2012**, *22*, 20535. [CrossRef]
51. Yan, G.; Mariyappan, S.; Rousse, G.; Jacquet, Q.; Deschamps, M.; David, R.; Mirvaux, B.; Freeland, J.W.; Tarascon, J.M. Higher Energy and Safer Sodium Ion Batteries via an Electrochemically Made Disordered Na3V2(PO4)2F3 Material. *Nat. Commun.* **2019**, *10*, 1–12. [CrossRef] [PubMed]
52. Bianchini, M.; Fauth, F.; Brisset, N.; Weill, F.; Suard, E.; Masquelier, C.; Croguennec, L. Comprehensive Investigation of the Na3V2(PO4)2F3–NaV2(PO4)2F3 System by Operando High Resolution Synchrotron X-ray Diffraction. *Chem. Mater.* **2015**, *27*, 3009–3020. [CrossRef]
53. Broux, T.; Bamine, T.; Simonelli, L.; Stievano, L.; Fauth, F.; Ménétrier, M.; Carlier, D.; Masquelier, C.; Croguennec, L. VIV Disproportionation Upon Sodium Extraction From Na3V2(PO4)2F3 Observed by Operando X-ray Absorption Spectroscopy and State NMR. *J. Phys. Chem. C* **2017**, *121*, 4103–4111. [CrossRef]
54. Broux, T.; Bamine, T.; Fauth, F.; Simonelli, L.; Olszewski, W.; Marini, C.; Ménétrier, M.; Carlier, D.; Masquelier, C.; Croguennec, L. Strong Impact of the Oxygen Content in Na3V2(PO4)2F3-YOy ($0 \leq y \leq 0.5$) on Its Structural and Electrochemical Properties. *Chem. Mater.* **2016**, *28*, 7683–7692. [CrossRef]
55. Park, Y.U.; Seo, D.H.; Kim, H.; Kim, J.; Lee, S.; Kim, B.; Kang, K. A Family of High-Performance Cathode Materials for Na-Ion Batteries, Na3(VO1-XPO4)2 F1+2x ($0 \leq x \leq 1$): Combined First-Principles and Experimental Study. *Adv. Funct. Mater.* **2014**, *24*, 4603–4614. [CrossRef]
56. Mariyappan, S.; Wang, Q.; Tarascon, J.M. Will Sodium Layered Oxides Ever Be Competitive for Sodium Ion Battery Applications? *J. Electrochem. Soc.* **2018**, *165*, 3714–3722. [CrossRef]
57. Park, Y.U.; Seo, D.H.; Kwon, H.S.; Kim, B.; Kim, J.; Kim, H.; Kim, I.; Yoo, H.I.; Kang, K. A New High-Energy Cathode for a Na-Ion Battery with Ultrahigh Stability. *J. Am. Chem. Soc.* **2013**, *135*, 13870–13878. [CrossRef]
58. Serras, P.; Alonso, J.; Sharma, N.; Miguel, J.; Kubiak, P.; Gubieda, M.-L.; Rojo, T. Electrochemical Na Extraction/Insertion of Na3V2O2x(PO4)2F3−2x. *Chem. Mater.* **2013**, *25*, 4917–4925. [CrossRef]
59. Morimoto, H.; Ito, D.; Ogata, Y.; Suzuki, T.; Sakamaki, K.; Tsuji, T.; Hirukawa, M.; Matsumoto, A.; Tobishima, S. Charge/Discharge Behavior of Triclinic LiTiOPO4 Anode Materials for Lithium Secondary Batteries. *Electrochem. Soc. Jpn.* **2016**, *84*, 878–881. [CrossRef]
60. Barker, J.; Saidi, M.Y.; Swoyer, J.L. Electrochemical Insertion Properties of the Novel Lithium Vanadium Fluorophosphate, LiVPO4F. *J. Electrochem. Soc.* **2003**, *150*, 1394–1398. [CrossRef]
61. Badraoui, A.E.; Pivan, J.-Y.; Maunaye, M.; Pena, O.; Louer, M.; Louer, D. Order-Disorder Phenomena in Vanadium Phosphates. Structures and Properties of Tetragonal and Monoclinic VPO4·H2O. *Ann. Chim. Sci. Matériaux* **1998**, *23*, 97–101. [CrossRef]
62. Lii, K.H.; Li, H.; Cheng, C.; Wang, S. Synthesis and Structural Characterization of Sodium Vanadyl (IV) Orthophosphate NaVOPO4. *Z. Krist. Cryst. Mater.* **2010**, *197*, 67–73. [CrossRef]
63. Bianchini, M.; Ateba-Mba, J.M.; Dagault, P.; Bogdan, E.; Carlier, D.; Suard, E.; Masquelier, C.; Croguennec, L. Multiple Phases in the ε-VPO4O–LiVPO4O–Li2VPO4O System: A Combined Solid State Electrochemistry and Diffraction Structural Study. *J. Mater. Chem. A* **2014**, *2*, 10182–10192. [CrossRef]
64. Song, Y.; Zavalij, P.Y.; Whittingham, M.S. ε-VOPO4: Electrochemical Synthesis and Enhanced Cathode Behavior. *J. Electrochem. Soc.* **2005**, *152*, 721–728. [CrossRef]
65. Boivin, E.; Chotard, J.-N.; Ménétrier, M.; Bourgeois, L.; Bamine, T.; Carlier, D.; Fauth, F.; Suard, E.; Masquelier, C.; Croguennec, L. Structural and Electrochemical Studies of a New Tavorite Composition LiVPO4OH. *J. Mater. Chem. A* **2016**, *4*, 11030–11045. [CrossRef]
66. Lee Harrison, K.; Bridges, C.; Segre, C.; Varnado, D.; Applestone, D.; Bielawski, C.; Paranthaman, M.; Manthiram, A. Chemical and Electrochemical Lithiation of LiVOPO4 Cathodes for Lithium-Ion Batteries. *Chem. Mater.* **2014**, *26*, 3849–3861. [CrossRef]

67. Lin, Y.; Wen, B.; Wiaderek, K.; Sallis, S.; Liu, H.; Lapidus, S.; Borkiewicz, O.; Quackenbush, N.; Chernova, N.; Karki, K.; et al. Thermodynamics, Kinetics and Structural Evolution of ε -LiVOPO4 over Multiple Lithium Intercalation. *Chem. Mater.* **2016**, *28*, 1794–1805. [CrossRef]
68. Barker, J.; Saidi, M.Y.; Swoyer, J.L. A Comparative Investigation of the Li Insertion Properties of the Novel Fluorophosphate Phases, NaVPO4F and LiVPO4F. *J. Electrochem. Soc.* **2004**, *151*, 1670–1677. [CrossRef]
69. Barker, J.; Gover, R.K.B.; Burns, P.; Bryan, A.; Saidi, M.Y.; Swoyer, J.L. Structural and Electrochemical Properties of Lithium Vanadium Fluorophosphate, LiVPO4F. *J. Power Sources* **2005**, *146*, 516–520. [CrossRef]
70. Ellis, B.L.; Ramesh, T.N.; Davis, L.J.M.; Goward, G.R.; Nazar, L.F. Structure and Electrochemistry of Two-Electron Redox Couples in Lithium Metal Fluorophosphates Based on the Tavorite Structure. *Chem. Mater.* **2011**, *23*, 5138–5148. [CrossRef]
71. Boivin, E. Crystal Chemistry of Vanadium Phosphates as Positive Electrode Materials for Li-Ion and Na-Ion Batteries. Ph.D. Thesis, University of Picardie Jules Verne, Amiens, France, 2017.
72. Ateba Mba, J.-M.; Croguennec, L.; Basir, N.I.; Barker, J.; Masquelier, C. Lithium Insertion or Extraction from/into Tavorite-Type LiVPO4F: An In Situ X-Ray Diffraction Study. *J. Electrochem. Soc.* **2012**, *159*, 1171–1175. [CrossRef]
73. Piao, Y.; Qin, Y.; Ren, Y.; Heald, S.M.; Sun, C.; Zhou, D.; Polzin, B.J.; Trask, S.E.; Amine, K.; Wei, Y.; et al. A XANES Study of LiVPO4F: A Factor Analysis Approach. *Phys. Chem. Chem. Phys.* **2014**, *16*, 3254–3260. [CrossRef] [PubMed]
74. Zhong, S.; Chen, W.; Li, Y.; Zou, Z.; Liu, C. Synthesis of LiVPO4F with High Electrochemical Performance by Sol-Gel Route. *Trans. Nonferrous Met. Soc. China* **2010**, *20*, 275–278. [CrossRef]
75. Rangaswamy, P.; Shetty, V.R.; Suresh, G.S.; Mahadevan, K.M.; Nagaraju, D.H. Enhanced Electrochemical Performance of LiVPO4F/f-Graphene Composite Electrode Prepared via Ionothermal Process. *J. Appl. Electrochem.* **2017**, *47*. [CrossRef]
76. Messinger, R.J.; Ménétrier, M.; Salager, E.; Boulineau, A.; Duttine, M.; Carlier, D.; Ateba Mba, J.-M.; Croguennec, L.; Masquelier, C.; Massiot, D.; et al. Revealing Defects in Crystalline Lithium-Ion Battery Electrodes by Solid-State NMR: Applications to LiVPO4F. *Chem. Mater.* **2015**, *27*, 5212–5221. [CrossRef]
77. Bamine, T.; Boivin, E.; Boucher, F.; Messinger, R.J.; Salager, E.; Deschamps, M.; Masquelier, C.; Croguennec, L.; Ménétrier, M.; Carlier, D. Understanding Local Defects in Li-Ion Battery Electrodes through Combined DFT/NMR Studies: Application to LiVPO4F. *J. Phys. Chem. C* **2017**, *121*, 3219–3227. [CrossRef]
78. Kim, M.; Lee, S.; Kang, B. Fast-Rate Capable Electrode Material with Higher Energy Density than LiFePO4: 4.2 V LiVPO4F Synthesized by Scalable Single-Step Solid-State Reaction. *Adv. Sci.* **2015**, *3*, 1500366. [CrossRef]
79. Kim, M.; Lee, S.; Kang, B. High Energy Density Polyanion Electrode Material: LiVPO4O1-XFx (x ≈ 0.25) with Tavorite Structure. *Chem. Mater.* **2017**, *29*, 4690–4699. [CrossRef]
80. Boivin, E.; David, R.; Chotard, J.N.; Bamine, T.; Iadecola, A.; Bourgeois, L.; Suard, E.; Fauth, F.; Carlier, D.; Masquelier, C.; et al. LiVPO4F1-YOy Tavorite-Type Compositions: Influence of the Vanadyl-Type Defects' Concentration on the Structure and Electrochemical Performance. *Chem. Mater.* **2018**, *30*, 5682–5693. [CrossRef]
81. Boivin, E.; Iadecola, A.; Fauth, F.; Chotard, J.N.; Masquelier, C.; Croguennec, L. Redox Paradox of Vanadium in Tavorite LiVPO4F1-YOy. *Chem. Mater.* **2019**, *31*, 7367–7376. [CrossRef]
82. Wurm, C.; Morcrette, M.; Rousse, G.; Dupont, L.; Masquelier, C. Lithium Insertion/Extraction into/from LiMX2O7 Compositions (M = Fe, V; X = P, As) Prepared via a Solution Method. *Chem. Mater.* **2002**, *14*, 2701–2710. [CrossRef]
83. Barker, J.; Gover, R.K.B.; Burns, P.; Bryan, A. LiVP2O7: A Viable Lithium-Ion Cathode Material? *Electrochem. Solid State Lett.* **2005**, *8*, 446–448. [CrossRef]
84. Deng, C.; Zhang, S. 1D Nanostructured Na7V4(P2O7)4(PO4) as High-Potential and Superior-Performance Cathode Material for Sodium-Ion Batteries. *Appl. Mater. Interfaces* **2014**, *6*, 9111–9117. [CrossRef] [PubMed]
85. Kovrugin, V.; Chotard, J.-N.; Fauth, F.; Jamali, A.; David, R.; Christian, M. Structural and Electrochemical Studies of Novel Batteries. *J. Mater. Chem. A* **2017**, *5*, 14365–14376. [CrossRef]
86. Kim, J.; Yoon, G.; Kim, H.; Park, Y.U.; Kang, K. Na3V(PO4)2: A New Layered-Type Cathode Material with High Water Stability and Power Capability for Na-Ion Batteries. *Chem. Mater.* **2018**, *30*, 3683–3689. [CrossRef]
87. Liu, R.; Liang, Z.; Xiang, Y.; Zhao, W.; Liu, H.; Chen, Y. Recognition of V3+/V4+/V5+ Multielectron Reactions in Na3V(PO4)2: A Potential High Energy Density Cathode for Sodium-Ion Batteries. *Molecules* **2020**, *25*, 1000. [CrossRef]
88. Sandineni, P.; Madria, P.; Ghosh, K.; Choudhury, A. A Square Channel Vanadium Phosphite Framework as High Voltage Cathode for Li- and Na- Ion Batteries. *Mater. Adv.* **2020**. [CrossRef]
89. Kuang, Q.; Xu, J.; Zhao, Y.; Chen, X.; Chen, L. Layered Monodiphosphate Li9V3(P2O7)3(PO4)2: A Novel Cathode Material for Lithium-Ion Batteries. *Electrochim. Acta* **2011**, *56*, 2201–2205. [CrossRef]
90. Deng, C.; Zhang, S.; Zhao, B. First Exploration of Ultra Fine Na7V3(P2O7)4 as a High-Potential Cathode Material for Sodium-Ion Battery. *Energy Storage Mater.* **2016**, *4*, 71–78. [CrossRef]
91. Reynaud, M.; Wizner, A.; Katcho, N.A.; Loaiza, L.C.; Galceran, M.; Carrasco, J.; Rojo, T.; Armand, M.; Casas-Cabanas, M. Sodium Vanadium Nitridophosphate Na3V(PO3)3N as a High-Voltage Positive Electrode Material for Na-Ion and Li-Ion Batteries. *Electrochem. Commun.* **2017**, *84*, 14–18. [CrossRef]
92. Boivin, E.; Chotard, J.-N.; Bamine, T.; Carlier, D.; Serras, P.; Veronica, P.; Rojo, T.; Iadecola, A.; Dupont, L.; Bourgeois, L.; et al. Vanadyl-Type Defects in Tavorite-like NaVPO4F: From the Average Long Range Structure to Local Environments. *J. Mater. Chem. A* **2017**, *5*, 25044–25055. [CrossRef]

93. Fedotov, S.S.; Khasanova, N.R.; Samarin, A.S.; Drozhzhin, O.A.; Batuk, D.; Karakulina, O.M.; Hadermann, J.; Abakumov, A.M.; Antipov, E.V. AVPO4F (A = Li, K): A 4 V Cathode Material for High-Power Rechargeable Batteries. *Chem. Mater.* **2016**, *28*, 411–415. [CrossRef]
94. Zhang, B.; Dugas, R.; Rousse, G.; Rozier, P.; Abakumov, A.M.; Tarascon, J.M. Insertion Compounds and Composites Made by Ball Milling for Advanced Sodium-Ion Batteries. *Nat. Commun.* **2016**, *7*, 1–9. [CrossRef] [PubMed]
95. Makimura, Y.; Cahill, L.S.; Iriyama, Y.; Goward, G.R.; Nazar, L.F. Layered Lithium Vanadium Fluorophosphate, Li 5 V(PO 4) 2 F 2: A 4 V Class Positive Electrode Material for Lithium-Ion Batteries. *Chem. Mater.* **2008**, *20*, 4240–4248. [CrossRef]
96. Liang, Z.; Zhang, X.; Liu, R.; Ortiz, G.F.; Zhong, G.; Xiang, Y.; Chen, S.; Mi, J.; Wu, S.; Yang, Y. New Dimorphs of Na5V(PO4)2F2 as an Ultrastable Cathode Material for Sodium-Ion Batteries. *ACS Appl. Energy Mater.* **2020**, *3*, 1181–1189. [CrossRef]
97. Barker, J.; Saidi, M.Y.; Swoyer, J.L. Electrochemical Properties of β-LiVOPO4 Prepared by Carbothermal Reduction. *J. Electrochem. Soc.* **2004**, *151*, 796–800. [CrossRef]
98. He, G.; Kan, W.; Manthiram, A. β-NaVOPO4 Obtained by a Low-Temperature Synthesis Process: A New 3.3 V Cathode for Sodium-Ion Batteries. *Chem. Mater.* **2016**, *28*, 1503–1512. [CrossRef]
99. He, G.; Bridges, C.A.; Manthiram, A. Crystal Chemistry of Electrochemically and Chemically Lithiated Layered α-LiVOPO4. *Chem. Mater.* **2015**, *27*, 6699–6707. [CrossRef]
100. Satya Kishore, M.; Pralong, V.; Caignaert, V.; Malo, S.; Hebert, S.; Varadaraju, U.V.; Raveau, B. Topotactic Insertion of Lithium in the Layered Structure Li4VO(PO4)2: The Tunnel Structure Li5VO(PO4)2. *J. Solid State Chem.* **2008**, *181*, 976–982. [CrossRef]
101. Kim, J.; Kim, H.; Lee, S. High Power Cathode Material Na4VO(PO4)2 with Open Framework for Na Ion Batteries. *Chem. Mater.* **2017**, *29*, 3363–3366. [CrossRef]
102. Kishore, M.S.; Pralong, V.; Caignaert, V.; Varadaraju, U.V.; Raveau, B. A New Lithium Vanadyl Diphosphate Li2VOP2O7: Synthesis and Electrochemical Study. *Solid State Sci.* **2008**, *10*, 1285–1291. [CrossRef]
103. Manthiram, A.; Goodenough, J.B. Lithium Insertion into Fe2(MO4)3 Frameworks: Comparison of M = W with M = MO. *J. Solid State Chem.* **1987**, *71*, 349–360. [CrossRef]
104. Ben Yahia, M.; Lemoigno, F.; Rousse, G.; Boucher, F.; Tarascon, J.-M.; Doublet, M.-L. Origin of the 3.6 V to 3.9 V Voltage Increase in the LiFeSO4F Cathodes for Li-Ion Batteries. *Energy Environ. Sci.* **2012**, *5*, 9584. [CrossRef]
105. Parapari, S.S.; Ateba Mba, J.M.; Tchernychova, E.; Mali, G.; Arčon, I.; Kapun, G.; Gülgün, M.A.; Dominko, R. Effects of a Mixed O/F Ligand in the Tavorite-Type LiVPO4O Structure. *Chem. Mater.* **2020**, *32*, 262–272. [CrossRef]
106. Boivin, E.; Chotard, J.N.; Ménétrier, M.; Bourgeois, L.; Bamine, T.; Carlier, D.; Fauth, F.; Masquelier, C.; Croguennec, L. Oxidation under Air of Tavorite LiVPO4F: Influence of Vanadyl-Type Defects on Its Electrochemical Properties. *J. Phys. Chem. C* **2016**, *120*, 26187–26198. [CrossRef]
107. Nguyen, L.H.B.; Iadecola, A.; Belin, S.; Olchowka, J.; Masquelier, C.; Carlier, D.; Croguennec, L. A Combined Operando Synchrotron X-ray Absorption Spectroscopy and First-Principles Density Functional Theory Study to Unravel the Vanadium Redox Paradox in the Na3V2(PO4)2F3−Na3V2(PO4)2FO2 Compositions. *J. Phys. Chem. C* **2020**, *124*, 23511–23522. [CrossRef]
108. Bamine, T.; Boivin, E.; Masquelier, C.; Croguennec, L.; Salager, E.; Carlier, D. Local Atomic and Electronic Structure in the LiVPO4(F,O) Tavorite-Type Materials from Solid State NMR Combined with DFT Calculations. *Magn. Reson. Chem.* **2020**, *58*, 1109–1117. [CrossRef]
109. Ni, S.; Liu, J.; Chao, D.; Mai, L. Vanadate-Based Materials for Li-Ion Batteries: The Search for Anodes for Practical Applications. *Adv. Energy Mater.* **2019**, *9*, 1803324. [CrossRef]

Article

Unusual Spin Exchanges Mediated by the Molecular Anion $P_2S_6^{4-}$: Theoretical Analyses of the Magnetic Ground States, Magnetic Anisotropy and Spin Exchanges of MPS_3 (M = Mn, Fe, Co, Ni)

Hyun-Joo Koo [1,*], Reinhard Kremer [2] and Myung-Hwan Whangbo [1,3,*]

[1] Department of Chemistry and Research Institute for Basic Sciences, Kyung Hee University, Seoul 02447, Korea
[2] Max Planck Institute for Solid State Research, Heisenbergstrasse 1, D-70569 Stuttgart, Germany; rekre@mpg.fkf.de
[3] Department of Chemistry, North Carolina State University, Raleigh, NC 27695-8204, USA
* Correspondence: hjkoo@khu.ac.kr (H.-J.K.); mike_whangbo@ncsu.edu (M.-H.W.)

Citation: Koo, H.-J.; Kremer, R.; Whangbo, M.-H. Unusual Spin Exchanges Mediated by the Molecular Anion $P_2S_6^{4-}$: Theoretical Analyses of the Magnetic Ground States, Magnetic Anisotropy and Spin Exchanges of MPS_3 (M = Mn, Fe, Co, Ni). *Molecules* **2021**, *26*, 1410. https://doi.org/10.3390/molecules26051410

Academic Editor: Takashiro Akitsu

Received: 27 January 2021
Accepted: 26 February 2021
Published: 5 March 2021

Publisher's Note: MDPI stays neutral with regard to jurisdictional claims in published maps and institutional affiliations.

Copyright: © 2021 by the authors. Licensee MDPI, Basel, Switzerland. This article is an open access article distributed under the terms and conditions of the Creative Commons Attribution (CC BY) license (https://creativecommons.org/licenses/by/4.0/).

Abstract: We examined the magnetic ground states, the preferred spin orientations and the spin exchanges of four layered phases MPS_3 (M = Mn, Fe, Co, Ni) by first principles density functional theory plus onsite repulsion (DFT + U) calculations. The magnetic ground states predicted for MPS_3 by DFT + U calculations using their optimized crystal structures are in agreement with experiment for M = Mn, Co and Ni, but not for $FePS_3$. DFT + U calculations including spin-orbit coupling correctly predict the observed spin orientations for $FePS_3$, $CoPS_3$ and $NiPS_3$, but not for $MnPS_3$. Further analyses suggest that the ||z spin direction observed for the Mn^{2+} ions of $MnPS_3$ is caused by the magnetic dipole–dipole interaction in its magnetic ground state. Noting that the spin exchanges are determined by the ligand p-orbital tails of magnetic orbitals, we formulated qualitative rules governing spin exchanges as the guidelines for discussing and estimating the spin exchanges of magnetic solids. Use of these rules allowed us to recognize several unusual exchanges of MPS_3, which are mediated by the symmetry-adapted group orbitals of $P_2S_6^{4-}$ and exhibit unusual features unknown from other types of spin exchanges.

Keywords: magnetic ground state; spin exchange; magnetic anisotropy; molecular anion; MPS_3; magnetic orbitals; qualitative rules

1. Introduction

In an extended solid, transition-metal magnetic cations M are surrounded by main-group ligands L to form ML_n (typically, n = 3–6) polyhedra, and the unpaired spins of M are accommodated in the singly occupied d-states (i.e., the magnetic orbitals) of ML_n. Each d-state has the metal d-orbital combined out-of-phase with the p-orbitals of the surrounding ligands L. The tendency for two adjacent magnetic ions to have a ferromagnetic (FM) or an antiferromagnetic (AFM) spin alignment is determined by the spin exchange between them, which takes place through the M-L-M or M-L . . . L-M exchange path [1–4]. Whereas the characteristics (e.g., the angular and distance dependence) of the M-L-M exchanges is conceptually well understood [5–8], the properties of the M-L . . . L-M exchanges involving several main-group ligands have only come into focus in the last two decades [1–4]. Furthermore, the character of a M-L . . . L-M exchange can be modified if the L . . . L contact is bridged by a d^0 metal cation A to form a L . . . A . . . L bridge [1–4]. What has not been well understood so far is the M-L . . . L-M exchange in which the L . . . L contact is an integral part of the covalent framework of a molecular anion made up of main group elements (e.g., the $P_2S_6^{4-}$ anion in MPS_3, where M = Mn, Fe, Co, Ni), which might be termed the M-(L-L)-M exchange to emphasize its difference from the M-L-M, M-L . . . L-M and M-L . . . A . . . L-M exchanges.

In the present work we examine the M-(L-L)-M spin exchanges in the layered phases MPS$_3$ (M = Mn [9–11], Fe [9–11], Co [10,11], Ni [10,11]), which crystallize with a monoclinic structure (space group $C2/m$, no. 12). Each layer of MPS$_3$ is made up of the molecular anions P$_2$S$_6^{4-}$ possessing the structure of staggered ethane (i.e., a trigonal antiprism structure) (Figure 1a,b). The molecular anions P$_2$S$_6^{4-}$ form a trigonal layer (Figure 1c) with the P-P bonds perpendicular to the layer, and a high-spin M^{2+} cation occupies every S$_6$ octahedral site (deviations from a trigonal symmetry caused by the monoclinic distortions are less than 1°). Thus, each MPS$_3$ layer consists of a honeycomb arrangement of M^{2+} cations. With the c^*-direction of the MPS$_3$ taken as the z-direction, the P-P bond of each P$_2$S$_6^{4-}$ is parallel to the z-direction (| | z), and each MS$_6$ octahedron is arranged with one of its three-fold rotational axes along the | | z-direction.

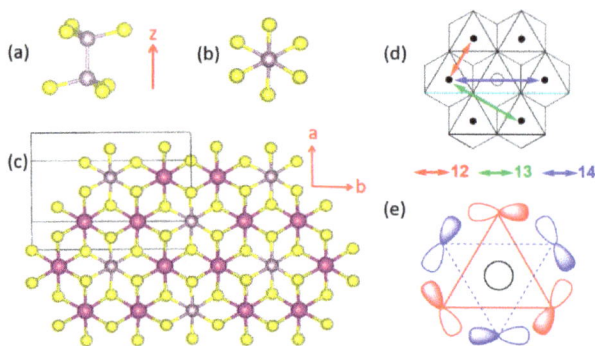

Figure 1. (a) Perspective and (b) projection views of a P$_2$S$_6^{4-}$ anion. (c) A projection view of a single MPS$_3$ layer along the c^*-direction (i.e., the z-direction), which is perpendicular to the MPS$_3$ layer. (d) Three kinds of the spin exchange paths in the MPS$_3$ honeycomb layers of MPS$_3$, where the labels 12, 13 and 14 refer to J$_{12}$, J$_{13}$ and J$_{14}$, respectively. (e) A group orbital of P$_2$S$_6^{4-}$ viewed along the P-P axis. The red triangle represents the three S atoms of the upper PS$_3$ pyramid, and the blue triangles those of the lower PS$_3$ pyramid.

To a first approximation, it may be assumed that each MPS$_3$ layer has a trigonal symmetry (see below for further discussion), so there are three types of spin exchanges to consider, i.e., the first nearest-neighbor (NN) spin exchange J$_{12}$, the second NN spin exchange J$_{13}$, and the third NN exchange J$_{14}$ (Figure 1d). J$_{12}$ is a spin exchange of the M-L-M type, in which the two metal ions share a common ligand, while J$_{13}$ and J$_{14}$ are nominally spin exchanges of the M-L ... L-M type, in which the two metal ions do not share a common ligand. In describing the magnetic properties of MPS$_3$ in terms of the spin exchanges J$_{12}$, J$_{13}$ and J$_{14}$, an interesting conceptual problem arises. Each P$_2$S$_6^{4-}$ anion is coordinated to the six surrounding M^{2+} cations simultaneously (Figure 1c,d), so one P$_2$S$_6^{4-}$ anion participates in all three different types of spin exchanges simultaneously with the surrounding six M^{2+} ions. Furthermore, the lone-pair orbitals of the S atoms of P$_2$S$_6^{4-}$, responsible for the coordination with M^{2+} ions, form symmetry-adapted group orbitals, in which all six S atoms participate (for example, see Figure 1e). Consequently, there is no qualitative argument with which to even guess the possible differences in J$_{12}$, J$_{13}$, and J$_{14}$. Over the past two decades, it became almost routine to quantitatively determine any spin exchanges of a magnetic solid by performing an energy-mapping analysis based on first principles DFT calculations. From a conceptual point of view, it would be very useful to have qualitative rules with which to judge whether the spin exchange paths involving complex intermediates are usual or unusual.

A number of experimental studies examined the magnetic properties of MPS$_3$ (M = Mn [9,11–14], Fe [9,11,15–18], Co [11,19], Ni [11,20]). The magnetic properties of MPS$_3$ (M = Mn, Fe, Co, Ni) monolayers were examined by DFT calculations to find their potential use as single-layer materials possessing magnetic order [21]. The present work is focused on the magnetic properties of bulk MPS$_3$. For the ordered AFM states of MPS$_3$, the neutron diffraction studies reported that the layers of MnPS$_3$ exhibits a honeycomb-type AFM spin arrangement, AF1 (Figure 2a), but those of FePS$_3$, CoPS$_3$ and NiPS$_3$ a zigzag-chain spin array, AF2 (Figure 2b), in which the FM chains running along the a-direction are antiferromagnetically coupled (hereafter, the ||a-chain arrangement). An alternative AFM arrangement, AF3 (Figure 2c), in which the FM zigzag chains running along the (a + b)-direction are antiferromagnetically coupled (hereafter, the ||(a + b)-chain arrangement), is quite similar in nature to the ||a-chain arrangement.

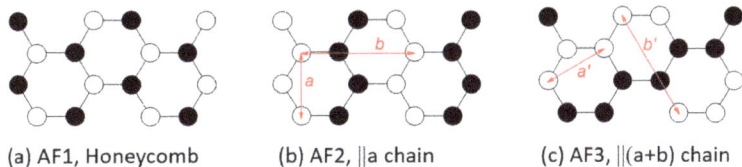

(a) AF1, Honeycomb (b) AF2, ||a chain (c) AF3, ||(a+b) chain

Figure 2. (a) The honeycomb AFM state, AF1. (b) The ||a-chain AFM state, AF2. (c) The ||(a + b)-chain AFM state, AF3.

At present, it is unclear why the spin arrangement of MnPS$_3$ differs from those of FePS$_3$, CoPS$_3$ and NiPS$_3$ and why FePS$_3$, CoPS$_3$ and NiPS$_3$ all adopt the ||a-chain arrangement rather than the ||(a + b)-chain arrangement. To explore these questions, it is necessary to examine the relative stabilities of a number of possible ordered spin arrangements of MPS$_3$ (M = Mn, Fe, Co, Ni) by electronic structure calculations and analyze the spin exchanges of their spin lattices.

Other quantities of importance for the magnetic ions M of an extended solid are the preferred orientations of their magnetic moments with respect to the local coordinates of the ML$_n$ polyhedra. These quantities, i.e., the magnetic anisotropy energies, are also readily determined by DFT calculations including spin orbit coupling (SOC). For the purpose of interpreting the results of these calculations, the selection rules for the preferred spin orientation of ML$_n$ were formulated [2,3,22–24] based on the SOC-induced interactions between the highest-occupied molecular orbital (HOMO) and lowest-unoccupied molecular orbital (LUMO) of ML$_n$. With the local z-axis of ML$_n$ taken along its n-fold rotational axis (n = 3, 4), the quantity needed for the selection rules is the minimum difference, $|\Delta L_z|$, in the magnetic quantum numbers L_z of the d-states describing the angular behaviors of the HOMO and LUMO. It is of interest to analyze the preferred spin orientations of the M^{2+} ions in MPS$_3$ (M = Mn, Fe, Co, Ni) from the viewpoint of the selection rules.

Our work is organized as follows: Section 2 describes simple qualitative rules governing spin exchanges. The details of our DFT calculations are presented in Section 3.1. The magnetic ground states of MPS$_3$ (M = Mn, Fe, Co, Ni) are discussed in Section 3.2, the preferred spin orientations of M^{2+} ions of MPS$_3$ in Section 3.3, and the quantitative values of the spin exchanges determined for MPS$_3$ in Section 3.4. We analyze the unusual features of the calculated spin exchanges via the P$_2$S$_6^{4-}$ anion in Section 3.5, and investigate in Section 3.6 the consequences of the simplifying assumption that the honeycomb spin lattice has a trigonal symmetry rather than a slight monoclinic distortion found experimentally. Our concluding remarks are summarized in Section 4.

2. Qualitative Rules Governing Spin Exchanges

2.1. Spin Exchange between Magnetic Orbitals

For clarity, we use the notation (φ_i, φ_j) to represent the spin exchange arising from the magnetic orbitals φ_i and φ_j at the magnetic ion sites A and B, respectively. It is well known that (φ_i, φ_j) consists of two competing terms [1–4,25]

$$(\varphi_i, \varphi_j) = J_F + J_{AF} \qquad (1)$$

The FM component J_F (>0) is proportional to the exchange repulsion,

$$J_F \propto K_{ij} \quad (2)$$

which increases with increasing the overlap electron density $\rho_{ij} = \varphi_i \varphi_j$. In case when the magnetic orbitals φ_i and φ_j are degenerate (e.g., between the t_{2g} states or between e_g states of the magnetic ions at octahedral sites), the AFM component J_{AF} (<0) is proportional to the square of the energy split Δe_{ij} between φ_i and φ_j induced by the interaction between them,

$$J_{AF} \propto -(\Delta e_{ij})^2 \propto -(S_{ij})^2 \quad (3)$$

The energy split Δe_{ij} is proportional to the overlap integral $S_{ij} = \langle \varphi_i | \varphi_j \rangle$, so that the magnitude of the AFM component J_{AF} increases with increasing that of $(S_{ij})^2$. If φ_i and φ_j are not degenerate (e.g., between the t_{2g} and e_g states of the magnetic ions), the magnitude of J_{AF} is approximately proportional to $-(S_{ij})^2$.

2.2. p-Orbital Tails of Magnetic Orbitals

The spin exchanges between adjacent transition-metal cations M are determined by the interactions between their magnetic orbitals, which in turn are governed largely by the overlap and the overlap electron density that are generated by the p-orbitals of the ligands present in the magnetic orbitals (the p-orbital tails, for short) [1–4]. Suppose that the metal ions M are surrounded by main-group ligands L to form ML_6 octahedra. In the t_{2g} and e_g states of an ML_6 octahedron (Figure 3a,b), the d-orbitals of M make σ and π antibonding combinations with the p-orbitals of the ligands L. Thus, the p-orbital tails of the t_{2g} and e_g states are represented as in Figure 4a,b, respectively, so that each M-L bond has the p_π and p_σ tails in the t_{2g} and e_g states, respectively, as depicted in Figure 4c. The triple-degeneracy of the t_{2g} and the double-degeneracy of the e_g states are lifted in a ML_5 square pyramid and a ML_4 square plane, both of which have a four-fold rotational symmetry; the t_{2g} states (xz, yz, xy) are split into (xz, yz) and xy, and the e_g states ($3z^2 - r^2$, $x^2 - y^2$) into $3z^2 - r^2$ and $x^2 - y^2$. Nevertheless, the description of the ligand p-orbital tails of the d-states depicted in Figure 4c remains valid.

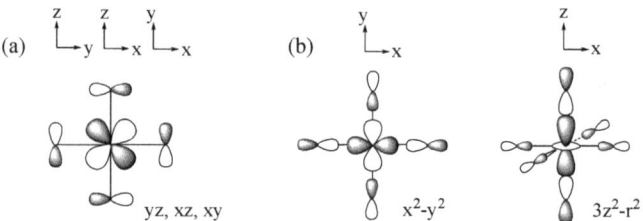

Figure 3. (a) The t_{2g} states and (b) the e_g states of a ML_6 octahedron.

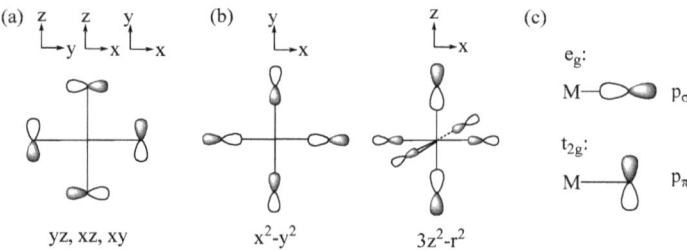

Figure 4. The p-orbital tails of (a) the t_{2g} and (b) the e_g states of a ML_6 octahedron. (c) The p_σ and p_π orbitals of the ligand p-orbital tails.

2.3. Spin Exchanges in Terms of the p-Orbital Tails

In this section, we generalize the qualitative rules of spin exchanges formulated for the magnetic solids of Cu^{2+} ions [4]. Each Cu^{2+} ion has only one magnetic orbital, i.e., the x^2-y^2 state in which each Cu-L bond has a p_σ tail. The d-electron configuration of the magnetic ion is $(t_{2g}\uparrow)^3(e_g\uparrow)^2(t_{2g}\downarrow)^0(e_g\downarrow)^0$ in MnPS$_3$, $(t_{2g}\uparrow)^3(e_g\uparrow)^2(t_{2g}\downarrow)^1(e_g\downarrow)^0$ in FePS$_3$, $(t_{2g}\uparrow)^3(e_g\uparrow)^2(t_{2g}\downarrow)^2(e_g\downarrow)^0$ in CoPS$_3$, and $(t_{2g}\uparrow)^3(e_g\uparrow)^2(t_{2g}\downarrow)^3(e_g\downarrow)^0$ in NiPS$_3$. Thus, the Mn^{2+}, Fe^{2+}, Co^{2+}, and Ni^{2+} ions possess 5, 4, 3, and 2 magnetic orbitals, respectively. For magnetic ions with several magnetic orbitals, the spin exchange J_{AB} between two such ions located at sites A and B is given by the sum of all possible individual exchanges (φ_i, φ_j),

$$J_{AB} = \frac{2}{n_A n_B} \sum_{i \in A} \sum_{j \in B} (\varphi_i, \varphi_j) \propto \sum_{i \in A} \sum_{j \in B} (\varphi_i, \varphi_j) \quad (4)$$

where n_A and n_B are the number of magnetic orbitals at the sites A and B, respectively. Each individual exchange (φ_i, φ_j) can be FM or AFM depending on which term, J_F or J_{AF}, dominates. Whether J_{AB} is FM or AFM depends on the sum of all individual (φ_i, φ_j) contributions.

2.3.1. M-L-M Exchange

As shown in Figure 5, there occur three types of M-L-M exchanges between the magnetic orbitals of t_{2g} and e_g states.

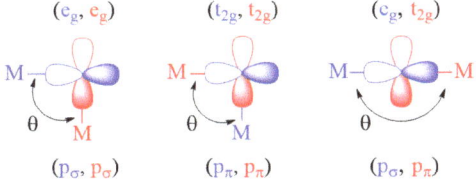

Figure 5. Three-types of M-L-M spin exchanges between t_{2g} and e_g magnetic orbitals.

If the M-L-M bond angle θ is 90° for the (e_g, e_g) and (t_{2g}, t_{2g}) exchanges, and also when θ is 180° for the (e_g, t_{2g}) exchange, the two p-orbital tails have an orthogonal arrangement so that $\langle \varphi_i | \varphi_j \rangle = 0$ (i.e., $J_{AF} = 0$). However, the overlap electron density $\varphi_i \varphi_j$ is nonzero (i.e., $J_F \neq 0$), hence predicting these spin exchanges to be FM. When the θ angles of the (e_g, e_g) and (t_{2g}, t_{2g}) exchanges increase from 90° toward 180°, and also when the angle θ of the (e_g, t_{2g}) exchange decreases from 180° toward 90°, both J_{AF} and J_F are nonzero so that the balance between the two determines if the overall exchange (φ_i, φ_j) becomes FM or AFM. These trends are what the Goodenough–Kanamori rules [5–8] predict.

2.3.2. M-L . . . L-M Exchange

There are two extreme cases of M-L . . . L-M exchange. When the p_σ-orbital tails are pointing toward each other (Figure 6a), the overlap integral, $\langle \varphi_i | \varphi_j \rangle$, can be substantial if the contact distance L . . . L lies in the vicinity of the van der Waals distance. However, the overlap electron density $\rho_{ij} = \varphi_i \varphi_j$ is practically zero because φ_i and φ_j do not have an overlapping region. Consequently, the in-phase and out-of-phase states Ψ_+ and Ψ_- are split in energy with a large separation Δe_{ij}. Thus, it is predicted that the M-L . . . L-M type exchange can only be AFM [1–4]. When the L . . . L linkage is bridged by a d^0 cation such as V^{5+} or W^{6+}, for example, only the out-of-phase state Ψ_- is lowered in energy by the d_π orbital of the cation A, reducing the Δe_{ij} so that the M-L . . . A . . . L-M exchange becomes weak (Figure 6b). Conversely, when the p-orbital tails of the M-L . . . L-M exchange path have an orthogonal arrangement (Figure 7a), the overlap integral, $\langle \varphi_i | \varphi_j \rangle$, is zero, making the M-L . . . L-M exchange weak. If the L . . . L linkage of such an exchange path is bridged by a d^0 cation, the out-of-phase state Ψ_- level is lowered in energy enlarging Δe_{ij} so that the M-L... A... L-M becomes strongly AFM (Figure 7b) [2–4].

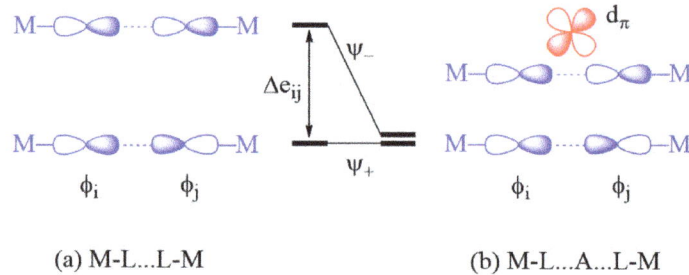

(a) M-L...L-M   (b) M-L...A...L-M

Figure 6. Interactions between the magnetic orbitals in the M-L... L-M exchange where their p_σ tails are pointing to each other. The large energy split Δe_{ij} of the M-L... L-M exchange in (a) is reduced by the d_π orbital of the d^0 cation A in the M-L... A... L-M exchange in (b).

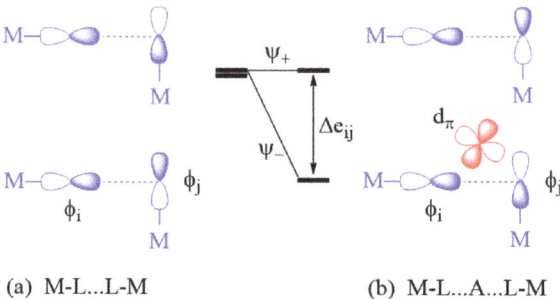

(a) M-L...L-M   (b) M-L...A...L-M

Figure 7. Interactions between the magnetic orbitals in the M-L... L-M exchange where their p_σ tails have an orthogonal arrangement. The small energy split Δe_{ij} of the M-L... L-M exchange in (a) is enlarged by the d_π orbital of the d^0 cation A in the M-L... A... L-M exchange in (b).

In the M-L... A... L-M exchange of Figure 7, the vanishingly small Δe_{ij} of the M-L... L-M exchange results because the two p_σ tails have an orthogonal arrangement. A very small Δe_{ij} for the M-L... L-M exchange occurs even if the two M-L bonds are pointing to each other as in Figure 6 when one M-L bond has a p_σ tail and the other has a p_π tail, and also when both M-L bonds have p_π tails. Such M-L... L-M spin exchanges become strong in the corresponding M-L... A... L-M exchanges.

2.3.3. Qualitative Rules Governing Spin Exchanges

The above discussions are based on the observation that the nature of a spin exchange, be it the M-L-M, M-L... L-M or M-L... A... L-M type, is governed by the ligand p-orbital tails present in the magnetic orbitals of the spin exchange path. The essential points of our discussions can be summarized as follows:

a. For an individual (φ_i, φ_j) exchange of a M-L-M type, the (t_{2g}, t_{2g}) and (e_g, e_g) exchanges are FM if the bond angle θ is 90°, and so is the (t_{2g}, e_g) exchange if the bond angle θ is 180°. These exchanges become AFM when the bond angles θ deviate considerably from these values.
b. An individual (φ_i, φ_j) exchange of a M-L... L-M or M-L... A... L-M type can only be AFM if not weak.
c. A strong individual (φ_i, φ_j) exchange of a M-L... L-M is weakened by the d^0 metal cation A in the M-L... A... L-M exchange, but a weak individual (φ_i, φ_j) exchange

of a M-L ... L-M is strengthened by the presence of a d^0 metal cation A in the M-L ... A ... L-M exchange.

d. When a magnetic ion has several unpaired spins, the spin exchange between two magnetic ions is given by the sum of all possible individual (φ_i, φ_j) exchanges.

These qualitative rules governing spin exchanges can serve as guidelines for exploring how the calculated spin exchanges are related to the structures of the exchange paths and also for ensuring that important exchange paths are included the set of spin exchanges to evaluate by the energy-mapping analysis.

3. Results and Discussion

3.1. Details of Calculations

We performed spin-polarized DFT calculations using the Vienna ab initio Simulation Package (VASP) [26,27], the projector augmented wave (PAW) method, and the PBE exchange-correlation functionals [28]. The electron correlation associated with the 3d states of M (M = Mn, Fe, Co, Ni) was taken into consideration by performing the DFT+U calculations [29] with the effective on-site repulsion $U_{eff} = U - J = 4$ eV on the magnetic ions. Our DFT + U calculations carried out for numerous magnetic solids of transition-metal ions showed that use of the U_{eff} values in the range of 3 − 5 eV correctly reproduce their magnetic properties (see the original papers cited in the review articles [1–3,22,24]). The primary purpose of using DFT + U calculations is to produce magnetic insulating states for magnetic solids. Use of $U_{eff} = 3 - 5$ eV in DFT + U calculations leads to magnetic insulating states for magnetic solids of Mn^{2+}, Fe^{2+}, Co^{2+}, and Ni^{2+} ions. The present work employed the representative U_{eff} value of 4 eV. We carried out DFT + U calculations (with $U_{eff} = 4$ eV) to optimize the structures of MPS_3 (M = Mn, Fe, Co, Ni) in their FM states by relaxing only the ion positions while keeping the cell parameters fixed and using a set of (4 × 2 × 6) k-points and the criterion of 5×10^{-3} eV/Å for the ionic relaxation. All our DFT + U calculations for extracting the spin-exchange parameters employed a (2a, 2b, c) supercell, the plane wave cutoff energy of 450 eV, the threshold of 10^{-6} eV for self-consistent-field energy convergence, and a set of (4 × 2 × 6) k-points. The preferred spin direction of the cation M^{2+} (M = Mn, Fe, Co, Ni) cation was determined by DFT + U + SOC calculations [30], employing a set of (4 × 2 × 6) k-points and the threshold of 10^{-6} eV for self-consistent-field energy convergence.

3.2. Magnetic Ground States of MPS_3

We probed the magnetic ground states of the MPS_3 phases by evaluating the relative energies, on the basis of DFT + U calculations, of the AF1, AF2 and AF3 spin configurations shown in Figure 2 as well as the FM, AF4, AF5, and AF6 states depicted in Supplementary Materials, Figure S1. As summarized in Table 1, our calculations using the experimental structures of MPS_3 show that the magnetic ground states of $MnPS_3$ and $NiPS_3$ adopt the honeycomb state AF1 and the ||a-chain state AF2, respectively, in agreement with experiment. In disagreement with experiment, however, the magnetic ground state is predicted to be the ||(a + b)-chain state AF3 for $FePS_3$, and the honeycomb state AF1 for $CoPS_3$. Since the energy differences between different spin ordered states are small, it is reasonable to speculate if they may be affected by small structural (monoclinic) distortion. Thus, we optimize the crystal structures of MPS_3 (M = Mn, Fe, Co, Ni) by performing DFT + U calculations to obtain the structures presented in the supporting material. Then, we redetermined the relative stabilities of the FM and AF1–AF6 states using these optimized structures. Results of these calculations are also summarized in Table 1. The optimized structures predict that the magnetic ground states of $MnPS_3$, $CoPS_3$ and $NiPS_3$ are the same as those observed experimentally, but that of $FePS_3$ is still the ||(a+b)-chain state AF3 rather than the ||a-chain state AF2 reported experimentally. This result is not a consequence of using the specific value of $U_{eff} = 4$ eV, because our DFT + U calculations for $FePS_3$ with $U_{eff} = 3.5$ and 4.5 eV lead to the same conclusion.

Table 1. Relative energies (in meV/formula unit) obtained for the seven ordered spin states of MPS$_3$ (M = Mn, Fe, Co, Ni) from DFT + U calculations with U$_{eff}$ = 4 eV. The numbers without parentheses are obtained by using the experimental structures, and those in parentheses by using the structures optimized by DFT + U calculations.

	Mn	Fe	Co	Ni
FM	33.77 (33.36)	31.25 (25.10)	71.46 (55.00)	45.00 (42.04)
AF1	0 (0)	12.24 (5.16)	0 (5.70)	6.50 (7.11)
AF2	15.54 (15.50)	12.92 (7.93)	45.05 (0)	0 (0)
AF3	14.25 (14.21)	0 (0)	34.02 (24.99)	0.35 (0.34)
AF4	14.72 (14.45)	20.85 (18.57)	22.16 (26.00)	52.40 (49.53)
AF5	12.77 (12.58)	15.79 (12.95)	157.25 (158.33)	33.62 (31.98)
AF6	17.24 (17.07)	10.57 (6.33)	140.58 (143.05)	16.43 (15.21)

To resolve the discrepancy between theory and experiment on the magnetic ground state of FePS$_3$, we note that the magnetic peak positions in the neutron diffraction profiles are determined by the repeat distances of the rectangular magnetic structures, namely, a and b for the AF2 state (Figure 2b), and a' and b' for the AF3 state (Figure 2c). In both the experimental and the optimized structures of FePS$_3$, it was found that $a = a'$ = 5.947 Å and $b = b'$ = 10.300 Å. Thus, for the neutron diffraction refinement of the magnetic structure for FePS$_3$, the AF2 and AF3 states provide an equally good model. In view of our computational results, we conclude that the AF3 state is the correct magnetic ground state for FePS$_3$.

The experimental and optimized structures of MPS$_3$ (M = Mn, Fe, Co, Ni) are very similar, as expected. The important differences between them affecting the magnetic ground state would be the M-S distances of the MS$_6$ octahedra, because the d-state splitting of the MS$_6$ octahedra is sensitively affected by them. The M-S distances of the MS$_6$ octahedra taken from the experimental and optimized crystal structures of MPS$_3$ are summarized in Table 2, and their arrangements in the honeycomb layer are schematically presented in Figure 8. All Mn-S bonds of MnS$_6$ in MnPS$_3$ are nearly equal in length, as expected for a high-spin d^5 ion (Mn^{2+}) environment. The Fe-S bonds of FeS$_6$ in the optimized structure of FePS$_3$ are grouped into two short and four long Fe-S bonds. This distinction is less clear in the experimental structure. The Co-S bonds of CoS$_6$ in the experimental and optimized structures of CoPS$_3$ are grouped into two short, two medium and two long Co-S bonds. However, the sequence of the medium and long Co-S bonds is switched between the two structures. In the experimental and optimized structures of NiPS$_3$, the Ni-S bonds of NiS$_6$ are grouped into two short, two medium and two long Ni-S bonds. This distinction is less clear in the experimental structure. Thus, between the experimental and optimized structures of MPS$_3$, the sequence of the two short, two medium and two long M-S bonds do not switch for M = Fe and Ni whereas it does for M = Co. The latter might be the cause for why the relative stabilities of the AF1 and AF2 states in CoPS$_3$ switches between the experimental and optimized structures.

Table 2. The M-S bond distances (in Å) of the MS$_6$ octahedra in MPS$_3$ (M = Mn, Fe, Co, Ni) obtained from the experimental and the optimized crystal structures, which are shown without and with parentheses, respectively.

Mn	Fe	Co	Ni
2.627 (2.632)	2.546 (2.525)	2.485 (2.492)	2.457 (2.453)
2.627 (2.632)	2.546 (2.526)	2.485 (2.492)	2.457 (2.453)
2.625 (2.635)	2.547 (2.571)	2.504 (2.525)	2.462 (2.457)
2.625 (2.635)	2.547 (2.572)	2.504 (2.525)	2.462 (2.457)
2.634 (2.639)	2.549 (2.572)	2.491 (2.537)	2.465 (2.461)
2.634 (2.639)	2.549 (2.573)	2.491 (2.537)	2.465 (2.461)

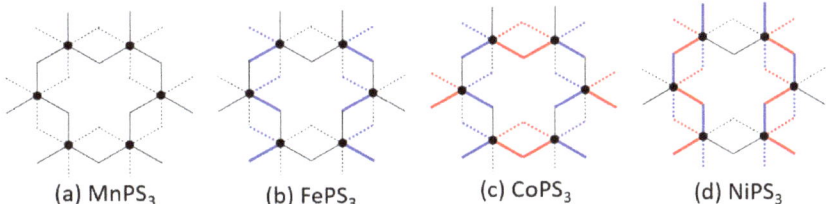

Figure 8. The arrangements of the M-S bond lengths of the MS$_6$ octahedra in MPS$_3$. The short M-S bonds are represented by blue lines, the medium M-S bonds by red lines, and the long M-S bonds by black lines.

3.3. Preferred Spin Orientation of MPS$_3$
3.3.1. Quantitative Evaluation

We determine the preferred spin orientations of the M^{2+} ions in MPS$_3$ (M = Mn, Fe, Co, Ni) phases by performing DFT + U + SOC calculations using their FM states with the ||z and ⊥z spin orientations. For the ⊥z direction we selected the ||a-direction. As summarized in Table 3, these calculations predict the preferred spin orientation to be the ||z direction for FePS$_3$, and the ||x direction for MnPS$_3$, CoPS$_3$ and NiPS$_3$. These predictions are in agreement with experiment for FePS$_3$ [9,18], CoPS$_3$ [19], and NiPS$_3$ [20], while this is not the case for MnPS$_3$ [9,12,14,31]. Our DFT + U + SOC calculations for the AF1 state of MnPS$_3$ show that the ||x spin orientation is still favored over the ||z orientation just as found from the calculations using the FM state of MnPS$_3$. The Mn^{2+} spins of MnPS$_3$ were reported to have the ||z orientation in the early studies [9,12], but were found to be slightly tilted away from the z-axis (by 8°) [14,31]. In our further discussion (see below), this small deviation is neglected.

Table 3. Relative energies (in K per formula unit) of the ||z and ⊥z spin orientations of the M^{2+} ions in the FM states of MPS$_3$ (M = Mn, Fe, Co, Ni) obtained by DFT + U + SOC calculations. The results calculated by using the optimized (experimental) structures are presented without (with) the parentheses.

	MnPS$_3$ [a]	FePS$_3$ [b]	CoPS$_3$	NiPS$_3$
⊥z	0 (0)	20.0 (21.8)	0 (0)	0 (0)
||z	0.3 (0.3)	0 (0)	3.8 (5.2)	0.8 (0.7)

[a] The same result is obtained by using the AF1 state, which is the magnetic ground state of MnPS$_3$. [b] The same results are obtained from our DFT+U calculations with U$_{eff}$ = 3.5 and 4.5 eV.

3.3.2. Qualitative Picture
Selection Rules of Spin Orientation and Implications

With the local z-axis of a ML_6 octahedron along its three-fold rotational axis (Figure 1a), the t_{2g} set is described by {1a, 1e'}, and the e_g set by {2e'}[22–24], where

$$1a = 3z^2 - r^2$$

$$\{1e'\} = \left\{\sqrt{\tfrac{2}{3}}xy - \sqrt{\tfrac{1}{3}}xz, \sqrt{\tfrac{2}{3}}(x^2 - y^2) - \sqrt{\tfrac{1}{3}}yz\right\}$$
$$\{2e'\} = \left\{\sqrt{\tfrac{1}{3}}xy + \sqrt{\tfrac{2}{3}}xz, \sqrt{\tfrac{1}{3}}(x^2 - y^2) + \sqrt{\tfrac{2}{3}}yz\right\}$$
(5)

Using these d-states, the electron configurations expected for the M^{2+} ions of MPS_3 (M = Mn, Fe, Co, Ni) are presented in Figure 9. In the spin polarized description of a magnetic ion, the up-spin d-states lie lower in energy than the down-spin states so that the HOMO and LUMO occur in the down-spin d-states for the M^{2+} ions with more than the d^5 electron count, so only the down-spin states are shown for $FePS_3$, $CoPS_3$, and $NiPS_3$ in Figure 9a–c. For $MnPS_3$ with d^5 Mn^{2+} ion, the HOMO is represented by the up-spin 1e', and the LUMO by the down-spin 1a and 2e' (Figure 9d).

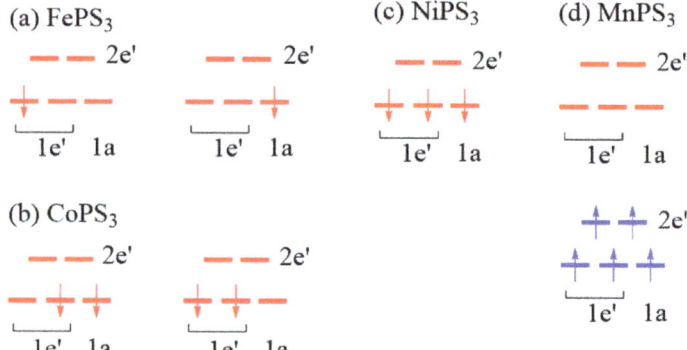

Figure 9. Electron configurations of the M^{2+} (M = Mn, Fe, Co, Ni) ions of (**a**) $FePS_3$, (**b**) $CoPS_3$, (**c**) $NiPS_3$, and (**d**) $MnPS_3$ in the spin polarized description. In (**a**–**c**), the up-spin d-states lying below the down-spin t_{2g} states are not shown for clarity.

In terms of the d-orbital angular states $|L, L_z\rangle$ ($L = 2$, $L_z = -2, -1, 0, 1, 2$), the 1e' state consists of the $|2, \pm 2\rangle$ and $|2, \pm 1\rangle$ sets in the weight ratio of 2:1, and the 2e' state in the weight ratio of 1:2 ratio. Consequently, the major component of the 1e' set is the $|2, \pm 2\rangle$ set, while that of the 2e' set is the $|2, \pm 1\rangle$ set.

The selection rules of the spin orientation are based on the $|\Delta L_z|$ value between the HOMO and LUMO of ML_n. If the HOMO and LUMO both occur in the up-spin state or in down-spin states (Figure 9a–c), the ||z spin orientation is predicted if $|\Delta L_z| = 0$, and the ⊥z spin orientation if $|\Delta L_z| = 1$. When $|\Delta L_z| > 1$, the HOMO and LUMO do not interact under SOC and hence do not affect the spin orientation. Between the 1a, 1e' and 2e' states, we note the following cases of values:

$$|\Delta L_z| = 0 \begin{cases} \text{between the major components of the 1e' set} \\ \text{between the major components of the 2e' set} \end{cases}$$
(6)

$$|\Delta L_z| = 1 \begin{cases} \text{between 1}a \text{ and the minor component of 1e'} \\ \text{between 1}a \text{ and the major component of 2e'} \\ \text{between the major components of 1e' and 2e'} \end{cases}$$
(7)

We now examine the preferred spin orientations of MPS$_3$ from the viewpoint of the selection rules and their electron configurations (Figure 9). The d-electron configuration of FePS$_3$ can be either (d↑)^5(1e'↓)^1 or (d↑)^5(1a↓)^1 (Figure 9a), where the notation (d↑)^5 indicates that all up-spin d-states are occupied. The (d↑)^5(1e'↓)^1 configuration, for which $|\Delta L_z| = 0$, predicts the ||z spin orientation, while the (d↑)^5(1a↓)^1 configuration, for which $|\Delta L_z| = 1$, predicts the ⊥z spin orientation. Thus, the (d↑)^5(1a↓)^1 configuration is correct for the Fe^{2+} ion of FePS$_3$. Since this configuration has the degenerate level 1e' unevenly occupied, it should possess uniaxial magnetism [2,3,22–24] and hence a large magnetic anisotropy energy. This is in support of the experimental finding of the Ising character of the spin lattice of FePS$_3$ [16] or the single-ion anisotropic character of the Fe^{2+} ion [17,18]. The d-electron configuration of CoPS$_3$ can be either (d↑)^5(1e'↓)^2 or (d↑)^5(1a↓)^1(1e'↓)^1 (Figure 9b). The (d↑)^5(1e'↓)^2 configuration, for which $|\Delta L_z| = 1$, predicts the ⊥z spin orientation, while the (d↑)^5(1a↓)^1(1e'↓)^1 configuration, for which $|\Delta L_z| = 0$, predicts the ||z spin orientation. Thus, the (d↑)^5(1e'↓)^2 configuration is correct for the Co^{2+} ion of CoPS$_3$. Since this configuration has the degenerate level 1e' evenly occupied, it does not possess uniaxial magnetism [2,3,22–24] and hence a small magnetic anisotropy energy. The d-electron configuration of NiPS$_3$ is given by (d↑)^5(1a)^1(1e'↓)^2 (Figure 9c), for which $|\Delta L_z| = 1$, so the ⊥z spin orientation is predicted in agreement with experiment.

Let us now consider the spin orientation of the Mn^{2+} ion of MnPS$_3$. First, it should be noted that, if the HOMO and LUMO occur in different spin states as in MnPS$_3$ (Figure 9d), the selection rules predict the opposite to those found for the case when the HOMO and LUMO occur all in up-spin states or all in down-spin states [2,3,22–24]. Namely, the preferred spin orientation is the ||z spin orientation if $|\Delta L_z| = 1$, but the ⊥z spin orientation if $|\Delta L_z| = 0$ [2,3,22–24]. According to Equation (7), $|\Delta L_z| = 1$ for the Mn^{2+} ion of MnPS$_3$, which predicts the ⊥z orientation as the preferred spin direction in agreement with the quantitative estimate of the magnetic anisotropy energy obtained from the DFT + U + SOC calculations, although this is in disagreement with experiment [5,8–10]. It has been suggested that the ||z spin orientation is caused by the magnetic dipole–dipole (MDD) interactions [13]. This subject will be probed in the following.

Magnetic Dipole–Dipole Interactions

Being of the order of 0.01 meV for two spin-1/2 ions separated by 2 Å, the MDD interaction is generally weak. For two spins located at sites i and j with the distance r_{ij} and the unit vector e_{ij} along the distance, the MDD interaction is defined as [32]

$$\left(\frac{g^2\mu_B^2}{a_0^3}\right)\left(\frac{a_0}{r_{ij}}\right)^3\left[-3(\vec{S}_i \cdot \vec{e}_{ij})(\vec{S}_j \cdot \vec{e}_{ij}) + (\vec{S}_i \cdot \vec{S}_j)\right] \qquad (8)$$

where a_0 is the Bohr radius (0.529177 Å), and $(g\mu_B)^2/(a_0)^3 = 0.725$ meV. The MDD effect on the preferred spin orientation of a given magnetic solid can be examined by comparing the MDD interaction energies calculated for a number of ordered spin arrangements. In summing the MDD interactions between various pairs of spin sites, it is necessary to employ the Ewald summation method [33–35]. Table 4 summarizes the MDD interaction energies calculated, by using the optimized structures of MPS$_3$ (M = Mn, Fe, Co, Ni), for the ||z and ||x spin directions in the AF1, AF2 and AF3 states. The corresponding results obtained by using the experimental structures of MPS$_3$ are summarized in Table S1.

Table 4. Relative energies (in K per formula unit) of the ||x and ||z spin orientations calculated by MDD calculations for the M^{2+} ions of MPS_3 (M = Mn, Fe, Co, Ni) in the AF1, AF2 and AF3 states using the optimized crystal structures.

	$MnPS_3$		$FePS_3$		$CoPS_3$		$NiPS_3$	
	||x	||z	||x	||z	||x	||z	||x	||z
AF1	0.48	0.17	0.36	0.12	0.21	0.07	0.09	0.03
AF2	0.00	0.35	0.00	0.26	0.00	0.15	0.00	0.07
AF3	0.55	0.38	0.38	0.27	0.22	0.15	0.10	0.07

These results can be summarized as follows: for the ||z spin orientation, the AF1 state is more stable than the AF2 and AF3 states. For the ||x spin orientation, the AF2 state is more stable than the AF1 and AF3 states. The ||x spin direction of the AF2 state is more stable than the ||z spin direction of the AF1 state. However, none of these results can reverse the relative stabilities of the ||z and ||x spin directions determined for $FePS_3$, $CoPS_3$, and $NiPS_3$ from the DFT + U + SOC calculations (Table 3). The situation is slightly different for $MnPS_3$, which adopts the AF1 state as the magnetic ground state. For $MnPS_3$ in this state, the MDD calculations predict that the ||z spin orientation is more stable than the ||x spin orientation by 0.3 K per formula unit (Table 4). Note that this prediction is the exact opposite to what the DFT + U + SOC calculations predict for $MnPS_3$ in the AF1 state (Table 3). Thus, the balance between these two opposing energy contributions will determine whether the ||z spin orientation is more stable than the ⊥z spin orientation in agreement with the experimental observation. Consequently, for $MnPS_3$ the MDD interaction dominates over the SOC effect which is plausible because of the half-filled shell electronic configuration. This is because the AF1 magnetic structure is forced on $MnPS_3$; in terms of purely MDD interactions alone, the ⊥z spin orientation in the AF2 state is most stable.

3.4. Quantitative Evaluations of Spin Exchanges

Due to the monoclinic crystal structure that MPS_3 adopts, each of the exchanges J_{12}, J_{13} and J_{14} (Figure 10a) are expected to split into two slightly different spin exchanges (Figure 10b) so that there are six spin exchanges J_1–J_6 to consider. To extract the values of the six spin exchanges J_1–J_6 (Figure 3), we employ the spin Hamiltonian expressed as:

$$H_{spin} = -\sum_{i>j} J_{ij} \hat{S}_i \cdot \hat{S}_j \quad (9)$$

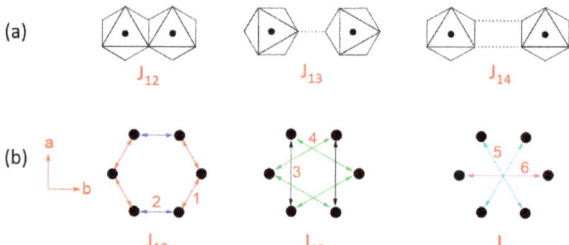

Figure 10. (a) Three kinds of spin exchange paths in each honeycomb layer of MPS_3. (b) Two kinds of the spin exchanges resulting from J_{12}, J_{13} and J_{14} due to the loss of the trigonal symmetry in the MPS_3 honeycomb layers. In (b), the numbers 1–6 refer to J_1–J_6, respectively.

Then, the energies of the FM and AF1–AF6 states of MPS$_3$ (M = Mn, Fe, Co, Ni) per 2 × 2 × 1 supercell are written as:

$$E_{FM} = (-16J_1 - 8J_2 - 16J_3 - 32J_4 - 16J_5 - 8J_6)S^2$$
$$E_{AF1} = (+16J_1 + 8J_2 - 16J_3 - 32J_4 + 16J_5 + 8J_6)S^2$$
$$E_{AF2} = (-16J_1 + 8J_2 - 16J_3 + 32J_4 + 16J_5 + 8J_6)S^2$$
$$E_{AF3} = (-8J_2 + 16J_3 + 16J_5 + 8J_6)S^2$$
$$E_{AF4} = (+16J_1 - 8J_2 - 16J_3 + 32J_4 - 16J_5 - 8J_6)S^2$$
$$E_{AF5} = (+8J_2 + 16J_3 - 16J_5 + 8J_6)S^2$$
$$E_{AF6} = (-8J_2 + 16J_3 + 16J_5 - 8J_6)S^2$$

where S is the spin on each M^{2+} ion (i.e., S = 5/2, 2, 3/2 and 1 for M = Mn, Fe, Co, and Ni, respectively). By mapping the relative energies of the FM and AF1–AF6 states determined in terms of the spin exchange J_1–J_6 onto the corresponding relative energies obtained from the DFT + U calculations (Table 1), we find the values of J_1–J_6 listed in Table 5. (The spin exchanges of MPS$_3$ determined by using their experimental crystal structures are summarized in Table S2)

Table 5. Spin exchanges J_1–J_6 obtained (for the optimized structures of MPS$_3$ (M = Mn, Fe, Co, Ni) from DFT + U calculations with U_{eff} = 4 eV) by simulating the relative energies of the FM and AF1–AF6 states with the six spin exchanges.

	Mn	Fe	Co	Ni
J_1	1.00	0.37	0.05	−0.25
J_2	0.87	−0.32	−0.91	−0.14
J_3	0.06	0.36	−0.55	0.04
J_4	0.05	0.07	0.04	−0.01
J_5	0.34	0.86	0.11	0.99
J_6	0.33	1.00	1.00	1.00
	J_1 = −16.0 K	J_6 = −18.4 K	J_6 = −608.7 K	J_6 = −172.4 K

With the sign convention adopted in Eq. 1, AFM exchanges are represented by J_{ij} < 0, and FM exchanges by J_{ij} > 0. From Table 5, the following can be observed:

a. In all MPS$_3$ (M = Mn, Fe, Co, Ni), $J_1 \neq J_2$, $J_3 \neq J_4$, and $J_5 \neq J_6$, reflecting that the exchange paths are different between J_1 and J_2, between J_3 and J_4, and between J_5 and J_6 (Figure 10).
b. $J_1 \approx J_2 < 0$, $J_3 \approx J_4 \approx 0$, and $J_5 \approx J_6 < 0$ for MnPS$_3$ while $J_1 \approx J_2 > 0$, $J_3 \approx J_4 \approx 0$, and $J_5 \approx J_6 < 0$ NiPS$_3$. To a first approximation, the electron configurations of MnPS$_3$ and NiPS$_3$ can be described by $(t_{2g})^3(e_g)^2$ and $(t_{2g})^6(e_g)^2$, respectively. That is, they do not possess an unevenly occupied degenerate state t_{2g}.
c. In FePS$_3$ and CoPS$_3$, J_1 and J_2 are quite different, and so are J_3 and J_4. While J_5 and J_6 are comparable in FePS$_3$, they are quite different in CoPS$_3$. The electron configurations of FePS$_3$ and NiPS$_3$ can be approximated by $(t_{2g})^4(e_g)^2$ and $(t_{2g})^5(e_g)^2$, respectively. Namely, they possess an unevenly occupied degenerate state t_{2g}.
d. The strongest exchange is J_1 in MnPS$_3$, but J_6 in other MPS$_3$ (M = Fe, Co, Ni).
e. The second NN exchange J_3 is strongly FM in CoPS$_3$, while the third NN exchange J_6 is very strongly AFM in CoPS$_3$ and NiPS$_3$.

From the viewpoints of the expected trends in spin exchanges, the observation (e) is quite unusual. This will be discussed in the next section.

3.5. Unusual Features of the M-L . . . L-M Spin Exchanges

3.5.1. Second Nearest-Neighbor Exchange

As pointed out in the previous section, the second NN exchange J_3 of CoPS$_3$ is strongly FM despite that it is a M-L . . . L-M exchange to a first approximation. This implies that the J_F component in some (φ_i, φ_j) exchanges is nonzero, namely, the overlap electron density associated with those exchanges is nonzero. This implies that the p-orbital tails of the two magnetic orbitals are hybridized with the group orbitals of the $P_2S_6^{4-}$ anion, i.e., they become delocalized into the whole $P_2S_6^{4-}$ anion. Each MS$_6$ octahedron has three mutually orthogonal "MS$_4$ square planes" containing the yz, xz and xy states (Figure 11a). At the four corners of these three square planes, the p-orbital tails of the d-states are present (Figure 3a).

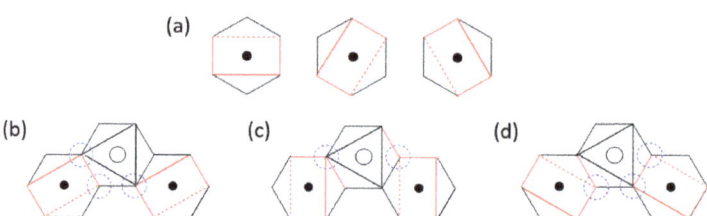

Figure 11. (a) Three MS$_4$ square planes of a MS$_6$ octahedron, containing the xy, xz and yz states of an MS$_6$ octahedron. (**b–d**) Three cases of the CoS$_4$ square planes containing the t_{2g} magnetic orbitals in the J_3 exchange path of CoPS$_3$.

The lone-pair orbitals of the S atoms are important for the formation of each MS$_6$ octahedron. Due to the bonding requirement of the $P_2S_6^{4-}$ anion, such lone pair orbitals become symmetry-adapted. An example in which the p-orbitals of all the S atoms are present is shown in Figure 1e.

With the $(t_{2g})^5(e_g)^2$ configuration, the Co^{2+} ion of CoPS$_3$ has five electrons in the t_{2g} level, namely, it has only one t_{2g} magnetic orbital. This magnetic orbital is contained in one of the three CoS$_4$ square planes presented in Figure 11b–d. When the S p-orbital at one corner of the $P_2S_6^{4-}$ anion interacts with a d-orbital of M, the S p-orbitals at the remaining corners are also mixed in. Thus, when $P_2S_6^{4-}$ anion shares corners with both MS$_4$ square planes of the J_3 exchange path, a nonzero overlap electron density is generated, thereby making the spin exchange FM. For convenience, we assume that the magnetic t_{2g} orbital of the Co^{2+} ion is the xy state. Then, there will be not only the (xy, xy) exchange, but also the (xy, x^2-y^2) and (x^2-y^2, xy) exchanges between the two Co^{2+} ions of the J_3 path. All these individual exchanges lead to nonzero overlap electron densities by the delocalization of the p-orbital tails with the group orbitals of the molecular anion $P_2S_6^{4-}$. In other words, the spin exchange J_3 in CoPS$_3$ is nominally a M-L . . . L-M, which is expected to be AFM by the qualitative rule, but it is strongly FM. It is clear that, if the L . . . L linkage is a part of the covalent framework of a molecular anion such as $P_2S_6^{4-}$, a nominal M-L . . . L-M exchange can become FM for a certain combination of the d-electron count of the metal M and the geometries of the exchange path.

3.5.2. Third Nearest-Neighbor Exchange

Unlike in MnPS$_3$ and FePS$_3$, the M-S . . . S-M exchange J_6 is unusually strong in CoPS$_3$ and NiPS$_3$ (Section 3.3). This is so despite that the S . . . S contact distances are longer in CoPS$_3$ and NiPS$_3$ than in MnPS$_3$ and FePS$_3$ (i.e., the S . . . S contact distance of the J_6 path in MPS$_3$ is 3.409, 3.416, 3.421 and 3.450 Å for M = Mn, Fe, Co and Ni, respectively). We note that a strong M-L . . . L-M exchange (i.e., a spin exchange leading to a large energy split Δe_{ij}) becomes weak, when the L . . . L contact is bridged by a d^0 cation like, e. g., V^{5+} and W^{6+} to form the M-L . . . A . . . L-M exchange path (Figure 6b), because the out-of-phase combination ψ_- is lowered in energy by interacting with the unoccupied d_π orbital of the cation A. Conversely, then, one may ask if the strength of a M-L ... L-M spin exchange can be enhanced by raising the ψ_- level. The latter can be achieved if the L ... L path provides an occupied level of π-symmetry that can interact with ψ_-. As depicted in Figure 12a, the J_6 path has the two MS_4 square planes containing the x^2-y^2 magnetic orbitals (Figure 12b). The lone-pair group orbital of the S_4 rectangular plane (Figure 12c) of the $P_2S_6^{4-}$ anion has the correct symmetry to interact with ψ_-, so that the ψ_- level is raised in energy thereby enlarging the energy split between ψ_+ and ψ_- and strengthening the J_6 exchange (Figure 12d). Although this reasoning applies equally to $MnPS_3$ and $FePS_3$, the latter do not have a strong J_6 exchange. This can be understood by considering Equation (1), which shows that a magnetic ion with several magnetic orbitals leads to several individual spin exchanges that can lead to FM contributions.

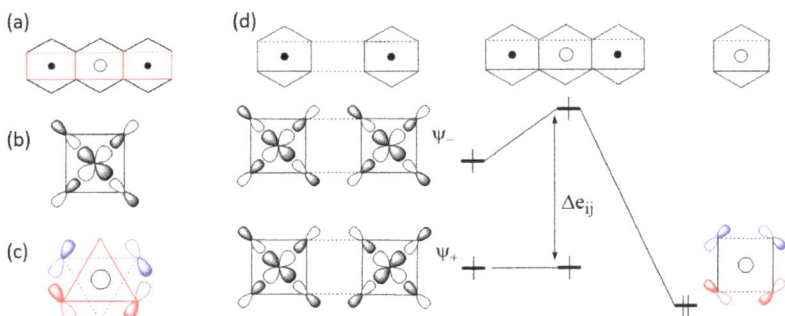

Figure 12. (**a**) The J_6 spin exchange path of MPS_3 (M = Co, Ni) viewed in terms of the MS_4 and P_2S_4 square planes. (**b**) The x^2-y^2 magnetic orbital of the MS_6 octahedron. (**c**) The S p-orbitals present at the corners of the P_2S_4 square plane. (**d**) How the M-S ... S-M spin exchange is enhanced by the through-bond effect of the intervening P_2S_6 octahedron.

In view of the above discussion, which highlights the unusual nature of the second and third NN spin exchanges mediated by a molecular anion such as $P_2S_6^{4-}$, we propose to use the notation M-(L-L)-M to distinguish it from M-L-M. M-L ... L-M and M-L ... A ... L-M type exchanges. The notation (L-L) indicates two different ligand sites of a multidentate molecular anion, each with lone pairs for the coordination with a cation M. Such M-(L-L)-M exchanges can be strongly FM or strongly AFM, as discussed above. Currently, there are no qualitative rules with which to predict whether they will be FM or AFM. A similar situation was found, for example, for the mineral Azurite $Cu_3(CO_3)_2(OH)_2$, in which every molecular anion CO_3^{2-} participates in three different Cu-(O-O)-Cu exchanges. DFT + U calculations show that one of these three is substantially AFM, but the remaining two are negligible. So far, this observation has not been understood in terms of qualitative reasoning.

3.6. Description Using Three Exchanges

Experimentally, the magnetic properties of MPS_3 have been interpreted in terms of three exchange parameters, namely, by assuming that $J_1 = J_2$ ($\equiv J_{12}$), $J_3 = J_4$ ($\equiv J_{13}$), and $J_5 = J_6$ ($\equiv J_{14}$). To investigate whether this simplified description is justified, we simulate the relative energies of the seven ordered spin states of MPS_3 by using the three exchanges J_{12}, J_{13} and J_{14} as parameters in terms of the least-square fitting analysis. Our results, summarized in Table 6, show that the standard deviations of J_{12}, J_{13} and J_{14} are small for $MnPS_3$ and $NiPS_3$, moderate in $FePS_3$, but extremely large in $CoPS_3$ (for details, see Figures S2–S5). The exchanges experimentally deduced for $FePS_3$ are $J_{12} = -17$ K, $J_{13} = -0.5$ K, and $J_{14} = 7$ K from neutron inelastic scattering measurements [17], -17 K $\leq J_{12} \leq -5.6$ K, -7.2 K $\leq J_{13} \leq 2.8$ K, and $0 \leq J_{14} \leq 10$ K from powder susceptibility measurements [9], and $J_{12} = -19.6$ K, $J_{13} = 10.3$ K, and $J_{14} = -2.2$ K from high field measurements [17]. These experimental estimates are dominated by J_{12}, but the theoretical estimates of Table 6 by J_{14}. One might note from Table 6 that the magnetic properties of MnPS$_3$, FePS$_3$ and NiPS$_3$ can be reasonably well approximated by two exchanges, that is, by J_{12} and J_{14} for MnPS$_3$, by J_{14} and J_{12} for NiPS$_3$, and by J_{14} and J_{13} for FePS$_3$. However, this three-parameter description leads to erroneous predictions for the magnetic ground states of MPS$_3$; it predicts the AF1 state to be the ground state for both MnPS$_3$ and CoPS$_3$. This prediction is correct for MnPS$_3$, but incorrect for CoPS$_3$. In addition, it predicts that the AF2 and AF3 states possess the same stability for all MPS$_3$ (M = Mn, Fe, Co, Ni), and are the ground states for FePS$_3$ and NiPS$_3$. These two predictions are both incorrect.

Table 6. Spin exchanges J_{12}, J_{13} and J_{14} in K obtained (for the optimized structures of MPS$_3$ (M = Mn, Fe, Co, Ni) from DFT + U calculations with U_{eff} = 4 eV) by simulating the relative energies of the FM and AF1–AF6 states with the three spin exchanges.

	Mn	Fe	Co	Ni
J_{12}	−15.5 ± 0.4	2.0 ± 7.7	−61.4 ± 119.0	36.3 ± 4.3
J_{13}	−0.9 ± 0.2	−7.7 ± 3.9	60.7 ± 55.3	0.0 ± 2.0
J_{14}	−5.3 ± 0.3	−20.9 ± 4.5	−59.1 ± 95.6	−186.0 ± 3.4

4. Concluding Remarks

Our DFT + U calculations for the optimized structures of MPS$_3$ (M = Mn, Fe, Co, Ni) reveal that, in agreement with experiment, the magnetic ground state of MnPS$_3$ is the AF1 state while those of CoPS$_3$ and NiPS$_3$ are the AF2 state. In disagreement with experiment, however, our calculations predict the AF2 state to be the magnetic ground state for FePS$_3$. Our DFT + U + SOC calculations show that, in agreement with experiment, the preferred spin orientation of FePS$_3$ is the ||z direction while those of CoPS$_3$ and NiPS$_3$ are the ⊥z direction, and the Fe^{2+} ion of FePS$_3$ exhibits uniaxial anisotropy. In disagreement with experiment, these calculations predict the preferred spin orientation for MnPS$_3$ to be the ⊥z direction. Our analyses suggest that the ||z spin direction experimentally observed for the Mn^{2+} ions arises from the magnetic dipole–dipole interactions in the AF1 magnetic state. We presented simple qualitative rules governing spin exchanges to be used as guidelines for gauging the nature of various spin exchanges. These rules allowed us to recognize several unusual exchanges of MPS$_3$; the second NN exchange J_3 of CoPS$_3$ is strongly FM while the third NN exchanges J_6 of CoPS$_3$ and NiPS$_3$ are very strongly AFM. These observations reflect the fact that the lone-pair orbitals of the P$_2$S$_6^{4-}$ ion, mediating the spin exchanges in MPS$_3$ are symmetry-adapted group orbitals, so the effect of coordinating one S atom to one M^{2+} ion is felt by all the remaining five S atoms of P$_2$S$_6^{4-}$. The spin exchanges mediated by molecular anions, termed the M-(L-L)-M type exchanges, differ in nature from the M-L-M, M-L … L-M and M-L … A … L-M type exchanges. To find qualitative trends in the M-(L-L)-M type exchanges, it is necessary to further study the spin exchanges involving various other molecular anions.

Supplementary Materials: The following are available online. Figure S1: Ordered spin states, FM, AF4, AF5 and AF6 employed together with the states AF1, AF2 and AF3 (see the text) to determine the magnetic ground states as well as the spin exchanges J_1–J_6 of MPS$_3$ (M = Mn, Fe, Co, Ni). Figure S2–S5. Results of the least-square fitting the relative energies of the seven ordered spin states (FM, AF1-AF6) of MnPS$_3$, CoPS$_3$ and FePS$_3$, and NiPS$_3$, Table S1: Relative energies (in K per formula unit) of the ||x and ||z spin orientations obtained by MDD calculations for the M^{2+} ions of MPS$_3$ (M = Mn, Fe, Co, Ni) in the AFM1, AF2 and AFM3 states using the experimental crystal structures. Table S2: Spin exchanges (in K) obtained for the experimental structures of MPS$_3$ (M = Mn, Fe, Co, Ni) from DFT + U calculations with U_{eff} = 4 eV.

Author Contributions: Conceptualization, M.-H.W.; formal analysis and investigation, H.-J.K., R.K. and M.-H.W.; resources, H.-J.K.; data curation, H.-J.K.; writing—original draft preparation, M.-H.W.; writing—review and editing, H.-J.K., R.K. and M.-H.W.; visualization, M.-H.W.; supervision, M.-

H.W.; funding acquisition, H.-J.K. All authors have read and agreed to the published version of the manuscript.

Funding: The work at Kyung Hee University was supported by the Basic Science Research Program through the National Research Foundation of Korea (NRFK) funded by the Ministry of Education (2020R1A6A1A03048004).

Institutional Review Board Statement: Not applicable.

Informed Consent Statement: Not applicable.

Acknowledgments: H.-J. K. thanks the NRFK for the fund 2020R1A6A1A03048004.

Conflicts of Interest: The authors declare no conflict of interest.

References

1. Whangbo, M.-H.; Koo, H.-J.; Dai, D. Spin exchange interactions and magnetic structures of extended magnetic solids with localized spins: Theoretical descriptions on formal, quantitative and qualitative levels. *J. Solid State Chem.* **2003**, *176*, 417–481. [CrossRef]
2. Xiang, H.J.; Lee, C.; Koo, H.-J.; Gong, X.; Whangbo, M.-H. Magnetic properties and energy-mapping analysis. *Dalton Trans.* **2013**, *42*, 823–853. [CrossRef]
3. Whangbo, M.-H.; Xiang, H.J. Magnetic Properties from the Perspectives of Electronic Hamiltonian: Spin Exchange Parameters, Spin Orientation and Spin-Half Misconception. In *Handbook in Solid State Chemistry, Volume 5: Theoretical Descriptions*; Wiley: New York, NY, USA, 2017; pp. 285–343.
4. Whangbo, M.-H.; Koo, H.-J.; Kremer, R.K. Spin Exchanges Between Transition-Metal Ions Governed by the Ligand p-Orbitals in Their Magnetic Orbitals. *Molecules* **2021**, *26*, 531. [CrossRef]
5. Goodenough, J.B.; Loeb, A.L. Theory of ionic ordering, crystal distortion, and magnetic exchange due to covalent forces in spinels. *Phys. Rev.* **1955**, *98*, 391–408. [CrossRef]
6. Goodenough, J.B. Theory of the role of covalence in the perovskite-type manganites [La, M(II)]MnO_3. *Phys. Rev.* **1955**, *100*, 564–573. [CrossRef]
7. Kanamori, J. Superexchange interaction and symmetry properties of electron orbitals. *J. Phys. Chem. Solids* **1959**, *10*, 87–98. [CrossRef]
8. Goodenough, J.B. *Magnetism and the Chemical Bond*; Interscience; Wiley: New York, NY, USA, 1963.
9. Kurosawa, K.; Saito, S.; Yamaguchi, Y. Neutron diffraction study on $MnPS_3$ and $FePS_3$. *J. Phys. Soc. Jpn.* **1983**, *52*, 3919–3926. [CrossRef]
10. Ouvrard, G.; Brec, R.; and Rouxel, J. Structural determination of some MPS_3 phases (M = Mn, Fe, Co, Ni and Cd). *Mater. Res. Bull.* **1985**, *20*, 1181–1189. [CrossRef]
11. Brec, R. Review on structural and chemical properties of transition metal phosphorus trisulfides MPS_3. *Solid State Ionics* **1986**, *22*, 3–30. [CrossRef]
12. Kuroda, K.; Kurosawa, K.; Shozo, S.; Honda, M.; Zhihong, Y.; Date, M. Magnetic-properties of layered compound $MnPS_3$. *J. Phys. Soc. Jpn.* **1986**, *55*, 4456–4463.
13. Hicks, T.J.; Keller, T.; Wildes, A.R. Magnetic dipole splitting of magnon bands in a two-dimensional antiferromagnet. *J. Magn. Magn. Mater.* **2019**, *474*, 512–516. [CrossRef]
14. Ressouche, E.; Loire, M.; Simonet, V.; Ballou, L.; Stunault, A.; Wildes, A. Magnetoelectric $MnPS_3$ as a candidate for ferrotoroicity. *Phys. Rev. B* **2010**, *82*, 100408(R). [CrossRef]
15. Okuda, K.; Kurosawa, K.; Saito, S. High field magnetization process in $FePS_3$. In *High Field Magnetism*; Date, M., Ed.; Elsevier: North Holland, Amsterdam, The Netherlands, 1983; pp. 55–58.
16. Rule, K.C.; McIntyre, G.J.; Kennedy, S.J.; Hicks, T.J. Single-crystal and powder neutron diffraction experiments on $FePS_3$: Search for the magnetic structure. *Phys. Rev. B* **2007**, *76*, 134402. [CrossRef]
17. Wildes, A.R.; Rule, K.C.; Bewley, R.I.; Enderle, M.; Hicks, T.J. The magnon dynamics and spin exchange parameters of $FePS_3$. *J. Phys. Condens. Matter* **2012**, *24*, 416004. [CrossRef]
18. Lançon, D.; Walker, H.C.; Ressouche, E.; Ouladiaf, B.; Rule, K.C.; McIntyre, G.J.; Hicks, T.J.; Rønnow, H.M.; Wildes, A.R. Magnetic structure and magnon dynamics of the quasi0two-dimensional antiferromagnet $FePS_3$. *Phys. Rev. B* **2016**, *94*, 214407. [CrossRef]
19. Wildes, A.R.; Simonet, V.; Ressouche, E.; Ballou, R.; McIntyre, G.J. The magnetic properties and structure of the quasi-two-dimensional antiferromagnet $CoPS_3$. *J. Phys. Condes. Matter* **2017**, *29*, 455801. [CrossRef]
20. Wildes, A.R.; Simonet, V.; Ressouche, E.; McIntyre, G.J.; Avdeev, M.; Suard, E.; Kimber, S.A.J.; Lançon, D.; Pepe, G.; Moubaraki, B.; et al. Magnetic structure of the quasi-two-dimensional antiferromagnet $NiPS_3$. *Phys. Rev. B* **2015**, *92*, 224408. [CrossRef]
21. Chittari, B.L.; Park, Y.J.; Lee, D.K.; Han, M.S.; MacDonal, A.H.; Hwang, E.H.; Jung, J.I. Electronic and magnetic properties of single-layer MPX_3 metal phosphorous trichalcogenides. *Phys. Rev. B* **2016**, *94*, 184428. [CrossRef]
22. Whangbo, M.-H.; Gordon, E.E.; Xiang, H.J.; Koo, H.-J.; Lee, C. Prediction of spin orientations in terms of HOMO-LUMO interactions using spin-orbit coupling as perturbation. *Acc. Chem. Res.* **2015**, *48*, 3080–3087. [CrossRef]

23. Gordon, E.E.; Xiang, H.J.; Köhler, J.; Whangbo, M.-H. Spin orientations of the spin-half Ir^{4+} ions in Sr_3NiIrO_6, Sr_2IrO_4 and Na_2IrO_3: Density functional, perturbation theory and Madelung potential analyses. *J. Chem. Phys.* **2016**, *144*, 114706. [CrossRef]
24. Whangbo, M.-H.; Xiang, H.J.; Koo, H.-J.; Gordon, E.E.; Whitten, J.L. Electronic and Structural Factors Controlling the Spin Orientations of Magnetic Ions. *Inorg. Chem.* **2019**, *58*, 11854–11874. [CrossRef]
25. Hay, P.J.; Thibeault, J.C.; Hoffmann, R. Orbital interactions in metal dimer complexes. *J. Am. Chem. Soc.* **1975**, *97*, 4884–4899. [CrossRef]
26. Kresse, G.; Furthmüller, J. Efficiency of ab-initio total energy calculations for metals and semiconductors using a plane-wave basis set. *Comput. Mater. Sci.* **1996**, *6*, 15–50. [CrossRef]
27. Kresse, G.; Joubert, D. From ultrasoft pseudopotentials to the projector augmented-wave method. *Phys. Rev. B* **1999**, *59*, 1758–1775. [CrossRef]
28. Perdew, J.P.; Burke, K.; Ernzerhof, M. Generalized Gradient Approximation Made Simple. *Phys. Rev. Lett.* **1996**, *77*, 3865. [CrossRef]
29. Dudarev, S.L.; Botton, G.A.; Savrasov, S.Y.; Humphreys, C.J.; Sutton, A.P. Electron-energy-loss spectra and the structural stability of nickel oxide: An LSDA+U study. *Phys. Rev. B* **1998**, *57*, 1505. [CrossRef]
30. Kuneš, K.; Novák, P.; Schmid, R.; Blaha, P.; Schwarz, K. Electronic structure of fcc Th: Spin-orbit calculation with $6p_{1/2}$ local orbital extension. *Phys. Rev. Lett.* **2001**, *64*, 153102.
31. Vaclavkova, D.; Delhomme, A.; Faugeras, C.; Potemski, M.; Bogucki, A.; Suffczyński, J.; Kossacki, P.; Wildes, A.R.; Grémaud, B.; Saúl, A. Magnetoelastic interaction in the two-dimensional magnetic material $MnPS_3$ studied by first principles calculations and Raman experiments. *2D Mater.* **2020**, *7*, 035030. [CrossRef]
32. Koo, H.-J.; Xiang, H.J.; Lee, C.; Whangbo, M.-H. Effect of Magnetic Dipole-Dipole Interactions on the Spin Orientation and Magnetic Ordering of the Spin-Ladder Compound $Sr_3Fe_2O_5$. *Inorg. Chem.* **2009**, *48*, 9051. [CrossRef]
33. Ewald, P.P. Die Berechnung Optischer und Elektroststischer Gitterpotentiale. *Ann. Phys.* **1921**, *64*, 253. [CrossRef]
34. Darden, T.; York, D.; Pedersen, L. Particle mesh Ewald: An N·log(N) method for Ewald sums in large systems. *J. Chem. Phys.* **1993**, *98*, 10089. [CrossRef]
35. Wang, H.; Dommert, F.; Holm, C. Optimizing working parameters of the smooth particle mesh Ewald algorithm in terms of accuracy and efficiency. *J. Chem. Phys.* **2010**, *133*, 034117. [CrossRef] [PubMed]

Review

Spin Hamiltonians in Magnets: Theories and Computations

Xueyang Li [1,2,†], Hongyu Yu [1,2,†], Feng Lou [1,2], Junsheng Feng [1,3], Myung-Hwan Whangbo [4] and Hongjun Xiang [1,2,*]

1. Key Laboratory of Computational Physical Sciences (Ministry of Education), State Key Laboratory of Surface Physics, Department of Physics, Fudan University, Shanghai 200433, China; 16110190006@fudan.edu.cn (X.L.); 20110190060@fudan.edu.cn (H.Y.); flou16@fudan.edu.cn (F.L.); fjs@hfnu.edu.cn (J.F.)
2. Shanghai Qi Zhi Institute, Shanghai 200232, China
3. School of Physics and Materials Engineering, Hefei Normal University, Hefei 230601, China
4. Department of Chemistry, North Carolina State University, Raleigh, NC 27695-8204, USA; whangbo@ncsu.edu
* Correspondence: hxiang@fudan.edu.cn
† These authors contributed equally to this work.

Abstract: The effective spin Hamiltonian method has drawn considerable attention for its power to explain and predict magnetic properties in various intriguing materials. In this review, we summarize different types of interactions between spins (hereafter, spin interactions, for short) that may be used in effective spin Hamiltonians as well as the various methods of computing the interaction parameters. A detailed discussion about the merits and possible pitfalls of each technique of computing interaction parameters is provided.

Keywords: spin Hamiltonian; magnetism; energy-mapping analysis; four-state method; Green's function method

Citation: Li, X.; Yu, H.; Lou, F.; Feng, J.; Whangbo, M.-H.; Xiang, H. Spin Hamiltonians in Magnets: Theories and Computations. *Molecules* 2021, 26, 803. https://doi.org/10.3390/molecules26040803

Academic Editor: Takashiro Akitsu
Received: 24 December 2020
Accepted: 2 February 2021
Published: 4 February 2021

Publisher's Note: MDPI stays neutral with regard to jurisdictional claims in published maps and institutional affiliations.

Copyright: © 2021 by the authors. Licensee MDPI, Basel, Switzerland. This article is an open access article distributed under the terms and conditions of the Creative Commons Attribution (CC BY) license (https://creativecommons.org/licenses/by/4.0/).

1. Introduction

The utilization of magnetism can date back to ancient China when the compass was invented to guide directions. Since the relationship between magnetism and electricity was revealed by Oersted, Lorentz, Ampere, Faraday, Maxwell, and others, more applications of magnetism have been invented, which include dynamos (electric generators), electric motors, cyclotrons, mass spectrometers, voltage transformers, electromagnetic relays, picture tubes, and sensing elements. During the information revolution, magnetic materials were extensively employed for information storage. The storage density, efficiency, and stability were substantially improved by the discovery and applications of the giant magnetoresistance effect [1,2], tunnel magnetoresistance [3–9], spin-transfer torques [10–13], etc. Recently, more and more novel magnetic states such as spin glasses [14,15], spin ice [16,17], spin liquid [18–22], and skyrmions [23–28] were found, revealing both theoretical and practical significance. For example, hedgehogs and anti-hedgehogs can be seen as the sources (monopoles) and the sinks (antimonopoles) of the emergent magnetic fields of topological spin textures [29], while magnetic skyrmions have shown promise as ultradense information carriers and logic devices [24].

To explain or predict the properties of magnetic materials, many models and methods have been invented. In this review, we will mainly focus on the effective spin Hamiltonian method based on first-principles calculations, and its applications in solid-state systems. In Section 2, we will introduce the effective spin Hamiltonian method. Firstly, in Section 2.1, the origin and the computing methods of the atomic magnetic moments are presented. Then, from Sections 2.2–2.6, different types of spin interactions that may be included in the spin Hamiltonians are discussed. Section 3 will discuss and compare various methods of computing the interaction parameters used in the effective spin Hamiltonians. In Section 4, we will give a brief conclusion of this review.

2. Effective Spin Hamiltonian Models

Though accurate, first-principles calculations are somewhat like black boxes (that is to say, they provide the final total results, such as magnetic moments and the total energy, but do not give a clear understanding of the physical results without further analysis), and have difficulties in dealing with large-scale systems or finite temperature properties. In order to provide an explicit explanation for some physical properties and improve the efficiency of thermodynamic and kinetic simulations, the effective Hamiltonian method is often adopted. In the context of magnetic materials where only the spin degree of freedom is considered, it can also be called the effective spin Hamiltonian method. Typically, the effective spin Hamiltonian models need to be carefully constructed and include all the possibly important terms; then the parameters of the models need to be calculated based on either first-principles calculations (see Section 3) or experimental data (see Section 3.3). Given the effective spin Hamiltonian and the spin configurations, the total energy of a magnetic system can be easily computed. Therefore, it is often adopted in Monte Carlo simulations [30] (or quantum Monte Carlo simulations) for assessing the total energy of many different configurations so that the finite temperature properties of magnetic materials can be studied. If the effects of atom displacements are taken into account, the effective Hamiltonians can also be applied to the spin molecular dynamics simulations [31–33], which is beyond the scope of this review.

In this review, we mainly focus on the classical effective spin Hamiltonian method where atomic magnetic moments (or spin vectors) are treated as classical vectors. In many cases, these classical vectors are assumed to be rigid so that their magnitudes keep constant during rotations. This treatment significantly simplifies the effective Hamiltonian models, and it is usually a good approximation, especially when atomic magnetic moments are large enough.

In this part, we shall first introduce the origin of the atomic magnetic moments as well as the methods of computing atomic magnetic moments. Then different types of spin interactions will be discussed.

2.1. Atomic Magnetic Moments

The origin of atomic magnetic moments is explained by quantum mechanics. Suppose the quantized direction is the z-axis, an electron with quantum numbers (n, l, m_l, m_s) leads to an orbital magnetic moment $\mu_l = -\mu_B l$ and a spin magnetic moment $\mu_s = -g_e \mu_B s$, with their z components $\mu_{lz} = -m_l \mu_B$ and $\mu_{sz} = -g_e m_s \mu_B$, where $\mu_B = \frac{|e|\hbar}{2m}$ is the Bohr magneton and $g_e \approx 2$ is the g-factor for a free electron. The energy of a magnetic moment μ in a magnetic field B (magnetic induction) along z-direction is $-\mu \cdot B = -\mu_z B$.

Considering the Russell-Saunders coupling (also referred to as L-S coupling), which applies to most multi-electronic atoms, the total orbital magnetic moment and the total spin magnetic moment are $\mu_L = -\mu_B L$ and $\mu_S = -g_e \mu_B S$, respectively, where $L = \sum_i l_i$ and $S = \sum_i s_i$ are the summation over electrons. The quantum numbers of each electron can usually be predicted by Hund's rules. Owing to the spin–orbit interaction, L and S both precess around the constant vector $J = L + S$. The time-averaged effective total magnetic moment is $\mu = -g_J \mu_B J$, where

$$g_J = 1 + \frac{J(J+1) + S(S+1) - L(L+1)}{2J(J+1)} \quad (1)$$

The atomic magnetic moments discussed above are based on the assumption that the atoms are isolated. Taking the influence of other atoms and external fields into account, the orbital interaction theory, the crystal field theory [34], or the ligand field theory [35,36] may be a better choice for theoretically predicting atomic magnetic moments. Notice that half-filled shells lead to a total $L = 0$, and that in solids and molecules, orbital moments of electrons are usually quenched, resulting in an effective $L = 0$ [37] (counterexamples may be found for 4f elements or for $3d^7$ configurations as in Co(II)). Therefore typically $\mu = \mu_S$, with its z component $\mu_z = -g_e S_z \mu_B$ (S_z is restricted to discrete values: $S, S-1, \ldots, -S$). Usually, a nonzero μ_S results from singly filled (localized) d or f orbitals, while s and p orbitals are typically either doubly filled or vacant so that they have no direct contribution to the atomic magnetic moments. Therefore, when referring to atomic magnetic moments, usually only the atoms in the d- and f-transition series need to be considered.

The atomic magnetic moments can also be predicted numerically employing first-principles calculations. Nevertheless, we should notice that traditional Kohn–Sham density functional theory (DFT) calculations [38,39] (based on single-electron approximation) are not reliable for predicting atomic magnetic moments, and hence require the consideration of strong correlation effect among electrons, especially when dealing with localized d or f orbitals. Based on the Hubbard model [40], such problem can often be remedied by introducing an intra-atomic interaction with effective on-site Coulomb and exchange parameters, U and J [41] (or only one parameter $U^{\text{eff}} = U - J$ in Dudarev's approach [42]). This approach is the DFT+U method [41–44], including LDA+U (LDA: Local density approximation), LSDA+U (LSDA: Local spin density approximation), GGA+U (GGA: Generalized gradient approximation), and so forth, where "+U" indicates the Hubbard "+U" correction. The parameters U and J can be estimated according to experience or semi-empirically by seeking agreement with experimental results of some specific properties, which is convenient but not very reliable. Considering how the values of the parameters U and J affect the prediction of atomic magnetic moments and other physical properties, we may need to compute these parameters more rigorously. A typical approach is constrained DFT calculations [43,45–47], where the local d or f charges are constrained to different values in several calculations, so that the parameters U and J can be obtained. Another approach based on constrained random-phase approximation (cRPA) [48–50] allows for considering the frequency (or energy) dependence of the parameters. More methods of computing U and J are summarized in Ref. [43]. There are more accurate approaches for dealing with strong correlated systems like DFT + Dynamical Mean Field Theory (DFT+DMFT) [51–55] and Reduced Density Matrix Functional Theory (RDMFT) [56,57], but they are much more sophisticated and computationally demanding so that they may be impractical for large-scale calculations. Wave function (WF) methods, such as Complete Active Space Self-Consistent Field (CASSCF) [58–61], Complete Active Space second-order Perturbation Theory (CASPT2) [62–64], Complete Active Space third-order Perturbation Theory (CASPT3) [65], and Difference Dedicated Configuration Interaction (DDCI) [66–68], are also widely adopted by theoretical chemists for studying magnetic properties of materials (especially molecules), including atomic magnetic moments and magnetic interactions. These WF methods are also more accurate but more computationally demanding than the DFT+U method, more detailed discussions of which can be found in Ref. [69].

2.2. Heisenberg Model

The simplest effective spin Hamiltonian model is the classical Heisenberg model, which can be reduced to Ising model or XY model. The classical Heisenberg model can be written as

$$H_{\text{spin}} = \sum_{i,j>i} J_{ij} \mathbf{S}_i \cdot \mathbf{S}_j, \quad (2)$$

where $\mathbf{S}_i$ and $\mathbf{S}_j$ indicate the total spin vectors on atoms i and j, and the summation is over all relevant pairs (ij). Its form was suggested by Heisenberg, Dirac, and Van Vleck. Such an interaction comes from the energy splitting between quantized parallel (ferromagnetic, FM; triplet state) and antiparallel (antiferromagnetic, AFM; singlet state) spin configurations. $J_{ij} > 0$ and $J_{ij} < 0$ prefer AFM and FM configurations, respectively. There may be a difference in the factor such as -1 and $\frac{1}{2}$ between different definitions of H_{spin}, which is also the case in other models as will be discussed.

The spatial wave function of two electrons should possess the form of $\psi_\pm = \frac{1}{\sqrt{2}}[\psi_a(r_1)\psi_b(r_2) \pm \psi_b(r_1)\psi_a(r_2)]$, where ψ_a and ψ_b are any single-electron spatial wave functions. A parallel triplet spin state and an antiparallel singlet spin state should correspond to an antisymmetric (ψ_-) and a symmetric (ψ_+) spatial wave function, respectively. For given ψ_a and ψ_b, the expectation value for the total energy of ψ_- can be different from that of ψ_+, which gives a preference to the AFM or FM spin configuration. Whether the AFM or FM spin configuration is preferred depends on the circumstances, and their energy difference can be described as a Heisenberg term $J_{12}S_1 \cdot S_2$.

In the simple case of an H_2 molecule, an AFM singlet spin state is preferred, whose symmetric spatial wave function ψ_+ corresponds to a bonding state [70,71]. However, this leads to the total magnetic moment of zero because the two antiparallel electrons share the same spatial state. In another simple case, where ψ_a and ψ_b stands for two degenerate and orthogonal orbitals of the same atom, an FM triplet spin state is preferred, which is in agreement with Hund's rules. Consider a set of orthogonal Wannier functions with $\phi_{n\lambda}(r - r_\alpha)$ resembling the nth atomic orbital with spin λ centered at the αth lattice site, and suppose there are Nh electrons each localized on one of the N lattice sites, each ion possessing h unpaired electrons. If these h electrons have the same exchange integrals with all the other electrons, the interaction resulting from the antisymmetrization of the wave functions can be expressed as

$$H_{\text{ex}} = \sum_{\substack{\alpha\alpha' \\ nn'}} J_{nn'}(r_\alpha, r_{\alpha'}) \left[\frac{1}{4} + S(r_\alpha) \cdot S(r_{\alpha'})\right] \quad (3)$$

which is called the Heisenberg exchange interaction [72]. After removing the constant terms, we can see such an interaction has the form of $H_{\text{spin}} = \sum_{i,j>i} J_{ij} S_i \cdot S_j$.

Based on molecular orbital analysis using ϕ_a and ϕ_b to denote the singly filled d orbitals of the two spin-$\frac{1}{2}$ magnetic ions (i.e., d^9 ions), Hay et al. [73] showed that the exchange interaction between the two ions can be approximately expressed as

$$J = -2K_{ab} + \frac{\Delta^2}{U^{\text{eff}}} = J_F + J_{AF} \quad (4)$$

where

$$K_{ab} \propto \left\langle \phi_a(1)\phi_b(2) \left| \frac{1}{r_{12}} \right| \phi_b(1)\phi_a(2) \right\rangle = \int \phi_a^*(r_1)\phi_b^*(r_2) \frac{1}{r_{12}} \phi_b(r_1)\phi_a(r_2) > 0 \quad (5)$$

$$U^{\text{eff}} = J_{aa} - J_{ab} \propto \left\langle \phi_a(1)\phi_a(2) \left| \frac{1}{r_{12}} \right| \phi_a(1)\phi_a(2) \right\rangle - \left\langle \phi_a(1)\phi_b(2) \left| \frac{1}{r_{12}} \right| \phi_a(1)\phi_b(2) \right\rangle > 0 \quad (6)$$

and Δ indicates the energy gap between the bonding state and antibonding state constructed by ϕ_a and ϕ_b. The two components of J have opposite signs, i.e., $J_F = -2K_{ab} < 0$ and $J_{AF} = \frac{\Delta^2}{U^{\text{eff}}} > 0$, which give preference to FM and AFM spin configurations, respectively. For general d^n cases, more orbitals should be considered. Therefore the expression of J will be more complicated, but the exchange interaction can still be similarly decomposed into FM and AFM contributions [73]. An application of this analysis is that when calculating exchange parameter J using Dudarev's approach of DFT+U with a parameter U^{eff}, the calculated value of J should vary with U^{eff} approximately as $J = J_F + \frac{\Delta^2}{U^{\text{eff}}}$ with J_F and Δ^2 to be fitted [74]. However, this is no longer correct when $U^{\text{eff}} \to 0$.

Another mechanism that leads to FM spin configurations is the double exchange, in which the interaction between two magnetic ions is induced by spin coupling to mobile electrons that travels from one ion to another. A mobile electron has lower energy if the localized spins are aligned. Such a mechanism is essential in metallic systems containing ions with variable charge states [75,76].

The superexchange is another important indirect exchange mechanism, where the interaction between two transition-metal (TM) ions is induced by spin coupling to two electrons on a non-magnetic ligand (L) ion that connects them, forming an exchange path of TM-L-TM type. Different mechanisms were proposed to explain the superexchange interaction. In Anderson's mechanism [77,78], the superexchange results from virtual processes in which an electron is transferred from the ligand to one of the neighboring magnetic ions, and then another electron on the ligand couples with the spin of the other magnetic ion through exchange interaction. In Goodenough's mechanism [79,80], the concept of semicovalent bonds was invented, where only one electron given by the ligand predominates in a semicovalent bond. Because of the exchange forces between the electrons on the magnetic ion and the electron given by the ligand, the ligand electron with its spin parallel to the net spin of the magnetic ion will spend more time on the magnetic ion than that with an antiparallel spin if the d orbital of the magnetic ion is less than half-filled, and vice versa. The magnetic atom and the ligand are supposed to be connected by a semicovalent bond or a covalent bond when they are near, or by an ionic bond (or possibly a metallic-like bond) otherwise. The superexchange interaction with semicovalent bonds existing is also called semicovalent exchange interaction. Kanamori summarized the dependence of the sign of the superexchange parameter (whether FM or AFM) on bond angle, bond type and number of d electrons (in different mechanisms), which is often referred as Goodenough–Kanamori (GK) rules [80–82]. For the 180° (bond angle) case, generally, AFM interaction is expected between cations of the same kind (counterexamples may exist for d^4 cases such as Mn^{3+}-Mn^{3+}, where the sign depends on the direction of the line of superexchange), and FM interaction is expected between two cations with more-than-half-filled and less-than-half-filled d-shells, respectively [81]. For the 90° case, the results are usually the opposite [81]. A schematic diagram of superexchange interactions (between cations both with more-than-half-filled d-shell) is given in Figure 1. More details of the discussions can be found in Ref. [81] and Ref. [82].

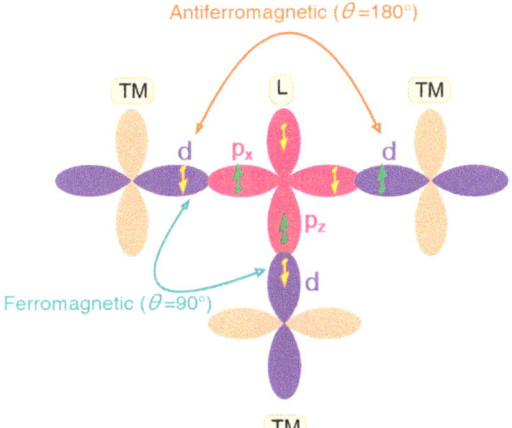

Figure 1. A schematic diagram of superexchange interactions between transition-metal (TM) ions both with more-than-half-filled d-shell. According to Goodenough–Kanamori (GK) rules, the 180° and the 90° cases favor antiferromagnetic (AFM) and ferromagnetic (FM) arrangements of TM ions, respectively. The main difference is whether the two electrons of L occupy the same p orbital, leading to different tendencies for the alignments of the two electrons of L that interact with two TM ions.

A counterexample of the GK rules can be found in the layered magnetic topological insulator MnBi$_2$Te$_4$, which possesses intrinsic ferromagnetism [83]. In contrast, the prediction of the GK rules leads to a weak AFM exchange interaction between Mn ions. In Ref. [84], the presence of Bi^{3+} was found to be essential for explaining this anomaly: d^5 ions in TM-L-TM spin-exchange paths would prefer FM coupling if the empty p orbitals of a nonmagnetic cation M (which is Bi^{3+} ion in the case of MnBi$_2$Te$_4$) hybridize strongly with those of the ligand L (but AFM coupling otherwise). Oleś et al. [85] pointed out that the GK rules may not be obeyed in transition metal compounds with orbital degrees of freedom (e.g., d^1 and d^2 electronic configurations) due to spin-orbital entanglement.

Exchange interactions between two TM ions also take place through the exchange paths of TM-L ... L-TM type [86], referred to as super-superexchanges, where TM ions do not share a common ligand. Each TM ion of a solid forms a TML$_n$ polyhedron (typically, n = 3–6) with the surrounding ligands L, and the unpaired spins of the TM ion are accommodated in the singly filled d-states of TML$_n$. Since each d-state has a d-orbital of TM combined out-of-phase with the p-orbitals of L, the unpaired spin of TM does not reside solely on the d-orbital of TM, as assumed by Goodenough and Kanamori, but is delocalized into the p-orbitals of the surrounding ligands L. Thus, TM-L ... L-TM type exchanges occur and can be strongly AFM when their L ... L contact distances are in the vicinity of the van der Waals distance so that the ligand p-orbitals overlap well across the L ... L contact.

Another mechanism is the indirect coupling of magnetic moments by conduction electrons, referred to as Ruderman–Kittel–Kasuya–Yosida (RKKY) interaction [87–90]. This kind of interaction between two spins S_1 and S_2 is also proportional to $S_1 \cdot S_2$ with an expression

$$H_{\text{RKKY}} \propto \sum_q \chi(q) e^{iq \cdot r_{21}} S_1 \cdot S_2. \tag{7}$$

The magnetic dipole–dipole interaction (between magnetic moments μ_1 and μ_2 located at different atoms) with energy

$$V = \frac{1}{R^3}\left[\mu_1 \cdot \mu_2 - 3(\hat{R} \cdot \mu_1)(\hat{R} \cdot \mu_2)\right] \tag{8}$$

also has contributions to the bilinear term, but it is typically much weaker than the exchange interactions in most solid-state materials such as iron and cobalt. The characteristic temperature of dipole–dipole interaction (or termed as "dipolar interaction") in magnetic materials is typically of the order of 1 K, above which no long-range order can be stabilized by such an interaction [37]. However, in some cases, such as in several single-molecule magnets (SMMs), the exchange interactions can be so weak that they are comparable to or weaker than dipolar interactions, thus the dipolar interactions must not be neglected [91].

For most magnetic materials, the Heisenberg interaction is the most predominant spin interaction. As a result, the simple classical Heisenberg model is able to explain the magnetic properties such as the ground states of spin configurations (FM or AFM) and the transition temperatures (Curie temperature for FM states or Néel temperature for AFM states) for many magnetic materials.

If some pairs of spins favor FM spin configurations while other pairs favor AFM configurations, frustration may occur, leading to more complicated and more interesting noncollinear spin configurations. For example, the FM effects of double exchange resulting from mobile electrons in some antiferromagnetic lattices give rise to a distortion of the ground-state spin arrangement and lead to a canted spin configuration [92]. A magnetic solid with moderate spin frustration lowers its energy by adopting a noncollinear superstructure (e.g., a cycloid or a helix) in which the moments of the ions are identical in magnitude but differ in orientation or a collinear magnetic superstructure (e.g., a spin density wave, SDW) in which the moments of the ions differ in magnitude but identical in orientation [93,94]. For a cycloid formed in a chain of magnetic ions, each successive spin rotates in one direction by a certain angle, so there are two opposite ways of rotating the successive spins hence producing two cycloids opposite in chirality but identical in energy. When these two cycloids occur with equal probability below a certain temperature, their superposition leads to a SDW [93,94]. On lowering the temperature further, the electronic structure of the spin-lattice relaxes to energetically favor one of the two chiral cycloids so that one can observe a cycloid state. The latter, being chiral, has no inversion symmetry and gives rise to ferroelectricity [95]. The spin frustration is also a potential driving force for topological states like skyrmions and hedgehogs [29].

2.3. The J Matrices and Single-Ion Anisotropy

The classical Heisenberg model can be generalized to a matrix form to include all the possible second-order interactions between two spins (or one spin itself):

$$H_{\text{spin}} = \sum_{i,j>i} S_i^T \mathbb{J}_{ij} S_j + \sum_i S_i^T \mathbb{A}_i S_i \qquad (9)$$

where $\mathbb{J}_{ij}$ and $\mathbb{A}_i$ are 3×3 matrices called the J matrix and single-ion anisotropy (SIA) matrix. The $\mathbb{J}_{ij}$ matrix can be decomposed into three parts: The isotropic Heisenberg exchange parameter $J_{ij} = (\mathbb{J}_{ij,xx} + \mathbb{J}_{ij,yy} + \mathbb{J}_{ij,zz})/3$ as in the classical Heisenberg model, the antisymmetric Dzyaloshinskii–Moriya interaction (DMI) matrix $\mathbb{D}_{ij} = (\mathbb{J}_{ij} - \mathbb{J}_{ij}^T)/2$ [96–98], and the symmetric (anisotropic) Kitaev-type exchange coupling matrix $\mathbb{K}_{ij} = (\mathbb{J}_{ij} + \mathbb{J}_{ij}^T)/2 - J_{ij}\mathbb{I}$ (where $\mathbb{I}$ denotes a 3×3 identity matrix). Thus $\mathbb{J}_{ij} = J_{ij}\mathbb{I} + \mathbb{D}_{ij} + \mathbb{K}_{ij}$ [99,100].

Now we analyze the possible origin of these terms by means of symmetry analysis. When considering interaction potential between (or among) spins, we should notice that the total interaction energy should be invariant under time inversion ($\{S_k\} \to \{-S_k\}$). Therefore, any odd order term in the spin Hamiltonian should be zero unless an external magnetic field is present when a term $-\sum_i \mu_i \cdot B = \sum_i g_e \mu_B B \cdot S_i$ should be added to the effective spin Hamiltonian. Ignoring the external magnetic field, the spin Hamiltonian should only contain even order terms, with the lowest order of significance being the second-order (the zeroth-order term is a constant and therefore not necessary). If the spin-orbit coupling (SOC) is negligible, the total effective spin Hamiltonian H_{spin} should be invariant under any global spin rotations, therefore H_{spin} should be expressed by only inner product terms of spins like terms proportional to $S_i \cdot S_j$, $(S_i \cdot S_j)(S_k \cdot S_l)$ and so on. That is to say, when SOC is negligible, the second-order terms in the H_{spin} should only include the classical Heisenberg term $\sum_{i,j>i} J_{ij} S_i \cdot S_j$, which implies that those interactions described by $\mathbb{A}_i$, $\mathbb{D}_{ij}$, and $\mathbb{K}_{ij}$ matrices all originate from SOC ($H_{\text{SO}} = \lambda \hat{S} \cdot \hat{L}$). What is more, if the spatial inversion symmetry is satisfied by the lattice, $\mathbb{J}_{ij}$ should be equal to $\mathbb{J}_{ij}^T$ so that there will be no DMI ($\mathbb{D}_{ij} = 0$). That is to say, the DMI can only exist where the spatial inversion symmetry is broken.

The SIA matrix $\mathbb{A}_i$ has only six independent components and is usually assumed to be symmetric. If we suppose the magnitude of the classical spin vector S_i to be independent of its direction, the isotropic part $\frac{1}{3}(\mathbb{A}_{i,xx} + \mathbb{A}_{i,yy} + \mathbb{A}_{i,zz})\mathbb{I}$ would be of no significance, and therefore $\mathbb{A}_i$ would have only five independent components after subtracting the isotropic part from itself. It is evident that the $S_i^T \mathbb{A}_i S_i$ prefers the direction of S_i along the eigenvector of $\mathbb{A}_i$ with the lowest eigenvalue. If this lowest eigenvalue is two-fold degenerate, the directions of S_i favored by SIA will be those belonging to the plane spanned by the two eigenvectors that share the lowest eigenvalue, in which case we say the ion i has easy-plane anisotropy. On the contrary, if the lowest eigenvalue is not degenerate while the higher eigenvalue is two-fold degenerate, we say the ion i has easy-axis anisotropy. In these two cases (easy-plane or easy-axis anisotropy), by defining the direction of z-axis parallel to the nondegenerate eigenvector, the $S_i^T \mathbb{A}_i S_i$ part would be simplified to $\mathbb{A}_{i,zz}(S_i^z)^2$ with only one independent component. The easy-axis anisotropy has been found to be helpful in stabilizing the long-range magnetic order and enhancing the Curie temperature in two-dimensional or quasi-two-dimensional systems [101]. The easy-plane anisotropy in three-dimensional ferromagnets can lead to the effect called "quantum spin reduction", where the mean spin at zero temperature has a value lower than the maximal one due to the quantum fluctuations [101,102]. Recently, several materials with unusually large easy-plane or easy-axis anisotropy were found [103–105], which, as single-ion magnets (SIMs), are promising for applications such as high-density information storage, spintronics, and quantum computing.

The DMI matrix $\mathbb{D}_{ij}$ is antisymmetric and therefore has only three independent components, which can be expressed by a vector $\boldsymbol{D}_{ij}$ with $\boldsymbol{D}_{ij,x} = \mathbb{D}_{ij,yz}$, $\boldsymbol{D}_{ij,y} = \mathbb{D}_{ij,zx}$, and $\boldsymbol{D}_{ij,z} = \mathbb{D}_{ij,xy}$. Thus, the DMI can be expressed by cross product: $\boldsymbol{S}_i^T \mathbb{D}_{ij} \boldsymbol{S}_j = \boldsymbol{D}_{ij} \cdot (\boldsymbol{S}_i \times \boldsymbol{S}_j)$. Such an interaction prefers the vectors $\boldsymbol{S}_i$ and $\boldsymbol{S}_j$ to be orthogonal to each other, with a rotation (of $\boldsymbol{S}_j$ relative to $\boldsymbol{S}_i$) around the direction of $-\boldsymbol{D}_{ij}$. Together with Heisenberg term $J_{ij} \boldsymbol{S}_i \cdot \boldsymbol{S}_j$, the preferred rotation angle between $\boldsymbol{S}_i$ and $\boldsymbol{S}_j$ relative to the collinear state preferred by the Heisenberg term would be $\arctan \frac{|\boldsymbol{D}_{ij}|}{|J_{ij}|}$. In Ref. [106], the DMI is shown to determine the chirality of the magnetic ground state of Cr trimers on Au(111). The DMI is also important in explaining the skyrmion states in many materials such as MnSi and FeGe [24,26,27,29,107–110]. Materials with skyrmion states induced by DMI usually have a large ratio of $\frac{|\boldsymbol{D}_1|}{|J_1|}$ (typically 0.1~0.2) where the subscript "1" means nearest pairs [24,100]. In Ref. [100], strong enough DMI for the existence of helical cycloid phases and skyrmionic states are predicted in Cr(I,X)$_3$ (X = Br or Cl) Janus monolayers (e.g., for Cr(I,Br)$_3$, supposing $|\boldsymbol{S}_i| = \frac{3}{2}$ for any i, the corresponding interaction parameters are computed as $J_1 = -1.800$ meV and $|\boldsymbol{D}_1| = 0.270$ meV, thus $\frac{|\boldsymbol{D}_1|}{|J_1|} = 0.150$), though monolayers such as CrI$_3$ only exhibit an FM state for lack of DMI. In Ref. [110], the nonreciprocal magnon spectrum (and the associated spectral weights) of MnSi, as well as its evolution as a function of magnetic field, is explained by a model including symmetric exchange, DMI, dipolar interactions, and Zeeman energy (related to the magnetic field).

The Kitaev matrix $\mathbb{K}_{ij}$ has five independent components as a symmetric matrix with zero trace. For the specific cases when $\boldsymbol{S}_i$ and $\boldsymbol{S}_j$ are parallel to each other (pointing in the same direction), $\boldsymbol{S}_i^T \mathbb{K}_{ij} \boldsymbol{S}_j$ would perform like $\boldsymbol{S}_i^T \mathbb{A}_i \boldsymbol{S}_i$ and show preference to the direction with the lowest eigenvalue of $\mathbb{K}_{ij}$; while when $\boldsymbol{S}_i$ and $\boldsymbol{S}_j$ are antiparallel to each other, $\boldsymbol{S}_i^T \mathbb{K}_{ij} \boldsymbol{S}_j$ would prefer the direction with the highest eigenvalue of $\mathbb{K}_{ij}$. The difference between the highest and lowest eigenvalue of $\mathbb{K}_{ij}$ can be defined as K_{ij} (a scalar), which characterizes the anisotropic contribution of $\mathbb{K}_{ij}$. Generally, the favorite direction of the spins is decided by both SIA and Kitaev interactions. The long-range ferromagnetic order in monolayer CrI$_3$ was explained by the anisotropic superexchange interaction since the Cr-I-Cr bond angle is close to 90° [111]. In Ref. [112], the interplay between the prominent Kitaev interaction and SIA was studied to explain the different magnetic behaviors of CrI$_3$ and CrGeTe$_3$ naturally. For CrI$_3$, supposing $|\boldsymbol{S}_i| = \frac{3}{2}$ for any i, the J_{ij} and K_{ij} parameters between nearest pairs are computed as -2.44 and 0.85 meV, respectively; while the only independent component $\mathbb{A}_{i,zz}$ of $\mathbb{A}_i$ is -0.26 meV. For CrGeTe$_3$, these three parameters are calculated to be -6.64, 0.36, and 0.25 meV, respectively. These two kinds of interactions are induced by SOC of the heavy ligands (I or Te) in these two materials (rather than the commonly believed Cr ions). Among different types of quantum spin liquids (QSLs), the exactly solvable Kitaev model with a ground state being QSL (with Majorana excitations) [113] has attracted much attention. Materials that achieve the realization of such Kitaev QSLs as α-RuCl$_3$ [114–116] and (Na$_{1-x}$Li$_x$)$_2$IrO$_3$ [117,118] (with an effective S = 1/2 spin value) with honeycomb lattices are discovered. A possible Kitaev QSL state is also predicted in epitaxially strained Cr-based monolayers with S = 3/2, e.g., CrSiTe$_3$ and CrGeTe$_3$ [119].

2.4. Fourth-Order Interactions without SOC

Sometimes, higher-order interactions are also crucial for explaining the magnetic properties of some materials, especially if the magnetic atoms have large magnetic moments or if the system is itinerant. As mentioned in Section 2.3, when SOC can be ignored, the effective spin Hamiltonian should only include inner product terms of spins. Besides second-order Heisenberg terms like $J_{ij}\mathbf{S}_i \cdot \mathbf{S}_j$, the terms with the next lowest order (which is fourth-order) are biquadratic (exchange) terms like $K_{ij}(\mathbf{S}_i \cdot \mathbf{S}_j)^2$, three-body fourth-order terms like $K_{ijk}(\mathbf{S}_i \cdot \mathbf{S}_j)(\mathbf{S}_i \cdot \mathbf{S}_k)$, and four-spin ring coupling terms like $K_{ijkl}(\mathbf{S}_i \cdot \mathbf{S}_j)(\mathbf{S}_k \cdot \mathbf{S}_l)$. That is to say, when SOC and external magnetic field are ignored, keeping the terms with orders no higher than fourth, the effective spin Hamiltonian can be expressed as

$$H_{\text{spin}} = \sum_{i,j>i} J_{ij}\mathbf{S}_i \cdot \mathbf{S}_j + \sum_{i,j>i} K_{ij}(\mathbf{S}_i \cdot \mathbf{S}_j)^2 + \sum_{\substack{i,j,\\k>j}} K_{ijk}(\mathbf{S}_i \cdot \mathbf{S}_j)(\mathbf{S}_i \cdot \mathbf{S}_k) + \sum_{\substack{i,j>i,\\k>i,\\l>k}} K_{ijkl}(\mathbf{S}_i \cdot \mathbf{S}_j)(\mathbf{S}_k \cdot \mathbf{S}_l) \quad (10)$$

The biquadratic terms have been found to be important in many systems, such as MnO [120,121], YMnO$_3$ [74], TbMnO$_3$ [122], iron-based superconductor KFe$_{1.5}$Se$_2$ [123], and 2D magnets [124]. In the case of TbMnO$_3$, besides the biquadratic terms, the four-spin couplings are also found to be important in explaining the non-Heisenberg behaviors [122]; the three-body fourth-order terms are also found to be important in simulating the total energies of different spin configurations [125] (a list of the fitted values of each important interaction parameter in TbMnO$_3$ is provided in the supplementary material of Ref. [125]). According to Ref. [126], in a Heisenberg chain system constructed from alternating $S > \frac{1}{2}$ and $S = \frac{1}{2}$ site spins, the additional isotropic three-body fourth-order terms are found to stabilize a variety of partially polarized states and two specific non-magnetic states including a critical spin-liquid phase and a critical nematic-like phase. In Ref. [127], the four-spin couplings were found to have a large effect on the energy barrier preventing skyrmions (or antiskyrmions) collapse into the ferromagnetic state in several transition-metal interfaces.

2.5. Chiral Magnetic Interactions Beyond DMI

Some high-order terms containing cross products of spins may also be necessary for fitting the models to the total energy or explaining some certain magnetic properties. Due to the chiral properties of these interactions like DMI, it is possible that they can also lead to, or explain, intriguing noncollinear spin textures such as skyrmions. In Ref. [128], topological–chiral interactions are found to be very prominent in MnGe, which includes chiral–chiral interactions (CCI) with the form

$$\kappa^{CC}_{ijki'j'k'}\left[\mathbf{S}_i \cdot (\mathbf{S}_j \times \mathbf{S}_k)\right]\left[\mathbf{S}_{i'} \cdot (\mathbf{S}_{j'} \times \mathbf{S}_{k'})\right] \quad (11)$$

whose local part has the form

$$\kappa^{CC}_{ijk}\left[\mathbf{S}_i \cdot (\mathbf{S}_j \times \mathbf{S}_k)\right]^2 \quad (12)$$

and spin–chiral interactions (SCI) with the form

$$\kappa^{SC}_{ijk}\left(\boldsymbol{\tau}_{ijk} \cdot \mathbf{S}_i\right)\left[\mathbf{S}_i \cdot (\mathbf{S}_j \times \mathbf{S}_k)\right] \quad (13)$$

where the unit vector $\boldsymbol{\tau}_{ijk} \propto (\mathbf{R}_j - \mathbf{R}_i) \times (\mathbf{R}_k - \mathbf{R}_i)$ is the surface normal of the oriented triangle spanned by the lattice sites $\mathbf{R}_i$, $\mathbf{R}_j$ and $\mathbf{R}_k$. The local scalar spin chirality $\chi_{ijk} = \mathbf{S}_i \cdot (\mathbf{S}_j \times \mathbf{S}_k)$ among triplets of spins can be interpreted as a fictitious effective magnetic field $\mathbf{B}^{\text{eff}} \propto \chi_{ijk}\boldsymbol{\tau}_{ijk}$, which leads to topological orbital moments $\mathbf{L}^{TO}$ (TOM) [129–134] arising from the orbital current of electrons hopping around the triangles. The TOM is defined as

$$\mathbf{L}^{TO}_i = \sum_{(jk)} \mathbf{L}^{TO}_{ijk} = \sum_{(jk)} \kappa^{TO}_{ijk} \chi_{ijk} \boldsymbol{\tau}_{ijk} \quad (14)$$

where κ^{TO}_{ijk} is the local topological orbital susceptibility. CCI corresponds to the interaction between pairs of topological orbital currents (or TOMs), whose local part can be interpreted as the orbital Zeeman interaction $\mathbf{L}^{TO}_i \cdot \mathbf{B}^{\text{eff}}$. SCI arises from the SOC, which couples the

TOM to single spin magnetic moments. An illustration of CCI and SCI is provided in Figure 2. Considerations of CCI and SCI improved the fitting of the total energy in MnGe (see details in Ref. [128]). Moreover, the authors showed the possibility that the CCI may lead to three-dimensional topological spin states and therefore may be vital in deciding the ground state of the spin configurations of MnGe, which was found to be a three-dimensional topological lattice (possibly built up with hedgehogs and anti-hedgehogs) experimentally [135].

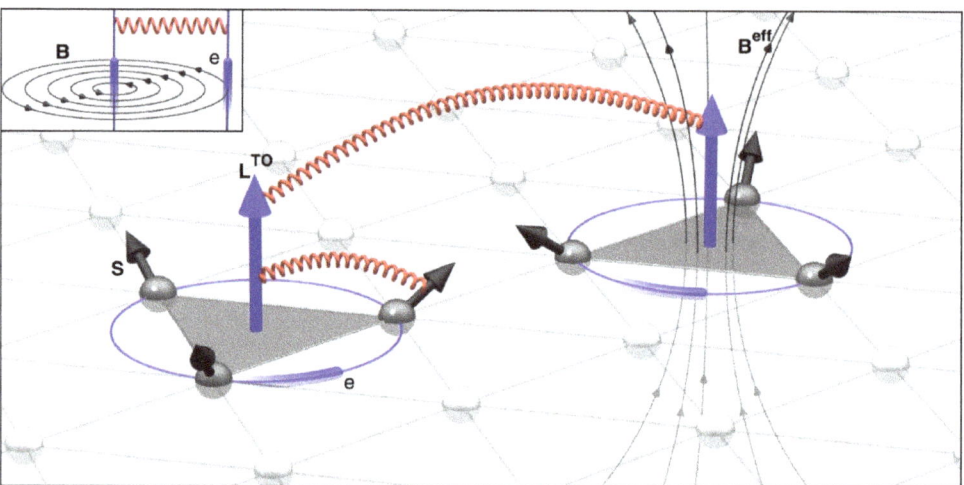

Figure 2. Schematic diagrams of chiral–chiral interactions (CCI) and spin–chiral interactions (SCI), as provided by S. Grytsiuk et al. in Ref. [128]. Spins and topological orbital moments (TOMs) are denoted as black arrows and blue arrows, respectively. CCI can be regarded as interactions between TOMs, while SCI can be interpreted as interactions between TOM and local spins.

A new type of chiral pair interaction $C_{ij} \cdot (S_i \times S_j)(S_i \cdot S_j)$, named as chiral biquadratic interaction (CBI), which is the biquadratic equivalent of the DMI, was derived from a microscopic model and demonstrated to be comparable in magnitude to the DMI in magnetic dimers made of 3d elements on Pt(111), Pt(001), Ir(111) and Re(0001) surface with strong SOC [136]. Similar but generalized chiral interactions such as $D_{ijjk} \cdot (S_j \times S_k)(S_i \cdot S_j)$ and $D_{ijkl} \cdot (S_k \times S_l)(S_i \cdot S_j)$, named four-spin chiral interactions, were discussed in Ref. [137], and they are found to be important in predicting a correct chirality for a spin spiral state of Fe chains deposited on the Re(0001) surface.

When there is a magnetic field, for a nonbipartite lattice, the magnetic field can couple with the spin and produce a new term of the form $J_{ijk} S_i \cdot (S_j \times S_k) = J_{ijk} \chi_{ijk}$ [138,139], which can be termed the three-spin chiral interaction (TCI) [140]. Such a chiral term can induce a gapless line in frustrated spin-gapped phases, and a critical chiral strength can change the ground state from spiral to Néel quasi-long-range-order phase [138]. This chiral term is also found to produce a chiral spin liquid state [141], where the time-reversal symmetry is broken spontaneously by the emergence of long-range order of scalar chirality [142].

2.6. Expansions of Magnetic Interactions

In general, a complete basis can be used to expand the spin interactions. One example is the spin-cluster expansion (SCE) [143–145], where unit vectors denoting the directions of spins are used as independent variables and spherical harmonic functions are used in the expressions of the basis functions. When SOC and the external magnetic field are not important, as mentioned in Section 2.3, only inner products of spins need to be considered. Consequently, the expansion can be

$$H_{\text{spin}} = E_0 + \sum_{n=1}^{\infty} J_n \left(\sum_{n\text{th nearest pairs } \langle k,l \rangle} e_k \cdot e_l \right) + \sum_{n'=1}^{\infty} J'_{n'} \sum_{n'\text{th type}} (e_k \cdot e_l)(e_m \cdot e_n) + \cdots \quad (15)$$

as used in Ref. [125]. When using expansions of spin vectors (or directions of the spins), suitable truncations on interaction distances and interaction orders are needed. Otherwise, the number of terms would be infinite, and as a result, the problem would be unsolvable.

3. Methods of Computing the Parameters of Effective Spin Hamiltonian Models

In this part, we mainly discuss the methods of computing the spin interaction parameters based on first-principles calculations of crystals, where periodic boundary conditions are tacitly supposed. These methods include different kinds of energy-mapping analysis (see Section 3.1) and Green's function method based on magnetic-force linear response theory (see Section 3.2). Discussions on the rigid spin rotation approximation and other assumptions are provided in Section 3.3. The cases of clusters, where periodic boundary conditions do not exist, will be briefly discussed in Section 3.3. Methods of obtaining spin interaction parameters from experiments will also be briefly mentioned in Section 3.3.

3.1. Energy-Mapping Analysis

In an energy-mapping analysis, we do several first-principles calculations to assess the total energies of different spin configurations. Then, we use the effective spin Hamiltonian to provide the expressions of the total energies of these spin configurations (with the expressions containing several undetermined parameters). By mapping the total energy expressions given by the effective spin Hamiltonian model to the results of first-principles calculations, the values of the undetermined parameters can be estimated. There are several types of energy-mapping analysis. For the first type, a minimal number of configurations are used, and a concrete expression for calculating the parameters can be given in advance. An example is by mapping between the eigenvalues and eigenfunctions of exact Hamiltonians and the effective spin Hamiltonian models (typically the Heisenberg model) to estimate the exchange parameters for relatively simple systems [146,147]. Several broken symmetry (BS) approaches are also of this type, where broken-symmetry states (instead of eigenstates of exact Hamiltonians) are adopted for energy mapping between the models and results of first-principles calculations [147,148]. A typical example of BS approach is the four-state method [148,149] where four special states are chosen for calculating each component of the parameters, which will be introduced in Section 3.1.1. For the second type, more configurations are used, and the parameters in the supposed effective spin Hamiltonian model are determined by employing least-squares fitting, which will be introduced in Section 3.1.2. The third type is similar to the second one, but the concrete form of the effective spin Hamiltonian model is not determined in advance. In the beginning, one includes many terms in the mode Hamiltonians. The relevance of each individual term depends on the fitting performance with respect to first-principles calculations. While selecting the relevant terms for a model Hamiltonian, it is important to search for the minimal Hamiltonian for a given magnetic system, namely, the one with the minimal number of parameters that capture its essential physics. This type of energy-mapping analysis will be introduced in Section 3.1.3.

In this section, we will mainly focus on the applications in solid state systems with periodic boundary conditions. The total energy of a designated configuration (which is usually a broken-symmetry state) is typically provided by first-principles calculations (e.g., DFT+U calculations) with constrained directions of magnetic moments.

3.1.1. Four-State Method

The energy-mapping analysis based on four ordered spin states [148,149], also referred to as the four-state method, assumes the effective spin Hamiltonian include only second-order terms (i.e., isotropic Heisenberg terms, DMI terms, Kitaev terms, and SIA terms). Each component of the parameters like J_{ij}, $\mathbf{D}_{ij,x}$, $\mathbb{A}_{i,xy}$, ($\mathbb{A}_{i,yy} - \mathbb{A}_{i,xx}$) and ($\mathbb{A}_{i,zz} - \mathbb{A}_{i,xx}$) can be obtained by first-principles calculations for four specified spin states [148]. Taking the isotropic Heisenberg parameter J_{ij} for example, with the spin-orbit coupling (SOC) switched off during the first-principles calculations, we use $E_{ij,\alpha\beta}$ ($\alpha, \beta = \uparrow, \downarrow$) to denote the energy of the configuration where spin i is parallel or antiparallel to the z direction (if $\alpha = \uparrow$ or $\downarrow$, respectively), spin j is parallel or antiparallel to the z direction (if $\beta = \uparrow$ or $\downarrow$, respectively), and all the spins except i and j are kept unchanged in the four states (which will be referred to as the "reference configuration", usually chosen to be a low-energy collinear state). Then J_{ij} can be expressed as

$$J_{ij} = \frac{E_{ij,\uparrow\uparrow} + E_{ij,\downarrow\downarrow} - E_{ij,\uparrow\downarrow} - E_{ij,\downarrow\uparrow}}{4S^2} \quad (16)$$

The schematic diagrams of these four states are shown in Figure 3.

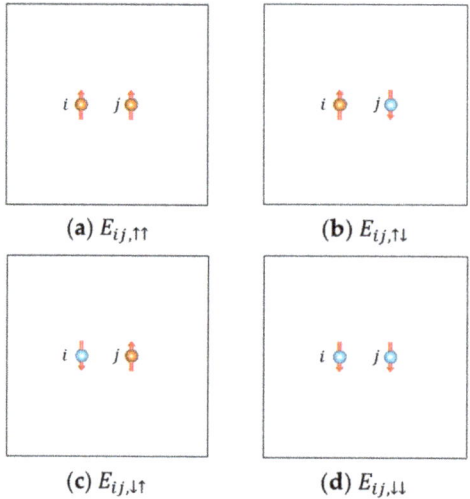

Figure 3. Schematic diagrams of the four states adopted in the four-state method for computing J_{ij}. Except atoms i and j, all the other atoms are omitted in this diagram (which are kept the same in the four states). Spins pointing up and down are indicated with orange and blue balls, respectively. The total energies given by first-principles calculations corresponding to these four states are denoted as (a) $E_{ij,\uparrow\uparrow}$, (b) $E_{ij,\uparrow\downarrow}$, (c) $E_{ij,\downarrow\uparrow}$, and (d) $E_{ij,\downarrow\downarrow}$, respectively.

In general, the second-order effective spin Hamiltonian takes the form of

$$H_{\text{spin}} = \sum_{i,j>i} S_i^T \mathbb{J}_{ij} S_j + \sum_i S_i^T \mathbb{A}_i S_i \quad (17)$$

and each component of the matrix $\mathbb{J}_{ij}$ and $\mathbb{A}_i$ can also be obtained by first-principles calculations for four specified spin states. To compute $\mathbb{J}_{ij,ab}$ ($a, b = x, y, z$), we use $E_{ij,ab,\alpha\beta}$ ($\alpha, \beta = \uparrow, \downarrow$) to denote the energy of the configuration where spin i is parallel or antiparallel to the a direction (if $\alpha = \uparrow$ or $\downarrow$, respectively), spin j is parallel or antiparallel to the b direction (if $\beta = \uparrow$ or $\downarrow$, respectively), and all the spins except for i and j are kept unchanged and parallel to the c-axis ($c = x, y$ or z, $c \neq a$, $c \neq b$) with an appropriate reference configuration and kept unchanged. Then, $\mathbb{J}_{ij,ab}$ can be expressed as

$$\mathbb{J}_{ij,ab} = \frac{E_{ij,ab,\uparrow\uparrow} + E_{ij,ab,\downarrow\downarrow} - E_{ij,ab,\uparrow\downarrow} - E_{ij,ab,\downarrow\uparrow}}{4S^2} \quad (18)$$

To compute $\mathbb{A}_{i,ab}$ ($a,b = x,y,z$ with $a \neq b$), we use $E_{i,ab,\alpha\beta}$ ($\alpha,\beta = \uparrow,\downarrow$) to denote the energy of the configuration where spin i is parallel to the direction whose a component is $\pm\frac{\sqrt{2}}{2}$ (for $\alpha = \uparrow$ or $\downarrow$, respectively), b component is $\pm\frac{\sqrt{2}}{2}$ (for $\beta = \uparrow$ or $\downarrow$, respectively), and the other component is 0. Here all the spins except for i and j are parallel to the c-axis ($c = x, y$ or z, $c \neq a$, $c \neq b$) with an appropriate reference configuration. Then, $\mathbb{A}_{i,ab}$ can be expressed as

$$\mathbb{A}_{i,ab} = \frac{E_{i,ab,\uparrow\uparrow} + E_{i,ab,\downarrow\downarrow} - E_{i,ab,\uparrow\downarrow} - E_{i,ab,\downarrow\uparrow}}{4S^2} \tag{19}$$

To compute $(\mathbb{A}_{i,aa} - \mathbb{A}_{i,bb})$ ($a,b = x,y,z$ with $a \neq b$), we use $E_{i,ab,\alpha\beta}$ ($\alpha,\beta = \uparrow,\downarrow$ or 0) to denote the energy of the configuration where spin i is parallel to the direction whose a component is ± 1 or 0 (if $\alpha = \uparrow$, $\downarrow$ or 0, respectively), b component is ± 1 or 0 (if $\beta = \uparrow$, $\downarrow$ or 0, respectively) and the other component is 0, while all the spins except i are parallel to the c-axis ($c = x, y$ or z, $c \neq a$, $c \neq b$) with an appropriate reference configuration. Then $(\mathbb{A}_{i,aa} - \mathbb{A}_{i,bb})$ can be expressed as

$$\mathbb{A}_{i,aa} - \mathbb{A}_{i,bb} = \frac{E_{i,ab,\uparrow 0} + E_{i,ab,\downarrow 0} - E_{i,ab,0\uparrow} - E_{i,ab,0\downarrow}}{4S^2} \tag{20}$$

It is easy to verify that, by employing this four-state method, each component of the $\mathbb{J}_{ij}$ and $\mathbb{A}_i$ can be obtained with the effects of other second-order terms entirely cancelled. Now we take the effects of fourth-order terms (without SOC) into account and check if the algorithms for computing $\mathbb{J}_{ij}$ and $\mathbb{A}_i$ are still rigorous. For $\mathbb{A}_i$, we can find out that the effects of all these terms are correctly cancelled. For $\mathbb{J}_{ij}$, the effects of biquadratic terms, three-body fourth-order terms, and most of the four-spin ring coupling terms are perfectly cancelled, while only the terms like $(\mathbf{S}_i \cdot \mathbf{S}_j)(\mathbf{S}_k \cdot \mathbf{S}_l)$ ($k,l \neq i,j$) will interfere with the calculation of J_{ij} because $(\mathbf{S}_k \cdot \mathbf{S}_l)$ is constant during the calculation and therefore mixed with the contribution of $J_{ij}(\mathbf{S}_i \cdot \mathbf{S}_j)$. The error of the calculated J_{ij} originated from four-spin ring coupling terms

$$\sum_{\substack{i,j>i,\\k>i,\\l>k}} K_{ijkl}(\mathbf{S}_i \cdot \mathbf{S}_j)(\mathbf{S}_k \cdot \mathbf{S}_l) \tag{21}$$

is given by

$$\sum_{\substack{k \neq i \text{ or } j\\l \neq i \text{ or } j\\(l > k)}} K_{ijkl}(\mathbf{S}_k \cdot \mathbf{S}_l) \tag{22}$$

while there is no easy way to get rid of this problem perfectly. Other parts of the $\mathbb{J}_{ij}$ (including $\mathbb{D}_{ij}$ and $\mathbb{K}_{ij}$) are not affected by these fourth-order terms (without SOC), but by other types of fourth-order terms (like four-spin chiral interactions) if SOC is taken into account (because of the similar reason).

In Ref. [122], the four-spin ring coupling interaction is found to be important in TbMnO$_3$, and therefore leads to instability in calculating the Heisenberg parameters using the four-state method when changing the reference configurations. This problem is remedied by calculating the Heisenberg parameter J_{ij} twice with the four-state method using the FM and A-type AFM (A-AFM, see Figure 4c) as the reference configurations, and use their average value as the final estimation of J_{ij}. The parameter of the vital ring coupling interaction is obtained by calculating the difference between the two calculated J_{ij} values (with FM and A-AFM reference configurations, respectively). This effective remedy is based on the assumption that only one kind of ring coupling interaction is essential. However, if there are more kinds of significant ring coupling or if we do not know which ring coupling is essential in advance, such a method of calculating J_{ij} is still not very trustworthy. Nevertheless, it is found that by taking the average of the calculated J_{ij} with four-state method with FM and G-type AFM (G-AFM, see Figure 4a) reference configurations, the influences of $K_{ijkl}(S_i \cdot S_j)(S_k \cdot S_l)$ with k and l being nearest pairs are eliminated. The terms $K_{ijkl}(S_i \cdot S_j)(S_k \cdot S_l)$ with non-nearest pairs of k and l still interfere with the calculation of J_{ij}, but they are generally very weak. Therefore, such a remedy to calculate J_{ij} using the four-state method should work well in most cases. In cases when a G-AFM state (in which all the nearest pairs of spins are antiparallelly aligned) cannot be defined (e.g., a triangular or a Kagomé lattice), there may be more than two reference configurations to use in the four-state method so as to eliminate the effects of $K_{ijkl}(S_i \cdot S_j)(S_k \cdot S_l)$ terms with non-nearest pairs of k and l. These reference configurations need to be designed carefully according to the specific circumstances to get rid of the effects of $K_{ijkl}(S_i \cdot S_j)(S_k \cdot S_l)$ terms as much as possible.

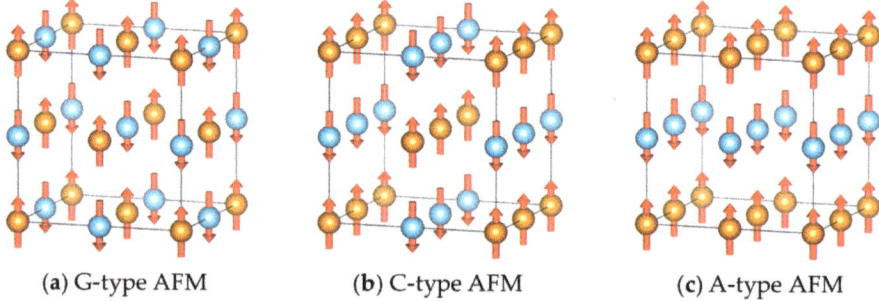

(a) G-type AFM   (b) C-type AFM   (c) A-type AFM

Figure 4. Schematic diagrams of (**a**) G-type AFM, (**b**) C-type AFM, and (**c**) C-type AFM states. Spins pointing to two opposite directions (e.g., up and down) are indicated with orange and blue balls, respectively.

The main advantages of the four-state method are its relatively small amount of first-principles calculations and its relatively good cancellations of other relevant terms. A weakness is that it cannot analyze the uncertainties of the parameters, so that we do not know how precise those estimated values are. Another weakness, which is also shared with other energy-mapping analysis methods, is that the computed $\mathbb{J}_{ij}$ between S_i and S_j is actually the sum over $\mathbb{J}_{ij'}$ with any lattice vector $\mathbf{R} = \mathbf{r}_{j'} - \mathbf{r}_j$. Therefore, to get rid of the effects of other spin pairs, a relatively large supercell is needed.

The four-state method can also be generalized to compute biquadratic parameters, where the SOC needs to be switched off during the first-principles calculations. For calculating K_{ij} in the term $K_{ij}(S_i \cdot S_j)^2$, we can let S_i pointing to the $(1,0,0)$ direction, S_j pointing to the $(1,0,0)$, $(-1,0,0)$, $\left(\frac{1}{\sqrt{2}}, \frac{1}{\sqrt{2}}, 0\right)$ and $\left(-\frac{1}{\sqrt{2}}, -\frac{1}{\sqrt{2}}, 0\right)$ directions, with other spins parallel to the z-axis. The corresponding total energies are denoted as E_1, E_2, E_3, and E_4, respectively. Then the K_{ij} can be expressed as

$$K_{ij} = E_1 + E_2 - E_3 - E_4 \tag{23}$$

It can be easily checked that the effects of other terms not higher than fourth order are totally eliminated. Therefore, this approach of calculating K_{ij} should be relatively rigorous theoretically.

Note that the four-state methods [148,149] could also give the derivatives of exchange interactions with respect to the atomic displacements without doing additional first-principles calculations due to the Hellmann-Feynman theorem. These derivatives are useful for the study of spin-lattice coupling related phenomena.

3.1.2. Direct Least Squares Fitting

Another type of energy-mapping analysis, instead of the four-state method, uses more first-principles calculations with different spin configurations and fits the results to the effective spin Hamiltonian using the least-squares method to estimate the parameters [74,122–124]. Ways of choosing spin configurations can depend on which parameters to estimate.

In Ref. [74], for the four Mn sites in a unit cell of YMnO$_3$, the polar and azimuthal angles (θ, φ) of their spins are given by $(0, 0)$, $(0, 0)$, $\left(\theta, \frac{3\pi}{2}\right)$, and $\left(\theta, \frac{\pi}{2}\right)$, respectively. By changing the θ from $0°$ to $180°$, different configurations are produced. If the effective spin Hamiltonian only contains Heisenberg terms, there will be a systematic deviation between the predicted value given by the effective Hamiltonian and the calculated value given by first-principles calculations. Such a deviation is well remedied by adding biquadratic exchange interactions into the effective Hamiltonian model. Thus, the biquadratic parameters can be fitted. Similar approaches were adopted by others to calculate and show the importance of biquadratic parameters and topological chiral–chiral contributions [122–124,128]. Apart from using an angular variable for generating spin configurations, using two or more variables is also practicable, or randomly chosen directions [106] can also be considered. Thus, more diverse configurations will be produced. The least-squares fitting will also work, but whether a systematic deviation exists will not be as apparent as the case when only one angular variable is used for generating configurations, and the fitting task may be more laborious.

The main virtues of this method are that the reliability of the model can be checked by the fitting performance and that the uncertainties of the parameters can be estimated if needed. This method is especially suitable for calculations of biquadratic parameters and topological CCI. However, when talking about calculations of Heisenberg parameters, this method needs more first-principles calculations and is thus less efficient. Furthermore, the fitted result of the Heisenberg parameter J_{ij} is vulnerable to the effects of other fourth-order interactions such as terms as $(S_i \cdot S_j)(S_i \cdot S_k)$ and $(S_i \cdot S_j)(S_k \cdot S_l)$. Therefore, the estimations of the Heisenberg parameters may not be very reliable if any of such fourth-order interactions are essential. This problem can be remedied by adding the related terms into the effective Hamiltonian model, while it is not easy to decide which terms to include in the model beforehand.

A possible way to get rid of the effects of other high-order terms is to perform artificial calculations where most of the magnetic ions are replaced by similar but nonmagnetic ions (e.g., substituting Fe^{3+} ions with nonmagnetic Al^{3+} ions) except for one or more ions to be studied [150]. For example, when calculating SIA, only one magnetic ion is not substituted, and by rotating this magnetic ion and calculating the total energy, the SIA can be studied. When studying two-body interactions between S_i and S_j, only two magnetic ions (at site i and j) are not substituted, and by rotating the spins (or magnetic moments) of these two magnetic ions, the interactions between them can be studied. Such a technique of substituting atoms can be applied to energy mapping analysis based on either the four-state method or least-squares fitting. In this way, effects from interactions involving other sites are effectively avoided. Nevertheless, this substitution method can make the chemical environments of the remaining magnetic ions different from those in the system with no substitution. This may make the calculations of the interaction parameters untrustworthy.

3.1.3. Methods Based on Expansions and Selecting Important Terms

The traditional energy-mapping analysis needs to construct an effective spin Hamiltonian first and then fit the undetermined parameters. However, it is not easy to give a perfect guess, especially when high-order interactions are essential. Such problems can be solved by considering almost all the possible terms utilizing some particular expansion with appropriate truncations. Usually, there are too many possible terms to be considered, so a direct fitting is impracticable; it requires at least as many first-principles calculations as the number of terms to determine, but leads to over-fitting problems due to too many parameters to determine. So, it is necessary to decide whether or not to include each term into the effective spin Hamiltonian on the basis of their contributions to the fitting performance.

In Ref. [145], SCE is adopted for the expansion of spin interactions of bcc and fcc Fe. After truncations based on the interacting distance and interaction orders, they considered 154 (179) possible different interaction terms in bcc (fcc) Fe. They randomly generated 3954 (835) different spin configurations in a 2 × 2 × 2 supercell for fitting. Their method of choosing terms is as follows: starting from the effective Hamiltonian model with only a constant term, try adding each possible term into the temporary model and accept the one providing the best fitting performance, in which way the terms are added to the model one by one. This method is the forward selection in variable selection problems, which is simple and straightforward, and it works well in most cases. However, this method may include unnecessary interactions to the effect Hamiltonian.

In Ref. [125], a machine learning method for constructing Hamiltonian (MLMCH) is proposed, which is more efficient and more reliable than the traditional forward selection method. Firstly, a testing set is used to avoid over-fitting problems. Secondly, not only adding terms but also deleting and substituting terms are considered during the search for the appropriate model. Thus, if an added term is judged to be unnecessary, it can still be removed from the model afterward. A penalty factor p^λ ($\lambda \geq 1$), where p is the number of parameters in the temporary model and λ is a given parameter, is used together with the loss function σ^2 (the fitting variance) to select models with fewer parameters. Several techniques are used to reduce the search space and enhance the search efficiency to select important terms out of tens of thousands of possible terms. The flow charts of this method of variable selection as well as the forward selection method are shown in Figure 5.

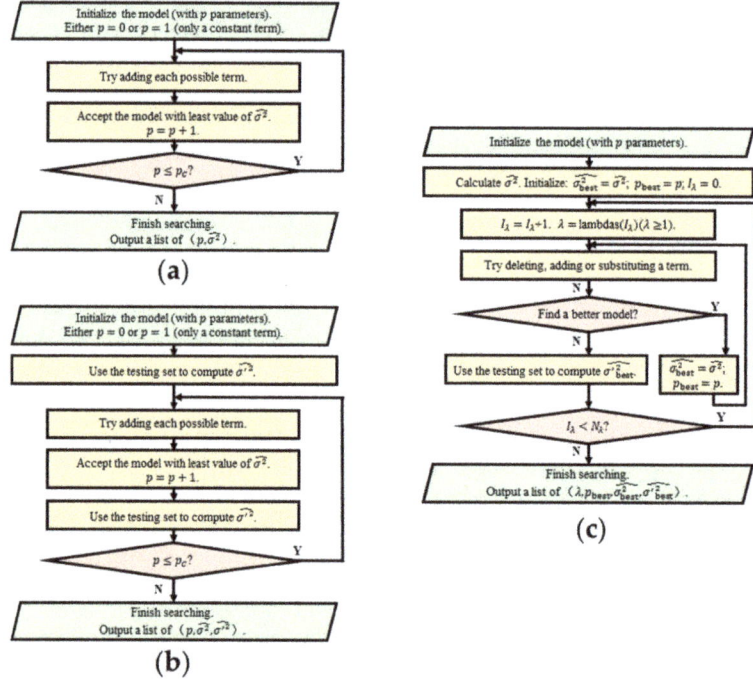

Figure 5. Flow charts of several methods of variable selection. In these flow charts, the fitting variance σ^2 estimated by the training set and the testing set are denoted as $\hat{\sigma}^2$ and $\widehat{\sigma'^2}$, respectively. (**a**) The traditional forward selection method as adopted in Ref. [145]. (**b**) The forward selection method using a testing set to check if over-fitting problems occur. (**c**) A simplified flow chart of the algorithm used in MLMCH [125], where some details are omitted. A testing set is used to check if over-fitting problems occur. The criterion for a better model is a smaller $(\sigma^2 \cdot \lambda^p)$ with $\lambda \geq 1$. There are N_λ values of λ, which are set in advance, saved in the array "lambdas(1: N_λ)", whose components are usually arranged in descending order of magnitude. For each value of λ, the best model is selected by using the criterion $(\sigma^2 \cdot \lambda^p)$.

This method is advantageous in two ways: (a) Constructing the effective spin Hamiltonians is carried out comprehensively, which makes it less likely to miss some critical interaction terms; (b) this method is general, so it can be applied to most magnetic materials. The least-squares fitting needed in this approach can also provide the estimations for the uncertainties of the parameters. The flaw is that it needs lots of (typically hundreds of) first-principles calculations, which could be impracticable when a very large supercell is needed (especially when the material is metallic so that long-range interactions are essential). The way to generate spin configurations (typically randomly distributed among all possible directions, sometimes deviating only moderately from the ground state) may have some room for improvement.

3.2. Green's Function Method Based on Magnetic-Force Linear Response Theory

In Green's function method based on magnetic-force linear response theory [151–159], we need localized basis functions $\psi_{im\sigma}(r)$ (i, m, σ indicating the site, orbital, and spin indices, respectively) based on the tight-binding model. The localized basis functions can be provided by DFT codes together with Wannier90 [160,161] or codes based on localized orbitals. By defining

$$\mathbb{H}_{imjm'\sigma\sigma'}(\boldsymbol{R}) = \langle \psi_{im\sigma}(r) | H | \psi_{im\sigma}(r+\boldsymbol{R}) \rangle \quad (24)$$

$$\mathbb{S}_{imjm'\sigma\sigma'}(\boldsymbol{R}) = \langle \psi_{im\sigma}(r) | \psi_{im\sigma}(r+\boldsymbol{R}) \rangle \quad (25)$$

$$\mathbb{H}(\boldsymbol{k}) = \sum_{R} \mathbb{H}(\boldsymbol{R}) e^{i\boldsymbol{k}\cdot\boldsymbol{R}} \quad (26)$$

$$\mathbb{S}(\boldsymbol{k}) = \sum_{R} \mathbb{S}(\boldsymbol{R}) e^{i\boldsymbol{k}\cdot\boldsymbol{R}} \quad (27)$$

the Green's function in reciprocal space and real space are defined as

$$\mathbb{G}(\boldsymbol{k}, \varepsilon) = (\varepsilon \mathbb{S}(\boldsymbol{k}) - \mathbb{H}(\boldsymbol{k}))^{-1} \quad (28)$$

and

$$\mathbb{G}(\boldsymbol{R}, \varepsilon) = \int_{BZ} \mathbb{G}(\boldsymbol{k}, \varepsilon) e^{-i\boldsymbol{k}\cdot\boldsymbol{R}} d\boldsymbol{k} \quad (29)$$

Based on the magnetic force theorem [162], the total energy variation due to a perturbation (which is the rotation of spins in this case) from the ground state equals the change of single-particle energies at the fixed ground-state potential:

$$\delta E = \int_{-\infty}^{E_F} \varepsilon \delta n(\varepsilon) d\varepsilon = -\int_{-\infty}^{E_F} \delta N(\varepsilon) d\varepsilon \quad (30)$$

where

$$n(\varepsilon) = -\frac{1}{\pi} Im \operatorname{Tr}(\mathbb{G}(\varepsilon)) \quad (31)$$

and

$$N(\varepsilon) = -\frac{1}{\pi} Im \operatorname{Tr}(\varepsilon - \mathbb{H}) \quad (32)$$

where traces are taken over orbitals. By defining

$$\mathbb{P}_i = \mathbb{H}_{ii}(\boldsymbol{R}=0) = \mathbb{1}_i^0 \ll + \vec{\mathbb{1}}_i \cdot \vec{\sigma} \quad (33)$$

with its component

$$\mathbb{P}_{imm'} = p_{imm'}^0 \ll + \vec{p}_{imm'} \cdot \vec{\sigma} \quad (34)$$

where $\vec{\sigma}$ is the vector composed of Pauli matrices. By defining

$$\mathbb{G}_{im,jm'} = G_{im,jm'}^0 \ll + \vec{G}_{im,jm'} \cdot \vec{\sigma} \quad (35)$$

the energy variation due to the two-spin interaction between sites i and j is

$$\delta E_{ij} = -\frac{2}{\pi} \int_{\infty}^{E_F} Im \text{Tr}\left(\delta \mathbb{H}_i \mathbb{G} \delta \mathbb{H}_j \mathbb{G}\right) d\varepsilon \tag{36}$$

with $\delta \mathbb{H}_i = \delta \boldsymbol{\phi}_i \times \boldsymbol{p}_i$. After mathematical simplification, the expression of δE_{ij} can be mapped to that given by the effective Hamiltonian model

$$H_{\text{spin}} = \sum_i S_i^T \mathbb{A}_i S_i + \sum_{i,j>i} \left[J_{ij} S_i \cdot S_j + \boldsymbol{D}_{ij} \cdot (S_i \times S_j) + S_i^T \mathbb{K}_{ij} S_j \right] + \sum_{i,j>i} K_{ij} (S_i \cdot S_j)^2 \tag{37}$$

(including all the second-order terms and a biquadratic term) to obtain the expressions of the parameters:

$$J_{ij} = Im\left(A_{ij}^{00} - A_{ij}^{xx} - A_{ij}^{yy} - A_{ij}^{zz} - 2 A_{ij}^{zz} S_i^{\text{ref}} \cdot S_j^{\text{ref}} \right) \tag{38}$$

$$\mathbb{K}_{ij}^{uv} = Im\left(A_{ij}^{uv} + A_{ij}^{vu} \right) \tag{39}$$

$$D_{ij}^u = Re\left(A_{ij}^{0u} - A_{ij}^{u0} \right) \tag{40}$$

$$B_{ij} = Im\left(A_{ij}^{zz} \right) \tag{41}$$

where

$$A_{ij}^{uv} = \frac{1}{\pi} \int_{-\infty}^{E_F} \text{Tr}\left\{ l_i^z \mathbb{G}_{ij}^u l_j^z \mathbb{G}_{ji}^v \right\} d\varepsilon \tag{42}$$

with $u, v \in \{0, x, y, z\}$ (the trace is also taken over orbitals), and S_i^{ref} indicates the unperturbed vector S_i. An xyz average strategy can be adopted so that some components inaccessible from one first-principles calculation can be obtained [157].

The main advantages of this method are that it only requires one or three DFT calculations to obtain all the parameters of second-order terms and biquadratic terms between different atoms, using only a small supercell (with a dense enough k-point sampling) to obtain interaction parameters between spins far away from each other. Therefore, it saves the computational cost compared to the energy-mapping analysis, especially when long-range interactions are essential. It also avoids the difficulties of reaching self-consistent-field convergence in DFT calculations for high-energy configurations, which may occur in the energy-mapping analysis. This method is good at describing states near the ground state but may not be so good at describing high-energy states. A limitation is that this method cannot obtain SIA parameters, and its calculations for biquadratic parameters are not very trustworthy [157]. The calculated Heisenberg parameter $\tilde{J}_{ij}$ is mixed with the contributions of other fourth-order interactions such as terms like $(S_i \cdot S_j)(S_i \cdot S_k)$ and $(S_i \cdot S_j)(S_k \cdot S_l)$. Therefore, the results may be unreliable if any of such fourth-order interactions are essential. Another little flaw is the noise of a typical order of magnitude of a few μeV introduced by the process of obtaining the Wannier orbitals [157]. In addition, the uncertainties of the parameters cannot be obtained by this method.

A recent study [140] considered the rotations of more than two spins as the perturbation and mapped the δE to the corresponding quantity given by an effective Hamiltonian with more types of interactions, including terms proportional to $S_i \cdot (S_j \times S_k)$, $(S_i \cdot S_j)(S_k \cdot S_l)$ and $(S_i \times S_j)(S_k \cdot S_l)$. This enables one to obtain the expressions needed for calculating the associated parameters. The derivations and forms of the expressions are much more complicated compared with those from the second-order interactions discussed above. This generalization of the traditional approach for calculating second-order interaction parameters remedied the problem of ignoring the effects of other high-order interactions to some extent. However, it is still a challenging task to get rid of the effects of high-order interactions on calculating the Heisenberg parameters (and other second-order parameters). The noise introduced by Wannier orbitals, the inability to determine SIA and the uncertainties of the resulting parameters are still the drawbacks of this approach. In addition, this method cannot give the derivatives of exchange interactions with respect to the atomic displacements, in contrast to the four-state method discussed above.

3.3. More Discussions on Calculating Spin Interaction Parameters

We should notice that, for all the methods discussed above, a rigid spin rotation approximation is used. The latter is equivalent to the supposition that the magnitudes of the spins should be constant in different configurations. However, this is not always true. For example, the magnitudes of the spins may be a little different in FM and AFM states. In Ref. [128], the energy-mapping analysis based on direct least-squares fitting (with configurations generated with different θ values) is adopted to study the spin interactions of MnGe and FeGe, to find that the agreement between the calculations and the model is enhanced by allowing the magnitudes of the spins to depend on the parameter θ (which decides the configurations) instead of using the fixed magnitudes (see Supplementary Materials of Ref. [128]). It is possible to obtain the relationship between the magnitudes of the spins and the parameter θ with an appropriate fitting or interpolation so that for a configuration with a new value of θ, the magnitudes of spins and the total energy can be predicted. Nevertheless, for a general spin configuration that cannot be described by a single θ, the prediction for the magnitudes of the spins can be very difficult. This is why one commonly employs the effective spin Hamiltonian by assuming that the magnitudes of the spins are constant.

Another perspective for the rigid spin rotation approximation, as implied in Ref. [143], is that even if the magnitudes of the spins are highly relevant to the configurations, the total energy can be fitted by using the directions of spins, instead of the spin vectors themselves, as the independent variables (which is mathematically equivalent to supposing the magnitudes of the spins to be constant) and considering an appropriate expansion of these variables (spin directions). For example, supposing SOC can be ignored (supposing SOC is switched off during first-principles calculations), the Heisenberg term $J_{ij}S_i \cdot S_j$ can be expressed as $J_{ij}S_i \cdot S_j = J_{ij}S_i e_i \cdot S_j e_j$; the magnitudes of the spins S_i and S_j depend on the angles between these two spins or neighboring spin directions (e.g., e_k). Therefore,

$$J_{ij}S_i \cdot S_j = \tilde{J}_{0,ij} e_i \cdot e_j + C_1 (e_i \cdot e_j)^2 + C_2 (e_i \cdot e_k)(e_i \cdot e_j) + C_3 (e_j \cdot e_{k'})(e_i \cdot e_j) + \cdots. \quad (43)$$

That is to say, the relevance of the magnitudes of spins to the configurations can be transferred to higher-order interactions when supposing the magnitudes of spins to be constant. These "artificial" higher-order terms, only emerging for compensating for such configuration dependency, are not very physical but can somewhat improve the fitting performances (when such dependency is prominent).

All the methods discussed above assumes, besides the rigid spin rotation approximation, that magnetic moments are localized on the atoms. In addition, we note that the DFT calculation results depend on the chosen exchange-correlation functional and the value of DFT+U parameters [157].

In the above discussions, we have supposed the periodic boundary conditions, for we mainly focus on the studies of crystals. When dealing with clusters (e.g., single-molecule magnets), we can still arrange a cluster in a crystal (using periodic boundary conditions) [159] with enough vacuum space to prevent the interactions between two clusters belonging to different periodic cells. If no periodic boundary conditions exist, the energy-mapping analysis can still work, while Green's function method based on magnetic-force linear response theory will fail because the reciprocal space is not defined. For the cases without periodic boundary conditions, theoretical chemists have developed several other approaches (such as wave-function based quantum chemical approaches) for studying magnetic interactions [69,146,147,163–165], detailed discussions of which are beyond the scope of this review.

The spin interaction parameters can also be obtained from comparing the experimental results of observable quantities such as transition temperatures, magnetization [166], specific heat [166], magnetic susceptibility [166,167], and magnon spectrum (given by inelastic neutron scattering measurements) [110,124,168–171] with the corresponding predictions given by the effective Hamiltonian model, which is similar to the idea of energy-mapping analysis based on least squares fitting. While the transition temperatures can only be used to roughly estimate the major interaction (typically the Heisenberg interaction between nearest pairs), the magnon spectrum can provide more detailed information and thus widely adopted for obtaining interaction parameters. These experimental results can also be used for checking the reliability of effective spin Hamiltonian models and the corresponding parameters obtained from first-principles calculations [172].

4. Conclusions

In this review, we summarized different types of spin interactions that an effective spin Hamiltonian may include. Recent studies have shown the importance of several kinds of high-order terms in some magnetic systems, especially biquadratic terms, four-spin ring interactions, topological chiral interactions, and chiral biquadratic interactions. In addition, we discussed in some detail the advantages and disadvantages of various methods of computing interaction parameters of the effective spin Hamiltonians. The energy-mapping analysis is easier to use, and it is less vulnerable to the effects of higher-order interactions (if carefully treated). Compared with the energy-mapping analysis, Green's function method requires less first-principle calculations and a relatively small supercell. The energy-mapping analysis usually gives a relatively good description of many kinds of states with diverse energies, while Green's function method provides a more accurate description of states close to the ground state (or the reference state). Both methods usually provide similar results and are both widely adopted in the studies of magnetic materials. We expect that first-principles based effective spin Hamiltonian will continue to play a key role in the investigation of novel magnetic states (e.g., quantum spin liquid and magnetic skyrmions).

Author Contributions: X.L. and H.Y. contributed equally to this work. Topic selection, X.L., F.L., J.F., and H.X.; literature search, X.L., H.Y., J.F., F.L., and H.X.; analysis and discussions, X.L., H.Y., J.F., H.X., and M.-H.W.; graphs, X.L., H.Y., and F.L.; writing-original draft preparation, X.L.; writing-review and editing, H.Y., F.L., H.X., M.-H.W., and J.F.; supervision, H.X. The manuscript was written through contributions of all authors. All authors have read and agreed to the published version of the manuscript.

Funding: This work is supported by NSFC (11825403, 11991061), the Program for Professor of Special Appointment (Eastern Scholar), the Qing Nian Ba Jian Program. J. S. Feng acknowledges the support from Anhui Provincial Natural Science Foundation (1908085MA10) and the Opening Foundation of State Key Laboratory of Surface Physics Fudan University (KF2019_07).

Conflicts of Interest: The authors declare no conflict of interest.

References

1. Baibich, M.N.; Broto, J.M.; Fert, A.; Vandau, F.N.; Petroff, F.; Eitenne, P.; Creuzet, G.; Friederich, A.; Chazelas, J. Giant magnetoresistance of (001)Fe/(001) Cr agnetic superlattices. *Phys. Rev. Lett.* **1988**, *61*, 2472–2475. [CrossRef] [PubMed]
2. Binasch, G.; Grunberg, P.; Saurenbach, F.; Zinn, W. Enhanced magnetoresistance in layered magnetic-structures with antiferromagnetic interlayer exchange. *Phys. Rev. B* **1989**, *39*, 4828–4830. [CrossRef] [PubMed]
3. Julliere, M. Tunneling between ferromagnetic-films. *Phys. Lett. A* **1975**, *54*, 225–226. [CrossRef]
4. Butler, W.H.; Zhang, X.G.; Schulthess, T.C.; MacLaren, J.M. Spin-dependent tunneling conductance of Fe vertical bar MgO vertical bar Fe sandwiches. *Phys. Rev. B* **2001**, *63*, 054416. [CrossRef]
5. Mathon, J.; Umerski, A. Theory of tunneling magnetoresistance of an epitaxial Fe/MgO/Fe(001) junction. *Phys. Rev. B* **2001**, *63*, 220403. [CrossRef]
6. Bowen, M.; Cros, V.; Petroff, F.; Fert, A.; Boubeta, C.M.; Costa-Kramer, J.L.; Anguita, J.V.; Cebollada, A.; Briones, F.; de Teresa, J.M.; et al. Large magnetoresistance in Fe/MgO/FeCo(001) epitaxial tunnel junctions on GaAs(001). *Appl. Phys. Lett.* **2001**, *79*, 1655–1657. [CrossRef]
7. Yuasa, S.; Nagahama, T.; Fukushima, A.; Suzuki, Y.; Ando, K. Giant room-temperature magnetoresistance in single-crystal Fe/MgO/Fe magnetic tunnel junctions. *Nat. Mater.* **2004**, *3*, 868–871. [CrossRef]

8. Parkin, S.S.P.; Kaiser, C.; Panchula, A.; Rice, P.M.; Hughes, B.; Samant, M.; Yang, S.H. Giant tunnelling magnetoresistance at room temperature with MgO (100) tunnel barriers. *Nat. Mater.* **2004**, *3*, 862–867. [CrossRef]
9. Ikeda, S.; Hayakawa, J.; Ashizawa, Y.; Lee, Y.M.; Miura, K.; Hasegawa, H.; Tsunoda, M.; Matsukura, F.; Ohno, H. Tunnel magnetoresistance of 604% at 300 K by suppression of Ta diffusion in CoFeB/MgO/CoFeB pseudo-spin-valves annealed at high temperature. *Appl. Phys. Lett.* **2008**, *93*, 082508. [CrossRef]
10. Ralph, D.C.; Stiles, M.D. Spin transfer torques. *J. Magn. Magn. Mater.* **2008**, *320*, 1190–1216. [CrossRef]
11. Jonietz, F.; Muehlbauer, S.; Pfleiderer, C.; Neubauer, A.; Muenzer, W.; Bauer, A.; Adams, T.; Georgii, R.; Boeni, P.; Duine, R.A.; et al. Spin Transfer Torques in MnSi at Ultralow Current Densities. *Science* **2010**, *330*, 1648–1651. [CrossRef]
12. Pai, C.-F.; Liu, L.; Li, Y.; Tseng, H.W.; Ralph, D.C.; Buhrman, R.A. Spin transfer torque devices utilizing the giant spin Hall effect of tungsten. *Appl. Phys. Lett.* **2012**, *101*, 122404. [CrossRef]
13. Mellnik, A.R.; Lee, J.S.; Richardella, A.; Grab, J.L.; Mintun, P.J.; Fischer, M.H.; Vaezi, A.; Manchon, A.; Kim, E.A.; Samarth, N.; et al. Spin-transfer torque generated by a topological insulator. *Nature* **2014**, *511*, 449–451. [CrossRef] [PubMed]
14. Edwards, S.F.; Anderson, P.W. Theory of Spin Glasses. *J. Phys. F-Met. Phys.* **1975**, *5*, 965–974. [CrossRef]
15. Binder, K.; Young, A.P. Spin-glasses-experimental facts, theoretical concepts, and open questions. *Rev. Mod. Phys.* **1986**, *58*, 801–976. [CrossRef]
16. Bramwell, S.T.; Gingras, M.J.P. Spin ice state in frustrated magnetic pyrochlore materials. *Science* **2001**, *294*, 1495–1501. [CrossRef]
17. Castelnovo, C.; Moessner, R.; Sondhi, S.L. Magnetic monopoles in spin ice. *Nature* **2008**, *451*, 42–45. [CrossRef] [PubMed]
18. Anderson, P.W. Resonating valence bonds: A new kind of insulator? *Mater. Res. Bull.* **1973**, *8*, 153–160. [CrossRef]
19. Shimizu, Y.; Miyagawa, K.; Kanoda, K.; Maesato, M.; Saito, G. Spin liquid state in an organic Mott insulator with a triangular lattice. *Phys. Rev. Lett.* **2003**, *91*, 107001. [CrossRef]
20. Balents, L. Spin liquids in frustrated magnets. *Nature* **2010**, *464*, 199–208. [CrossRef]
21. Yan, S.; Huse, D.A.; White, S.R. Spin-Liquid Ground State of the S = 1/2 Kagome Heisenberg Antiferromagnet. *Science* **2011**, *332*, 1173–1176. [CrossRef]
22. Han, T.-H.; Helton, J.S.; Chu, S.; Nocera, D.G.; Rodriguez-Rivera, J.A.; Broholm, C.; Lee, Y.S. Fractionalized excitations in the spin-liquid state of a kagome-lattice antiferromagnet. *Nature* **2012**, *492*, 406–410. [CrossRef] [PubMed]
23. Skyrme, T.H.R. A unifield field theory of mesons and baryons. *Nucl. Phys.* **1962**, *31*, 556–569. [CrossRef]
24. Fert, A.; Cros, V.; Sampaio, J. Skyrmions on the track. *Nat. Nanotechnol.* **2013**, *8*, 152–156. [CrossRef]
25. Muehlbauer, S.; Binz, B.; Jonietz, F.; Pfleiderer, C.; Rosch, A.; Neubauer, A.; Georgii, R.; Boeni, P. Skyrmion Lattice in a Chiral Magnet. *Science* **2009**, *323*, 915–919. [CrossRef]
26. Neubauer, A.; Pfleiderer, C.; Binz, B.; Rosch, A.; Ritz, R.; Niklowitz, P.G.; Boeni, P. Topological Hall Effect in the A Phase of MnSi. *Phys. Rev. Lett.* **2009**, *102*, 186602. [CrossRef]
27. Pappas, C.; Lelievre-Berna, E.; Falus, P.; Bentley, P.M.; Moskvin, E.; Grigoriev, S.; Fouquet, P.; Farago, B. Chiral Paramagnetic Skyrmion-like Phase in MnSi. *Phys. Rev. Lett.* **2009**, *102*, 197202. [CrossRef]
28. Roessler, U.K.; Bogdanov, A.N.; Pfleiderer, C. Spontaneous skyrmion ground states in magnetic metals. *Nature* **2006**, *442*, 797–801. [CrossRef]
29. Fujishiro, Y.; Kanazawa, N.; Tokura, Y. Engineering skyrmions and emergent monopoles in topological spin crystals. *Appl. Phys. Lett.* **2020**, *116*, 090501. [CrossRef]
30. Hastings, W.K. Monte-carlo sampling methods using markov chains and their applications. *Biometrika* **1970**, *57*, 97–109. [CrossRef]
31. Ma, P.-W.; Woo, C.H.; Dudarev, S.L. Large-scale simulation of the spin-lattice dynamics in ferromagnetic iron. *Phys. Rev. B* **2008**, *78*, 024434. [CrossRef]
32. Ma, P.-W.; Dudarev, S.L.; Woo, C.H. Spin-lattice-electron dynamics simulations of magnetic materials. *Phys. Rev. B* **2012**, *85*, 184301. [CrossRef]
33. Omelyan, I.P.; Mryglod, I.M.; Folk, R. Algorithm for molecular dynamics simulations of spin liquids. *Phys. Rev. Lett.* **2001**, *86*, 898–901. [CrossRef] [PubMed]
34. Van Vleck, J.H. Theory of the variations in paramagnetic anisotropy among different salts of the iron group. *Phys. Rev.* **1932**, *41*, 208–215. [CrossRef]
35. Ballhausen, C.J. *Introduction to Ligand Field Theory*; McGraw-Hill: New York, NY, USA, 1962.
36. Griffith, J.S. *The Theory of Transition-Metal Ions*; Cambridge University Press: Cambridge, UK, 1964.
37. Mattis, D.C. *The Theory of Magnetism Made Simple: An. Introduction to Physical Concepts and to Some Useful Mathematical Methods*; World Scientific Publishing Company: Singapore, 2006.
38. Hohenberg, P.; Kohn, W. Inhomogeneous electron gas. *Phys. Rev. B* **1964**, *136*, B864. [CrossRef]
39. Kohn, W.; Sham, L.J. Self-consistent equations including exchange and correlation effects. *Phys. Rev.* **1965**, *140*, A1133. [CrossRef]
40. Hubbard, J. Electron correlations in narrow energy bands. *Proc. R. Soc. Lond. Ser. A-Math. Phys. Sci.* **1963**, *276*, 238–257.
41. Liechtenstein, A.I.; Anisimov, V.I.; Zaanen, J. Density-functional theory and strong-interactions-orbital ordering in mott-hubbard insulators. *Phys. Rev. B* **1995**, *52*, R5467–R5470. [CrossRef]
42. Dudarev, S.L.; Botton, G.A.; Savrasov, S.Y.; Humphreys, C.J.; Sutton, A.P. Electron-energy-loss spectra and the structural stability of nickel oxide: An LSDA+U study. *Phys. Rev. B* **1998**, *57*, 1505–1509. [CrossRef]
43. Himmetoglu, B.; Floris, A.; de Gironcoli, S.; Cococcioni, M. Hubbard-Corrected DFT Energy Functionals: The LDA + U Description of Correlated Systems. *Int. J. Quantum Chem.* **2014**, *114*, 14–49. [CrossRef]

44. The VASP Manual—Vaspwiki. Available online: https://www.vasp.at/wiki/index.php/The_VASP_Manual (accessed on 3 February 2021).
45. Dederichs, P.H.; Blugel, S.; Zeller, R.; Akai, H. Ground-states of constrained systems-application to cerium impurities. *Phys. Rev. Lett.* **1984**, *53*, 2512–2515. [CrossRef]
46. Hybertsen, M.S.; Schluter, M.; Christensen, N.E. Calculation of coulomb-interaction parameters for la2cuo4 using a constrained-density-functional approach. *Phys. Rev. B* **1989**, *39*, 9028–9041. [CrossRef]
47. Gunnarsson, O.; Andersen, O.K.; Jepsen, O.; Zaanen, J. Density-functional calculation of the parameters in the anderson model-application to Mn in CdTe. *Phys. Rev. B* **1989**, *39*, 1708–1722. [CrossRef]
48. Springer, M.; Aryasetiawan, F. Frequency-dependent screened interaction in Ni within the random-phase approximation. *Phys. Rev. B* **1998**, *57*, 4364–4368. [CrossRef]
49. Kotani, T. Ab initio random-phase-approximation calculation of the frequency-dependent effective interaction between 3d electrons: Ni, Fe, and MnO. *J. Phys. Condens. Matter* **2000**, *12*, 2413–2422. [CrossRef]
50. Aryasetiawan, F.; Imada, M.; Georges, A.; Kotliar, G.; Biermann, S.; Lichtenstein, A.I. Frequency-dependent local interactions and low-energy effective models from electronic structure calculations. *Phys. Rev. B* **2004**, *70*, 195104. [CrossRef]
51. Metzner, W.; Vollhardt, D. Correlated lattice fermions in dinfinity dimensions. *Phys. Rev. Lett.* **1989**, *62*, 324–327. [CrossRef] [PubMed]
52. Georges, A.; Kotliar, G. Hubbard-model in infinite dimensions. *Phys. Rev. B* **1992**, *45*, 6479–6483. [CrossRef] [PubMed]
53. Georges, A.; Kotliar, G.; Krauth, W.; Rozenberg, M.J. Dynamical mean-field theory of strongly correlated fermion systems and the limit of infinite dimensions. *Rev. Mod. Phys.* **1996**, *68*, 13–125. [CrossRef]
54. Pavarini, E.; Koch, E.; Vollhardt, D.; Lichtenstein, A. The LDA+ DMFT approach to strongly correlated materials. *Reihe Modeling Simul.* **2011**, *1*, 2–13.
55. Embedded DMFT Functional Tutorials. Available online: http://hauleweb.rutgers.edu/tutorials/ (accessed on 3 February 2021).
56. Yang, W.T.; Zhang, Y.K.; Ayers, P.W. Degenerate ground states and a fractional number of electrons in density and reduced density matrix functional theory. *Phys. Rev. Lett.* **2000**, *84*, 5172–5175. [CrossRef] [PubMed]
57. Sharma, S.; Dewhurst, J.K.; Lathiotakis, N.N.; Gross, E.K.U. Reduced density matrix functional for many-electron systems. *Phys. Rev. B* **2008**, *78*, 201103. [CrossRef]
58. Roos, B.O.; Taylor, P.R.; Siegbahn, P.E.M. A complete active space scf method (casscf) using a density-matrix formulated super-ci approach. *Chem. Phys.* **1980**, *48*, 157–173. [CrossRef]
59. Siegbahn, P.E.M.; Almlof, J.; Heiberg, A.; Roos, B.O. The complete active space scf (casscf) method in a newton-raphson formulation with application to the HNO molecule. *J. Chem. Phys.* **1981**, *74*, 2384–2396. [CrossRef]
60. Knowles, P.J.; Sexton, G.J.; Handy, N.C. Studies using the casscf wavefunction. *Chem. Phys.* **1982**, *72*, 337–347. [CrossRef]
61. Snyder, J.W., Jr.; Parrish, R.M.; Martinez, T.J. alpha-CASSCF: An Efficient, Empirical Correction for SA-CASSCF To Closely Approximate MS-CASPT2 Potential Energy Surfaces. *J. Phys. Chem. Lett.* **2017**, *8*, 2432–2437. [CrossRef] [PubMed]
62. Roos, B.O.; Linse, P.; Siegbahn, P.E.M.; Blomberg, M.R.A. A simple method for the evaluation of the 2nd-order perturbation energy from external double-excitations with a casscf reference wavefunction. *Chem. Phys.* **1982**, *66*, 1–2, 197–207. [CrossRef]
63. Andersson, K.; Malmqvist, P.A.; Roos, B.O.; Sadlej, A.J.; Wolinski, K. Second-order perturbation theory with a CASSCF reference function. *J. Phys. Chem.* **1990**, *94*, 5483–5488. [CrossRef]
64. Andersson, K.; Malmqvist, P.A.; Roos, B.O. 2nd-order perturbation-theory with a complete active space self-consistent field reference function. *J. Chem. Phys.* **1992**, *96*, 1218–1226. [CrossRef]
65. Werner, H.J. Third-order multireference perturbation theory—The CASPT3 method. *Mol. Phys.* **1996**, *89*, 645–661. [CrossRef]
66. Miralles, J.; Daudey, J.P.; Caballol, R. Variational calculation of small energy differences—The singlet-triplet gap in Cu2Cl6 2. *Chem. Phys. Lett.* **1992**, *198*, 555–562. [CrossRef]
67. Miralles, J.; Castell, O.; Caballol, R.; Malrieu, J.P. Specific ci calculation of energy differences—transition energies and bond-energies. *Chem. Phys.* **1993**, *172*, 33–43. [CrossRef]
68. Malrieu, J.P.; Daudey, J.P.; Caballol, R. Multireference self-consistent size-consistent singles and doubles configuration-interaction for ground and excited-states. *J. Chem. Phys.* **1994**, *101*, 8908–8921. [CrossRef]
69. Malrieu, J.P.; Caballol, R.; Calzado, C.J.; de Graaf, C.; Guihery, N. Magnetic Interactions in Molecules and Highly Correlated Materials: Physical Content, Analytical Derivation, and Rigorous Extraction of Magnetic Hamiltonians. *Chem. Rev.* **2014**, *114*, 429–492. [CrossRef] [PubMed]
70. Heitler, W.; London, F. Wechselwirkung neutraler Atome und homöopolare Bindung nach der Quantenmechanik. *Z. Für Phys.* **1927**, *44*, 6–7, 455–472. [CrossRef]
71. Levine, D.S.; Head-Gordon, M. Clarifying the quantum mechanical origin of the covalent chemical bond. *Nat. Commun.* **2020**, *11*, 4893. [CrossRef]
72. White, R.M.; White, R.M.; Bayne, B. *Quantum Theory of Magnetism*; Springer: Berlin/Heidelberg, Germany, 1983.
73. Hay, P.J.; Thibeault, J.C.; Hoffmann, R. Orbital interactions in metal dimer complexes. *J. Am. Chem. Soc.* **1975**, *97*, 4884–4899. [CrossRef]
74. Novak, P.; Chaplygin, I.; Seifert, G.; Gemming, S.; Laskowski, R. Ab-initio calculation of exchange interactions in YMnO(3). *Comput. Mater. Sci.* **2008**, *44*, 79–81. [CrossRef]

75. Zener, C. Interaction between the d-shells in the transition metals. 2. ferromagnetic compounds of manganese with perovskite structure. *Phys. Rev.* **1951**, *82*, 403–405. [CrossRef]
76. Anderson, P.W.; Hasegawa, H. Considerations on double exchange. *Phys. Rev.* **1955**, *100*, 675–681. [CrossRef]
77. Anderson, P.W. Antiferromagnetism—Theory of superexchange interaction. *Phys. Rev.* **1950**, *79*, 350–356. [CrossRef]
78. Kramers, H. L'interaction entre les atomes magnétogènes dans un cristal paramagnétique. *Physica* **1934**, *1*, 182–192. [CrossRef]
79. Goodenough, J.B.; Loeb, A.L. Theory of ionic ordering, crystal distortion, and magnetic exchange due to covalent forces in spinels. *Phys. Rev.* **1955**, *98*, 391–408. [CrossRef]
80. Goodenough, J.B. Theory of the role of covalence in the perovskite-type manganites la,m(ii) mno3. *Phys. Rev.* **1955**, *100*, 564–573. [CrossRef]
81. Kanamori, J. Superexchange interaction and symmetry properties of electron orbitals. *J. Phys. Chem. Solids* **1959**, *10*, 87–98. [CrossRef]
82. Goodenough, J.B. *Magnetism and the Chemical Bond*; Interscience Publishers: New York, NY, USA, 1963; Volume 1.
83. Gong, C.; Li, L.; Li, Z.; Ji, H.; Stern, A.; Xia, Y.; Cao, T.; Bao, W.; Wang, C.; Wang, Y.; et al. Discovery of intrinsic ferromagnetism in two-dimensional van der Waals crystals. *Nature* **2017**, *546*, 265–269. [CrossRef] [PubMed]
84. Li, J.; Ni, J.Y.; Li, X.Y.; Koo, H.J.; Whangbo, M.H.; Feng, J.S.; Xiang, H.J. Intralayer ferromagnetism between S = 5/2 ions in MnBi2Te4: Role of empty Bi p states. *Phys. Rev. B* **2020**, *101*, 201408. [CrossRef]
85. Oles, A.M.; Horsch, P.; Feiner, L.F.; Khaliullin, G. Spin-orbital entanglement and violation of the Goodenough-Kanamori rules. *Phys. Rev. Lett.* **2006**, *96*, 147205. [CrossRef] [PubMed]
86. Whangbo, M.H.; Koo, H.J.; Dai, D. Spin exchange interactions and magnetic structures of extended magnetic solids with localized spins: Theoretical descriptions on formal, quantitative and qualitative levels. *J. Solid State Chem.* **2003**, *176*, 417–481. [CrossRef]
87. Ruderman, M.A.; Kittel, C. Indirect exchange coupling of nuclear magnetic moments by conduction electrons. *Phys. Rev.* **1954**, *96*, 99–102. [CrossRef]
88. Zener, C. Interaction between the d-shells in the transition metals. *Phys. Rev.* **1951**, *81*, 440–444. [CrossRef]
89. Kasuya, T. A Theory of metallic ferromagnetism and antiferromagnetism on zeners model. *Prog. Theor. Phys.* **1956**, *16*, 45–57. [CrossRef]
90. Yosida, K. Magnetic properties of cu-mn alloys. *Phys. Rev.* **1957**, *106*, 893–898. [CrossRef]
91. Zhang, L.; Zhang, Y.-Q.; Zhang, P.; Zhao, L.; Guo, M.; Tang, J. Single-Molecule Magnet Behavior Enhanced by Synergic Effect of Single-Ion Anisotropy and Magnetic Interactions. *Inorg. Chem.* **2017**, *56*, 7882–7889. [CrossRef]
92. Degennes, P.G. Effects of double exchange in magnetic crystals. *Phys. Rev.* **1960**, *118*, 141–154. [CrossRef]
93. Gordon, E.E.; Derakhshan, S.; Thompson, C.M.; Whangbo, M.-H. Spin-Density Wave as a Superposition of Two Magnetic States of Opposite Chirality and Its Implications. *Inorg. Chem.* **2018**, *57*, 9782–9785. [CrossRef] [PubMed]
94. Koo, H.-J.; PN, R.S.; Orlandi, F.; Sundaresan, A.; Whangbo, M.-H. On Ferro- and Antiferro-Spin-Density Waves Describing the Incommensurate Magnetic Structure of NaYNiWO6. *Inorg. Chem.* **2020**, *59*, 17856–17859. [CrossRef]
95. Xiang, H.J.; Whangbo, M.H. Density-functional characterization of the multiferroicity in spin spiral chain cuprates. *Phys. Rev. Lett.* **2007**, *99*, 257203. [CrossRef] [PubMed]
96. Dzyaloshinsky, I. A thermodynamic theory of "weak" ferromagnetism of antiferromagnetics. *J. Phys. Chem. Solids* **1958**, *4*, 241–255. [CrossRef]
97. Moriya, T. Anisotropic superexchange interaction and weak ferromagnetism. *Phys. Rev.* **1960**, *120*, 91–98. [CrossRef]
98. Moriya, T. New mechanism of anisotropic superexchange interaction. *Phys. Rev. Lett.* **1960**, *4*, 228–230. [CrossRef]
99. Xu, C.; Xu, B.; Dupe, B.; Bellaiche, L. Magnetic interactions in BiFeO3: A first-principles study. *Phys. Rev. B* **2019**, *99*, 104420. [CrossRef]
100. Xu, C.; Feng, J.; Prokhorenko, S.; Nahas, Y.; Xiang, H.; Bellaiche, L. Topological spin texture in Janus monolayers of the chromium trihalides Cr(I, X)(3). *Phys. Rev. B* **2020**, *101*, 060404. [CrossRef]
101. Spirin, D.V. Magnetization of two-layer ferromagnets with easy-plane and/or easy-axis single-site anisotropy. *J. Magn. Magn. Mater.* **2003**, *264*, 121–129. [CrossRef]
102. Onufrieva, F.P. Magnetic property anomalies of ferromagnets with mono-ionic anisotropy. *Fiz. Tverd. Tela* **1984**, *26*, 3435–3437.
103. Zadrozny, J.M.; Xiao, D.J.; Atanasov, M.; Long, G.J.; Grandjean, F.; Neese, F.; Long, J.R. Magnetic blocking in a linear iron(I) complex. *Nat. Chem.* **2013**, *5*, 577–581. [CrossRef] [PubMed]
104. Liu, J.; Chen, Y.-C.; Liu, J.-L.; Vieru, V.; Ungur, L.; Jia, J.-H.; Chibotaru, L.F.; Lan, Y.; Wernsdorfer, W.; Gao, S.; et al. A Stable Pentagonal Bipyramidal Dy(III) Single-Ion Magnet with a Record Magnetization Reversal Barrier over 1000 K. *J. Am. Chem. Soc.* **2016**, *138*, 5441–5450. [CrossRef] [PubMed]
105. Deng, Y.-F.; Wang, Z.; Ouyang, Z.-W.; Yin, B.; Zheng, Z.; Zheng, Y.-Z. Large Easy-Plane Magnetic Anisotropy in a Three-Coordinate Cobalt(II) Complex Li(THF)(4) Co(NPh2)(3). *Chem. A Eur. J.* **2016**, *22*, 14821–14825. [CrossRef]
106. Antal, A.; Lazarovits, B.; Udvardi, L.; Szunyogh, L.; Ujfalussy, B.; Weinberger, P. First-principles calculations of spin interactions and the magnetic ground states of Cr trimers on Au(111). *Phys. Rev. B* **2008**, *77*, 174429. [CrossRef]
107. Bak, P.; Jensen, M.H. Theory of helical magnetic-structures and phase-transitions in mnsi and fege. *J. Phys. C Solid State Phys.* **1980**, *13*, L881–L885. [CrossRef]
108. Yu, X.Z.; Kanazawa, N.; Onose, Y.; Kimoto, K.; Zhang, W.Z.; Ishiwata, S.; Matsui, Y.; Tokura, Y. Near room-temperature formation of a skyrmion crystal in thin-films of the helimagnet FeGe. *Nat. Mater.* **2011**, *10*, 106–109. [CrossRef]

109. Huang, S.X.; Chien, C.L. Extended Skyrmion Phase in Epitaxial FeGe(111) Thin Films. *Phys. Rev. Lett.* **2012**, *108*, 267201. [CrossRef]
110. Weber, T.; Waizner, J.; Tucker, G.S.; Georgii, R.; Kugler, M.; Bauer, A.; Pfleiderer, C.; Garst, M.; Boeni, P. Field dependence of nonreciprocal magnons in chiral MnSi. *Phys. Rev. B* **2018**, *97*, 224403. [CrossRef]
111. Lado, J.L.; Fernandez-Rossier, J. On the origin of magnetic anisotropy in two dimensional CrI3. *2d Materials* **2017**, *4*, 035002. [CrossRef]
112. Xu, C.; Feng, J.; Xiang, H.; Bellaiche, L. Interplay between Kitaev interaction and single ion anisotropy in ferromagnetic CrI3 and CrGeTe3 monolayers. *Npj Comput. Mater.* **2018**, *4*, 57. [CrossRef]
113. Kitaev, A. Anyons in an exactly solved model and beyond. *Ann. Phys.* **2006**, *321*, 2–111. [CrossRef]
114. Banerjee, A.; Bridges, C.A.; Yan, J.Q.; Aczel, A.A.; Li, L.; Stone, M.B.; Granroth, G.E.; Lumsden, M.D.; Yiu, Y.; Knolle, J.; et al. Proximate Kitaev quantum spin liquid behaviour in a honeycomb magnet. *Nat. Mater.* **2016**, *15*, 733–740. [CrossRef] [PubMed]
115. Baek, S.H.; Do, S.H.; Choi, K.Y.; Kwon, Y.S.; Wolter, A.U.B.; Nishimoto, S.; van den Brink, J.; Buechner, B. Evidence for a Field-Induced Quantum Spin Liquid in alpha-RuCl3. *Phys. Rev. Lett.* **2017**, *119*, 037201. [CrossRef]
116. Zheng, J.; Ran, K.; Li, T.; Wang, J.; Wang, P.; Liu, B.; Liu, Z.-X.; Normand, B.; Wen, J.; Yu, W. Gapless Spin Excitations in the Field-Induced Quantum Spin Liquid Phase of alpha-RuCl3. *Phys. Rev. Lett.* **2017**, *119*, 227208. [CrossRef] [PubMed]
117. Cao, G.; Qi, T.F.; Li, L.; Terzic, J.; Cao, V.S.; Yuan, S.J.; Tovar, M.; Murthy, G.; Kaul, R.K. Evolution of magnetism in the single-crystal honeycomb iridates (Na1-xLix)(2)IrO3. *Phys. Rev. B* **2013**, *88*, 220414. [CrossRef]
118. Manni, S.; Choi, S.; Mazin, I.I.; Coldea, R.; Altmeyer, M.; Jeschke, H.O.; Valenti, R.; Gegenwart, P. Effect of isoelectronic doping on the honeycomb-lattice iridate A(2)IrO(3). *Phys. Rev. B* **2014**, *89*, 245113. [CrossRef]
119. Xu, C.; Feng, J.; Kawamura, M.; Yamaji, Y.; Nahas, Y.; Prokhorenko, S.; Qi, Y.; Xiang, H.; Bellaiche, L. Possible Kitaev Quantum Spin Liquid State in 2D Materials with S = 3/2. *Phys. Rev. Lett.* **2020**, *124*, 087205. [CrossRef]
120. Harris, E.A.; Owen, J. Biquadratic exchange between Mn2+ ions in MgO. *Phys. Rev. Lett.* **1963**, *11*, 9. [CrossRef]
121. Rodbell, D.S.; Harris, E.A.; Owen, J.; Jacobs, I.S. Biquadratic exchange and behavior of some antiferromagnetic substances. *Phys. Rev. Lett.* **1963**, *11*, 10. [CrossRef]
122. Fedorova, N.S.; Ederer, C.; Spaldin, N.A.; Scaramucci, A. Biquadratic and ring exchange interactions in orthorhombic perovskite manganites. *Phys. Rev. B* **2015**, *91*, 165122. [CrossRef]
123. Zhu, H.-F.; Cao, H.-Y.; Xie, Y.; Hou, Y.-S.; Chen, S.; Xiang, H.; Gong, X.-G. Giant biquadratic interaction-induced magnetic anisotropy in the iron-based superconductor A(x)Fe(2-y)Se(2). *Phys. Rev. B* **2016**, *93*, 024511. [CrossRef]
124. Kartsev, A.; Augustin, M.; Evans, R.F.L.; Novoselov, K.S.; Santos, E.J.G. Biquadratic exchange interactions in two-dimensional magnets. *Npj Comput. Mater.* **2020**, *6*, 150. [CrossRef]
125. Li, X.-Y.; Lou, F.; Gong, X.-G.; Xiang, H. Constructing realistic effective spin Hamiltonians with machine learning approaches. *New J. Phys.* **2020**, *22*, 053036. [CrossRef]
126. Ivanov, N.B.; Ummethum, J.; Schnack, J. Phase diagram of the alternating-spin Heisenberg chain with extra isotropic three-body exchange interactions. *Eur. Phys. J. B* **2014**, *87*, 226. [CrossRef]
127. Paul, S.; Haldar, S.; von Malottki, S.; Heinze, S. Role of higher-order exchange interactions for skyrmion stability. *Nat. Commun.* **2020**, *11*, 4756. [CrossRef]
128. Grytsiuk, S.; Hanke, J.P.; Hoffmann, M.; Bouaziz, J.; Gomonay, O.; Bihlmayer, G.; Lounis, S.; Mokrousov, Y.; Bluegel, S. Topological-chiral magnetic interactions driven by emergent orbital magnetism. *Nat. Commun.* **2020**, *11*, 511. [CrossRef] [PubMed]
129. Hoffmann, M.; Weischenberg, J.; Dupe, B.; Freimuth, F.; Ferriani, P.; Mokrousov, Y.; Heinze, S. Topological orbital magnetization and emergent Hall effect of an atomic-scale spin lattice at a surface. *Phys. Rev. B* **2015**, *92*, 020401. [CrossRef]
130. Hanke, J.P.; Freimuth, F.; Nandy, A.K.; Zhang, H.; Bluegel, S.; Mokrousov, Y. Role of Berry phase theory for describing orbital magnetism: From magnetic heterostructures to topological orbital ferromagnets. *Phys. Rev. B* **2016**, *94*, 121114. [CrossRef]
131. Dias, M.D.S.; Bouaziz, J.; Bouhassoune, M.; Bluegel, S.; Lounis, S. Chirality-driven orbital magnetic moments as a new probe for topological magnetic structures. *Nat. Commun.* **2016**, *7*, 13613. [CrossRef] [PubMed]
132. Hanke, J.-P.; Freimuth, F.; Bluegel, S.; Mokrousov, Y. Prototypical topological orbital ferromagnet gamma-FeMn. *Sci. Rep.* **2017**, *7*, 41078. [CrossRef] [PubMed]
133. Dias, M.D.S.; Lounis, S. Insights into the orbital magnetism of noncollinear magnetic systems. *Proc. SPIE* **2017**, *10357*, 103572A.
134. Lux, F.R.; Freimuth, F.; Bluegel, S.; Mokrousov, Y. Engineering chiral and topological orbital magnetism of domain walls and skyrmions. *Commun. Phys.* **2018**, *1*, 60. [CrossRef]
135. Tanigaki, T.; Shibata, K.; Kanazawa, N.; Yu, X.; Onose, Y.; Park, H.S.; Shindo, D.; Tokura, Y. Real-Space Observation of Short-Period Cubic Lattice of Skyrmions in MnGe. *Nano Lett.* **2015**, *15*, 5438–5442. [CrossRef]
136. Brinker, S.; Dias, M.D.S.; Lounis, S. The chiral biquadratic pair interaction. *New J. Phys.* **2019**, *21*, 083015. [CrossRef]
137. Laszloffy, A.; Rozsa, L.; Palotas, K.; Udvardi, L.; Szunyogh, L. Magnetic structure of monatomic Fe chains on Re(0001): Emergence of chiral multispin interactions. *Phys. Rev. B* **2019**, *99*, 184430. [CrossRef]
138. Parihari, D.; Pati, S.K. Effect of chiral interactions in frustrated magnetic chains. *Phys. Rev. B* **2004**, *70*, 180403. [CrossRef]
139. Kostyrko, T.; Bulka, B.R. Canonical perturbation theory for inhomogeneous systems of interacting fermions. *Phys. Rev. B* **2011**, *84*, 035123. [CrossRef]
140. Mankovsky, S.; Polesya, S.; Ebert, H. Extension of the standard Heisenberg Hamiltonian to multispin exchange interactions. *Phys. Rev. B* **2020**, *101*, 174401. [CrossRef]

141. Wen, X.G.; Wilczek, F.; Zee, A. Chiral spin states and superconductivity. *Phys. Rev. B* **1989**, *39*, 11413–11423. [CrossRef]
142. Bauer, B.; Cincio, L.; Keller, B.P.; Dolfi, M.; Vidal, G.; Trebst, S.; Ludwig, A.W.W. Chiral spin liquid and emergent anyons in a Kagome lattice Mott insulator. *Nat. Commun.* **2014**, *5*, 5137. [CrossRef]
143. Drautz, R.; Fahnle, M. Spin-cluster expansion: Parametrization of the general adiabatic magnetic energy surface with ab initio accuracy. *Phys. Rev. B* **2004**, *69*, 104404. [CrossRef]
144. Singer, R.; Faehnle, M. Construction of basis functions for the spin-cluster expansion of the magnetic energy on the atomic scale in rotationally invariant systems. *J. Math. Phys.* **2006**, *47*, 113503. [CrossRef]
145. Singer, R.; Dietermann, F.; Faehnle, M. Spin Interactions in bcc and fcc Fe beyond the Heisenberg Model. *Phys. Rev. Lett.* **2011**, *107*, 017204. [CrossRef] [PubMed]
146. Reinhardt, P.; Habas, M.P.; Dovesi, R.; Moreira, I.D.; Illas, F. Magnetic coupling in the weak ferromagnet CuF2. *Phys. Rev. B* **1999**, *59*, 1016–1023. [CrossRef]
147. Illas, F.; Moreira, I.D.R.; de Graaf, C.; Barone, V. Magnetic coupling in biradicals, binuclear complexes and wide-gap insulators: A survey of ab initio wave function and density functional theory approaches. *Theor. Chem. Acc.* **2000**, *104*, 265–272. [CrossRef]
148. Xiang, H.; Lee, C.; Koo, H.-J.; Gong, X.; Whangbo, M.-H. Magnetic properties and energy-mapping analysis. *Dalton Trans.* **2013**, *42*, 823–853. [CrossRef] [PubMed]
149. Xiang, H.J.; Kan, E.J.; Wei, S.-H.; Whangbo, M.H.; Gong, X.G. Predicting the spin-lattice order of frustrated systems from first principles. *Phys. Rev. B* **2011**, *84*, 224429. [CrossRef]
150. Weingart, C.; Spaldin, N.; Bousquet, E. Noncollinear magnetism and single-ion anisotropy in multiferroic perovskites. *Phys. Rev. B* **2012**, *86*, 094413. [CrossRef]
151. Liechtenstein, A.I.; Katsnelson, M.I.; Antropov, V.P.; Gubanov, V.A. Local spin-density functional-approach to the theory of exchange interactions in ferromagnetic metals and alloys. *J. Magn. Magn. Mater.* **1987**, *67*, 65–74. [CrossRef]
152. Katsnelson, M.I.; Lichtenstein, A.I. First-principles calculations of magnetic interactions in correlated systems. *Phys. Rev. B* **2000**, *61*, 8906–8912. [CrossRef]
153. Katsnelson, M.I.; Kvashnin, Y.O.; Mazurenko, V.V.; Lichtenstein, A.I. Correlated band theory of spin and orbital contributions to Dzyaloshinskii-Moriya interactions. *Phys. Rev. B* **2010**, *82*, 100403. [CrossRef]
154. Lounis, S.; Dederichs, P.H. Mapping the magnetic exchange interactions from first principles: Anisotropy anomaly and application to Fe, Ni, and Co. *Phys. Rev. B* **2010**, *82*, 180404. [CrossRef]
155. Szilva, A.; Costa, M.; Bergman, A.; Szunyogh, L.; Nordstrom, L.; Eriksson, O. Interatomic Exchange Interactions for Finite-Temperature Magnetism and Nonequilibrium Spin Dynamics. *Phys. Rev. Lett.* **2013**, *111*, 127204. [CrossRef] [PubMed]
156. Korotin, D.M.; Mazurenko, V.V.; Anisimov, V.I.; Streltsov, S.V. Calculation of exchange constants of the Heisenberg model in plane-wave-based methods using the Green's function approach. *Phys. Rev. B* **2015**, *91*, 224405. [CrossRef]
157. He, X.; Helbig, N.; Verstraete, M.J.; Bousquet, E. TB2J: A Python Package for Computing Magnetic Interaction Parameters. Available online: https://arxiv.org/abs/2009.01910 (accessed on 3 February 2021).
158. Wan, X.; Yin, Q.; Savrasov, S.Y. Calculation of magnetic exchange interactions in Mott-Hubbard systems. *Phys. Rev. Lett.* **2006**, *97*, 266403. [CrossRef]
159. Boukhvalov, D.W.; Dobrovitski, V.V.; Katsnelson, M.I.; Lichtenstein, A.I.; Harmon, B.N.; Kogerler, P. Electronic structure and exchange interactions in V-15 magnetic molecules: LDA+U results. *Phys. Rev. B* **2004**, *70*, 054417. [CrossRef]
160. Mostofi, A.A.; Yates, J.R.; Lee, Y.-S.; Souza, I.; Vanderbilt, D.; Marzari, N. wannier90: A tool for obtaining maximally-localised Wannier functions. *Comput. Phys. Commun.* **2008**, *178*, 685–699. [CrossRef]
161. Pizzi, G.; Vitale, V.; Arita, R.; Bluegel, S.; Freimuth, F.; Geranton, G.; Gibertini, M.; Gresch, D.; Johnson, C.; Koretsune, T.; et al. Wannier90 as a community code: New features and applications. *J. Phys. Condens. Matter* **2020**, *32*, 165902. [CrossRef]
162. Wang, X.D.; Wang, D.S.; Wu, R.Q.; Freeman, A.J. Validity of the force theorem for magnetocrystalline anisotropy. *J. Magn. Magn. Mater.* **1996**, *159*, 337–341. [CrossRef]
163. Flygare, W.H. Magnetic-interactions in molecules and an analysis of molecular electronic charge distribution from magnetic parameters. *Chem. Rev.* **1974**, *74*, 653–687. [CrossRef]
164. Brown, J.M.; Sears, T.J. Reduced form of the spin-rotation hamiltonian for asymmetric-top molecules, with applications to ho2 and nh2. *J. Mol. Spectrosc.* **1979**, *75*, 111–133. [CrossRef]
165. Lahti, P.M.; Ichimura, A.S. Semiempirical study of electron exchange interaction in organic high-spin pi-systems—classifying structural effects in organic magnetic molecules. *J. Org. Chem.* **1991**, *56*, 3030–3042. [CrossRef]
166. Chattopadhyay, S.; Lenz, B.; Kanungo, S.; Panda, S.K.; Biermann, S.; Schnelle, W.; Manna, K.; Kataria, R.; Uhlarz, M.; Skourski, Y.; et al. Pronounced 2/3 magnetization plateau in a frustrated S = 1 isolated spin-triangle compound: Interplay between Heisenberg and biquadratic exchange interactions. *Phys. Rev. B* **2019**, *100*, 094427. [CrossRef]
167. Takano, K.; Sano, K. Determination of exchange parameters from magnetic susceptibility. *J. Phys. Soc. Jpn.* **1997**, *66*, 1846–1847. [CrossRef]
168. Turek, I.; Kudrnovsky, J.; Drchal, V.; Bruno, P. Exchange interactions, spin waves, and transition temperatures in itinerant magnets. *Philos. Mag.* **2006**, *86*, 1713–1752. [CrossRef]
169. Lancon, D.; Ewings, R.A.; Guidi, T.; Formisano, F.; Wildes, A.R. Magnetic exchange parameters and anisotropy of the quasi-two-dimensional antiferromagnet NiPS3. *Phys. Rev. B* **2018**, *98*, 134414. [CrossRef]

170. Chen, L.; Chung, J.-H.; Gao, B.; Chen, T.; Stone, M.B.; Kolesnikov, A.I.; Huang, Q.; Dai, P. Topological Spin Excitations in Honeycomb Ferromagnet CrI3. *Phys. Rev. X* **2018**, *8*, 041028. [CrossRef]
171. Oba, M.; Nakamura, K.; Akiyama, T.; Ito, T.; Weinert, M.; Freeman, A.J. Electric-Field-Induced Modification of the Magnon Energy, Exchange Interaction, and Curie Temperature of Transition-Metal Thin Films. *Phys. Rev. Lett.* **2015**, *114*, 107202. [CrossRef] [PubMed]
172. Nishino, M.; Yoshioka, Y.; Yamaguchi, K. Effective exchange interactions and magnetic phase transition temperatures in Prussian blue analogs: A study by density functional theory. *Chem. Phys. Lett.* **1998**, *297*, 51–59. [CrossRef]

Review

Spin Exchanges between Transition Metal Ions Governed by the Ligand p-Orbitals in Their Magnetic Orbitals

Myung-Hwan Whangbo [1,2,*], Hyun-Joo Koo [1] and Reinhard K. Kremer [3]

1. Department of Chemistry and Research Institute for Basic Sciences, Kyung Hee University, Seoul 02447, Korea; hjkoo@khu.ac.kr
2. Department of Chemistry, North Carolina State University, Raleigh, NC 27695-8204, USA
3. Max Planck Institute for Solid State Research, Heisenbergstrasse 1, D-70569 Stuttgart, Germany; rekre@fkf.mpg.de
* Correspondence: whangbo@ncsu.edu; Tel.: +0f1-919-515-3464

Citation: Whangbo, M.-H.; Koo, H.-J.; Kremer, R.K. Spin Exchanges between Transition Metal Ions Governed by the Ligand p-Orbitals in Their Magnetic Orbitals. *Molecules* 2021, 26, 531. https://doi.org/10.3390/molecules26030531

Academic Editors: Alessandro Stroppa and Igor Djerdj
Received: 19 December 2020
Accepted: 15 January 2021
Published: 20 January 2021

Publisher's Note: MDPI stays neutral with regard to jurisdictional claims in published maps and institutional affiliations.

Copyright: © 2021 by the authors. Licensee MDPI, Basel, Switzerland. This article is an open access article distributed under the terms and conditions of the Creative Commons Attribution (CC BY) license (https://creativecommons.org/licenses/by/4.0/).

Abstract: In this review on spin exchanges, written to provide guidelines useful for finding the spin lattice relevant for any given magnetic solid, we discuss how the values of spin exchanges in transition metal magnetic compounds are quantitatively determined from electronic structure calculations, which electronic factors control whether a spin exchange is antiferromagnetic or ferromagnetic, and how these factors are related to the geometrical parameters of the spin exchange path. In an extended solid containing transition metal magnetic ions, each metal ion M is surrounded with main-group ligands L to form an ML_n polyhedron (typically, n = 3–6), and the unpaired spins of M are represented by the singly-occupied d-states (i.e., the magnetic orbitals) of ML_n. Each magnetic orbital has the metal d-orbital combined out-of-phase with the ligand p-orbitals; therefore, the spin exchanges between adjacent metal ions M lead not only to the M–L–M-type exchanges, but also to the M–L . . . L–M-type exchanges in which the two metal ions do not share a common ligand. The latter can be further modified by d^0 cations A such as V^{5+} and W^{6+} to bridge the L . . . L contact generating M–L . . . A . . . L–M-type exchanges. We describe several qualitative rules for predicting whether the M–L . . . L–M and M–L . . . A . . . L–M-type exchanges are antiferromagnetic or ferromagnetic by analyzing how the ligand p-orbitals in their magnetic orbitals (the ligand p-orbital tails, for short) are arranged in the exchange paths. Finally, we illustrate how these rules work by analyzing the crystal structures and magnetic properties of four cuprates of current interest: α-CuV_2O_6, $LiCuVO_4$, $(CuCl)LaNb_2O_7$, and $Cu_3(CO_3)_2(OH)_2$.

Keywords: Keywords: spin exchange; magnetic orbitals; ligand p-orbital tails; M–L–M exchange; M–L . . . L–M exchange; α-CuV_2O_6; $LiCuVO_4$; $(CuCl)LaNb_2O_7$; $Cu_3(CO_3)_2(OH)_2$

1. Introduction

An extended solid consisting of transition metal magnetic ions has closely packed energy states (Figure 1a,b) so that, at a given non-zero temperature, the ground state as well as a vast number of the excited states can be thermally occupied. The thermodynamic properties such as the magnetic susceptibility and the specific heat of a magnetic system represents the weighted average of the properties associated with all thermally occupied states, with their Boltzmann factors as the weights. Such a quantity is difficult to calculate if all states were to be determined by first principle electronic structure calculations.

To generate the states of a given magnetic system and subsequently calculate the thermally-averaged physical property, a model Hamiltonian (also called a toy Hamiltonian) is invariably employed [1–3]. A typical model Hamiltonian used for this purpose is the Heisenberg-type spin Hamiltonian, H_{spin}, expressed as:

$$H_{spin} = \sum_{i>j} J_{ij} \hat{S}_i \cdot \hat{S}_j \qquad (1)$$

where the energy spectrum of a magnetic system as the sum of the pairwise spin exchange interactions $J_{ij}\hat{S}_i \cdot \hat{S}_j$ is approximated. The spin operators $\hat{S}_i$ and $\hat{S}_j$ (at the spin sites i and j, respectively) can be treated as the spin vectors $\vec{S}_i$ and $\vec{S}_j$, respectively, unless they operate on spin states. If all magnetic ions of a given system are identical with spin S, each term $\vec{S}_i \cdot \vec{S}_j$ can be written as $\vec{S}_i \cdot \vec{S}_j = S^2 \cos\theta_{ij}$ where θ_{ij} is the angle between the two spin vectors. In such a case, Equation (1) is rewritten as:

$$H_{spin} = \sum_{i>j} J_{ij} S^2 \cos\theta_{ij} \qquad (2)$$

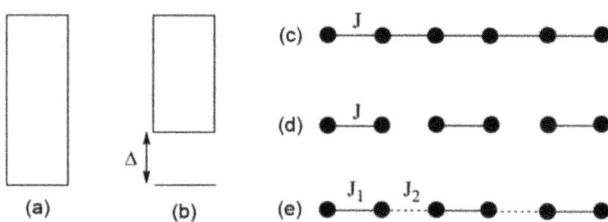

Figure 1. (**a**,**b**) Allowed energy states of a magnetic solid. Between the lowest-lying excited state and the ground state, there is no energy gap in (**a**), but a non-zero energy gap in (**b**). (**c**–**e**) Examples of simple spin lattices: a uniform chain in (**c**); isolated spin dimers in (**d**); and an alternating chain in (**e**). Here, all nearest-neighbor spins are antiferromagnetically coupled.

In a collinearly ordered spin arrangement of a magnetic solid, every two spin arrangements are either ferromagnetic (FM, i.e., parallel ($\theta_{ij} = 0°$)) or antiferromagnetic (AFM, i.e., antiparallel ($\theta_{ij} = 180°$)). With the definition of a spin Hamiltonian as in Equation (1), AFM and FM spin exchanges are represented by positive and negative J_{ij} values, respectively. For any collinearly ordered spin arrangement, the total energy is readily written as a function of the various spin exchanges J_{ij}.

The energy spectrum allowed for a magnetic system, and hence its magnetic properties, depend on its spin lattice. The latter refers to the repeat pattern of predominant spin exchange paths, i.e., those with large $|J_{ij}|$ values. For example, between the ground and the excited states, a uniform half-integer spin AFM chain (Figure 1c) has no energy gap (Figure 1a), whereas an isolated AFM dimer (Figure 1d) and an alternating AFM chain (Figure 1e) have a non-zero energy gap (Figure 1b). The spin lattice of a given magnetic solid is determined by its electronic structure, which makes it interesting how to identify the spin lattice of a magnetic solid on the basis of its atomic and electronic structures. Strong AFM exchanges between magnetic ions are often termed magnetic bonds, in contrast to chemical bonds determined by strong chemical bonding. Thus, Figure 1c represents a uniform AFM chain, and Figure 1d represents isolated AFM dimers. The magnetic bonds do not necessarily follow the geometrical pattern of the magnetic ion arrangement dictated by chemical bonding.

In a magnetic solid, transition metal ions M often share a common ligand L to form M–L–M bridges (Figure 2a). The spin exchange between the magnetic ions in an M–L–M bridge has been termed "superexchange" [4–7]. Whether the spin arrangement between two metal ions becomes FM or AFM, as described by the Goodenough–Kanamori rules formulated in the late 1950s, depends on the geometry of the M–L–M bridge [5–9]. Ever since, the Goodenough–Kanamori rules have greatly influenced the thinking of inorganic and solid-state chemists dealing with magnetic systems. Since the early 2000s, it has become increasingly clear that the magnetic properties of certain compounds cannot be conclusively explained unless also one takes into consideration the spin exchanges of the M–L . . . L–M (Figure 2b) or the M–L . . . A . . . L–M (Figure 2c) types, termed super-superexchanges [1–3], in which the metal ions do not share a common ligand. In the late 1950s, it was impossible to imagine that spin exchanges could take place in such paths in which the M... M distances were very long, because the prevailing concept of chemical bonding at that time, mainly based on the valence bond picture [10], suggested that an unpaired electron of a magnetic ion is accommodated in a pure d-orbital of M. In an extended solid, each transition metal ion M is typically surrounded with main-group ligand atoms L to form an ML_n polyhedron (n = 3–6), and each unpaired electron of ML_n resides in a singly-occupied d-state (referred to as a magnetic orbital) of ML_n, in which the d-orbital of M is combined out-of-phase with the p-orbitals of L. In this molecular orbital picture, the unpaired spin density is already delocalized from the d-orbital of M (the "magnetic orbital head") to the p-orbitals of L (the "magnetic orbital tails") [1–3]. Thus, it is quite natural to think that the spin exchange between the metal ions in an M–L... L–M path takes place through the overlap between the ligand p-orbital tails present in the L... L contact. This interaction between the p-orbital tails can be modified by the empty d_π orbitals of the d^0 cation A (e.g., V^{5+} and W^{6+}) in an M–L... A... L–M exchange.

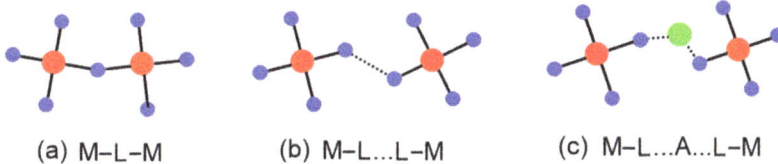

(a) M–L–M   (b) M–L...L–M   (c) M–L...A...L–M

Figure 2. Three types of spin exchange paths associated with two magnetic ions: (**a**) M–L–M, (**b**) M–L... L–M, and (**c**) M–L... A... L–M, where A represents a d^0 cation such as V^{5+} or W^{6+}. The M, L, and A are represented by red, blue, and green circles, respectively.

Given that spin exchanges between magnetic ions are determined by the p-orbital tails of their magnetic orbitals, it is not surprising that spin exchanges of the M–L... L–M or M–L... A... L–M type can be stronger than those of the M–L–M type, and that spin exchanges are generally strong only along a certain direction of the crystal structure. Although this aspect has been repeatedly pointed out in review articles over the years [1–3], it is still not infrequent to observe that experimental results are incorrectly interpreted simply because spin lattices have been deduced considering only the M–L–M type spin exchanges. Such an unfortunate mishap is akin to providing solutions in search of a problem. In this review, written as a tribute to John B. Goodenough for his long and illustrious scientific career culminating with the Nobel Prize in 2019, we review what electronic factors govern the nature of the M–L–M, M–L... L–M and M–L... A... L–M type spin exchanges, with an ultimate goal to provide several qualitative rules useful for finding the spin lattice relevant for any given magnetic system.

Our work is organized as follows: Section 2 examines how to quantitatively determine the values of spin exchanges by carrying out the energy-mapping analysis based on electronic structure calculations. Section 3 explores the electronic factors controlling whether a spin exchange is AFM or FM. In Section 4, we derive several qualitative rules that enable one to predict whether the M–L... L–M and M–L... A... L–M type spin exchanges are AFM or FM by analyzing how their ligand p-orbitals are arranged in the exchange paths. To illustrate how the resulting structure–property relationships operate, in Section 5 we examine the crystal structures and magnetic properties of α-CuV_2O_6, $LiCuVO_4$, $(CuCl)LaNb_2O_7$ and $Cu_3(CO_3)_2(OH)_2$. Our concluding remarks are presented in Section 6.

2. Energy Mapping Analysis for Quantitative Evaluation of Spin Exchanges

This section probes how to define the relevant spin Hamiltonian for a given magnetic system on a quantitative level. This requires the determination of the spin exchanges to include in the spin Hamiltonian. For various collinearly ordered spin states of a given magnetic system, one finds the expressions for their relative energies in terms of the spin exchange parameters J_{ij} to evaluate, performs DFT+U [11] or DFT+hybrid [12] electronic structure calculations for the ordered spin states to determine the numerical values for their relative energies, and finally maps the two sets of relative energies to find the numerical values of the exchanges.

2.1. Using Eigenstates

To gain insight into the meaning and the nature of a spin exchange, we examine a spin dimer made up of two S = 1/2 ions (Figure 3a). The spin Hamiltonian describing the energies of this dimer is given by:

$$H_{spin} = J\hat{S}_i \cdot \hat{S}_j, \qquad (3)$$

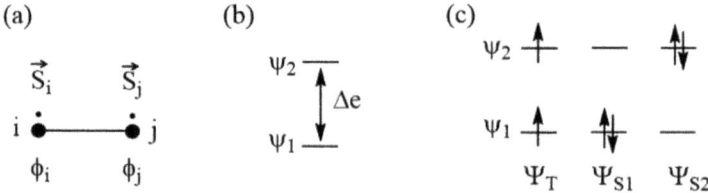

Figure 3. (a) Spin dimer made up of two S = 1/2 ions at sites i and j. Each site has one unpaired spin. The magnetic orbitals at the sites i and j are represented by ϕ_i and ϕ_j, respectively. (b) The bonding and antibonding states, Ψ_1 and Ψ_2, resulting from the interactions between ϕ_i and ϕ_j, are split in energy by Δe. (c) The singlet and triplet electron configurations resulting from Ψ_1 and Ψ_2.

The spin states allowed for this dimer are the singlet and triplet states, Ψ_S and Ψ_T, respectively.

$$\begin{array}{c}\Psi_T = |\uparrow\uparrow\rangle,\ |\downarrow\downarrow\rangle,\ (|\uparrow\downarrow\rangle + |\downarrow\uparrow\rangle)/\sqrt{2} \\ \Psi_S = (|\uparrow\downarrow\rangle - |\downarrow\uparrow\rangle)/\sqrt{2}\end{array} \qquad (4)$$

It can be readily shown that these states are the eigenstates of the spin Hamiltonian by rewriting Equation (3) as:

$$H_{spin} = J\hat{S}_{iz}\hat{S}_{jz} + (J/2)[\hat{S}_{i+}\hat{S}_{j-} + \hat{S}_{i-}\hat{S}_{j+}] \qquad (5)$$

where $\hat{S}_{mz}$ (m = i, j) is the z-component of $\hat{S}_m$, while $\hat{S}_{m+}$ and $\hat{S}_{m-}$ are the raising and lowering operators associated with $\hat{S}_m$, respectively. Then, it is found that:

$$\begin{array}{c}H_{spin}\Psi_T = (J/4)\Psi_T \\ H_{spin}\Psi_S = (-3J/4)\Psi_S\end{array} \qquad (6)$$

Thus:

$$\Delta E_{eigen} = E_T - E_S = \langle\Psi_T|H_{spin}|\Psi_T\rangle - \langle\Psi_S|H_{spin}|\Psi_S\rangle = J \qquad (7)$$

Therefore, the spin exchange J is related to the energy difference between the singlet and triplet states of the spin dimer, as illustrated in Figure 4a.

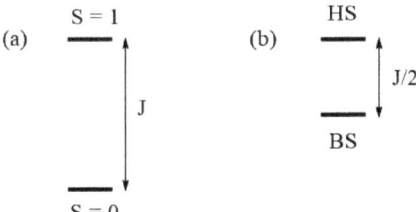

Figure 4. Relationships of the spin exchange J to the energy difference between two spin states of a spin dimer made up of two S = 1/2 ions in terms of (**a**) the eigenstates and (**b**) the broken-symmetry states. The legends HS and BS in (**b**) refer to the high-symmetry and broken-symmetry states, respectively.

2.2. Using Broken-Symmetry States

In general, it is not an easy task to find the eigenstates of a general spin Hamiltonian (e.g., Equation (1)), which makes it difficult to relate the spin exchanges to the energy differences between the eigenstates of a magnetic system. However, the energies for the broken-symmetry (BS) states (i.e., the non-eigenstates) of a spin Hamiltonian are easy to evaluate. For the magnetic dimer in Section 2.1, the high-spin (HS) state, $\Psi_{HS} = |\uparrow\uparrow\rangle$ or $|\downarrow\downarrow\rangle$, which has an FM spin arrangement, is an eigenstate of the spin Hamiltonian (Equation (3)). However, the low-spin (LS) state can be expressed as:

$$\Psi_{LS} = |\uparrow\downarrow\rangle \text{ or } |\downarrow\uparrow\rangle$$

This state has an AFM spin arrangement and is the BS state of the spin Hamiltonian. Using Equation (5), it is found that:

$$H_{spin}|\uparrow\downarrow\rangle = (-J/4)|\uparrow\downarrow\rangle + (J/2)|\downarrow\uparrow\rangle \\ H_{spin}|\downarrow\uparrow\rangle = (-J/4)|\downarrow\uparrow\rangle + (J/2)|\uparrow\downarrow\rangle \qquad (8)$$

Therefore:

$$\langle\uparrow\downarrow|H_{spin}|\uparrow\downarrow\rangle = \langle\downarrow\uparrow|H_{spin}|\downarrow\uparrow\rangle = -J/4 \qquad (9)$$

From Equations (6) and (9), we obtain:

$$\Delta E_{BS} = \langle\Psi_{HS}|H_{spin}|\Psi_{HS}\rangle - \langle\Psi_{BS}|H_{spin}|\Psi_{BS}\rangle = J/2 \qquad (10)$$

The spin exchange J is related to the energy difference between the states including the BS states (Figure 4b).

2.3. Energy Mapping

For a general spin lattice, the energy difference between any two states (involving BS and HS states) can be readily determined by using the spin Hamiltonian of Equation (3), which is expressed as a function of unknown constants J_{ij}. To evaluate the spin exchanges J_{ij} of a general spin Hamiltonian, it is necessary to numerically determine the relative energies of various ordered spin states. This is done by performing DFT+U [11] or DFT+hybrid [12] electronic structure calculations for the collinearly ordered spin states of a magnetic system on the basis of the electronic Hamiltonian, H_{elec}. These types of calculations ensure that the electronic structures calculated for various ordered spin states have a bandgap as expected for a magnetic insulator. Suppose that N different spin exchange paths J_{ij} are considered to describe a given magnetic solid. If one considers N + 1 ordered spin states, for example, one can determine N different relative energies $\Delta E_{spin}(i)$ (i = 1, 2,···, N) expressed in terms of N different spin exchanges J_{ij} (in principle, one could use more spin states, but at least N+1 are necessary; using more would produce error bars and increase the precision of the analysis). By carrying out electronic structure calculations for the N+1 ordered spin states of the magnetic system, one obtains the numerical values for the N different relative energies $\Delta E_{elec}(i)$ (i = 1, 2,···, N). Then, by equating the $\Delta E_{spin}(i)$ (i = 1, 2, ···, N) values to the corresponding $\Delta E_{elec}(i)$ (i = 1, 2, ⋯, N) values, the N different spin exchanges J_{ij} are obtained.

$$\Delta E_{spin}(i) \ (i = 1, 2, \cdots, N) \leftrightarrow \Delta E_{elec}(i) \ (i = 1, 2, \cdots, N) \tag{11}$$

The spin exchanges discussed so far are known as Heisenberg exchanges. There are other variants of interactions between spins which, though weaker than Heisenberg exchanges in strength, are needed to explain certain magnetic properties not covered by the symmetrical Heisenberg exchanges. They include Dzyaloshinskii–Moriya exchanges (or antisymmetric exchanges) and asymmetric exchanges [2,3]. The energy-mapping analysis based on collinearly ordered spin states allows one to determine the Heisenberg spin exchanges only. To evaluate the Dzyaloshinskii–Moriya and asymmetric spin exchanges, the energy-mapping analysis employs the four-state method [2,13], in which non-collinearly ordered broken-symmetry states are used. A further generalization of this energy-mapping method was developed to enable the evaluation of other energy terms that one might include in a model spin Hamiltonian [14].

3. Qualitative Features of Spin Exchange

A spin Hamiltonian appropriate for a given magnetic system is one that consists of the predominant spin exchange paths. Such a spin Hamiltonian can be determined by evaluating the values of various possible spin exchanges for the magnetic system by performing the energy-mapping analysis as described in Section 2. If the spin lattice is chosen without quantitatively evaluating its spin exchanges, one might inadvertently choose a spin lattice irrelevant for the interpretation of the experimental data. When simulating the thermodynamic properties using a chosen set of spin exchanges, the values of the spin exchanges are optimized until they provide the best possible simulation even if the chosen spin lattice is incorrect from the viewpoint of electronic structure. Thus, in principle, more than one spin lattice might provide an equally good simulation. In interpreting the experimental results of a magnetic system with a correct spin lattice, it is crucial to know what electronic and structural factors control the signs and the magnitudes of spin exchanges.

3.1. Parameters Affecting Spin Exchanges

To examine what energy parameters govern the sign and magnitude of a spin exchange, we revisit the spin exchange of the spin dimer (Figure 3a) by explicitly considering the electronic structures of its singlet and triplet states. For simplicity, we represent each spin site with one magnetic orbital. As will be discussed in Section 5, the nature of the magnetic orbital plays a crucial role in determining the sign and magnitude of a spin exchange. We label the magnetic orbitals located at the spin sites i and j as ϕ_i and ϕ_j, respectively (Figure 3a). These orbitals overlap weakly, and hence interact weakly, to form the in-phase and out-of-phase states, Ψ_1 and Ψ_2, respectively, with energy split Δe between the two (Figure 3b). The overlap integral $\langle \phi_i | \phi_j \rangle$ between ϕ_i and ϕ_j is small for magnetic systems, so Ψ_1 and Ψ_2 are well approximated by:

$$\Psi_1 \approx \frac{(\phi_i + \phi_j)}{\sqrt{2}}, \ \Psi_2 \approx \frac{(\phi_i - \phi_j)}{\sqrt{2}} \tag{12}$$

For simplicity, it is assumed here that the in-phase combination (i.e., the bonding combination) is described by the plus combination, which amounts to the assumption that the overlap integral $\langle \phi_i | \phi_j \rangle$ is positive. The energy split Δe is approximately proportional to the overlap integral:

$$\Delta e \propto \langle \phi_i | \phi_j \rangle \tag{13}$$

In understanding the qualitative features describing the electronic energy difference between the singlet and triplet states, ΔE_{elec}, and hence the spin exchange J, it is necessary to consider two other quantities. One is the on-site repulsion U_{ii}:

$$U_{ii} = \langle \phi_i(1)\phi_i(2)|1/r_{12}|\phi_i(1)\phi_i(2)\rangle \quad (i = 1, 2) \tag{14}$$

where the product $\phi_i(m)\phi_i(m)$ is the electron density $\rho_{ii}(m)$ associated with the orbital $\phi_i(m)$ occupied by electron m (=1, 2). Thus, U_{ii} is the self-repulsion when the orbital ϕ_i is occupied by two electrons. If the spin sites i and j are identical in nature, the on-site repulsion U_{jj} at the site j is the same as U_{ii}, therefore it is convenient to use the symbol U to represent both U_{ii} and U_{jj} (i.e., $U = U_{ii} = U_{jj}$). The other quantity of interest is the exchange repulsion K_{ij} between ϕ_i and ϕ_j:

$$K_{ij} = \langle \phi_i(1)\phi_j(2)|1/r_{12}|\phi_j(1)\phi_i(2)\rangle \tag{15}$$

This is the self-repulsion arising from the overlap electron density:

$$\rho_{ij}(m) = \phi_i(m)\phi_j(m) \quad (m = 1, 2) \tag{16}$$

To illustrate the difference between the overlap integral and overlap electron density, we consider the p_x and p_y atomic orbitals located at a same atomic site (Figure 5a,b). The product $p_x p_y$ represents the overlap electron density ρ_{xy}, which consists of four overlapping regions (Figure 5c); two regions of positive electron density (colored in pink) and two regions of negative electron density (colored in cyan). The overlap integral $\langle p_x | p_y \rangle$ is the sum of these four overlap electron densities, which adds up to zero.

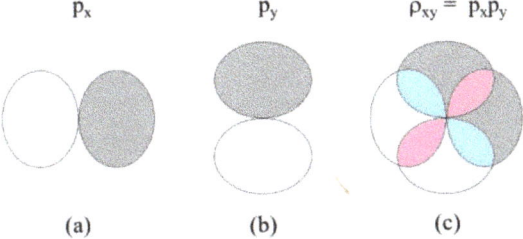

Figure 5. The overlap density resulting from two p-orbital at a given atomic site: (**a**) a p_x orbital; (**b**) a p_y orbital; and (**c**) the overlap density between the two orbitals, $\rho_{xy} = p_x p_y$. The pink and cyan regions have positive and negative values, respectively.

The exchange repulsion between p_x and p_y is written as:

$$K_{xy} = \langle p_x(1)p_x(2)|1/r_{12}|p_y(1)p_y(2)\rangle \tag{17}$$

Equation (17) is given by the sum of the self-repulsion resulting from each overlapping region of the overlap electron density ρ_{xy} (Figure 5c). Each overlapping region, be it positive or negative, leads to a positive repulsion; therefore, the exchange repulsion K_{xy} is positive.

3.2. Two Competing Components of Spin Exchange

In Figure 3c, the configuration Ψ_T represents the triplet state of the dimer while the configurations Ψ_{S1} and Ψ_{S2} each represent a singlet state. For a magnetic system, for which Δe is very small, the singlet state is not well described by Ψ_{S1} alone and is represented by a linear combination of Ψ_{S1} and Ψ_{S2}, i.e., $\Psi_S = c_1 \Psi_{S1} + c_2 \Psi_{S2}$, where the mixing coefficients c_1 and c_2 are determined by the interaction between Ψ_{S1} and

Ψ_{S2}, namely, $\langle\Psi_{S1}|H_{elec}|\Psi_{S2}\rangle = K_{ij}$, as well as by the energies of Ψ_{S1} and Ψ_{S2}, that is, $\langle\Psi_{S1}|H_{elec}|\Psi_{S1}\rangle = E_1$ and $\langle\Psi_{S2}|H_{elec}|\Psi_{S2}\rangle = E_2$. After some lengthy manipulations under the condition, $K_{ij} \gg |E_1-E_2|$, which is satisfied for magnetic systems, the energy difference between the triplet and singlet states, and hence the spin exchange J, is expressed as [15]:

$$J = E_T - E_S \approx -2K_{ij} + \frac{(\Delta e)^2}{U} \quad (18)$$

Note that the spin exchange J consists of two components, $J = J_F + J_{AF}$, where:

$$\begin{aligned}J_F &= -2K_{ij} < 0 \\ J_{AF} &= \frac{(\Delta e)^2}{U} > 0\end{aligned} \quad (19)$$

The magnitude of the FM component $|J_F|$ increases as the exchange repulsion K_{ij} increases, namely, as the overlap density $|\rho_{ij}| = |\phi_i\phi_j|$ increases. The strength of the AFM component J_{AF} increases with increasing the energy split Δe, i.e., with increasing the overlap integral $\langle\phi_i|\phi_j\rangle$, while J_{AF} decreases as the on-site repulsion U increases.

As already discussed in Section 2, the quantitative values of spin exchanges can be accurately determined by the energy-mapping analysis based on first-principles DFT+U or DFT+hybrid calculations. The purpose of Equations (18) and (19) is not to determine the numerical value of any spin exchange, but to show that each spin exchange J consists of two competing components, J_F and J_{AF}, that the overall sign of J is determined by which component dominates, and which electronic parameters govern the strength of each component.

4. Spin Exchanges Determined by the Ligand p-Orbitals in the Magnetic Orbitals

To illustrate how spin exchanges of transition metal magnetic ions are controlled by the ligand p-orbitals in their magnetic orbitals, we consider various spin exchanges involving Cu^{2+} (d^9, S = 1/2) ions as an example, which typically form axially elongated CuL_6 octahedra (Figure 6a). The energies of their d-states are split as (xz, yz) < xy < $3z^2-r^2$ < x^2-y^2 (Figure 6b) so that the magnetic orbital of each Cu^{2+} ion is represented by the x^2-y^2 state, in which the Cu x^2-y^2 orbital induces σ-antibonding with the p-orbitals of four equatorial ligands L (Figure 6c). In this section, we examine how the various types of spin exchanges associated with Cu^{2+} ions are controlled by the ligand p-orbitals of their x^2-y^2 states. The major component of the magnetic orbital of a Cu^{2+} ion (Figure 6d) is the Cu d-orbital (i.e., the magnetic orbital "head"), and the minor component the ligand p-orbitals (i.e., the magnetic orbital "tail"). In this section, we probe how the nature and strengths of M–L–M, M–L . . . L–M and M–L . . . A . . . L–M-type exchanges are determined by how the ligand p-orbital tails of their magnetic orbitals are arranged in their exchange paths.

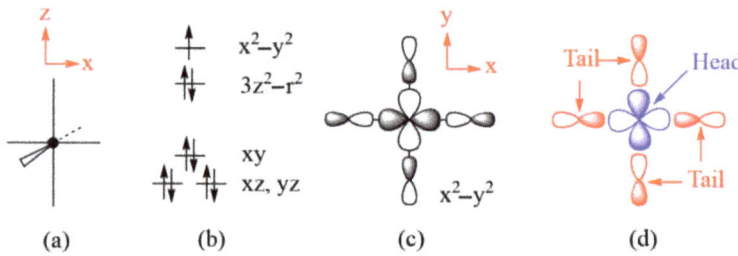

Figure 6. (a) An axially elongated CuL_6 octahedron. (b) The electron configuration of a $Cu^{2+}(d^9)$ ion at an axially elongated octahedral site. (c) The magnetic orbital of the $Cu^{2+}(d^9)$ ion at an axially elongated octahedral site, which is contained in the CuL_4 equatorial plane. (d) The head and tails of the magnetic orbital.

4.1. M–L ... L–M and M–L ... A ... L–M Spin Exchanges

The next-nearest-neighbor (nnn) spin exchange J_{nnn} (Figure 7a) that occurs in a CuL_2 (L = O, Cl, Br) ribbon chain, is obtained by sharing the opposite edges of CuL_4 square planes. This spin exchange is an example of a strong Cu–L ... L–Cu exchange (Figure 7b), when the L ... L contact distance is in the vicinity of the van der Waals distance. The latter will be assumed to be the case in what follows. The two magnetic orbitals interact across the L ... L contacts through the overlap of their p-orbital tails. This through-space interaction leads to the in-phase and out-of-phase combinations (Ψ_+ and Ψ_-, respectively) of the magnetic orbitals (Figure 7c), with energy split Δe between the two (Figure 7d). This makes the AFM component J_{AF} non-zero. The overlap electron density associated with the interacting p-orbital tails is nearly zero, so the FM component J_F is practically zero. As a result, J_{nnn} becomes AFM.

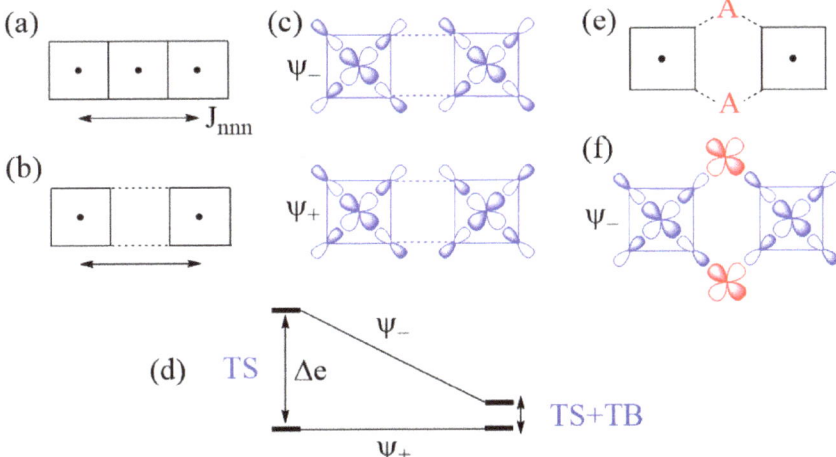

Figure 7. (a) The next-nearest-neighbor spin exchange J_{nnn} in a CuL_2 ribbon chain made up of edge-sharing CuL_4 square planes. (b) A case of strong Cu–L ... L–Cu exchange, which represents the next-nearest-neighbor exchange J_{nnn}. (c) The in-phase and out-of-phase combinations of the two magnetic orbitals, Ψ_+ and Ψ_-, respectively, associated with the Cu–L ... L–Cu exchange. (d) The large energy split Δe resulting from the through-space (TS) interaction in the Cu–L ... L–Cu exchange becomes small in the Cu–L ... A ... L–Cu exchange as a result of the through-bond (TB) interaction that occurs with the Ψ_- state. (e) A Cu–L ... A ... L–Cu exchange generated when each L ... L contact is bridged by a d^0 metal cation A. (f) The bonding interaction of the Ψ_- state with the d_π orbital of A.

In the M–L ... L–M spin exchange J_{nnn} discussed above, there are two equivalent exchange paths due to the ribbon structure. Suppose that each L ... L contact of the M–L ... L–M exchange path is bridged by a d^0 transition metal cation A (e.g., V^{5+} and W^{6+}) to form an M–L ... A ... L–M spin exchange (Figure 7e). We now analyze the relative strengths of the M–L ... L–M and M–L ... A ... L–M spin exchanges. Across the L ... L contact of the M–L ... L–M path, the p-orbital tails of L are combined in-phase in Ψ_+, but out-of-phase in Ψ_-. Thus, the empty d_π orbital of the cation A interacts in-phase with Ψ_- (Figure 7f) to lower the energy of Ψ_-, but it does not interact with Ψ_+, therefore the energy of Ψ_+ is unaffected. This selective interaction of the bridging d^0 cation A with the L ... L contact of the M–L ... L–M spin exchange has a dramatic consequence on the strength of an M–L ... A ... L–M spin exchange. When the M–L ... L–M exchange has a strong through-space interaction, the through-bond interaction reduces the large energy split Δe to a small value, thereby weakening the overall M–L ... A ... L–M spin exchange (Figure 7d).

Another example of a strong M–L ... L–M exchange occurs when the p-orbital tails of L are pointed to each other along the L ... L contact (Figure 8a). The through-space interaction between the magnetic orbitals leads to the in-phase and out-of-phase combinations, Ψ_+ and Ψ_-, respectively (Figure 8b), with a large energy split Δe between the two (Figure 8c) and a negligible overlap electron density between the interacting p-orbital tails. As a result, the M–L ... L–M spin exchange becomes AFM. If the L ... L contact of an M–L ... L–M exchange path is bridged by a d^0 transition metal cation A to form an M–L ... A ... L–M spin exchange (Figure 8d), only the Ψ_- state of the M–L ... L–M path interacts effectively with one of the empty d_π orbitals of A (Figure 8e). Thus, a strong through-space interaction in the M–L ... L–M exchange leads to a weak overall M–L ... A ... L–M spin exchange due to the effect of the through-bond interaction (Figure 8c).

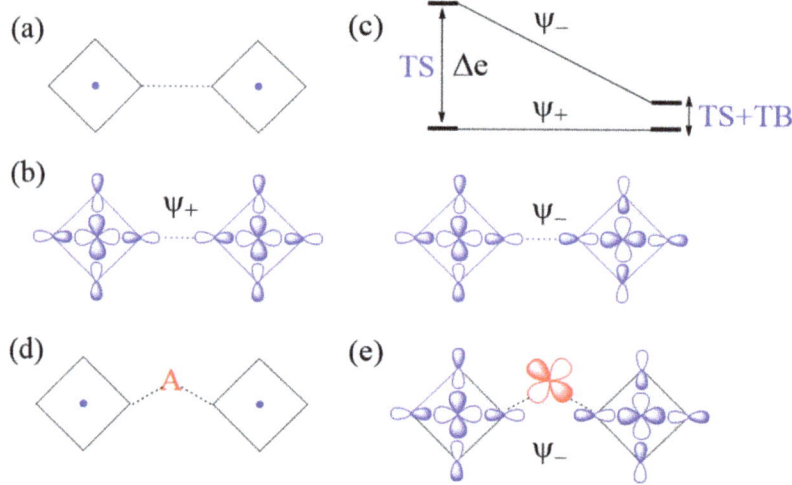

Figure 8. (**a**) A case of strong Cu–L ... L–Cu spin exchange where the two Cu–L bonds leading to the L ... L contacts are linear. (**b**) The in-phase and out-of-phase combinations (Ψ_+ and Ψ_-, respectively) of the two magnetic orbitals, resulting from the through-space (TS) interactions. (**c**) The energy split Δe between Ψ_+ and Ψ_- is large when the overlap between the p-orbital tails is large. (**d**) A Cu–L ... A ... L–Cu exchange generated when the L ... L contact is bridged by a d^0 metal cation A. (**e**) The bonding interaction of the Ψ_- state with the d_π orbital of A. (**f**) The large energy split Δe resulting from the through-space (TS) interaction in the Cu–L ... L–Cu exchange becomes small in the Cu–L ... A ... L–Cu exchange as a result of the through-bond (TB) interaction that occurs primarily with the Ψ_- state.

An example of very weak Cu–L ... L–Cu exchange is shown in Figure 9a, in which the two magnetic orbitals are arranged such that the p-orbital tails are orthogonal to each other (Figure 9b), and their overlap vanishes so that the energy split between Ψ_+ and Ψ_- vanishes (i.e., $\Delta e = 0$) (Figure 9c) so that $J_{AF} = 0$. In addition, J_F should vanish because the overlap electron density resulting from the p-orbital tails will be practically zero. Then, the spin exchange J would be zero. If the L ... L contact of such an M–L ... L–M exchange path is bridged by a d^0 transition metal cation A to form an M–L ... A ... L–M spin exchange (Figure 9d), the Ψ_- state of the M–L ... L–M path interacts with the empty d_π orbital of A, thereby lowering its energy (Figure 9e) while that of the Ψ_+ state is unchanged. Thus, when the M–L ... L–M exchange has a very weak through-space interaction, the through-bond interaction induces the large energy split Δe, so that the overall M–L ... A ... L–M spin exchange becomes strong (Figure 9f).

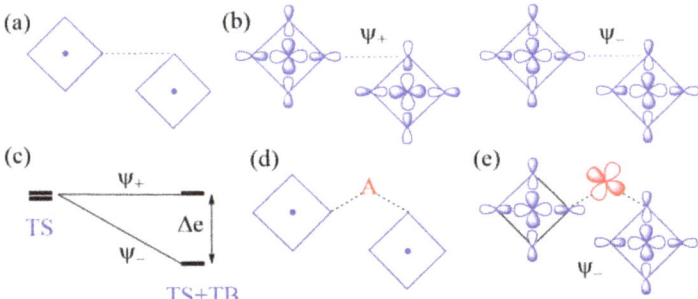

Figure 9. (**a**) A case of weak Cu–L ... L–Cu spin exchange, where the two Cu–L bonds leading to the L ... L contacts are orthogonal. (**b**) The in-phase and out-of-phase combinations (Ψ_+ and Ψ_-, respectively) of the two magnetic orbitals. (**c**) The vanishing energy split Δe resulting from the through-space (TS) interaction in the Cu–L ... L–Cu exchange is enhanced in the Cu–L ... A ... L–Cu exchange as a result of the through-bond (TB) interaction that occurs primarily with the Ψ_- state. (**d**) A Cu–L ... A ... L–Cu exchange generated when the L ... L contact is bridged by a d^0 metal cation A. (**e**) The in-phase interaction of the Ψ_- state with the d_π orbital of A.

In short, a strong M–L ... L–M exchange becomes a weak M–L ... A ... L–M exchange when the L ... L linkage is bridged by d^0 cations, while a weak M–L ... L–M exchange becomes a strong M–L ... A ... L–M exchange when the L ... L linkage is bridged by a d^0 cation.

4.2. M–L–M Spin Exchanges

The Goodenough–Kanamori rules cover these types of spin exchanges [5–9]. For the sake of completeness, we discuss these types of spin exchanges from the viewpoint of the ligand p-orbital tails on the basis of Equations (18) and (19). Let us consider a Cu_2L_6 dimer resulting from two CuL_4 square planes obtained by sharing an edge (Figure 10a), where the ligand L can be O, Cl, or Br. The two magnetic orbitals associated with the nearest-neighbor (nn) spin exchange J_{nn}, presented in Figure 10b, interact at the bridging ligands L of the M–L–M paths. If the CuL_4 units have an ideal square planar shape, the ∠M–L–M angle becomes 90° so that the two p-orbital tails at the bridging ligands L are orthogonal to each other. Thus, as discussed in Section 3, the overlap integral between them is zero. Therefore, the in-phase and out-of-phase combinations of the two magnetic orbitals (Figure 10b) is not split in energy, so $\Delta e = 0$ (Figure 10c) and $J_{AF} = 0$. However, the overlap electron density between the two p-orbital tails at the bridging ligand L is not zero, i.e., $J_F > 0$. Thus, the M–L–M spin exchange becomes FM. When the ∠M–L–M angle deviates from 90° (Figure 10d), the two p-orbital tails at the bridging ligands L are no longer orthogonal to each other so the overlap integral between them is non-zero. Therefore, the in-phase and out-of-phase combinations of the two magnetic orbitals (Figure 10e) differ in energy, so $\Delta e > 0$ (Figure 10f) and J_{AF} is non-zero. The overlap electron density between the two p-orbital tails is non-zero, so J_F is non-zero. Thus, whether the spin exchange is FM or AFM depends on which component, J_F or J_{AF}, dominates, which in turn depends on the ∠M–L–M angle ϕ. Typically, the angle ϕ where FM changes to AFM is slightly greater than 90° due to the involvement of the ligand s-orbital [15], which is commonly neglected for simplicity.

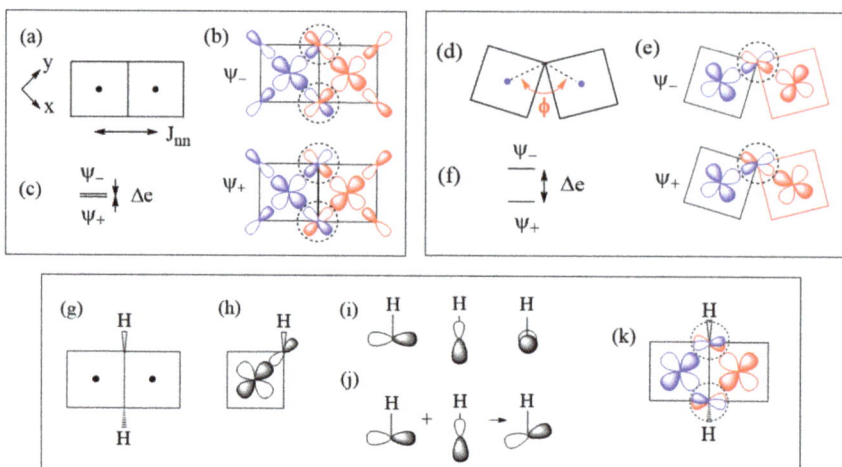

Figure 10. (**a**–**c**) A Cu–L–Cu spin exchange with ∠M–L–M angle ϕ = 90°: This occurs in a Cu$_2$L$_6$ dimer shown in (**a**), which is made up of two coplanar CuL$_4$ square planes by sharing an edge. The in-phase and out-of-phase combinations of the two magnetic orbitals in (**b**), and the energy split Δe between the two in (**c**). (**d**–**f**) A Cu–L–Cu spin exchange with ∠M–L–M angle ϕ > 90°: This occurs when two non-coplanar CuL$_4$ square planes are corner-shared as in (**d**). The in-phase and out-of-phase combinations of the two magnetic orbitals in (**e**), and the energy split Δe between the two in (**f**). (**g**–**k**) Effect of the molecular anions OH$^-$ on the spin exchange of the edge-sharing Cu$_2$O$_6$ dimer: the shared edge consists of two OH$^-$ ions in (**g**); the p-orbital tail of one OH$^-$ ligand in a CuO$_4$ square plane expected if the OH$^-$ is treated as O^{2-} in (**h**); the three oxygen lone pairs associated with an isolated OH$^-$ in (**i**); the tilting of the O 2p-orbital tail by mixing two oxygen lone pairs in (**j**); and the arrangement of the tilted O 2p-tails at the bridging O atoms of the Cu$_2$O$_6$ dimer in (**k**).

So far in our discussion, it has been implicitly assumed that each main-group ligand L exists as a spherical anion (e.g., each O as an O^{2-} anion, and each Cl atom as a Cl$^-$ anion). However, this picture is not quite accurate when the ligand atom makes a strong covalent bonding with another main-group element to form a molecular anion such as OH$^-$. Suppose that each O atom on the shared edge of the CuO$_6$ (L = O) dimer is not a O^{2-} but an OH$^-$ anion (Figure 10g). Then, the ligand p-orbital tail on that O cannot be the p-orbital pointed along one lobe of the Cu x^2-y^2 orbital (Figure 10h) because it is incompatible with the O–H bonding, which has three directional O lone pairs depicted in Figure 10i. To satisfy both the strong covalent-bonding with H and the weak covalent-bonding with Cu, the O lone pair of OH$^-$ tilts slightly toward one lobe of the x^2-y^2 orbital (Figure 10j). As a result, the ligand p-orbital tails, arising from the two magnetic orbitals at the bridging O atoms, are not orthogonal as in Figure 10b but become more parallel to each other (Figure 10k). As a result, the spin exchange between the two Cu^{2+} ions in Figure 10g becomes AFM (see below). Another molecular anion of interest is the carbonate ion CO$_3^{2-}$, in which each O atom makes a strong covalent bond with C; therefore, the O atoms of CO$_3^{2-}$ should not be treated as isolated O^{2-} anions in their coordination with transition metal cations. In general, the presence of molecular anions such as OH$^-$ and CO$_3^{2-}$ in a magnetic solid makes it difficult to deduce, on a qualitative reasoning, what its spin lattice would be. This is where the quantitative energy-mapping analysis is indispensable, because it does not require any qualitative reasoning.

4.3. Qualitative Rules for Spin Exchanges Based on the p-Orbital Tails of Magnetic Orbitals

In a magnetic orbital of an ML$_n$ polyhedron, the d-orbital of M dictates by its symmetry which p-orbitals of the ligands L become the p-orbital tails. From the viewpoint of orbital interaction, a spin exchange between magnetic ion is none other than the interaction between their magnetic orbitals. The latter is caused by the interaction between their p-orbital tails, not by that between their d-orbital heads. In other words, a spin exchange is not a "head-to-head" interaction but a "tail-to-tail" interaction. By considering these tail-to-tail interactions described above, we arrive at the following four qualitative rules governing the nature and strengths of the M–L ... L–M and M–L ... A ... L–M type exchanges under the assumption that the L ... L contact distance is in the vicinity of the van der Waals distance:

(1) When the p-orbital tails generate a large overlap integral but a small overlap electron density, the M–L ... L–M exchange is AFM.
(2) When the p-orbital tails generate a small overlap integral but a large overlap electron density, the M–L ... L–M exchange is FM.
(3) When the p-orbital tails generate neither a non-zero overlap integral nor a non-zero overlap electron density, the M–L ... L–M exchange vanishes.
(4) When the M–L ... L–M exchange is strongly AFM, the corresponding M–L ... A ... L–M becomes a weak exchange. When the M–L ... L–M exchange is a weak exchange, the corresponding M–L ... A ... L–M becomes strongly AFM.

These rules on the M–L ... L–M and M–L ... A ... L–M exchanges should be used together with the Goodenough–Kanamori rules in choosing a proper set of spin exchanges to evaluate using the energy-mapping analysis based on DFT+U or DFT+hybrid calculations. In principle, this analysis can provide quantitative values for any possible exchanges of a given magnetic system. However, even this quantitative tool cannot determine the value of any spin exchange unless it is included in the set of spin exchanges for the energy-mapping analysis. It is paramount to consider in detail the structural features governing the strengths of spin exchanges in order not to miss exchange paths crucial for defining the correct spin lattice of a given solid.

5. Representative Examples

In Section 4, we analyzed the structural features governing the nature of the three types of spin exchanges, i.e., M–L–M, M–L ... L–M and M–L ... A ... L–M, which occur in various magnetic solids. This section will discuss the occurrence of these exchanges in actual magnetic solids by analyzing the crystal structures and magnetic properties of four representative magnetic solids, α-CuV$_2$O$_6$, LiCuVO$_4$, (CuCl)LaNb$_2$O$_7$ and Cu$_3$(CO$_3$)$_3$(OH)$_3$. α-CuV$_2$O$_6$, LiCuVO$_4$, and (CuCl)LaNb$_2$O$_7$ were chosen to show that correct spin lattices can be readily predicted by the qualitative rules of Section 4.3, although they have to be confirmed by performing the energy-mapping analyses. Azurite Cu$_3$(CO$_3$)$_2$(OH)$_2$ was chosen to demonstrate that the spin lattice of a certain magnetic system cannot be convincingly deduced solely on the basis of the qualitative rules. The magnetic ions of such a system are coordinated with molecular anions in which the first-coordinate main-group ligands L make strong covalent bonds with other main-group elements (e.g., H in the OH$^-$ ion, and C in the CO$_3^{2-}$ ion). In such a case, use of the energy-mapping analysis is the only recourse with which to find the spin lattice correct for a given system.

5.1. Two-Dimensional Behavior of α-CuV$_2$O$_6$

The magnetic properties of α-CuV$_2$O$_6$ were initially analyzed in terms of a one-dimensional (1D) spin S = 1/2 Heisenberg chain model with uniform nearest-neighbor AFM spin exchange [16–18]. However, a rather high Néel temperature of ~22.4 K indicated the occurrence of a substantial interchain spin exchange of the order of 50% of the intrachain exchange, casting serious doubts on the applicability of a simple chain description. α-CuV$_2$O$_6$ consists of CuO$_4$ chains, made up of edge-sharing CoO$_6$ octahedra, which run along the a-direction (Figure 11a). If the two axially elongated Cu–O bonds are removed from each CuO$_6$ octahedron to identify its CuO$_4$ equatorial plane containing the magnetic orbital, one finds that each CuO$_4$ chain of edge-sharing CoO$_6$ octahedra becomes a chain of stacked CuO$_4$ square planes (Figure 11b). In each stack-chain along the a-direction, adjacent CuO$_4$ square planes are parallel to each other such that the adjacent magnetic orbitals generate neither a non-zero overlap nor a non-zero overlap electron density. The same is true between adjacent CuO$_4$ square planes along the b-direction (Figures 11c and 12a). However, between adjacent CuO$_4$ square planes along the c-direction, an almost linear Cu–O ... O–Cu contact occurs with O ... O distance of 2.757 Å (Figures 11c and 12b), slightly shorter than the van der Waals distance of 2.80 Å. Thus, this Cu–O ... O–Cu spin exchange along the c-direction (J_c) should be substantial.

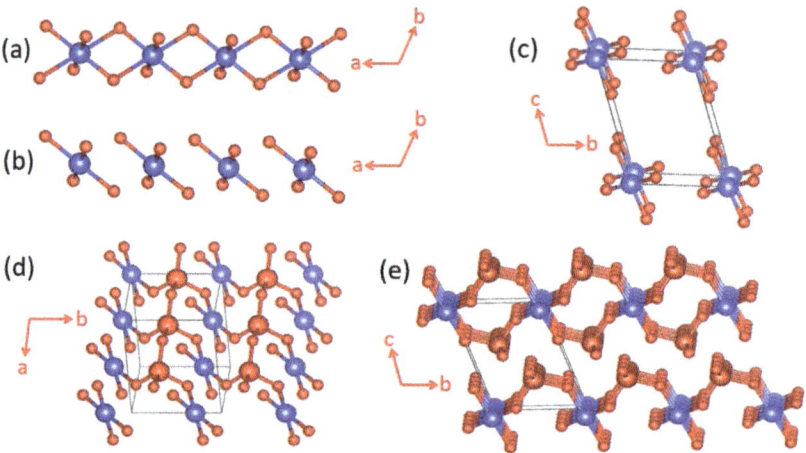

Figure 11. Crystal structure of α-CuV$_2$O$_6$, where the blue spheres represent the Cu atoms, and the large and small red spheres the V and O atoms, respectively: (**a**) A CuO$_4$ chain along the a-direction, which is made up of edge-sharing, axially-elongated, CuO$_6$ octahedra. (**b**) A stack of CuO$_4$ square planes along the a-direction, which results from the chain of edge-sharing CuO$_6$ octahedra by removing the axial Cu-O bonds. (**c**) Stacks of CuO$_4$ square planes running along the c-direction. (**d**) One sheet of CuO$_4$ stack chains parallel to the a–b plane condensed with chains of corner-sharing VO$_4$ tetrahedra on one side of the sheet. (**e**) Stacking of CuV$_2$O$_6$ layers forming α-CuV$_2$O$_6$.

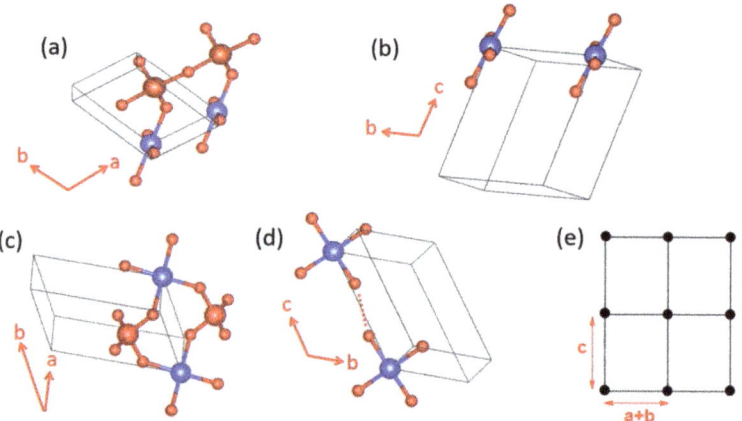

Figure 12. Spin exchange paths and spin lattice of α-CuV$_2$O$_6$: (**a**) Spin exchange along the b-direction, J_b; (**b**) Spin exchange along the c-direction, J_c; (**c**) Spin exchange along the a-direction, J_a; (**d**) Spin exchange along the (a + b)-direction, J_{a+b}; (**e**) Rectangular spin lattice made up of J_c and J_{a+b}.

Each sheet of CuO_4 stack-chains parallel to the a–b plane is corner-shared with V_2O_6 chains, above and below the sheet (Figure 11d), to form layers of composition CuV_2O_6 (Figure 11e). Each V_2O_6 chain is made up of corner-sharing VO_4 tetrahedra containing V^{5+} (d^0, S = 0) ions. As a consequence, adjacent CuO_4 square planes are corner-shared with V_2O_6 chains (Figure 12c). The adjacent CuO_4 square planes are not bridged by a single VO_4 tetrahedron; therefore, the spin exchange along the a-direction, J_a, is expected to be weak. Along the (a + b)-direction, two adjacent CuO_4 square planes are bridged by VO_4 to form two Cu–O . . . V^{5+} . . . O–Cu paths, in which the two Cu–O bonds in each path have a near-orthogonal arrangement.

As already discussed (Figure 10), this Cu–O . . . V^{5+} . . . O–Cu spin exchange should be substantial due to the through-bond effect of the V^{5+} cation. Consequently, the spin lattice of α-CuV_2O_6 must be described by a two-dimensional (2D) rectangular lattice defined by J_{a+b} and J_c. In support of this analysis, the energy-mapping analysis based on DFT+U calculations with U_{eff} = 4 eV show that J_{a+b} and J_c are the two dominant spin exchanges, and are nearly equal in magnitude, namely, J_{a+b} = 86.8 K and J_c/J_{a+b} = 0.88 [19]. In agreement with this finding, one re-investigation of the magnetic properties of α-CuV_2O_6 clearly attested a 2D S = 1/2 rectangular spin lattice model with an anisotropy ratio of 0.7 [19]. The magnetic structure determined from neutron powder diffraction data was in best agreement with these DFT+U calculations. In terms of chemical bonding, α-CuV_2O_6 consists of CuV_2O_6 layers stacked along the c-direction. There is no chemical bonding between adjacent layers; only van der Waals interactions. In terms of magnetic bonding, however, α-CuV_2O_6 consists of 2D spin lattices parallel to the (a + b)-c plane. There is negligible magnetic bonding perpendicular to this plane.

5.2. One-Dimensional Chain Behavior of $LiCuVO_4$

$LiCuVO_4$ consists of axially elongated CuO_6 octahedra, which form edge-sharing CuO_4 chains along the b-direction, which are corner-shared with VO_4 tetrahedra containing V^{5+} (d^0, S = 0) ions (Figure 13a). When the axial Cu–O bonds are deleted, one finds $CuVO_4$ layers in which the CuO_2 ribbon chains are corner-shared by VO_4 tetrahedra (Figure 13b). A perspective view of a single $CuVO_4$ layer approximately along the c-direction (Figure 13c) shows that each VO_4 tetrahedron bridges two neighboring CuO_2 ribbon chains. Thus, the spin exchanges of interest are the nearest-neighbor exchanges J_{nn} of the Cu–O–Cu type and the next-nearest-neighbor spin exchange J_{nnn} of the Cu–O . . . O–Cu type within each CuO_2 ribbon chain as well as the interchain spin exchange J_a along the a-direction of the Cu–O . . . V^{5+} . . . O-Cu type (Figure 13d).

In the absence of the V^{5+} ion, the exchange J_a would be similar in strength to J_{nnn}. However, the O . . . V^{5+} . . . O bridges will weaken the strength of J_a, as discussed in Figure 7. Furthermore, there is no spin exchange path between adjacent $CuVO_4$ layers. Consequently, the spin lattice of $LiCuVO_4$ is a 1D chain running along the b-direction, as is the 1D ribbon chain. The major cause for the occurrence of the 1D chain character is the Cu–O . . . V^{5+} . . . O–Cu spin exchange, which nearly vanishes because the effect of the through-space interaction is canceled by that of the through-bond interaction. In agreement with this reasoning, DFT+U calculations with U_{eff} = 4 eV show that J_{nnn} is strongly AFM (i.e., 208.7 K), while J_{nn} and J_a are weakly FM (i.e., J_{nn}/J_{nnn} = −0.12, and J_a/J_{nnn} = −0.08) [20].

As in the case of α-CuV_2O_6, the magnetic properties of $LiCuVO_4$ were initially analyzed in terms of a 1D chain with uniform nearest-neighbor Heisenberg spin exchange [21–23]. The magnetic susceptibility of $LiCuVO_4$ showed a typical broad maximum at about 28 K, characteristic for 1D behavior with strong intrachain spin exchanges, whereas long-range AFM ordering was detected only below ~2.2 K. The need to modify this simple 1D description was brought about by Gibson et al., who determined the magnetic structure of $LiCuVO_4$ from single crystal neutron diffraction [24]. They found that the spins of each CuO_2 ribbon chain had a cycloid structure (Figure 14a) with adjacent Cu^{2+} moments making an angle of slightly less than 90°. The incommensurate cycloid structure is explained by spin frustration due to competing exchange J_{nn}, which is FM, and J_{nnn}, which is AFM [20,24–26]. If all the spins of a CuO_2 ribbon chain were to be collinear, the ribbon chain cannot satisfy all nearest-neighbor exchanges FM and all next-nearest-neighbors AFM simultaneously. Thus, the spin arrangement of the CuO_2 ribbon chain is spin-frustrated. To reduce the extent of this spin frustration, the spins of the ribbon chain adopt a noncollinear spin arrangement. The noncollinear spin arrangement observed for $LiCuVO_4$ has a cycloid structure in each ribbon chain, in which the nearest-neighbor spins are nearly orthogonal to each other, while the next-nearest-neighbor spins generate a near-AFM arrangement (Figure 14a) [24,25]. In a cycloid, each successive spin in the CuO_2 ribbon chain rotates in one direction by a certain angle. Thus, a cycloid structure is chiral in nature, which means that the alternative cycloid structure opposite in chirality but identical in energy is equally probable (Figure 14b) [27].

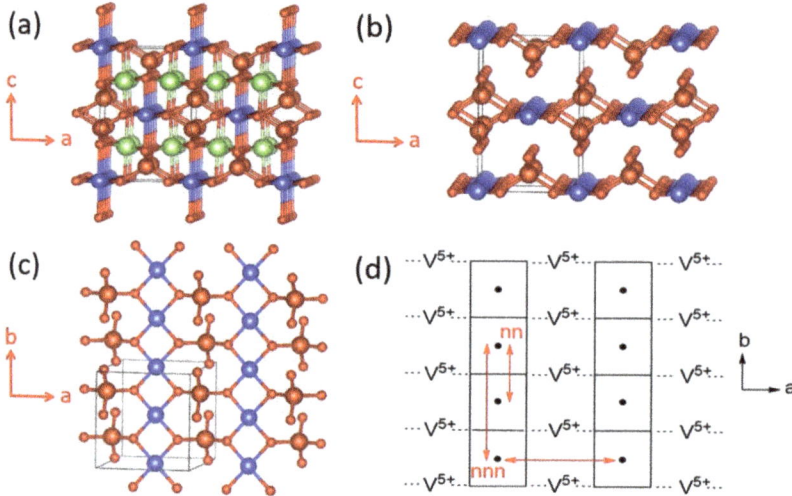

Figure 13. (a) The crystal structure of $LiCuVO_4$ viewed approximately along the b-direction, where the blue and green spheres represent the Cu and Li atoms, respectively, and the large and small red spheres represent the V and O atoms, respectively. (b) The $CuVO_4$ lattice resulting from $LiCuVO_4$ by removing the axial Cu–O bonds and Li atoms. (c) A perspective view of a single $CuVO_4$ layer approximately along the c-direction. (d) The spin exchange paths present in a single $CuVO_4$ layer, where the labels nn, nnn and a represent the spin exchanges J_{nn}, J_{nnn} and J_a, respectively.

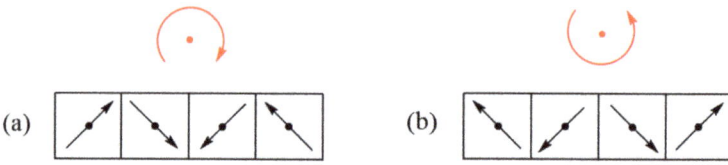

Figure 14. (a,b) Two cycloids, which are opposite in chirality but are identical in energy.

In general, when the temperature is lowered below a certain temperature, T_{SDW}, a moderately spin-frustrated magnetic system gives rise to two cycloids of opposite chirality with equal probability. The resulting superposition of the two (Figure 15a–c) leads to a state known as a spin density wave (SDW) [27,28]. The latter becomes transverse if the preferred spin orientation at each magnetic ion is perpendicular to the SDW propagation direction, but becomes longitudinal if the spin orientation prefers the SDW propagation direction

(Figure 15d–f). When the temperature is lowered further below T_{SDW}, the electronic structure of the spin-lattice may relax to energetically favor one of the two chiral cycloids so that one can observe a cycloid state at a temperature slightly below T_{SDW}. The repeat unit of a cycloid is determined by the spin frustration present in the magnetic system, therefore a cycloid phase is typically incommensurate.

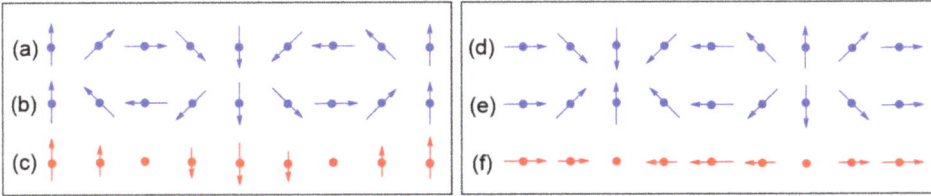

Figure 15. The superposition of the two chiral cycloids in (**a**); (**b**) leads to the transverse SDW in (**c**). The superposition of the two chiral cycloids in (**d**); (**e**) leads to the longitudinal SDW in (**f**). For ease of illustration, the angle of the spin rotation was taken to be 45°.

In the cycloid state, LiCuVO$_4$ exhibits ferroelectricity [29–31], because a cycloid structure lacks inversion symmetry. The polarization of LiCuVO$_4$ can be switched with magnetic and electric fields [32–35]. LiCuVO$_4$ has attracted special attention for the possibility of inducing new phases by applying external magnetic fields. Saturation of the Cu moments occurs above ~45 T, depending on the orientation of the crystal [36]. The competition between J_{nn} and J_{nnn} in LiCuVO$_4$ is considered a promising setting for unusual bond nematicity and spin nematic phases close to the full magnetic saturation [37–45].

5.3. Spin Gap Behavior from the CuO$_2$Cl$_2$ Perovskite Layer of (CuCl)LaNb$_2$O$_7$

(CuCl)LaNb$_2$O$_7$ consists of CuClO$_2$ and LaNb$_2$O$_7$ layers, which alternate along the c-direction by sharing their O corners [46]. Each LaNb$_2$O$_7$ layer represents two consecutive layers (Figure 16a) of LaNbO$_3$ perovskite. The building blocks of the CuClO$_2$ layer are the CuCl$_4$O$_2$ octahedra with their O atoms at the apical positions forming a linear O–Cu–O bond aligned along the c-direction. Suppose that the CuCl$_4$ equatorial planes are square in shape with four-fold rotational symmetry around the O–Cu–O axis (Figure 16b,c), and such CuCl$_4$O$_2$ octahedra corner-share their Cl atoms to form a perovskite layer CuClO$_2$. Then, the highest-occupied d-states of each CuCl$_4$O$_2$ octahedron are degenerate and have three electrons to accommodate, causing a Jahn–Teller instability of each CuCl$_4$O$_2$ octahedron. In the CuClO$_2$ layer, the CuCl$_4$O$_2$ octahedra must undergo a cooperative Jahn–Teller distortion. In the tetragonal structure (SG, P4/mmm) of (CuCl)LaNb$_2$O$_7$ [46], each Cl site is split into four positions (Figure 16d). A Jahn–Teller distortion available to such a CuCl$_4$O$_2$ octahedron is an axial-elongation, in which one linear Cl–Cu–Cl bond is shortened while lengthening the other linear Cl–Cu–Cl bond, as shown in Figure 16e, which can be simplified as in Figure 16f. By choosing one of the four split positions from each Cl site, it is possible to construct the CuClO$_2$ layer with a cooperative Jahn–Teller distortion, as presented in Figure 17a, which shows the Cu–Cl–Cu–Cl zigzag chains running along the b-direction with the plane of each CuCl$_2$O$_2$ rhombus perpendicular to the a–b plane. Each Cl site has four split positions when all these four possibilities occur equally.

With the local coordinate axes of each CuCl$_2$O$_2$ rhombus taken as in Figure 17a, then the magnetic orbital of the Cu^{2+} ion can be described as the x^2-z^2 state in which Cu x^2-z^2 orbital makes σ-antibonding interactions with the p-orbitals of O and Cl. Extended Hückel tight binding calculations [47] for the CuCl$_2$O$_2$ rhombus show that in the magnetic orbital of the $CuCl_2O_2$ rhombus, the Cu^{2+} ion is described by $0.646\,(3z^2-r^2) - 0.272\,(x^2-y^2)$. The latter is rewritten as:

$$0.646(3z^2-r^2) - 0.272(x^2-y^2) \approx [2(3z^2-r^2)-(x^2-y^2)]/3$$
$$\propto (z^2-x^2) + 0.25(x^2-y^2) \approx (z^2-x^2)$$

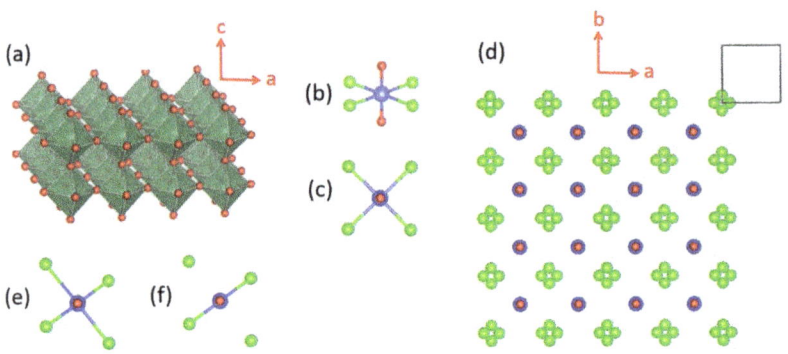

Figure 16. (a) One $LaNb_2O_7$ layer of tetragonal $(CuCl)LaNb_2O_7$, where the La atoms at the 12 coordinate sites formed by eight corner-sharing NbO_6 octahedra are not shown for simplicity. (b) Perspective and (c) projection views of a $CuCl_4O_2$ octahedron with 4-fold rotational symmetry around the O–Cu–O axis aligned along the c-direction. (d) A projection view of a $CuClO_2$ layer of tetragonal $(CuCl)LaNb_2O_7$ in which every Cl site is split into four positions. (e,f) Projection views of the Jahn–Teller distorted $CuCl_4O_2$ octahedron, where the axially elongated Cu–Cl bonds are shown in (e), but are not shown in (f) for simplicity.

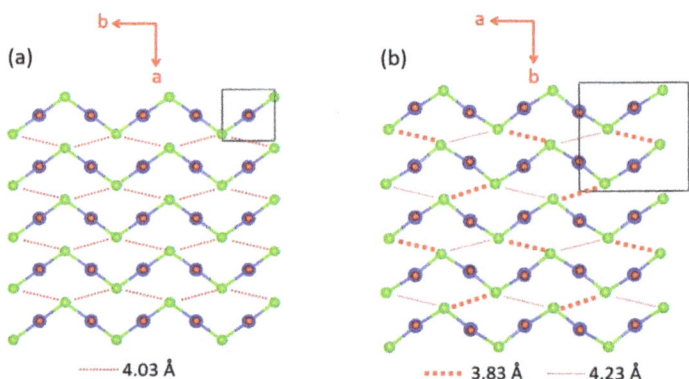

Figure 17. The structures of the $CuClO_2$ layer with cooperative Jahn–Teller distortion in (a) the tetragonal and (b) orthorhombic phases of $(CuCl)LaNb_2O_7$. The Cl–Cu–Cl unit is linear in the tetragonal structure, but is slightly bent in the orthorhombic structure. The latter has an important consequence on the spin exchanges.

Namely, it is dominated by the (z^2-x^2) character. This is consistent with the NMR/NQR study of $(CuCl)LaNb_2O_7$, which showed that the d-state of the Cu^{2+} ion is mostly characterized by $3z^2-r^2$ with some contribution of x^2-y^2 [48].

We note that the $CuCl_2O_2$ rhombuses of each Cu–Cl–Cu–Cl zigzag chain make Cu–Cl... Cl–Cu contacts of Cl... Cl = 4.03 Å with its adjacent zigzag chains (Figure 17a). The Cl p-orbitals of the (z^2-x^2) magnetic orbital (Figure 18a) are pointed approximately along the Cl ... Cl contact, so their overlap is substantial. In all other spin exchange paths, the Cl p-orbitals in their magnetic orbitals are not arranged to overlap well. Thus, only the exchange along the Cu–Cl ... Cl–Cu direction is expected to be substantially AFM. Then the spin lattice of the CuClO$_2$ layer would be 1D Heisenberg uniform AFM chains running along the Cu–Cl ... Cl–Cu directions, e.g., the (a + 2b)- and (a − 2b)-directions (Figure 17a). Uniform AFM chains do not have a spin gap, but the magnetic properties of (CuCl)LaNb$_2$O$_7$ reveal a spin gap behavior [49]. Thus, the tetragonal structure of (CuCl)LaNb$_2$O$_7$ is inconsistent with experiment. If one discards the possibility of cooperative Jahn-Teller distortions, one can generate several nonuniform clusters made up of distorted CuCl$_4$O$_2$ octahedra [50]. However, the latter would lead to several different spin gaps rather than that observed by Kageyama et al. [49], who showed that the magnetic susceptibility of (CuCl)LaNb$_2$O$_7$ can be approximated by an isolated spin dimer model with the intradimer distance of approximately 8.8 Å, which corresponds to the fourth-nearest-neighbor Cu ... Cu distance. The spin-gap behavior of (CuCl)LaNb$_2$O$_7$ was surprising, given the belief of the square lattice arrangement of Cu^{2+} ions is spin-frustrated [49,51,52]. This led to several DFT studies designed to find the precise crystal structure of (CuCl)LaNb$_2$O$_7$ [53–55], leading to the conclusion that an orthorhombic structure of space group Pbam is correct for (CuCl)LaNb$_2$O$_7$ [54,55].

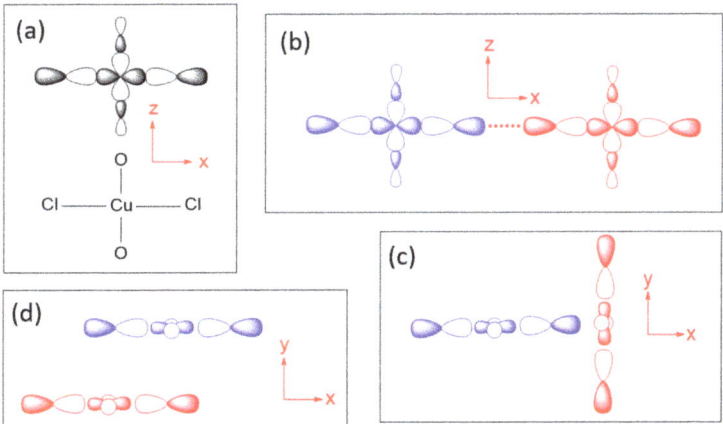

Figure 18. (**a**) The magnetic orbital lying in the CuCl$_2$O$_2$ rhombus plane, which is best described as the x^2–z^2 state. (**b**–**d**) Three different types of the magnetic orbital arrangements found in the CuClO$_2$ layers of orthorhombic CuCl(LaV$_2$O$_7$). In this idealized representation of the CuCl$_2$O$_2$ rhombus, the slight bending of the Cl–Cu–Cl linkage is neglected for clarity.

The cause for the spin gap behavior of (CuCl)LaNb$_2$O$_7$ is found from its orthorhombic structure (SG, *Pbam*) (Figure 17b) [56,57], in which the arrangement of the Cu^{2+} ions are no longer tetragonal so that adjacent Cu–Cl–Cu–Cl zigzag chains have two kinds of Cl ... Cl contacts (i.e., 3.83 and 4.23 Å), and the Cu–Cl ... Cl–Cu chains become alternating with shorter and longer Cl ... Cl contacts. Furthermore, although the Cl–Cu–Cl unit of each CuCl$_2$O$_2$ rhombus is slightly bent in the orthorhombic structure, the latter has an important consequence on the spin exchanges (see below).

The six spin exchange paths J$_1$–J$_6$ of the CuClO$_2$ layer are depicted in Figure 19a. In the spin exchanges J$_1$ and J$_2$, the Cl p-orbital tails are approximately pointed toward each other (Figure 18b). J$_3$ is a Cu–Cl–Cu exchange with ∠Cu–Cl–Cu angle somewhat greater than 90° (namely, 109.0°). In J$_4$ and J$_5$, the Cl p-orbital tails are approximately orthogonal to each other (Figure 18c), but they differ due to the bending in the Cl–Cu–Cl units. In the J$_6$ path, the Cl p-orbital tails are not pointed toward to each other but are approximately parallel to each other (Figure 18d). In the exchange paths J_1 and J_2, the Cu–Cl . . . Cl linkage is more linear and the Cl . . . Cl contact is shorter in J_1. This suggests that J_1 is more strongly AFM than J_2, thus the spin lattice of the $CuClO_2$ layer is an alternating AFM chain. In agreement with this argument, the energy-mapping analysis based on DFT+U calculations shows that $J_1 = 87.5$ K and $J_2/J_1 = 0.18$. In addition, this analysis reveals that J_3–J_6 are all FM with $J_3/J_1 = -0.39$, $J_4/J_1 = -0.38$, $J_5/J_1 = -0.14$ and $J_5/J_1 = -0.04$ [56]. It is of interest to note that the strongest AFM exchange J_1 is the fourth-nearest-neighbor spin exchange, with a Cu . . . Cu distance of 8.53 Å [49]. As shown in Figure 19b, the spins of the $CuClO_2$ layer form alternating AFM chains. Chemically, the Cu–Cl zigzag chains run along the a-direction. In terms of magnetic bonding, however, the spins of the $CuClO_2$ layer consist of J_1-J_2 alternating AFM chains not only along the (a + 2b)-direction but also along the (−a + 2b)-direction. This explains why the magnetic susceptibility of $(CuCl)LaNb_2O_7$ exhibits a spin gap behavior. Due to the bending of the Cl–Cu–Cl units, the Cl p-orbital tail of one $CuCl_2O_2$ rhombus is pointed toward one Cl atom (away from both Cl atoms) of the other rhombus in the J_4 (J_5) path. This makes J_4 more strongly FM than J_5 is. J_3 is strongly FM despite the fact that the ∠Cu–Cl–Cu angle is somewhat greater than 90°, probably because the Cl 3p orbital tails are more diffuse than the 2p-orbital tails of the second-row ligand (e.g., O). If the spin lattice of the $CuClO_2$ layer is described by using the three strongest spin exchanges, namely, the AFM exchange J_1 as well as the FM exchanges J_3 and J_4, then the resulting spin lattice is topologically equivalent to the Shastry–Sutherland spin lattice (Figure 19c) [56,58].

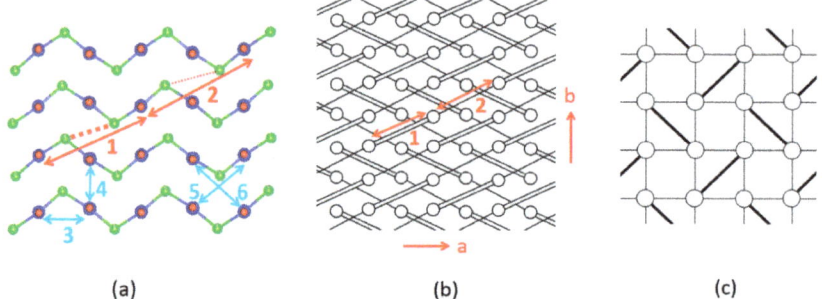

Figure 19. (a) Six spin exchange paths of the $CuClO_2$ layer in orthorhombic $CuCl(LaV_2O_7)$, and the numbers 1–6 refer to J_1–J_6, respectively. (b) The J_1–J_2 AFM alternating chains the $CuClO_2$ layer. Only the Cu^{2+} ions are shown as empty circles, with J_1 and J_2 paths represented by cylinders and lines, respectively. (c) The simplified spin lattice of the $CuClO_2$ layer generated by the three strongest spin exchanges, which is the Shastry–Sutherland spin lattice.

5.4. Two Dimensional Magnetic Character of Azurite $Cu_3(CO_3)_2(OH)_2$

Early interests in the magnetic properties of the mineral Azurite, $Cu_3(CO_3)_2(OH)_2$, in which Cu^{2+} ions are coordinated with molecular anions CO_3^{2-} and OH^-, focused mainly on the paramagnetic AFM ordering transition of the Cu^{2+} moments that occurs at about 1.86 K [59–63]. Renewed interests in the properties of $Cu_3(CO_3)_2(OH)_2$ arose from low-temperature high-field magnetization measurements by Kikuchi et al. [64], who detected a magnetization plateau extending over a wide field interval between 16 and 26 T or 11 and 30 T, depending on the crystal orientation. Only one-third of the Cu magnetic moments saturate in these field ranges whereas complete saturation of all Cu moments occurs above 32.5 T [64]. In $Cu_3(CO_3)_2(OH)_2$, the Cu^{2+} ions form CuO_4 square planar units with the CO_3^{2-} and OH^- ions. In each CuO_4 unit, two O atoms come from two CO_3^{2-} ions, and the remaining two O atoms from two OH^- ions. These CuO_4 units form Cu_2O_6 edge-sharing dimers, which alternate with CuO_4 monomers by corner-sharing to make a diamond chain (Figure 20a,b). Guided by the crystal structure, Kikuchi et al. [64] explained their results on $Cu_3(CO_3)_2(OH)_2$ by considering spin frustration in the diamond chains (Figure 20b), to conclude that all spin exchange constants (J_1, J_2 and J_3) in the diamond chains are AFM with J_2 being the dominant exchange, and that the moment of the Cu^{2+} ion of the monomer is susceptible to external magnetic fields because its spin exchanges with the two adjacent dimers (i.e., $2J_1 + 2J_3$) are nearly canceled. In questioning this scenario, Gu et al. and Rule et al. suggested that one of the monomer–dimer spin exchanges is FM, implying the absence of spin frustration [65–67]. The diamond chain picture had to be revised when the spin exchanges of $Cu_3(CO_3)_2(OH)_2$, evaluated using the energy-mapping analysis [68], showed that, although Kikuchi et al.'s description of the diamond chain was correct, $Cu_3(CO_3)_2(OH)_2$ is a 2D spin lattice made up of inter-linked diamond chains.

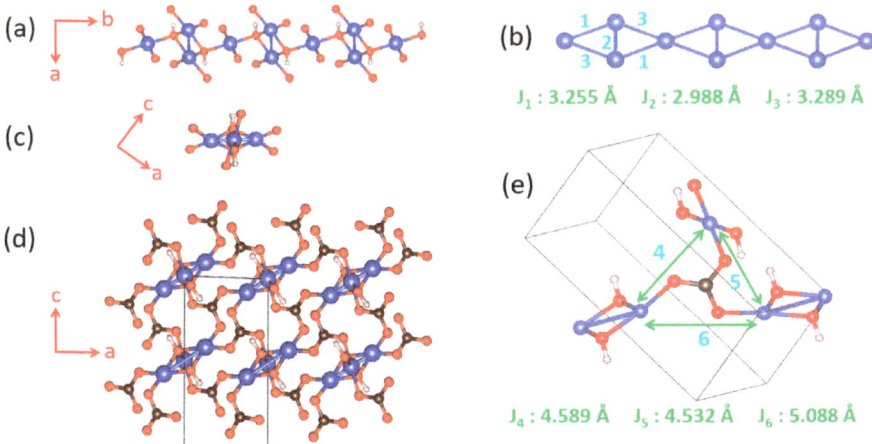

Figure 20. (**a**) The diamond chain made up of CuO_4 square planar units. (**b**) Definition of the three spin exchanges J_1–J_3 in a diamond chain. (**c**) A projection view of the diamond chain along the chain direction. (**d**) The structure of Azurite viewed along the diamond chain direction. (**e**) Definition of the three spin exchanges J_4–J_6 between adjacent diamond chains.

In the three-dimensional (3D) structure of $Cu_3(CO_3)_2(OH)_2$, the diamond chains are interconnected by the CO_3^{2-} ions. Using the projection view of the diamond chain along the chain direction (Figure 20c), the 3D structure of $Cu_3(CO_3)_2(OH)_2$ can be represented as in Figure 20d, which shows that each CO_3^{2-} ion bridges three different diamond chains. The spin exchange paths of interest for $Cu_3(CO_3)_2(OH)_2$ are J_1–J_3 in each diamond chain (Figure 20b), as well as J_4–J_6 between diamond chains (Figure 20e). In the diamond unit of Azurite (Figure 21a), the two bridging O atoms of the edge-sharing dimer Cu_2O_6 form O–H bonds. Thus, the spin exchange J_2 (Figure 20b) between the two Cu^{2+} ions in the edge-sharing dimer Cu_2O_6 are expected to be strongly AFM, as discussed in Section 4.2. The spin exchanges J_1 and J_3 of the diamond (Figure 20b) would be similar in strength because their two Cu–O–Cu exchange paths are nearly equivalent due to the near perpendicular arrangement the CuO_4 monomer plane to the Cu_2O_6 dimer plane (Figure 21a). Due to the perpendicular arrangement of the two planes, J_1 and J_3 are expected to be weakly AFM and smaller than J_2. What is difficult to predict without quantitative calculations is the relative strengths of the inter-chain exchanges J_4–J_6 (Figure 20e). For example, we consider the J_4 exchange path shown in Figure 21b. The O p_π and O p_σ orbitals of the CO_3^{2-} ion interact with the p-orbital tail of the Cu^{2+} ion magnetic orbital (Figure 21c), which depend not only on the $\angle$C–O–Cu bond angle, but also on the dihedral angles associated with the spin exchange path (e.g., $\angle$O–C–O–Cu dihedral angle). It is necessary to resort to the energy-mapping analysis based on DFT+U calculations to find the relative strengths of the spin exchanges J_1–J_6. The arrangement of these spin exchange paths in Azurite is presented in Figure 22a. Results of our analysis for J_1–J_4 are summarized in Table 1.

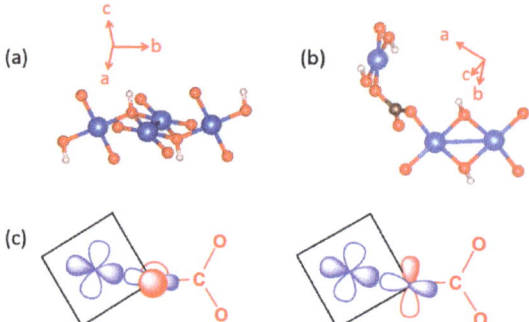

Figure 21. (a) A diamond unit of Azurite. (b) Arrangement of the CuO$_4$ monomer and Cu$_2$O$_6$ dimer associated with the spin exchange path J$_4$. (c) Interaction of the O p$_\pi$ and O p$_\sigma$ orbitals of the CO$_3^{2-}$ ion with the p-orbital tail of the Cu^{2+} ion magnetic orbital.

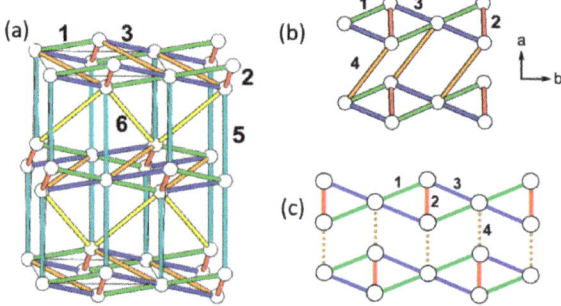

Figure 22. (a) Arrangement of the spin exchange paths J$_1$–J$_6$ in Azurite. (b) Arrangement of the four strongest spin exchanges J$_1$–J$_4$ in Azurite. (c) 2D spin lattice describing Azurite.

Table 1. Values of the spin exchanges J$_1$–J$_6$ determined by DFT+U calculations with U$_{eff}$ = 4 and 5 eV.

	U$_{eff}$ = 4	U$_{eff}$ = 5
J$_2$	241.5 K	194.6 K
J$_3$/J$_2$	0.28	0.27
J$_1$/J$_2$	0.21	0.19
J$_4$/J$_2$	0.17	0.16
J$_5$/J$_2$	−0.01	−0.02
J$_6$/J$_2$	0.04	0.03

Our energy-mapping analysis shows that J$_5$ and J$_6$ are negligibly weak compared with J$_1$–J$_4$. The latter four exchanges are all AFM; J$_2$ is the strongest while J$_1$, J$_3$ and J$_4$ are comparable in magnitude, which is in support of Kikuchi et al.'s deduction of the spin exchanges J$_1$–J$_3$ for the diamond chain. The values of J$_1$–J$_4$ are smaller from the DFT+U calculations with U$_{eff}$ = 5 eV than from those with U$_{eff}$ = 4 eV. This is understandable because the J$_{AF}$ component decreases with increasing the on-site repulsion U (Equation (19)). The strengths of the exchanges J$_1$–J$_4$ decrease in the order, J$_2$ >> J$_3$ ≈ J$_1$ > J$_4$. Thus, in the spin lattice of Azurite, the diamond chains defined by the intrachain exchanges J$_2$, J$_3$ and J$_1$ interact by the interchain exchange J$_4$ (Figure 22b). Therefore, the spin lattice of Azurite is a 2D spin lattice described by the exchanges J$_1$–J$_4$ depicted in Figure 22c. J$_3$ and J$_1$ are practically equal in strength; therefore, use of a symmetrical diamond chain (i.e., with the approximation $J_3 \approx J_1$) would be a good approximation. In any event, it is crucial not to neglect the interchain exchange J_4 because it is comparable in magnitude to J_3 and J_1. Alternatively, the spin lattice can be described in terms of the alternating AFM chains defined by J_2 and J_4. The 2D spin lattice consists of these alternating AFM chains that are spin-frustrated by the interchain exchanges J_1 and J_3.

The importance of interchain exchange indicating that the correct spin lattice of Azurite is not a diamond chain but a 2D net in which the diamond chains are interconnected by the spin exchange spin J_4 was recognized by Kang et al. in 2009 [68]. This finding, though controversial and vigorously disputed in the beginning, is now accepted as a prerequisite for correctly describing Azurite [69]. The presence of an interchain exchange naturally allows one to understand a long-range AFM order and explain gapped modes in the spin dynamics along the diamond chains [70], and the magnetic contribution to the thermal conductivity [71].

6. Concluding Remarks

In this review, we discussed the theoretical foundations of the concept of spin exchanges and analyzed which electronic factors affect their signs and strengths. Noting that a spin exchange between two magnetic ions is mediated by the ligand p-orbital tails, we derived several qualitative rules for predicting whether a given M–L ... L–M or M–L ... A ... L–M exchange would be AFM or FM by inspecting the arrangement of their ligand p-orbital tails in the exchange paths. As long as the L ... L distance is in the vicinity of the van der Waals distance, the M–L ... L–M or M–L ... A ... L–M spin exchange can be strong and often stronger than the M–L–M exchanges. In searching for the spin lattice relevant for a given magnetic solid, therefore, it is crucial not to omit the M–L ... L–M and M–L ... A ... L–M spin exchanges when present. The qualitative rules on the M–L ... L–M and M–L ... A ... L–M exchanges, described in Section 4.3, can be used together with the Goodenough–Kanamori rules on M–L–M spin exchanges in selecting a proper set of spin exchanges to evaluate using the energy-mapping analysis.

The important aspect emerging from our discussions is that the nature of a spin exchange is determined by the interactions between the magnetic orbitals. These are governed by the ligand p-orbitals, not by the metal d-orbitals. The essential role that the metal d-orbitals play in any spin exchange is rather indirect. In a magnetic orbital of an ML_n polyhedron, the metal d-orbital selects with which ligand p-orbitals it combines and hence determines the nature of the p-orbital tails in the magnetic orbital. The spin exchange between magnetic ions, namely, the interaction between their magnetic orbitals, rests upon the interaction between their p-orbital tails.

Author Contributions: Conceptualization, M.-H.W.; methodology, M.-H.W.; validation, M.-H.W., H.-J.K. and R.K.K.; formal analysis, M.-H.W.; investigation, M.-H.W., H.-J.K. and R.K.K.; writing—original draft preparation, M.-H.W.; writing—review and editing, M.-H.W., H.-J.K. and R.K.K.; visualiza-tion, M.-H.W.; supervision, M.-H.W.; project administration, M.-H.W.; funding acquisition, H.-J.K. All authors have read and agreed to the published version of the manuscript.

Funding: The work at Kyung Hee University was supported by the Basic Science Research Program through the National Research Foundation of Korea (NRFK) funded by the Ministry of Education (2020R1A6A1A03048004).

Institutional Review Board Statement: Not applicable.

Informed Consent Statement: Not applicable.

Acknowledgments: H.-J.K. thanks the NRFK for the fund 2020R1A6A1A03048004.

Conflicts of Interest: The authors declare no conflict of interest.

References

1. Whangbo, M.-H.; Koo, H.-J.; Dai, D. Spin exchange interactions and magnetic structures of extended magnetic solids with localized spins: Theoretical descriptions on formal, quantitative and qualitative levels. *J. Solid State Chem.* **2003**, *176*, 417–481. [CrossRef]
2. Xiang, H.J.; Lee, C.; Koo, H.-J.; Gong, X.; Whangbo, M.-H. Magnetic properties and energy-mapping analysis. *Dalton Trans.* **2013**, *42*, 823–853. [CrossRef] [PubMed]
3. Whangbo, M.-H.; Xiang, H.J. Magnetic Properties from the Perspectives of Electronic Hamiltonian: Spin Exchange Parameters, Spin Orientation and Spin-Half Misconception. In *Handbook in Solid State Chemistry, Volume 5: Theoretical Descriptions*; Wiley: New York, NY, USA, 2017; pp. 285–343.
4. Anderson, P.W. Antiferromagnetism-Theory of superexchange. *Phys. Rev.* **1950**, *79*, 350–356. [CrossRef]
5. Goodenough, J.B.; Loeb, A.L. Theory of ionic ordering, crystal distortion, and magnetic exchange due to covalent forces in spinels. *Phys. Rev.* **1955**, *98*, 391–408. [CrossRef]
6. Goodenough, J.B. Theory of the role of covalence in the perovskite-type manganites [La, M(II)]MnO$_3$. *Phys. Rev.* **1955**, *100*, t564–573. [CrossRef]
7. Kanamori, J. Superexchange interaction and symmetry properties of electron orbitals. *J. Phys. Chem. Solids* **1959**, *10*, 87–98. [CrossRef]
8. Goodenough, J.B. *Magnetism and the Chemical Bond*; Interscience, Wiley: New York, NY, USA, 1963.
9. Martin, R.L. Metal-metal interaction in paramagnetic clusters. In *New Pathways of Inorganic Chemistry*; Ebsworth, E.A.V., Maddock, A.G., Sharp, A.G., Eds.; Cambridge University Press: Cambridge, UK, 1968; Chapter 9.
10. Pauling, L. *The Nature of the Chemical Bond and the Structure of Molecules and Solids: The Introduction to Mode*, 3rd ed.; Cornell University: Ithaca, NY, USA, 1960.
11. Dudarev, S.L.; Botton, G.A.; Savrasov, S.Y.; Humphreys, C.J.; Sutton, A.P. Electron-energy-loss spectra and the structural stability of nickel oxide: An LSDA+U study. *Phys. Rev. B* **1998**, *57*, 1505–1509. [CrossRef]
12. Heyd, J.; Scuseria, G.E.; Ernzerhof, M. Hybrid functionals based on a screened Coulomb potential. *J. Chem. Phys.* **2003**, *118*, 8207. [CrossRef]
13. Xiang, H.J.; Kan, E.J.; Wei, S.-H.; Whangbo, M.-H.; Gong, X.G. Predicting the spin-lattice order of frustrated systems from first principles. *Phys. Rev. B* **2011**, *84*, 224429. [CrossRef]
14. Li, X.-Y.; Yu, H.Y.; Lou, F.; Feng, J.-S.; Whangbo, M.-H.; Xiang, H.J. Spin Hamiltonians in Magnets: Theories and Computations. *Molecules*. submitted for Publication.
15. Hay, P.J.; Thibeault, J.C.; Hoffmann, R. Orbital interactions in metal dimer complexes. *J. Am. Chem. Soc.* **1975**, *97*, 4884–4899. [CrossRef]
16. Vasil'ev, A.N.; Ponomarenko, L.A.; Smirnov, A.I.; Antipov, E.V.; Velikodny, Y.A.; Isobe, M.; Ueda, Y. Short-range and long-range magnetic ordering in α-CuV$_2$O$_6$. *Phys. Rev. B* **1999**, *60*, 3021–3024. [CrossRef]
17. Kikuchi, J.; Ishiguchi, K.; Motoya, K.; Itoh, M.; Inari, K.; Eguchi, N.; Akimitsu, J. NMR and neutron scattering studies of quasi one-dimensional magnet CuV$_2$O$_6$. *J. Phys. Soc. Jpn.* **2000**, *69*, 2660–2668. [CrossRef]
18. Prokofiev, A.V.; Kremer, R.K.; Assmus, W. Crystal growth and magnetic properties of α-CuV$_2$O$_6$. *J. Cryst. Growth* **2001**, *231*, 498–505. [CrossRef]
19. Golubev, A.M.; Nuss, J.; Kremer, R.K.; Gordon, E.E.; Whangbo, M.-H.; Ritter, C.; Weber, L.; Wessel, S. Two-dimensional magnetism in α-CuV$_2$O$_6$. *Phys. Rev. B* **2020**, *102*, 014436. [CrossRef]
20. Koo, H.-J.; Lee, C.; Whangbo, M.-H.; McIntyre, G.J.; Kremer, R.K. On the nature of the spin frustration in the CuO$_2$ ribbon chains of LiCuVO$_4$: Crystal structure determination at 1.6 K, magnetic susceptibility analysis and density functional evaluation of the spin exchange constants. *Inorg. Chem.* **2011**, *50*, 3582–3588. [CrossRef]
21. Yamaguchi, M.; Furuta, T.; Ishikawa, M. Calorimetric study of several cuprates with restricted dimensionality. *J. Phys. Soc. Jpn.* **1996**, *65*, 2998–3006. [CrossRef]
22. Vasil'ev, A.N.; Ponomarenko, L.A.; Manaka, H.; Yamada, I.; Isobe, M.; Ueda, Y. Quasi-one-dimensional antiferromagnetic spinel compound LiCuVO$_4$. *Physica B* **2000**, *284–288*, 1619–1620. [CrossRef]
23. Vasil'ev, A.N.; Ponomarenko, L.A.; Manaka, H.; Yamada, I.; Isobe, M.; Ueda, Y. Magnetic and resonant properties of quasi-one-dimensional antiferromagnet LiCuVO$_4$. *Phys. Rev. B* **2001**, *64*, 024419. [CrossRef]
24. Gibson, B.J.; Kremer, R.K.; Prokofiev, A.V.; Assmus, W.; McIntyre, G.J. Incommensurate antiferromagnetic order in the S=1/2 quantum chain compound LiCuVO$_4$. *Physica B* **2004**, *350*, e253–e256. [CrossRef]
25. Dai, D.; Koo, H.-J.; Whangbo, M.-H. Investigation of the incommensurate and commensurate magnetic superstructures of LiCuVO$_4$ and CuO on the basis of the isotropic spin exchange and classical spin approximations. *Inorg. Chem.* **2004**, *43*, 4026–4035. [CrossRef] [PubMed]
26. Enderele, M.; Mukherjee, C.; Fåk, B.; Kremer, R.K.; Broto, J.-M.; Rosner, H.; Drechsler, S.-L.; Richter, J.; Malek, J.; Prokofiev, A.; et al. Quantum helimagnetism of the frustrated spin-1/2 chain LiCuVO$_4$. *Europhys. Lett.* **2005**, *70*, 237–243. [CrossRef]
27. Gordon, E.E.; Derakhshan, S.; Thompson, C.M.; Whangbo, M.-H. Spin-density wave as a superposition of two magnetic states of opposite chirality and its implications. *Inorg. Chem.* **2018**, *57*, 9782–9785. [CrossRef] [PubMed]
28. Koo, H.-J.; N, R.S.P.; Orlandi, F.; Sundaresan, A.; Whangbo, M.-H. On Ferro and antiferro spin density waves describing the incommensurate magnetic structure of NaYNiWO$_6$. *Inorg. Chem.* **2020**, *59*, 17856–17859. [CrossRef]

29. Naito, Y.; Sato, K.; Yasui, Y.; Kobayashi, Y.; Kobayashi, Y.; Sato, M. Relationship between magnetic structure and ferroelectricity of LiVCuO$_4$. *J. Phys. Soc. Jpn.* **2007**, *76*, 023708. [CrossRef]
30. Xiang, H.J.; Whangbo, M.-H. Density-functional characterization of the multiferroicity in spin spiral chain cuprates. *Phys. Rev. Lett.* **2007**, *99*, 257203. [CrossRef] [PubMed]
31. Yasui, Y.; Naito, Y.; Sato, K.; Moyoshi, T.; Sato, M.; Kakurai, K. Studies of Multiferroic System LiCu$_2$O$_2$: I. Sample characterization and relationship between magnetic properties and multiferroic nature. *J. Phys. Soc. Jpn.* **2008**, *77*, 023712. [CrossRef]
32. Schrettle, F.; Krohns, S.; Lunkenheimer, P.; Hemberger, J.; Büttgen, N.; von Nidda, H.-A.K.; Prokofiev, A.V.; Loidl, A. Switching the ferroelectric polarization in the S = 1/2 chain cuprate LiCuVO$_4$ by external magnetic fields. *Phys. Rev. B* **2008**, *77*, 144101. [CrossRef]
33. Ruff, A.; Krohns, S.; Lunkenheimer, P.; Prokofiev, A.; Loidl, A. Dielectric properties and electrical switching behaviour of the spin-driven multiferroic LiCuVO$_4$. *J. Phys. Condens. Matter* **2014**, *26*, 485901. [CrossRef]
34. Ruff, A.; Lunkenheimer, P.; von Nidda, H.-A.K.; Widmann, S.; Prokofiev, A.; Svistov, L.; Loidl, A.; Krohns, S. Chirality-driven ferroelectricity in LiCuVO$_4$. *NPJ Quantum Mater.* **2019**, *4*, 24. [CrossRef]
35. Mourigal, M.; Enderle, M.; Kremer, R.K.; Law, J.M.; Fåk, B. Ferroelectricity from spin supercurrents in LiCuVO$_4$. *Phys. Rev. B* **2014**, *83*, 100409(R). [CrossRef]
36. Banks, M.G.; Heidrich-Meisner, F.; Honecker, A.; Rakoto, H.; Broto, J.-M.; Kremer, R.K. High field magnetization of the frustrated one-dimensional quantum antiferromagnet LiCuVO$_4$. *J. Phys. Condens. Matter* **2007**, *19*, 145227. [CrossRef]
37. Hikihara, T.; Kecke, L.; Momoi, T.; Furusaki, A. Vector chiral and multipolar orders in the spin-1/2 frustrated ferromagnetic chain in magnetic field. *Phys. Rev. B* **2008**, *78*, 144404. [CrossRef]
38. Mourigal, M.; Enderle, M.; Fåk, B.; Kremer, R.K.; Law, J.M.; Schneidewind, A.; Hiess, A.; Prokofiev, A. Evidence of a bond-nematic phase in LiCuVO$_4$. *Phys. Rev. Lett.* **2012**, *109*, 027203. [CrossRef] [PubMed]
39. Orlova, A.; Green, E.L.; Law, J.M.; Gorbunov, D.I.; Chanda, G.; Krämer, S.; Horvatić, M.; Kremer, R.K.; Wosnitza, J.; Rikken, G.L.J.A. Nuclear magnetic resonance signature of the spin-nematic phase in LiCuVO$_4$ at high magnetic fields. *Phys. Rev. Lett.* **2017**, *118*, 247201. [CrossRef]
40. Zhitomirsky, M.E.; Tsunetsugu, H. Magnon pairing in quantum spin nematic. *Europhys. Lett.* **2010**, *92*, 37001. [CrossRef]
41. Svistova, L.E.; Fujita, T.; Yamaguchi, H.; Kimura, S.; Omura, K.; Prokofiev, A.; Smirnova, A.I.; Honda, Z.; Hagiwara, M. New high magnetic field phase of the frustrated S = 1/2 Chain compound LiCuVO$_4$. *JETP Lett.* **2011**, *93*, 21. [CrossRef]
42. Ueda, H.T.; Momoi, T. Nematic phase and phase separation near saturation field in frustrated ferromagnets. *Phys. Rev. B* **2013**, *87*, 144417. [CrossRef]
43. Sato, M.; Hikihara, T.; Momoi, T. Spin-nematic and spin-density-wave orders in spatially anisotropic frustrated magnets in a magnetic field. *Phys. Rev. Lett.* **2013**, *110*, 077206. [CrossRef]
44. Starykh, O.A.; Balents, L. Excitations and quasi-one-dimensionality in field-induced nematic and spin density wave states. *Phys. Rev. B* **2014**, *89*, 104407. [CrossRef]
45. Büttgen, N.; Nawa, K.; Fujita, T.; Hagiwara, M.; Kuhns, P.; Prokofiev, A.; Reyes, A.P.; Svistov, L.E.; Yoshimura, K.; Takigawa, M. Search for a spin-nematic phase in the quasi-one-dimensional frustrated magnet LiCuVO$_4$. *Phys. Rev. B* **2014**, *90*, 134401. [CrossRef]
46. Caruntu, G.; Kodenkandath, T.A.; Wiley, J.B. Neutron diffraction study of the oxychloride layered perovskite, (CuCl)LaNb$_2$O$_7$. *Mater. Res. Bull.* **2002**, *37*, 593–598. [CrossRef]
47. Hoffmann, R. An extended Hückel theory. I. Hydrocarbons. *J. Chem. Phys.* **1963**, *39*, 1397. [CrossRef]
48. Yoshida, M.; Ogata, N.; Takigawa, M.; Yamaura, J.-I.; Ichihara, M.; Kitano, T.; Kageyama, H.; Ajiro, Y.; Yoshimura, K. Magnetic and structural studies of the quasi-two-dimensional spin-gap system (CuCl)LaNb$_2$O$_7$. *J. Phys. Soc. Jpn.* **2007**, *76*, 104703. [CrossRef]
49. Kageyama, H.; Kitano, T.; Oba, N.; Nishi, M.; Nagai, S.; Hirota, K.; Viciu, L.; Wiley, J.B.; Yasuda, J.; Baba, Y.; et al. Spin-singlet ground state in two-dimensional S = 1/2 frustrated square lattice: (CuCl)LaNb$_2$O$_7$. *J. Phys. Soc. Jpn.* **2005**, *74*, 1702–1705. [CrossRef]
50. Whangbo, M.-H.; Dai, D. On the Disorder of the Cl atom position in and its probable effect on the magnetic properties of (CuCl)LaNb$_2$O$_7$. *Inorg. Chem.* **2006**, *45*, 6227–6234. [CrossRef]
51. Kageyama, H.; Yasuda, J.; Kitano, T.; Totsuka, K.; Narumi, Y.; Hagiwara, M.; Kindo, K.; Baba, Y.; Oba, N.; Ajiro, Y.; et al. Anomalous magnetization of two-dimensional S = 1/2 frustrated square-lattice antiferromagnet (CuCl)LaNb$_2$O$_7$. *J. Phys. Soc. Jpn.* **2005**, *74*, 3155–3158. [CrossRef]
52. Hiroi, A.K.Z.; Tsujimoto, Y.; Kitano, T.; Kageyama, H.; Ajiro, Y.; Yoshimura, K. Bose-Einstein condensation of quasi-two-dimensional frustrated quantum magnet (CuCl)LaNb$_2$O$_7$. *J. Phys. Soc. Jpn.* **2007**, *76*, 093706. [CrossRef]
53. Tsirlin, A.A.; Rosner, H. Structural distortion and frustrated magnetic interactions in the layered copper oxychloride (CuCl)LaNb$_2$O$_7$. *Phys. Rev. B* **2009**, *79*, 214416. [CrossRef]
54. Ren, C.-Y.; Cheng, C. Atomic and magnetic structures of (CuCl)LaNb$_2$O$_7$ and (CuBr)LaNb$_2$O$_7$: Density functional calculations. *Phys. Rev. B* **2010**, *82*, 024404. [CrossRef]
55. Tsirlin, A.A.; Abakumov, A.M.; van Tendeloo, G.; Rosner, H. Interplay of atomic displacements in the quantum magnet, (CuCl)LaNb$_2$O$_7$. *Phys. Rev. B* **2010**, *82*, 054107. [CrossRef]

56. Tassel, C.; Kang, J.; Lee, C.; Hernandez, O.; Qiu, Y.; Paulus, W.; Collet, E.; Lake, B.; Guidi, T.; Whangbo, M.-H.; et al. Ferromagnetically coupled Shastry-Sutherland quantum spin singlets in (CuCl)LaNb$_2$O$_7$. *Phys. Rev. Lett.* **2010**, *105*, 167205. [CrossRef] [PubMed]
57. Hernandez, O.J.; Tassel, C.; Nakano, K.; Paulus, W.; Ritter, C.; Collet, E.; Kitada, A.; Yoshimura, K.; Kageyama, H. First single-crystal synthesis and low-temperature structural determination of the quasi-2D quantum spin compound (CuCl)LaNb$_2$O$_7$. *Dalton Trans.* **2011**, *40*, 4605–4613. [CrossRef] [PubMed]
58. Furukawa, S.; Dodds, T.; Kim, Y.B. Ferromagnetically coupled dimers on the distorted Shastry-Sutherland lattice: Application to (CuCl)LaNb$_2$O$_7$. *Phys. Rev. B* **2011**, *84*, 054432. [CrossRef]
59. Spence, R.D.; Ewing, R.D. Evidence for antiferromagnetism in Cu$_3$(CO$_3$)$_2$(OH)$_2$. *Phys. Rev.* **1958**, *112*, 1544–1545. [CrossRef]
60. Forstat, H.; Taylor, G.; King, B.R. Low-temperature heat capacity of Azurite. *J. Chem. Phys.* **1959**, *31*, 929–931. [CrossRef]
61. van der Lugt, W.; Poulis, N.J. Proton magnetic resonance in Azurite. *Physica* **1959**, *25*, 1313–1320. [CrossRef]
62. Garber, M.; Wagner, R. The susceptibility of Azurite. *Physica* **1960**, *26*, 777. [CrossRef]
63. Love, N.D.; Duncan, T.K.; Bailey, P.T.; Forstat, H. Spin flopping in Azurite. *Phys. Lett. A* **1970**, *33*, 290–291. [CrossRef]
64. Kikuchi, H.; Fujii, Y.; Mitsudo, S.; Idehara, T. Magnetic properties of the frustrated diamond chain compound Cu$_3$(CO$_3$)$_2$(OH)$_2$. *Phys. B* **2003**, *329*, 967–968. [CrossRef]
65. Gu, B.; Su, G. Comment on "Experimental Observation of the 1/3 Magnetization Plateau in the Diamond-Chain Compound Cu$_3$(CO$_3$)$_2$(OH)$_2$. *Phys. Rev. Lett.* **2006**, *97*, 089701. [CrossRef] [PubMed]
66. Gu, B.; Su, G. Magnetism and thermodynamics of spin-1/2 Heisenberg diamond chains in a magnetic field. *Phys. Rev. B* **2007**, *75*, 174437. [CrossRef]
67. Rule, K.C.; Wolter, A.U.B.; Süllow, S.; Tennant, D.A.; Brühl, A.; Köhler, S.; Wolf, B.; Lang, M.; Schreuer, J. Nature of the spin dynamics and 1/3 magnetization plateau in Azurite. *Phys. Rev. Lett.* **2008**, *100*, 117202. [CrossRef] [PubMed]
68. Kang, J.; Lee, C.; Kremer, R.K.; Whangbo, M.-H. Consequences of the intrachain dimer-monomer spin frustration and the interchain dimer-monomer spin exchange in the diamond-chain compound azurite Cu$_3$(CO$_3$)$_2$(OH)$_2$. *J. Phys. Condens. Matter* **2009**, *21*, 392201. [CrossRef] [PubMed]
69. Jeschke, H.; Opahle, I.; Kandpal, H.; Valentí, R.; Das, H.; Saha-Dasgupta, T.; Janson, O.; Rosner, H.; Brühl, A.; Wolf, B.; et al. Multistep approach to microscopic models for frustrated quantum magnets: The case of the natural mineral Azurite. *Phys. Rev. Lett.* **2011**, *106*, 217201. [CrossRef]
70. Rule, K.C.; Tennant, D.A.; Caux, J.-S.; Gibson, M.C.R.; Telling, M.T.F.; Gerischer, S.; Süllow, S.; Lang, M. Dynamics of azurite Cu$_3$(CO$_3$)$_2$(OH)$_2$ in a magnetic field as determined by neutron scattering. *Phys. Rev. B* **2011**, *84*, 184419. [CrossRef]
71. Wu, C.; Song, J.D.; Zhao, Z.Y.; Shi, J.; Xu, H.S.; Zhao, J.Y.; Liu, X.G.; Zhao, X.; Sun, X.F. Thermal conductivity of the diamond-chain compound Cu$_3$(CO$_3$)$_2$(OH)$_2$. *J. Phys. Condens. Matter* **2016**, *28*, 056002. [CrossRef]

Article

WO₃ Nanowire/Carbon Nanotube Interlayer as a Chemical Adsorption Mediator for High-Performance Lithium-Sulfur Batteries

Sang-Kyu Lee [1], Hun Kim [1], Sangin Bang [1], Seung-Taek Myung [2,*] and Yang-Kook Sun [1,*]

1. Department of Energy Engineering, Hanyang University, Seoul 04763, Korea; chemiphile7@gmail.com (S.-K.L.); rlagns0829@nate.com (H.K.); b930403@gmail.com (S.B.)
2. Department of Nanotechnology and Advanced Materials Engineering & Sejong Battery Institute, Sejong University, Seoul 05006, Korea
* Correspondence: smyung@sejong.ac.kr (S.-T.M.); yksun@hanyang.ac.kr (Y.-K.S.); Tel.: +82-2-2220-0524 (Y.-K.S.)

Abstract: We developed a new nanowire for enhancing the performance of lithium-sulfur batteries. In this study, we synthesized WO₃ nanowires (WNWs) via a simple hydrothermal method. WNWs and one-dimensional materials are easily mixed with carbon nanotubes (CNTs) to form interlayers. The WNW interacts with lithium polysulfides through a thiosulfate mediator, retaining the lithium polysulfide near the cathode to increase the reaction kinetics. The lithium-sulfur cell achieves a very high initial discharge capacity of 1558 and 656 mAh g^{-1} at 0.1 and 3 C, respectively. Moreover, a cell with a high sulfur mass loading of 4.2 mg cm^{-2} still delivers a high capacity of 1136 mAh g^{-1} at a current density of 0.2 C and it showed a capacity of 939 mAh g^{-1} even after 100 cycles. The WNW/CNT interlayer maintains structural stability even after electrochemical testing. This excellent performance and structural stability are due to the chemical adsorption and catalytic effects of the thiosulfate mediator on WNW.

Keywords: lithium-sulfur batteries; tungsten oxide nanowire; interlayer; thiosulfate mediator

Citation: Lee, S.-K.; Kim, H.; Bang, S.; Myung, S.-T.; Sun, Y.-K. WO₃ Nanowire/Carbon Nanotube Interlayer as a Chemical Adsorption Mediator for High-Performance Lithium-Sulfur Batteries. *Molecules* **2021**, *26*, 377. https://doi.org/10.3390/molecules26020377

Academic Editor: Stephane Jobic
Received: 18 December 2020
Accepted: 8 January 2021
Published: 13 January 2021

Publisher's Note: MDPI stays neutral with regard to jurisdictional claims in published maps and institutional affiliations.

Copyright: © 2021 by the authors. Licensee MDPI, Basel, Switzerland. This article is an open access article distributed under the terms and conditions of the Creative Commons Attribution (CC BY) license (https://creativecommons.org/licenses/by/4.0/).

1. Introduction

Next-generation batteries have been in the spotlight to meet the demand for batteries with higher energy densities than commercial lithium-ion batteries [1–6]. Among them, lithium-sulfur batteries have been actively investigated owing to their attractive features: (1) ultrahigh theoretical energy density (2660 Wh kg^{-1}), (2) worldwide abundant sulfur resources as low-cost active materials, and (3) lower environmental impact because their main constituents are sulfur and carbon [7–10].

Lithium-sulfur batteries are based on a general reaction mechanism; during discharge, high-order lithium polysulfides (Li₂S$_n$, when $n \geq 4$) are generated and gradually converted into low-order lithium polysulfides (Li₂S$_n$, when $n < 4$) during lithiation (reduction) to finally form Li₂S. Subsequent recharging drives delithiation (oxidation), so the cycle can be repeated [11]. High-order lithium polysulfides are soluble in the electrolyte, causing a shuttle phenomenon that induces the loss of active materials during the charging process and, consequently, incomplete battery charging [12,13]. By contrast, the use of volatile electrolytes exacerbates the electrolyte deficiency phenomenon, which results in lithium dendrite growth and deterioration in the stability of the cells [14]. In addition, sulfur (as active material) and lithium sulfides (as discharge products) are insulators that encourage the compositization of sulfur with electro-conducting materials [11]. For instance, porous carbons are capable of supporting large amounts of sulfur, and graphene and carbon nanotubes (CNTs) are proven applications in the fabrication of conductive carbon-sulfur composites [7,15]. Metal oxides, doped carbon, and modified membranes inhibit the migration of polysulfides to improve capacity retention during cycling [16,17].

Recent improvements in the performance of lithium-sulfur cells are attributed to film-type interlayers between the separator and the cathode. These interlayers facilitate electron transfer and prevent the migration of the produced lithium polysulfide toward the anode, thereby minimizing the loss of active materials to significantly improve capacity retention during cycling [16–19]. Inspired by metal–air batteries, metal oxides such as TiO_2, MnO_2, V_2O_5, and ZnO are of interest because of their ability to capture polysulfides as thiosulfate species [20–24]. More recently, Lin et al. introduced the catalytic effect of oxygen-deficient $WO_{3-\delta}$ ($\delta = 0.1$) nanoplates on the reduction of polysulfides to Li_2S_2 or Li_2S [25]. Choi et al. applied a ~100 nm-thick WO_3 layer onto a graphene-sulfur composite, achieving capacity retention of 95% after 500 cycles [26]. Yang et al. demonstrated the ability of WO_3 and carbon nanofibers, i.e., WO_3-decorated N, S co-doped carbon nanofibers, to trap polysulfides and form an electro-conducting network [27]. These studies demonstrate the catalytic efficacy of WO_3 in reducing high-order polysulfides to low-order Li_2S.

It is anticipated that this catalytic effect of WO_3 can be significantly enhanced by maximizing the active surface area through the formation of WO_3 nanowires (WNW). Herein, we introduce a viable WNW/carbon nanotube (CNT) composite interlayer between the S electrode and separator. As discussed, WNW and lithium polysulfide interact to produce thiosulfate, which combines with high-order polysulfides to form low-order polysulfide and polythionate complexes. This mediating reaction enhances the kinetics of Li_2S formation during discharge and S formation during charge. The loss of active material is mitigated by the facile conversion of S to Li_2S, which suppresses the migration of high-order polysulfides toward the anode. Owing to the thiosulfate mediation on the surface of the WNW/CNT interlayer, the sulfur electrode exhibits excellent electrochemical capacity and rate capability even at high sulfur mass loadings of 4.2 mg cm^{-2}.

2. Results and Discussion

During the fabrication of the WNW/CNT composite, a simple hydrothermal reaction occurred between $Na_2WO_4 \cdot 2H_2O$ and HCl; namely, $Na_2WO_4 \cdot 2H_2O + 2HCl \rightarrow WO_3 + 2NaCl + 3H_2O$ (1), resulting in the formation of WO_3 with a P6/mm space group (a = 7.298 Å and c = 3.899 Å) (Figure 1a). Images obtained through SEM and TEM indicate that the synthesized WO_3 has a long nanowire shape with a thickness of approximately 20 nm (Figure 1b,c), which is similar to the morphology of CNTs (Figure S1a). Reaction (1) is likely responsible for the nucleation of WO_3, while the subsequently added ammonium sulfate determines the one-directional growth of the WO_3 particles. The homogeneous distribution of W and O on the nanowires is evident in the inset of Figure 1c [28]. The WNW has a high surface area of 111.31 m^2 g^{-1} and a total pore volume of 0.48 cm^3 g^{-1} according to nitrogen adsorption/desorption analysis (Figure S1b). The pore diameter distribution curve reveals that the size of the mesopores is centered at ~5.0 nm (Figure S1c). The produced WNWs were mixed with CNTs to fabricate WNW/CNT paper, which is prepared by the same method to fabricate the CNT paper (Figure 1d). The XRD patterns of the CNT and WNW/CNT composite reflect a CNT-derived peak at approximately 26° (2 θ) (Figure 1a). Despite the similarity in morphology between the nanowires, element maps show the presence of C, W, and O elements, indicating the successful fabrication of WNW/CNT paper, as summarized in Figure 1e.

To confirm the thiosulfate mediation effect on the surface of WNW, a 2 mL lithium polysulfide solution (0.0025 M Li_2S_8 in DME/DOL) was added to vials containing 10 mg of CNT and WNW, respectively (Figure 2a). After aging for 12 h, the transparent yellow-colored WNW-Li_2S_8 solution turns colorless. In contrast, the transparent yellow-colored CNT-Li_2S_8 solution shows no color change after aging for 12 h, indicating that the lithium polysulfide (Li_2S_8) has not been adsorbed by the CNT.

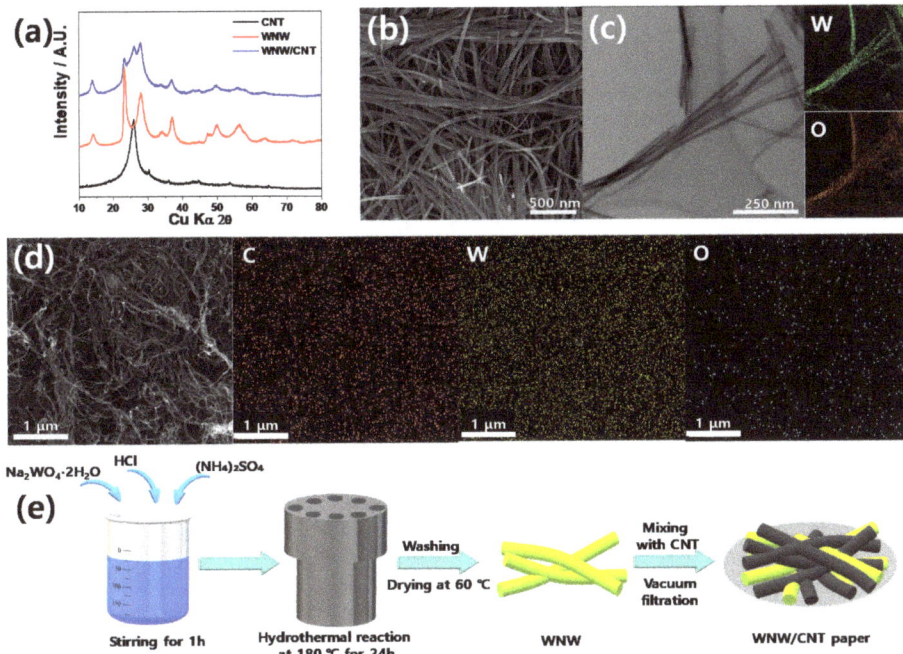

Figure 1. (a) XRD patterns of carbon nanotube (CNT), WO$_3$ nanowires (WNW), and WNW/CNT. (b) SEM and (c) TEM image of WNW. (d) SEM images of WNW/CNT interlayer with EDS mapping images of carbon, tungsten, and oxygen. (e) Synthesis of the WNW/CNT interlayer.

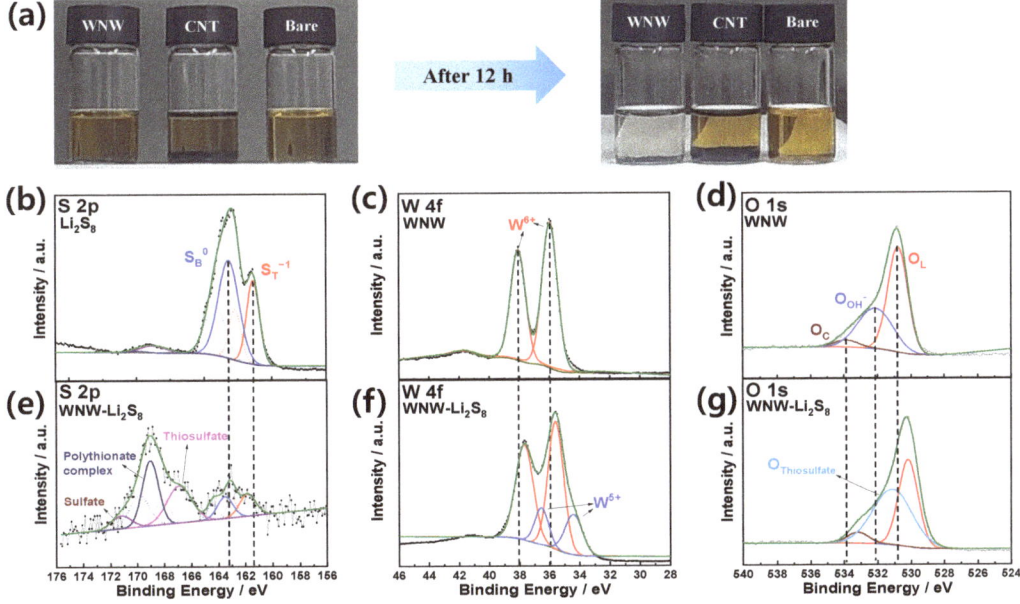

Figure 2. (a) Photographs of lithium polysulfide (0.0025 M Li$_2$S$_8$ 2 mL) electrolyte adsorption test. (b) XPS S 2p spectrum of Li$_2$S$_8$. XPS spectra of WNW: (c) W 4f and (d) O 1s. XPS spectra of WNW after Li$_2$S$_8$ adsorption test (e) S 2p, (f) W 4f, and (g) O 1s.

As shown in Figure S1b,c, the large specific surface area and pores of WNW maximize the reaction sites in contact with the lithium polysulfide. The chemical states of the surface elements of WNW in the aged WNW-Li_2S_8 solution were observed using XPS (Figure 2b–g). Li_2S_8 shows two S $2p_{3/2}$ peaks at 161.4 and 163.2 eV, assigned to the terminal (S_T^{-1}) and bridging sulfur (S_B^0) atoms, respectively (Figure 2b) [29]. WNW displays a pair of peaks at 36.0 and 38.1 eV, ascribed to W $4f_{7/2}$ and W $4f_{5/2}$, respectively, indicating that the synthesized WO_3 nanowire is composed of W^{6+} (Figure 2c) [30]. The O 1s XPS spectrum shows a dominant peak at 530.8 eV that is attributed to the lattice oxygen of WNW, while the broad shoulder at 532.1 and 533.8 eV are due to contaminants such as hydroxyl groups or carbonates on the surface of WNW (Figure 2d) [31]. Figure 2e–g show the XPS spectra of the lithium polysulfide adsorbed on WNW (in the WNW-Li_2S_8 solution after aging for 12 h). While Figure 2b reflects dominant peaks corresponding to the terminal (161.4 eV) and bridging (163.2 eV) sulfur atoms in Li_2S_8, the S 2p spectrum of WNW-Li_2S_8 in Figure 2e shows dominant S $2p_{3/2}$ peaks at 166.6 and 169.0 eV. The S $2p_{3/2}$ peak at 166.6 eV corresponds to the thiosulfate binding energy, while the peak at 169.0 eV is ascribed to the polythionate complex [29]. Interestingly, in the W 4f spectrum of WNW-Li_2S_8, the binding energies of the two peaks corresponding to W^{6+} are 0.4 eV less than those in Figure 2c, and are accompanied by the emergence of peaks at 34.4 and 36.5 eV (Figure 2f). These peaks are attributed to W^{5+} as a result of the formation of WS_xO_y oxysulfide species [32–34]; that is, the oxidation of high-order lithium polysulfide to thiosulfate is accompanied by the reduction of W^{6+} to W^{5+}. The peaks of the O 1s spectrum of WNW-Li_2S_8 appear at lower binding energies than those of the O 1s spectrum of WNW, while the area of the center peak at 531.1 eV is greater than that of the corresponding peak in Figure 2d owing to the formation of thiosulfate. Based on the XPS data, it is reasonable to conclude that thiosulfate is produced, since polysulfides provide electrons to the WNWs at the interface, thereby reducing W^{6+} to W^{5+}. The produced thiosulfate combines with high-order polysulfides to form polythionate complexes and low-order polysulfides. The thiosulfate formed on the WNW surface acts as a redox mediator and facilitates lithium polysulfide conversion. This process is summarized in Figure 3.

Cyclic voltammetry (CV) data demonstrate the improved reversibility of the sulfur electrode (S: 3 mg cm^{-2}) in the presence of a WNW/CNT interlayer (Figure 4). For the cell containing a WNW/CNT interlayer (Figure 4a), a reduction peak is observed at 2.23 V during the first cathodic scan, associated with the conversion of sulfur to long-chain lithium polysulfides. A cathodic peak, associated with a change in short-chain lithium polysulfides (Li_2S_2, Li_2S), is observed at 2.04 V. Two anodic peaks appear at 2.36 and 2.42 V, which are associated with the recovery of sulfur from the lithium polysulfides [35,36]. Similarly, in the case of the CNT interlayer (Figure 4b), cathodic peaks appear at ~2.2 and ~1.95 V, and the reverse reactions occur at ~2.38 and ~2.46 V during the anodic scan. The cathodic peaks of the first and second CV curves differ slightly owing to the sulfur changing from alpha to beta phase [18,37]. The slightly higher voltage of the redox peaks for the cell containing the WNW/CNT interlayer suggests that the thiosulfate mediator on WNW enhances the low-order lithium polysulfide conversion kinetics [29,38]. In contrast, the CV curves of the cell without an interlayer show poor electrochemical activity, compared to the cells with interlayers (Figure 4c).

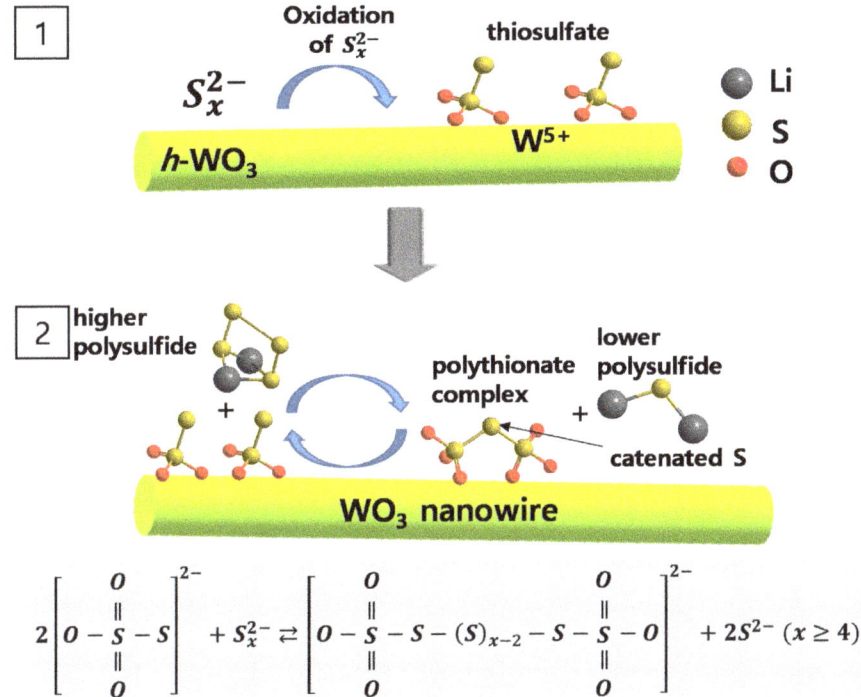

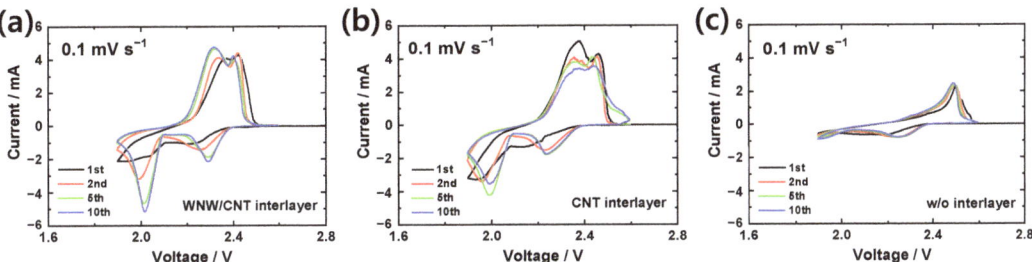

Figure 3. Schematic of lithium polysulfide conversion catalyzed on WO$_3$ nanowire.

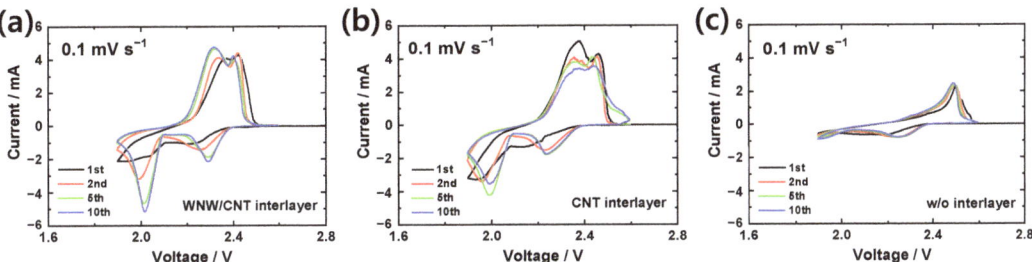

Figure 4. Cyclic voltammetry curves of lithium-sulfur cells with (**a**) WNW/CNT interlayer, (**b**) CNT interlayer, and (**c**) without an interlayer.

Figure 5 compares the effect of the WNW/CNT interlayer on the long-term cell capacity at different currents. As anticipated in the CV evaluation (Figure 4), a disappointing performance is observed for the sulfur cell without an interlayer (Figure S2). With the CNT interlayer (S-CNT cell), the lithium-sulfur cell delivers the highest capacity of 1224 mAh g^{-1} with a capacity fading rate of 0.09% per cycle for the first 160 cycles. After 160 cycles, the cell capacity decreases drastically until, at the 300th cycle, only 27.6% of the initial capacity is retained (Figure 5a,c). For the lithium-sulfur cell with the WNW/CNT interlayer (S-WNW/CNT cell), the delivered initial capacity of 1225 mAh g^{-1} fades at a rate of 0.11% per cycle (Figure 5b,c). Notably, ~68.3% (836.6 mAh g^{-1}) of the initial capacity is retained after 300 cycles at 0.5 C, which is ~40% higher than the retained capacity of the S-CNT cell. After 300 cycles, the WNW/CNT cell demonstrates an areal capacity exceeding

2.5 mAh cm^{-2}; moreover, it maintains a Coulombic efficiency (CE) of 97% for the duration of the cycling (Figure 5c).

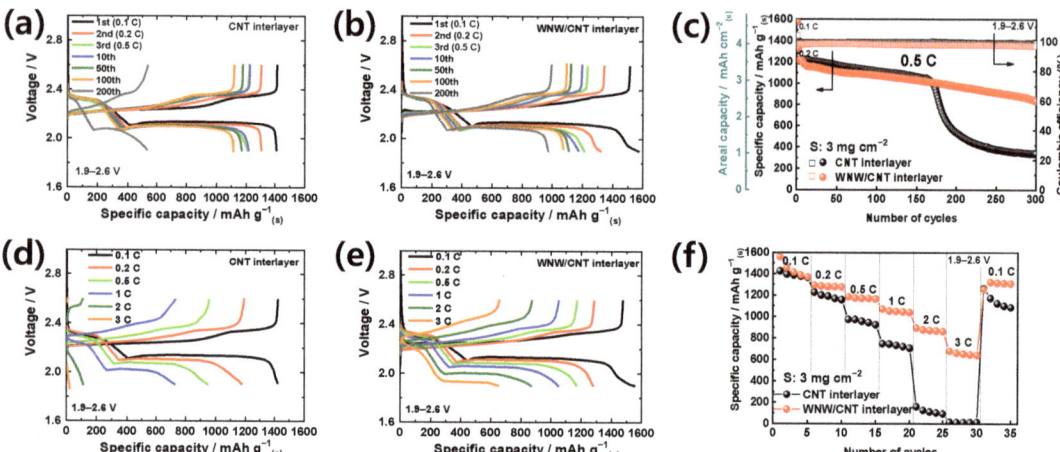

Figure 5. Voltage profiles of the lithium-sulfur cells, at 0.5 C, containing (**a**) CNT interlayer and (**b**) WNW/CNT interlayer, respectively. (**c**) Cycling performance of the lithium-sulfur cells. Voltage profiles of the lithium-sulfur cells, at different current densities, containing a (**d**) CNT interlayer and a (**e**) WNW/CNT interlayer, respectively. (**f**) Rate capability of the lithium-sulfur cells. The sulfur loading of all cells is 3 mg cm^{-2}.

The efficacy of the WNW/CNT interlayer is further evaluated in terms of rate capability. The S-CNT cell shows discharge capacities of 1424.6 mAh g^{-1} at 0.1 C and 734.7 mAh g^{-1} at 1 C, while the capacity is limited at 3 C (Figure 5d,f). When the current returns to a rate of 0.1 C, the capacity increases to 1263 mAh g^{-1}; however, the cell is not able to maintain this capacity (Figure 5f). In contrast, the S-WNW/CNT cell can deliver a capacity of 656.0 mAh g^{-1} at a rate of 3 C (Figure 5e) and shows stable cycling performance after subsequent recovery of the current to 0.1 C (1316.2 mAh g^{-1}). Environmental tests employing symmetric cells with Li$_2$S$_8$ electrolytes also demonstrate better electrochemical reversibility at high rates (Figure S3). The S-WNW/CNT cell shows a higher redox current density than the S-CNT cell at increasing scan rates, which verifies that the redox kinetics are significantly improved in the presence of liquid-phase polysulfides.

The high-rate cycling performance of S-WNW/CNT cells was further tested at a rate of 2 C. When the mass loading of the active material is increased from 1.5 to 3.0 mg cm^{-2}, the cell delivers a discharge capacity of 572 mAh g^{-1} after 400 cycles (Figure 6a).

The presence of the WNW/CNT interlayer enables the delivery of capacity even at increased sulfur loading densities of 4.2, 4.5, and 5.3 mg cm^{-2} (Figure 6b–d). It is worth highlighting that these superior cell performances at high sulfur loadings are achievable in the presence of WNW/CNT interlayers and may be associated with their electrochemical activity. Therefore, we conducted electrochemical tests on WNW/CNT and CNT interlayers without sulfur electrodes (Figure S4). The WNW/CNT interlayer has a discharge at the first cycle (127.8 mAh g^{-1}), whereas the capacity during subsequent cycles is negligible. Evidently, the WNW/CNT interlayer does not participate in the electrochemical reaction of the sulfur electrode but facilitates electron transfer during cycling.

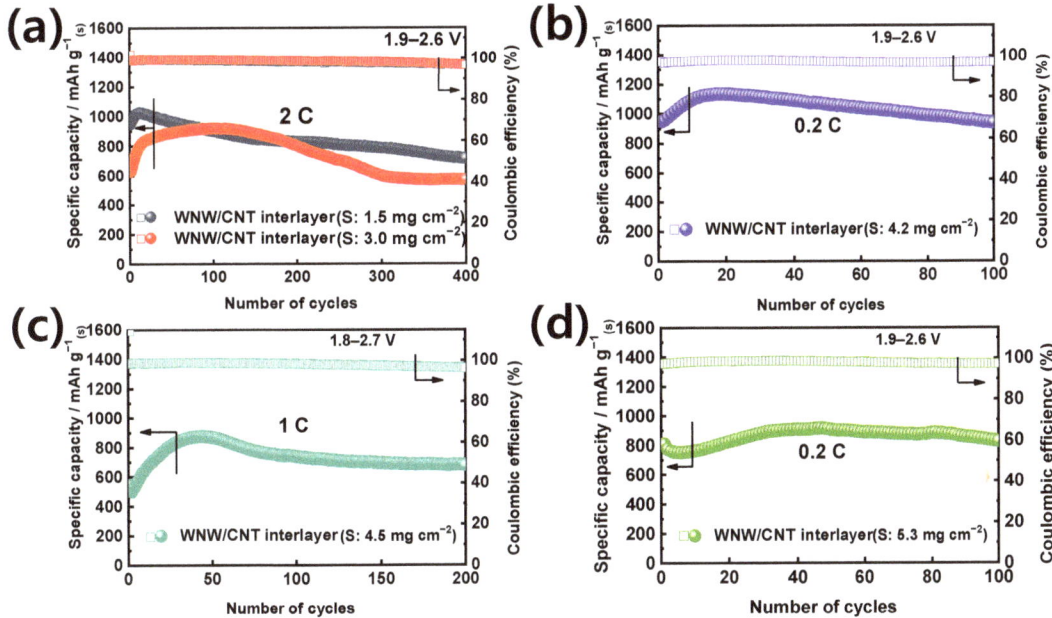

Figure 6. (**a**) Long-term cycling performance of lithium-sulfur cells with a WNW/CNT interlayer at 2 C (sulfur mass loading: 1.5 and 3.0 mg cm^{-2}, respectively). Cycling performance of lithium-sulfur cells with sulfur mass loadings of (**b**) 4.2 mg cm^{-2} (0.2 C), (**c**) 4.5 mg cm^{-2} (1 C), and (**d**) 5.3 mg cm^{-2} (0.2 C).

To gain insight into the superior electrode performance of S-WNW/CNT cells, we observed the interlayers with charged states after extensive cycling (Figure 7). Compared to the morphology of the as-fabricated WNW/CNT interlayer (Figure 1d), the morphology of the WNW/CNT interlayer (Figure 7a,b) after cycling appears unchanged. However, the morphology of the CNT interlayer (Figure 7c,d) after cycling does differ from that of the as-fabricated CNT layer (Figure S1a), which could be attributed to the localized accumulation of lithium polysulfides since carbon has a low affinity for lithium polysulfides [23]. This heterogeneity of the CNT interlayer is likely to impede the facile conversion of sulfur to lithium polysulfides during long-term cycling, illustrated by the undesired capacity fading observed for the S-CNT cell; see Figure 5c. The WNW/CNT interlayer is capable of chemically adsorbing lithium polysulfides and promoting the redox reaction of lithium polysulfide owing to the catalytic effect of WNW; this suppresses the accumulation of lithium polysulfides on the interlayer.

Although cycling does not alter the morphology of the WNW/CNT interlayer, the XPS results indicate a change in the chemical state of the interlayer (Figure 7e). After cycling, the binding energies of the two peaks associated with W^{6+} are ~0.2 eV less than those of the corresponding peaks before cycling, while additional peaks appear at 32.9, 34.3, 35.0, and 36.8 eV. The new peaks are associated with W^{5+} (34.3 and 36.8 eV) and W^{4+} (32.9 and 35.0 eV) [39], which form when WO_3 is reduced during charge transfers between lithium polysulfide and WNW. The TEM-EDS mapping images support this relationship (Figure 7f). The homogeneous elemental distribution on the WNW results from the adherence of lithium polysulfide via thiosulfate mediation, which occurs more actively on the WNW surface than on that of the CNT, which has a weak affinity for lithium polysulfide [36,40]. These results are consistent with the interaction between the thiosulfate mediator and lithium polysulfide on WNW, as shown in Figure 2a, confirming that WO_3 and polysulfides can be chemically bonded. Therefore, the WNW plays a pivotal role in capturing lithium polysulfide during electrochemical reactions and promoting the redox reaction of lithium polysulfide through catalytic action, thereby enhancing the electrochemical performance of sulfur electrodes in lithium-sulfur cells. The electrochemical performance of lithium-sulfur batteries incorporating a WNW/CNT interlayer is compared with that of lithium-sulfur batteries incorporating other types of interlayer or functional separators in Figure 8 and Table S1, highlighting the superiority of the present WNW/CNT interlayer.

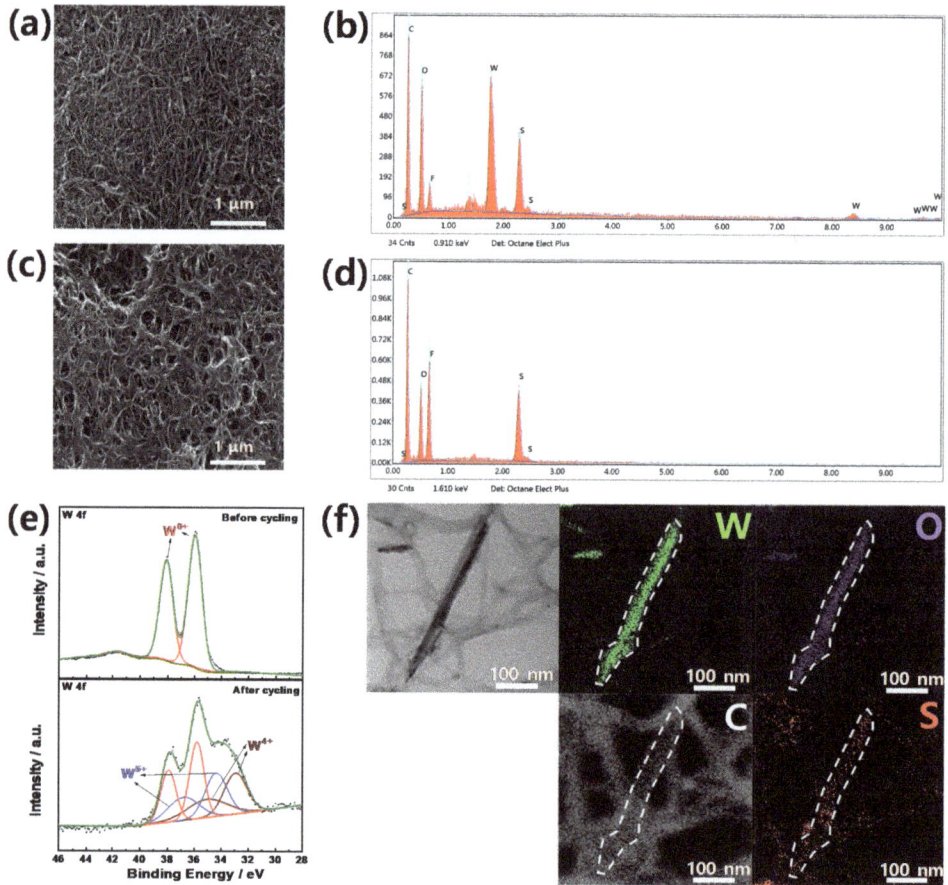

Figure 7. SEM images and area EDS measurements of WNW/CNT and CNT interlayers and their charged states after 100 cycles: (**a**,**b**) WNW/CNT interlayer and (**c**,**d**) CNT interlayer. (**e**) XPS spectra (W 4f) of the WNW/CNT interlayer before and after cycling. (**f**) TEM images of a cycled WNW/CNT interlayer with EDS mapping images of tungsten, oxygen, sulfur, and carbon.

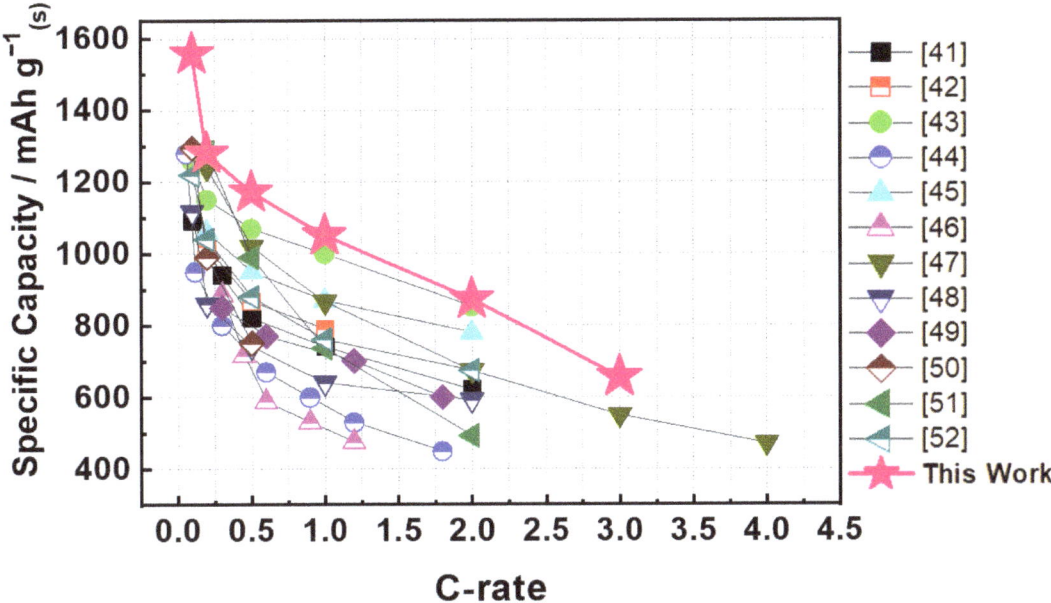

Figure 8. Comparison of the rate capabilities of lithium-sulfur batteries incorporating different types of interlayer or functional separators [41–52].

3. Materials and Methods

3.1. Synthesis of WNW

WNWs were synthesized using a hydrothermal method [53]. Sodium tungstate dihydrate powder (3.29 g, Sigma Aldrich, St. Louis, MO, USA) was dissolved in deionized water (76 mL), and an aqueous HCl solution (3 M) was used to adjust the pH to 1.0. Ammonium sulfate (Sigma Aldrich, 2.64 g) was subsequently added to the solution to control the morphology of the WO_3 product and stirred for 1 h. The solution was transferred into a Teflon-lined, stainless-steel autoclave (capacity: 100 mL) and heated at 180 °C for 24 h. After the reaction, the autoclave was cooled to room temperature. The WO_3 precipitate, with a bright emerald-green color, was collected through filtration and washed with distilled water. To remove reactant residues, the product was rinsed with ethanol three times. Finally, the product was dried in a vacuum oven at 60 °C for 12 h.

3.2. Preparation of CNT Paper and WNW/CNT Paper

To produce CNT paper, multi-walled carbon nanotubes (MWCNTs, Hanwha Chem., Seoul, Korea, 100 mg) were dispersed in deionized water (200 mL) and isopropyl alcohol (8 mL) through ultrasonication for 10 min. The MWCNTs were retrieved through vacuum filtration using membrane paper (Advantec, Tokyo, Japan, No.2) and dried in an oven at 60 °C for 12 h. To fabricate WNW/CNT paper, WNWs (50 mg) and MWCNTs (50 mg) were mixed with a mixer-mill (Retsch, Haan, Germany, MM 400) at a frequency of 5 Hz for 15 min. The WNW/CNT paper was subsequently prepared following a process that is similar to making CNT paper.

After drying, the WNW/CNT paper was peeled from the membrane and cut into 16 mm-diameter circular disks for electrochemical testing. The weight of the interlayer was adjusted for each sulfur mass loading to achieve the required sulfur content ratio. If the sulfur mass loading was 3.0 mg cm^{-2}, the weight per area of the interlayer was approximately 2.0 mg cm^{-2} (the overall sulfur content was ~48%).

3.3. Preparation of Electrodes

A sulfur cathode, for use with an interlayer, was prepared by applying a slurry, consisting of sulfur powder (Sigma Aldrich, 70 wt.%), carbon black Super-P (20 wt.%), and polyacrylic acid (PAA, 10 wt.%) in ethanol, onto carbon-coated aluminum foil, and then drying it overnight at 60 °C in an oven. A sulfur cathode, for use without an interlayer, was made of 48 wt.% sulfur, 32 wt.% Super-P, and 20 wt.% PAA binder. A WNW cathode, used to evaluate the electrochemical reactivity of WNW, was prepared in a similar way to the above sulfur cathodes. A slurry containing 80 wt.% WNW, 10 wt.% Super-P, and 10 wt.% PAA was applied to the gas diffusion layer (GDL).

3.4. Material Characterization

The morphology of the as-synthesized WNW and interlayers was investigated using scanning electron microscopy (SEM; Nova NanoSEM 450, FEI, Hillsboro, OR, USA) and transmission electron microscopy (TEM; JEM-2100F, JEOL, Tokyo, Japan). The crystal structures of the obtained products were characterized by X-ray diffractometry (MiniFlex600, Rigaku, Japan) using Cu Kα radiation. The surface area and pore size distribution of the WNWs were measured using a Quantachrome Autosorb iQ MP automated gas adsorption system with liquid nitrogen (at 77 K). X-ray photoelectron spectroscopy (XPS; Thermo Fisher Scientific, Waltham, MA, USA) was used to investigate the surface and chemical states of the WNWs; the binding energy values were calibrated based on the C 1s peak at 284.5 eV.

3.5. Electrochemical Measurements

Electrochemical tests were performed with coin-type cells (R2032) constructed with sulfur electrodes. WNW/CNT or CNT interlayers were sandwiched between the cathode and a microporous polypropylene film membrane (Celgard 2400) separator; the assembly was countered with a Li-foil anode. The prepared sulfur cathodes were cut into circular disks with diameters of 10 mm, and the sulfur mass loading levels on the electrodes ranged from ~1.5 to ~5.3 mg cm^{-2}. The overall sulfur content in the electrodes was ~48%. The electrolyte solution contained 1 M lithium bis(trifluoromethanesulfonyl)imide (LiTFSI) and 0.4 M lithium nitrate (LiNO$_3$) in a 1:1 (v/v) mixture of 1,3-dioxolane (DOL) and 1,2-dimethoxyethane (DME). The electrolyte/sulfur ratio (E/S ratio) was 20:1. All cells were assembled in an Ar-filled glove box (MBRAUN). Galvanostatic charge-discharge tests were conducted in constant current mode from 0.1 (167.5 mA g^{-1}) to 3 C (5.025 A g^{-1}) within a voltage range of 1.9 to 2.6 V at 30 °C, monitored using a battery testing system (TOSCAT-3100, Toyo System Co., Iwaki, Japan). For sulfur electrodes with high mass loading (4.5 mg cm^{-2}), the cells were tested at 1 C currents in a voltage range of 1.8 to 2.7 V. Cyclic voltammetry (CV) measurements were carried out on a multi-channel potentiostat (VMP-3, Biologic) across the voltage range of 1.9 to 2.6 V at a scan rate of 0.1 mV s^{-1}. The CV of the WNW cathode was tested at 0.05 mV s^{-1} using 1 M LiTFSI in a 1:1 (v/v) mixture of DOL and DME. Symmetric cells were assembled into an R2032 coin-type cell with a polypropylene separator, two identical electrodes (WNW/CNT interlayer, or CNT interlayer), and Li$_2$S$_8$ electrolyte (20.0 mL containing 0.25 M Li$_2$S$_8$ and 1 M LiTFSI in a 1:1 (v/v) mixture of DOL/DME). CV measurements of the symmetrical cells were conducted at different scan rates (-0.8 to 0.8 V).

4. Conclusions

We introduced a 1D-structured WNW/CNT interlayer into a lithium-sulfur battery to capture lithium polysulfides. As intended by the design, the presence of the WNW/CNT interlayer improves the electrode performance in terms of cyclability and rate capability even when the loading density of the active material is increased. The interlayer structure remains intact even after long-term cycling. We believe that this simply made WNW/CNT interlayer presents a promising approach to improve the performance of lithium-sulfur batteries. Its excellent performance is due to the trapping of lithium polysulfide in the interlayer by chemical adsorption and catalytic effect of the WNW as thiosulfate mediators.

Supplementary Materials: The following are available online. Figure S1: (a) SEM images of CNT interlayer with EDS mapping images of carbon, tungsten, oxygen. (b) Isotherm profiles and (c) pore size distributions of WNW. Figure S2: (a) Cycling performance and (b) voltage profiles of a lithium-sulfur cell without an interlayer. Figure S3: Cyclic voltammetry curves of the lithium-sulfur cells with a (a) WNW/CNT interlayer and (b) CNT interlayer, respectively, at different scan rates. Figure S4: (a) Cycling performances of the interlayers without sulfur electrodes. (b) Cyclic voltammetry (CV) profiles of WNW (1.5–3.8 V). Table S1: Comparison of overall electrochemical performances previously reported lithium-sulfur batteries incorporating different types of interlayer or functional separators.

Author Contributions: Conceptualization, S.-K.L.; formal analysis, S.-K.L.; investigation, S.-K.L., H.K. and S.B.; data curation, S.-K.L., H.K. and S.B.; writing—original draft preparation, S.-K.L. and S.-T.M.; writing—review and editing, S.-T.M. and Y.-K.S.; visualization, S.-K.L.; supervision, S.-T.M. and Y.-K.S. All authors have read and agreed to the published version of the manuscript.

Funding: This work was supported by the Technology Innovation Program (20012330, Development of manufacturing technology for high performance cathode electrode with an areal capacity of 6 mAh/cm^2 for all solid state battery) funded By the Ministry of Trade, Industry & Energy (MOTIE, Korea). This work was also supported by National Research Foundation of Korea (NRF) grant funded by the Korea government Ministry of Education and Science Technology (MEST) (NRF-2018R1A2B3008794).

Institutional Review Board Statement: Not applicable.

Informed Consent Statement: Not applicable.

Data Availability Statement: The data presented in this study are available on request from the corresponding author.

Conflicts of Interest: The authors declare no conflict of interest.

Sample Availability: Samples of the compounds are available from the authors.

References

1. Xue, W.; Shi, Z.; Suo, L.; Wang, C.; Wang, Z.; Wang, H.; So, K.P.; Maurano, A.; Yu, D.; Chen, Y.; et al. Intercalation-conversion hybrid cathodes enabling Li–S full-cell architectures with jointly superior gravimetric and volumetric energy densities. *Nat. Energy* **2019**, *4*, 374–382. [CrossRef]
2. Bruce, P.G.; Freunberger, S.A.; Hardwick, L.J.; Tarascon, J.M. Li–O$_2$ and Li–S batteries with high energy storage. *Nat. Mater.* **2012**, *11*, 19–29. [CrossRef] [PubMed]
3. Ji, X.; Lee, K.T.; Nazar, L.F. A highly ordered nanostructured carbon-sulphur cathode for lithium-sulphur batteries. *Nat. Mater.* **2009**, *8*, 500–506. [CrossRef] [PubMed]
4. Hwang, J.-Y.; Myung, S.-T.; Sun, Y.-K. Sodium-ion batteries: Present and future. *Chem. Soc. Rev.* **2017**, *46*, 3529–3614. [CrossRef]
5. Lu, J.; Li, L.; Park, J.-B.; Sun, Y.-K.; Wu, F.; Amine, K. Aprotic and Aqueous Li–O$_2$ Batteries. *Chem. Rev.* **2014**, *114*, 5611–5640. [CrossRef]
6. Liu, D.; Zhang, C.; Zhou, G.; Lv, W.; Ling, G.; Zhi, L.; Yang, Q.H. Catalytic Effects in Lithium–Sulfur Batteries: Promoted Sulfur Transformation and Reduced Shuttle Effect. *Adv. Sci.* **2018**, *5*, 1700270.
7. Kumar, R.; Liu, J.; Hwang, J.-Y.; Sun, Y.-K. Recent research trends in Li–S batteries. *J. Mater. Chem. A* **2018**, *6*, 11582–11605. [CrossRef]
8. Wang, H.Q.; Zhang, W.C.; Xu, J.Z.; Guo, Z.P. Advances in Polar Materials for Lithium-Sulfur Batteries. *Adv. Funct. Mater.* **2018**, *28*, 1707520. [CrossRef]
9. Huang, S.; Lim, Y.V.; Zhang, X.; Wang, Y.; Zheng, Y.; Kong, D.; Ding, M.; Yang, S.A.; Yang, H.Y. Regulating the polysulfide redox conversion by iron phosphide nanocrystals for high-rate and ultrastable lithium-sulfur battery. *Nano Energy* **2018**, *51*, 340–348. [CrossRef]
10. Wu, P.; Chen, L.H.; Xiao, S.S.; Yu, S.; Wang, Z.; Li, Y.; Su, B.L. Insight into the positive effect of porous hierarchy in S/C cathodes on the electrochemical performance of Li-S batteries. *Nanoscale* **2018**, *10*, 11861–11868. [CrossRef]
11. Manthiram, A.; Fu, Y.; Chung, S.H.; Zu, C.; Su, Y.S. Rechargeable lithium-sulfur batteries. *Chem. Rev.* **2014**, *114*, 11751–11787. [CrossRef] [PubMed]
12. Mikhaylik, Y.V.; Akridge, J.R. Polysulfide shuttle study in the Li/S battery system. *J. Electrochem. Soc.* **2004**, *151*, A1969–A1976. [CrossRef]
13. Chung, S.-H.; Chang, C.-H.; Manthiram, A. Progress on the Critical Parameters for Lithium-Sulfur Batteries to be Practically Viable. *Adv. Funct. Mater.* **2018**, *28*, 1801188. [CrossRef]
14. Mikhaylik, Y.V.; Kovalev, I.; Schock, R.; Kumaresan, K.; Xu, J.; Affinito, J. High Energy Rechargeable Li-S Cells for EV Application: Status, Remaining Problems and Solutions. *ECS Trans.* **2010**, *25*, 23–34. [CrossRef]

15. Zhang, Y.; Gao, Z.; Song, N.; He, J.; Li, X. Graphene and its derivatives in lithium–sulfur batteries. *Mater. Today Energy* **2018**, *9*, 319–335. [CrossRef]
16. Jeong, Y.C.; Kim, J.H.; Nam, S.; Park, C.R.; Yang, S.J. Rational Design of Nanostructured Functional Interlayer/Separator for Advanced Li-S Batteries. *Adv. Funct. Mater.* **2018**, *28*, 1707411. [CrossRef]
17. Huang, J.-Q.; Zhang, Q.; Wei, F. Multi-functional separator/interlayer system for high-stable lithium-sulfur batteries: Progress and prospects. *Energy Storage Mater.* **2015**, *1*, 127–145. [CrossRef]
18. Luo, L.; Qin, X.; Wu, J.; Liang, G.; Li, Q.; Liu, M.; Kang, F.; Chen, G.; Li, B. An interwoven MoO_3@CNT scaffold interlayer for high-performance lithium–sulfur batteries. *J. Mater. Chem. A* **2018**, *6*, 8612–8619. [CrossRef]
19. Yue, X.-Y.; Li, X.-L.; Meng, J.-K.; Wu, X.-J.; Zhou, Y.-N. Padding molybdenum net with Graphite/MoO_3 composite as a multi-functional interlayer enabling high-performance lithium-sulfur batteries. *J. Power Sources* **2018**, *397*, 150–156. [CrossRef]
20. Xiao, Z.; Yang, Z.; Wang, L.; Nie, H.; Zhong, M.; Lai, Q.; Xu, X.; Zhang, L.; Huang, S. A Lightweight TiO_2/Graphene Interlayer, Applied as a Highly Effective Polysulfide Absorbent for Fast, Long-Life Lithium-Sulfur Batteries. *Adv. Mater.* **2015**, *27*, 2891–2898. [CrossRef]
21. Dong, W.; Meng, L.; Hong, X.; Liu, S.; Shen, D.; Xia, Y.; Yang, S. MnO_2/rGO/CNTs Framework as a Sulfur Host for High-Performance Li-S Batteries. *Molecules* **2020**, *25*, 1989. [CrossRef]
22. Zhao, T.; Ye, Y.; Peng, X.; Divitini, G.; Kim, H.-K.; Lao, C.-Y.; Coxon, P.R.; Xi, K.; Liu, Y.; Ducati, C.; et al. Advanced Lithium-Sulfur Batteries Enabled by a Bio-Inspired Polysulfide Adsorptive Brush. *Adv. Funct. Mater.* **2016**, *26*, 8418–8426. [CrossRef]
23. Yang, L.; Li, G.; Jiang, X.; Zhang, T.; Lin, H.; Lee, J.Y. Balancing the chemisorption and charge transport properties of the interlayer in lithium–sulfur batteries. *J. Mater. Chem. A* **2017**, *5*, 12506–12512. [CrossRef]
24. Liu, F.; Xiao, Q.; Wu, H.B.; Sun, F.; Liu, X.; Li, F.; Le, Z.; Shen, L.; Wang, G.; Cai, M.; et al. Regenerative Polysulfide-Scavenging Layers Enabling Lithium-Sulfur Batteries with High Energy Density and Prolonged Cycling Life. *ACS Nano* **2017**, *11*, 2697–2705. [CrossRef]
25. Lin, H.B.; Zhang, S.L.; Zhang, T.R.; Ye, H.L.; Yao, Q.F.; Zheng, G.W.; Lee, J.Y. Elucidating the Catalytic Activity of Oxygen Deficiency in the Polysulfide Conversion Reactions of Lithium-Sulfur Batteries. *Adv. Energy Mater.* **2018**, *8*, 1801868. [CrossRef]
26. Choi, S.; Seo, D.H.; Kaiser, M.R.; Zhang, C.; van der Laan, T.; Han, Z.J.; Bendavid, A.; Guo, X.; Yick, S.; Murdock, A.T.; et al. WO_3 nanolayer coated 3D-graphene/sulfur composites for high performance lithium/sulfur batteries. *J. Mater. Chem. A* **2019**, *7*, 4596–4603. [CrossRef]
27. Yang, X.; Zu, H.; Luo, L.; Zhang, H.; Li, J.; Yi, X.; Liu, H.; Wang, F.; Song, J. Synergistic tungsten oxide/N, S co-doped carbon nanofibers interlayer as anchor of polysulfides for high-performance lithium-sulfur batteries. *J. Alloys Compd.* **2020**, *833*, 154969. [CrossRef]
28. Zhang, J.; Tu, J.-P.; Xia, X.-H.; Wang, X.-L.; Gu, C.-D. Hydrothermally synthesized WO_3 nanowire arrays with highly improved electrochromic performance. *J. Mater. Chem.* **2011**, *21*, 5492–5498. [CrossRef]
29. Liang, X.; Hart, C.; Pang, Q.; Garsuch, A.; Weiss, T.; Nazar, L.F. A highly efficient polysulfide mediator for lithium-sulfur batteries. *Nat. Commun.* **2015**, *6*, 5682. [CrossRef]
30. Naumkin, A.V.; Kraut-Vass, A.; Gaarenstroom, S.W.; Powell, C.J. *NIST Standard Reference Database 20, Version 4.1*; NIST: Gaithersburg, ML, USA, 2012.
31. Kondalkar, V.V.; Mali, S.S.; Kharade, R.R.; Khot, K.V.; Patil, P.B.; Mane, R.M.; Choudhury, S.; Patil, P.S.; Hong, C.K.; Kim, J.H.; et al. High performing smart electrochromic device based on honeycomb nanostructured h-WO_3 thin films: Hydrothermal assisted synthesis. *Dalton Trans.* **2015**, *44*, 2788–2800. [CrossRef]
32. Hur, Y.G.; Lee, D.-W.; Lee, K.-Y. Hydrocracking of vacuum residue using $NiWS_{(x)}$ dispersed catalysts. *Fuel* **2016**, *185*, 794–803. [CrossRef]
33. Ishutenko, D.; Minaev, P.; Anashkin, Y.; Nikulshina, M.; Mozhaev, A.; Maslakov, K.; Nikulshin, P. Potassium effect in K-Ni(Co)PW/Al_2O_3 catalysts for selective hydrotreating of model FCC gasoline. *Appl. Catal. B* **2017**, *203*, 237–246. [CrossRef]
34. Nikulshina, M.; Mozhaev, A.; Lancelot, C.; Marinova, M.; Blanchard, P.; Payen, E.; Lamonier, C.; Nikulshin, P. MoW synergetic effect supported by HAADF for alumina based catalysts prepared from mixed $SiMonW12-n$ heteropolyacids. *Appl. Catal. B* **2018**, *224*, 951–959. [CrossRef]
35. Hwang, J.-Y.; Kim, H.M.; Lee, S.-K.; Lee, J.-H.; Abouimrane, A.; Khaleel, M.A.; Belharouak, I.; Manthiram, A.; Sun, Y.-K. High-Energy, High-Rate, Lithium-Sulfur Batteries: Synergetic Effect of Hollow TiO_2-Webbed Carbon Nanotubes and a Dual Functional Carbon-Paper Interlayer. *Adv. Energy Mater.* **2016**, *6*, 1501480. [CrossRef]
36. Hwang, J.-Y.; Kim, H.M.; Shin, S.; Sun, Y.-K. Designing a High-Performance Lithium-Sulfur Batteries Based on Layered Double Hydroxides-Carbon Nanotubes Composite Cathode and a Dual-Functional Graphene-Polypropylene-Al_2O_3 Separator. *Adv. Funct. Mater.* **2018**, *28*, 1704294. [CrossRef]
37. Walus, S.; Barchasz, C.; Colin, J.F.; Martin, J.F.; Elkaim, E.; Lepretre, J.C.; Alloin, F. New insight into the working mechanism of lithium-sulfur batteries: In situ and operando X-ray diffraction characterization. *Chem. Commun.* **2013**, *49*, 7899–7901. [CrossRef]
38. Liang, X.; Kwok, C.Y.; Lodi-Marzano, F.; Pang, Q.; Cuisinier, M.; Huang, H.; Hart, C.J.; Houtarde, D.; Kaup, K.; Sommer, H.; et al. Tuning Transition Metal Oxide-Sulfur Interactions for Long Life Lithium Sulfur Batteries: The "Goldilocks" Principle. *Adv. Energy Mater.* **2016**, *6*, 1501636. [CrossRef]
39. Xie, F.Y.; Gong, L.; Liu, X.; Tao, Y.T.; Zhang, W.H.; Chen, S.H.; Meng, H.; Chen, J. XPS studies on surface reduction of tungsten oxide nanowire film by Ar^+ bombardment. *J. Electron Spectrosc. Relat. Phenom.* **2012**, *185*, 112–118. [CrossRef]

40. Pang, Q.; Kundu, D.; Cuisinier, M.; Nazar, L.F. Surface-enhanced redox chemistry of polysulphides on a metallic and polar host for lithium-sulphur batteries. *Nat. Commun.* **2014**, *5*, 4759. [CrossRef]
41. Liang, G.; Wu, J.; Qin, X.; Liu, M.; Li, Q.; He, Y.B.; Kim, J.K.; Li, B.; Kang, F. Ultrafine TiO_2 Decorated Carbon Nanofibers as Multifunctional Interlayer for High-Performance Lithium-Sulfur Battery. *ACS Appl. Mater. Interfaces* **2016**, *8*, 23105–23113. [CrossRef]
42. Chang, C.-H.; Chung, S.-H.; Manthiram, A. Ultra-lightweight PANiNF/MWCNT-functionalized separators with synergistic suppression of polysulfide migration for Li–S batteries with pure sulfur cathodes. *J. Mater. Chem. A* **2015**, *3*, 18829–18834. [CrossRef]
43. Lee, C.L.; Kim, I.D. A hierarchical carbon nanotube-loaded glass-filter composite paper interlayer with outstanding electrolyte uptake properties for high-performance lithium-sulphur batteries. *Nanoscale* **2015**, *7*, 10362–10367. [CrossRef] [PubMed]
44. Kim, H.M.; Hwang, J.Y.; Manthiram, A.; Sun, Y.-K. High-Performance Lithium-Sulfur Batteries with a Self-Assembled Multiwall Carbon Nanotube Interlayer and a Robust Electrode-Electrolyte Interface. *ACS Appl. Mater. Interfaces* **2016**, *8*, 983–987. [CrossRef] [PubMed]
45. Song, R.; Fang, R.; Wen, L.; Shi, Y.; Wang, S.; Li, F. A trilayer separator with dual function for high performance lithium–sulfur batteries. *J. Power Sources* **2016**, *301*, 179–186. [CrossRef]
46. Lin, C.; Zhang, W.; Wang, L.; Wang, Z.; Zhao, W.; Duan, W.; Zhao, Z.; Liu, B.; Jin, J. A few-layered Ti_3C_2 nanosheet/glass fiber composite separator as a lithium polysulphide reservoir for high-performance lithium–sulfur batteries. *J. Mater. Chem. A* **2016**, *4*, 5993–5998. [CrossRef]
47. Lu, Y.; Gu, S.; Guo, J.; Rui, K.; Chen, C.; Zhang, S.; Jin, J.; Yang, J.; Wen, Z. Sulfonic Groups Originated Dual-Functional Interlayer for High Performance Lithium-Sulfur Battery. *ACS Appl. Mater. Interfaces* **2017**, *9*, 14878–14888. [CrossRef]
48. Kim, J.H.; Seo, J.; Choi, J.; Shin, D.; Carter, M.; Jeon, Y.; Wang, C.; Hu, L.; Paik, U. Synergistic Ultrathin Functional Polymer-Coated Carbon Nanotube Interlayer for High Performance Lithium-Sulfur Batteries. *ACS Appl. Mater. Interfaces* **2016**, *8*, 20092–20099. [CrossRef]
49. Guo, P.; Liu, D.; Liu, Z.; Shang, X.; Liu, Q.; He, D. Dual functional MoS_2/graphene interlayer as an efficient polysulfide barrier for advanced lithium-sulfur batteries. *Electrochim. Acta* **2017**, *256*, 28–36. [CrossRef]
50. Chang, C.-H.; Chung, S.-H.; Manthiram, A. Transforming waste newspapers into nitrogen-doped conducting interlayers for advanced Li–S batteries. *Sustain. Energy Fuels* **2017**, *1*, 444–449. [CrossRef]
51. Shi, H.; Zhao, X.; Wu, Z.-S.; Dong, Y.; Lu, P.; Chen, J.; Ren, W.; Cheng, H.-M.; Bao, X. Free-standing integrated cathode derived from 3D graphene/carbon nanotube aerogels serving as binder-free sulfur host and interlayer for ultrahigh volumetric-energy-density lithium sulfur batteries. *Nano Energy* **2019**, *60*, 743–751. [CrossRef]
52. Zheng, B.; Yu, L.; Li, N.; Xi, J. Efficiently immobilizing and converting polysulfide by a phosphorus doped carbon microtube textile interlayer for high-performance lithium-sulfur batteries. *Electrochim. Acta* **2020**, *345*, 136186. [CrossRef]
53. Wang, H.P.; Ma, G.F.; Tong, Y.C.; Yang, Z.R. Biomass carbon/polyaniline composite and WO_3 nanowire-based asymmetric supercapacitor with superior performance. *Ionics* **2018**, *24*, 3123–3131. [CrossRef]

Article

A Bifunctional Hybrid Electrocatalyst for Oxygen Reduction and Oxygen Evolution Reactions: Nano-Co$_3$O$_4$-Deposited La$_{0.5}$Sr$_{0.5}$MnO$_3$ via Infiltration

Seona Kim [1], Guntae Kim [2,*] and Arumugam Manthiram [1,*]

[1] Materials Science and Engineering Program & Texas Materials Institute, The University of Texas at Austin, Austin, TX 78712, USA; seona0623@gmail.com
[2] Department of Energy Engineering, Ulsan National Institute of Science and Technology (UNIST), Ulsan 44919, Korea
* Correspondence: gtkim@unist.ac.kr (G.K.); manth@austin.utexas.edu (A.M.)

Citation: Kim, S.; Kim, G.; Manthiram, A. A Bifunctional Hybrid Electrocatalyst for Oxygen Reduction and Oxygen Evolution Reactions: Nano-Co$_3$O$_4$-Deposited La$_{0.5}$Sr$_{0.5}$MnO$_3$ via Infiltration. *Molecules* **2021**, *26*, 277. http://doi.org/10.3390/molecules 26020277

Academic Editor: Myung-Hwan Whangbo
Received: 5 December 2020
Accepted: 5 January 2021
Published: 8 January 2021

Publisher's Note: MDPI stays neutral with regard to jurisdictional claims in published maps and institutional affiliations.

Copyright: © 2021 by the authors. Licensee MDPI, Basel, Switzerland. This article is an open access article distributed under the terms and conditions of the Creative Commons Attribution (CC BY) license (https:// creativecommons.org/licenses/by/ 4.0/).

Abstract: For rechargeable metal–air batteries, which are a promising energy storage device for renewable and sustainable energy technologies, the development of cost-effective electrocatalysts with effective bifunctional activity for both oxygen reduction reaction (ORR) and oxygen evolution reaction (OER) has been a challenging task. To realize highly effective ORR and OER electrocatalysts, we present a hybrid catalyst, Co$_3$O$_4$-infiltrated La$_{0.5}$Sr$_{0.5}$MnO$_{3-\delta}$ (LSM@Co$_3$O$_4$), synthesized using an electrospray and infiltration technique. This study expands the scope of the infiltration technique by depositing ~18 nm nanoparticles on unprecedented ~70 nm nano-scaffolds. The hybrid LSM@Co$_3$O$_4$ catalyst exhibits high catalytic activities for both ORR and OER (~7 times, ~1.5 times, and ~1.6 times higher than LSM, Co$_3$O$_4$, and IrO$_2$, respectively) in terms of onset potential and limiting current density. Moreover, with the LSM@Co$_3$O$_4$, the number of electrons transferred reaches four, indicating that the catalyst is effective in the reduction reaction of O$_2$ via a direct four-electron pathway. The study demonstrates that hybrid catalysts are a promising approach for oxygen electrocatalysts for renewable and sustainable energy devices.

Keywords: bifunctional catalyst; hybrid catalyst; oxygen reduction reaction; oxygen evolution reaction; four-electron pathway

1. Introduction

Renewable and sustainable energy storage and conversion systems such as fuel cells, electrolysis cells, and secondary batteries have attracted a great deal of attention in recent years to address the environmental challenges we face today [1–4]. Specifically, metal–air batteries are considered to be promising systems due to their extremely high theoretical energy density [4–9]. Metal–air batteries consist of a pure metal (e.g., Li, Na, Zn, Al) anode and an external cathode of ambient air with an aqueous or aprotic electrolyte. To be suitable for rechargeable batteries, the oxygen electrocatalysts should exhibit effective catalytic activity, not only for the oxygen reduction reaction (ORR), but also for the oxygen evolution reaction (OER)—that is, they should exhibit bifunctionality [10–15]. However, due to the sluggish reaction kinetics of the ORR and OER (which respectively correspond to the discharge and charge processes), the development of highly efficient and cost-effective oxygen electrocatalysts is considered to be an essential challenge. Although Pt, IrO$_2$, RuO$_2$, and their alloys and composites show superior activities for ORR or OER, they suffer from the drawbacks of poor stability, scarcity, and high cost.

The focus in the literature has been on transition-metal oxides because oxygen electrocatalysts possess bifunctionality due to their operational stability in alkaline solutions and promising catalytic activities for ORR and OER. Among them, Co-containing oxides such as Co$_3$O$_4$, Ln$_x$Sr$_{1-x}$CoO$_{3-\delta}$, and Ba$_x$Sr$_{1-x}$Co$_y$Fe$_{1-y}$O$_{3-\delta}$ have especially been reported as efficient electrocatalysts for OER [13,16–21]. Mn-containing oxides (e.g., Mn_2O_3, $NiMnO_2$, $CoMnO_2$, and $La_{0.5}Sr_{0.5}MnO_{3-\delta}$) have demonstrated excellent catalytic activity for ORR [22–26]. Although the above transition-metal oxides exhibit some bifunctionality, their outstanding catalytic activity is generally specific to either ORR or OER, but not both.

Hybrid catalysts composed of more than two materials have been suggested as an approach to secure the desired bifunctionality. For example, Cu-nanoparticle-loaded Co_3O_4 microspheres, Co_3O_4–graphene composite, $LaNiO_3$-Pt/C core–corona structure catalyst, and Co_3O_4 nanocrystals grown on graphene have been demonstrated [8,17,27–31]. Based on earlier studies, hybrid catalysts (A@B) show enhanced ORR/OER activity compared to the individual catalysts (A or B) and/or their physical mixtures (A + B) due to the synergistic effects between the constituent materials [17,27,32]. Therefore, in this study, a hybrid catalyst consisting of $La_{0.5}Sr_{0.5}MnO_{3-\delta}$ (LSM) and Co_3O_4 was designed to achieve high catalytic activities for both ORR and OER.

Nanosizing fabrications were employed to maximize the performances of each constituent material by endowing enlarged reaction sites and shortened ion-/charge-transport path [33–35]. Among the various fabrication techniques, electrospray and infiltration techniques have been used due to their simplicity, effectiveness, and reproducibility. In addition, the stoichiometric ratios can be precisely controlled because the fabrication techniques rely on the precursor mixing at the molecular level.

LSM nanoparticles were prepared through electrospray, and Co_3O_4 was deposited on the LSM surface through infiltration to synthesize the hybrid catalyst (LSM@Co_3O_4). We discovered that 20 wt.% Co_3O_4-infiltrated LSM exhibited a comparable catalytic activity to benchmark bifunctional oxygen electrocatalysts. Furthermore, because the calculated number of transferred electrons is four, it indicates that O_2 follows a direct reduction pathway with less thermodynamic reaction potential.

2. Results and Discussion

2.1. Structural and Morphological Characterization

Figure 1a illustrates the electrospray fabrication of $La_{0.5}Sr_{0.5}MnO_{3-\delta}$ (LSM) with the precursor solution including La-, Sr-, and Mn nitrates, PVP, and DMF. A syringe needle tip and aluminum foil collector were connected to the anode and cathode of a high-voltage power supply, respectively. After a heat treatment at 850 °C, X-ray diffraction (XRD) was conducted in the range of $20° < 2\theta < 60°$ to identify the crystal structure of the prepared samples. As shown in Figure 1b, LSM was successfully synthesized with a single-phase structure of simple perovskite without any significant impurities. The scanning electron microscopy (SEM) images in Figure 1c,d illustrate that spherical LSM nanoparticles 60–100 nm in size were formed and organically connected.

Employing the infiltration technique for these nanoparticle scaffolds is a new approach and a unique attempt. Typically, micro-scale pore/particles have been used as the scaffold for infiltration technique and the nanoparticles deposited on the scaffold surface have a size of 60 nm or more. However, the nano-sized LSM scaffold has low interactive force with the precursor solution due to the reduced contact area. To overcome this problem and increase the wettability, ethanol was used as the solvent in this study due to its low boiling point, low surface tension (22.3 mN m^{-1} at 20 °C), and good solubility of cobalt nitrate. The contact angle of the precursor solution was measured to check the wettability by dropping 3 µL of the precursor solution on the polished LSM pellet. The cobalt precursor solution had a low contact angle of 15.8° (inset in Figure 2a), which indicates good wettability, and the use of ethanol facilitated the deposition of precursors on the surface of LSM nanoparticles. Table 1 summarizes the abbreviations for the samples.

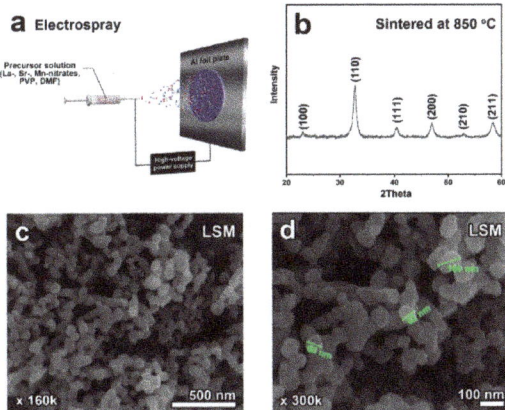

Figure 1. (**a**) Schematic illustration of the electrospray fabrication of $La_{0.5}Sr_{0.5}MnO_{3-\delta}$ (LSM). (**b**) X-ray diffraction patterns and (**c,d**) scanning electron microscopy images of LSM sintered at 850 °C for 4 h in air.

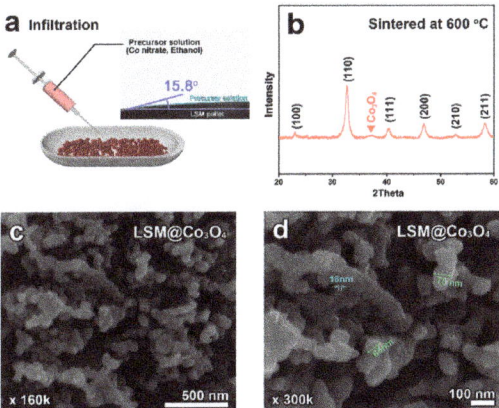

Figure 2. (**a**) Schematic illustration of the infiltration technique for LSM@Co_3O_4 with the contact angle of the precursor solution (inset). (**b**) X-ray diffraction pattern and (**c,d**) scanning electron microscopy images of LSM@Co_3O_4 sintered at 600 °C for 4 h in air.

Table 1. Chemical compositions and abbreviations of the samples.

	Abbreviations	Chemical Composition
	LSM	$La_{0.5}Sr_{0.5}MnO_{3-\delta}$
LSM@Co_3O_4	LSM@Co_3O_4-20	20 wt.% Co_3O_4-infiltrated LSM
	LSM@Co_3O_4-25	25 wt.% Co_3O_4-infiltrated LSM

To obtain an effective Co_3O_4 electrocatalyst, the infiltrated powder was sintered at 600 °C and the sintered powder is denoted as LSM@Co_3O_4. The XRD pattern of LSM@Co_3O_4 (Figure 2b) shows the phase of LSM and Co_3O_4 without any detectable impurity phases, indicating the successful synthesis of Co_3O_4. The broad peak due to Co_3O_4 can be attributed to the nano size of the deposited particles. The high-magnification SEM images of LSM@Co_3O_4 in Figure 2c,d confirm that the smaller nanoparticles (~18 nm) of Co_3O_4 were uniformly deposited on the entire LSM nano-scaffold (~70 nm). Furthermore, LSM@Co$_3$O$_4$ retained the organically connected microstructures, which is effective for the transfer of electrons.

2.2. Rotating Ring-Disk Electrode (RRDE) Test

With different loading amounts of Co$_3$O$_4$, the catalytic activities of LSM@Co$_3$O$_4$ were measured for OER and ORR in 0.1 M KOH. The catalytic activities were compared with the benchmark catalysts IrO$_2$ and Pt/C for OER and ORR to evaluate LSM@Co$_3$O$_4$ as a bifunctional electrocatalyst. For OER, the catalytic activities of LSM, LSM@Co$_3$O$_4$-20, and LSM@Co$_3$O$_4$-25 were compared with Co$_3$O$_4$ and the benchmark catalyst IrO$_2$ (Figure 3a). LSM has a nano-sized microstructure, but its OER current density showed a low value of 1.85 mA cm^{-2} at 1.70 V vs. reversible hydrogen electrode (RHE). After conducting Co$_3$O$_4$ infiltration, the current density was significantly enhanced to 13.01 mA cm^{-2} for LSM@Co$_3$O$_4$-20 and 10.63 mA cm^{-2} for LSM@Co$_3$O$_4$-25, indicating that the surface material was the dominant factor determining the catalytic activity for the OER. Especially, the current density of LSM@Co$_3$O$_4$-20 was ~7 times, ~1.5 times, and ~1.6 times higher, respectively, than those of LSM, Co$_3$O$_4$, and IrO$_2$. The onset potential was also enhanced, as could be deduced from the voltage at 1 mA cm^{-2} of LSM@Co$_3$O$_4$-20 (1.55 V) and LSM@Co$_3$O$_4$-25 (1.55 V), compared to those of LSM (1.65 V) and Co$_3$O$_4$ (1.57 V). Moreover, the values are comparable to those of benchmark electrocatalysts such as Pt/C (1.61 V) and IrO$_2$ (1.55 V). To compare OER kinetics, Tafel plots of LSM, LSM@Co$_3$O$_4$-20, and LSM@Co$_3$O$_4$-25 were derived from their OER polarization curves, and are presented in Figure 3c. The Tafel slope of LSM@Co$_3$O$_4$-25 was 73 mV decade^{-1}, which is lower than that of LSM (84 mV decade^{-1}), LSM@Co$_3$O$_4$-20 (89 mV decade^{-1}), and IrO$_2$ (121 mV decade^{-1}). These improved catalytic activities of LSM@Co$_3$O$_4$-20 imply the existence of synergistic effects between LSM and Co$_3$O$_4$ and the potential for the use of LSM@Co$_3$O$_4$-20 as a bifunctional electrocatalyst.

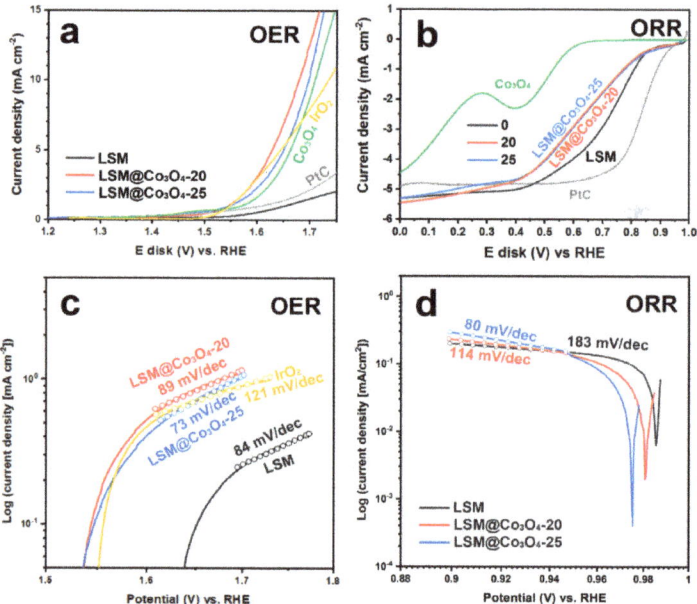

Figure 3. Polarization curves and Tafel slopes according to the loading amount of Co$_3$O$_4$ for (**a**,**c**) oxygen evolution reaction (OER) and (**b**,**d**) oxygen reduction reaction (ORR) in O$_2$-saturated 0.1 M KOH at a rotation rate of 1600 rpm and a scan rate of 0.01 V s^{-1}.

Figure 3b,d presents the polarization curves of LSM@Co$_3$O$_4$ for the ORR with the benchmark samples Co$_3$O$_4$ and Pt/C. LSM on its own exhibited a good limiting current density (−5.30 mA cm^{-2}) compared to Pt/C (−4.90 mA cm^{-2}). However, with the Co$_3$O$_4$ infiltration, the limiting current density was enhanced to −5.45 and −5.30 mA cm^{-2}, respectively, for LSM@Co$_3$O$_4$-20 and LSM@Co$_3$O$_4$-25. The ORR onset potentials at a current density of −0.3 mA cm^{-2} were observed to be 0.87, 0.87, and 0.89 V vs. RHE, respectively, for LSM, LSM@Co$_3$O$_4$-20, and LSM@Co$_3$O$_4$-25. On its own, Co$_3$O$_4$ showed a lower ORR activity, with the values of −4.45 mA cm^{-2} for limiting current density and 0.59 V vs. RHE for onset potential. Considering that a hybrid catalyst (A@B) has the properties of both constituent materials (A and B), the reduced activity in the range of 0.10–0.89 V may originate from the low catalytic activity of Co$_3$O$_4$, covering the surface of LSM. In Figure 3d, the Tafel slopes are plotted from the ORR polarization curves near the onset potential. Tafel slopes were obtained from the Koutecky–Levich (K-L) equation: LSM (80 mV decade^{-1}), LSM@Co$_3$O$_4$-20 (114 mV decade^{-1}), and LSM@Co$_3$O$_4$-25 (184 mV decade^{-1}). The lower Tafel slope for LSM@Co$_3$O$_4$ compared to that for LSM indicates that the infiltration of Co$_3$O$_4$ facilitated oxygen diffusion on the catalyst surface and improved the ORR performance.

2.3. Four-Electron Pathway

As shown in Figure 4a,b, the electrochemical reduction of O$_2$ can occur via two pathways, either a direct 4-electron pathway or a 2+2-electron pathway with peroxide intermediates. For oxygen electrocatalysis, the 4-electron pathway is most desirable because the direct pathway requires less thermodynamic reaction potential (Table 2). The 2+2-electron pathway is preferred in the industry for H$_2$O$_2$ production rather than for ORR catalysis.

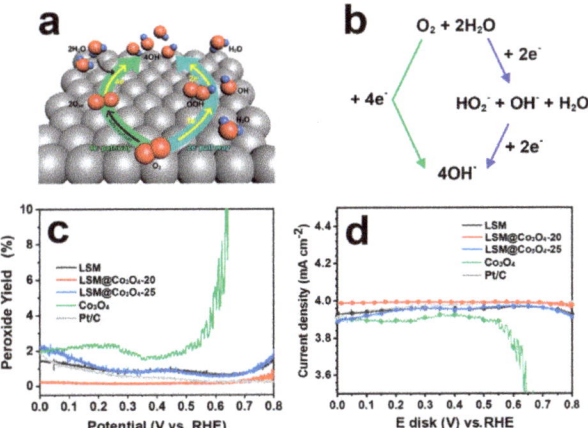

Figure 4. (a) Electron transfer number (n) depending on the loading amount of Co$_3$O$_4$ for the oxygen reduction reaction (ORR). (b) Schematic representation of the ORR mechanism by direct 4-electron and indirect 2+2 electron pathways. (c) Assessment of peroxide yield calculated by the (d) ring current in O$_2$-saturated 0.1 M KOH at a rotation rate of 1600 rpm and a scan rate of 0.01 V s^{-1}.

Table 2. Thermodynamic electrode potential of the O$_2$ reduction pathway.

	Reaction Process	Thermodynamic Electrode Potential at Standard Conditions, V vs. SHE
4-electron pathway	$O_2 + 2H_2O + 4e^- \rightarrow 4OH^-$	0.401
2+2-electron pathway	$O_2 + H_2O + 2e^- \rightarrow HO_2^- + OH^-$ $HO_2^- + H_2O + 2e^- \rightarrow 3OH^-$	−0.065 0.867

To verify the ORR catalytic pathways, ring disk current was employed to detect the current of HO_2^- concomitantly produced via a two-electron pathway. The peroxide yield and electron transfer number (n) were calculated from the following equations with the measured current of the ring electrode and disk electrode for LSM, LSM@Co$_3$O$_4$-20, and LSM@Co$_3$O$_4$-25. The values of commercial Pt/C and Co$_3$O$_4$ were also compared as a benchmark.

$$\% \, HO_2^- = 200 \times (I_r/N)/(I_d + I_r/N) \quad (1)$$

$$n = 4 \times I_d/(I_d + I_r/N) \quad (2)$$

where I_d represents the disk current, I_r indicates the ring current, and N is the current collection efficiency of the Pt ring. N was experimentally determined to be 0.41 from the reduction of K$_3$Fe[CN]$_6$.

The measured H$_2$O$_2$ yield was below 3% for LSM, LSM@Co$_3$O$_4$-20, and LSM@Co$_3$O$_4$-25, which is comparable to that for Pt/C and much lower than that for Co$_3$O$_4$ (Figure 4c). The number of electrons transferred for LSM@Co$_3$O$_4$-20 was calculated to be 3.96–4.00 (Figure 4d), which is higher than that for LSM (3.90–3.96) and LSM@Co$_3$O$_4$-25 (3.90–3.97). These results confirm that LSM, LSM@Co$_3$O$_4$-20, and LSM@Co$_3$O$_4$-25 followed a 4-electron reduction pathway rather than the 2+2-electron pathway. Especially, LSM@Co$_3$O$_4$-20 formed a lower quantity of peroxides than Pt/C, allowing the ORR to proceed through an almost ideal 4-electron reduction pathway.

2.4. X-ray Photoelectron Spectroscopy (XPS) Anaylsis

Based on the RRDE results, LSM@Co$_3$O$_4$ exhibited good catalytic activity for both ORR and OER, comparable to the benchmark catalysts (Pt/C, Co$_3$O$_4$, and IrO$_2$). Since the catalytic activities of LSM@Co$_3$O$_4$-20 and LSM@Co$_3$O$_4$-25 were similar and the n-value of LSM@Co$_3$O$_4$-20 was closer to four, we selected LSM@Co$_3$O$_4$-20 as a representative sample of LSM@Co$_3$O$_4$ for X-ray photoelectron spectroscopy (XPS) characterization.

XPS was performed to analyze the surface electronic state of the catalysts (i.e., LSM, Co$_3$O$_4$, and LSM@Co$_3$O$_4$-20) for O 1s, Co 2p, and Mn 2p. The XPS results were calibrated with the binding energy (BE) of the C 1s peak at 284.3 eV. In Figure 5a–c, the O 1s spectrum can be deconvoluted into three main peaks of lattice oxygen (O$_{lattice}$, red), surface-adsorbed oxygen species (O$_{ad}$, green), and adsorbed molecular water (H$_2$O$_{ad}$, blue). The relative contents of the oxygen species were calculated from the relative area of the three peaks, and are listed in Table 3. The calculated O$_{ad}$/O$_{lattice}$ ratios were 1.68, 2.02, and 2.39, respectively, for LSM, Co$_3$O$_4$, and LSM@Co$_3$O$_4$-20. Considering that O$_{ad}$ can easily be converted to O$_2$, the higher O$_{ad}$/O$_{lattice}$ ratio of LSM@ Co$_3$O$_4$-20 is in good agreement with the OER results (i.e., indicating an enhanced catalytic activity of LSM@ Co$_3$O$_4$-20).

According to previous studies, Co^{3+} cations at the surface play a significant role in the OER due to their unique electronic state, favorable as both an electron donor and an electron acceptor for O$_2$ and electron capturing. The Co^{3+}/Co^{2+} ratio was 0.45 for Co$_3$O$_4$ and 0.68 for LSM@Co$_3$O$_4$-20, respectively. The highly concentrated Co^{3+} of LSM@Co$_3$O$_4$-20 indicates the high number of donor–acceptor reduction sites, supporting the enhanced OER catalytic activity. As presented in Figure 5f,g, the Mn^{3+}/Mn^{4+} ratio was reduced from 0.83 for LSM to 0.77 for LSM@Co$_3$O$_4$-20, indicating ligand effects between LSM and the infiltrated Co$_3$O$_4$ layer. The ligand effect, a type of interfacial effect that leads to synergistic effects on the catalytic activity, refers to the electron transfer derived by the different electronic configurations of two adjacent materials. It has good agreement with the XPS results for Co 2p, in which LSM@Co$_3$O$_4$-20 had a higher Co^{3+} concentration compared to Co$_3$O$_4$. Based on these results, we speculate that the electrons in the LSM layer were transferred to the infiltrated Co$_3$O$_4$ layer by ligand effects, resulting in an electron-rich state on the surface. Correspondingly, the bond strength between the metal oxide and oxygen species at the surface of an electrocatalyst can be altered and the catalytic activity for ORR and OER can be tuned.

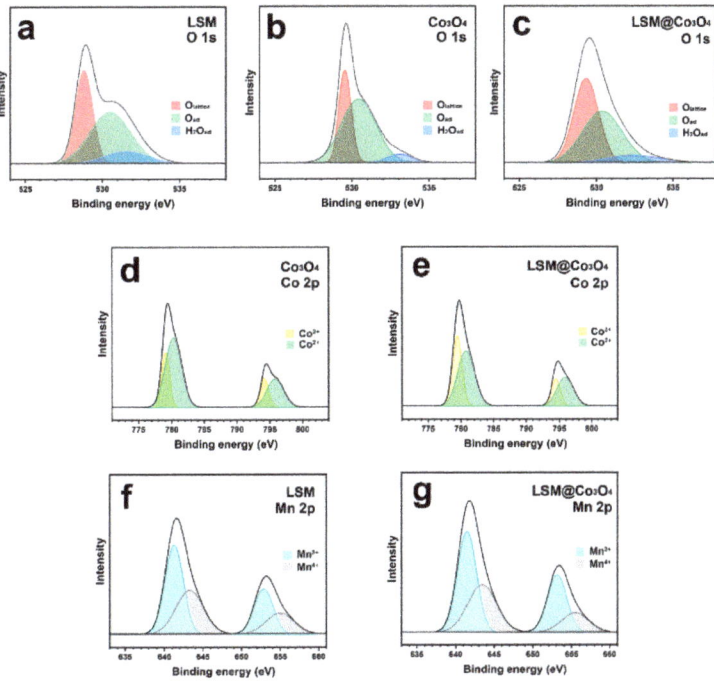

Figure 5. (a–c) O 1s, (d,e) Co 2p, and (f,g) Mn 2p XPS peaks of, respectively, LSM, Co_3O_4, and LSM@Co_3O_4-20.

Table 3. XPS peak deconvolution results of LSM, CO_3O_4, and LSM@Co_3O_4-20.

Sample		Species		BE (eV)	Ratio (Co^{3+}/Co^{2+})	Ratio (Mn^{3+}/Mn^{4+})	Ratio ($O_{ad}/O_{lattice}$)
LSM	Mn	$2p_{3/2}$	Mn^{3+}	641.49		0.83	
			Mn^{4+}	643.36			
		$2p_{1/2}$	Mn^{3+}	653.26			
			Mn^{4+}	656.21			
	O 1s		$O_{lattice}$	528.77			1.68
			O_{ad}	528.98			
			H_2O_{ad}	530.96			
		C 1s		284.3			
Co_3O_4	Co	$2p_{3/2}$	Co^{3+}	779.32	0.45		
			Co^{2+}	780.56			
		$2p_{1/2}$	Co^{3+}	794.34			
			Co^{2+}	795.77			
	O 1s		$O_{lattice}$	529.57			2.02
			O_{ad}	530.48			
			H_2O_{ad}	533.08			
		C 1s		284.3			
LSM@Co_3O_4-20	Co	$2p_{3/2}$	Co^{3+}	779.53	0.68		
			Co^{2+}	780.84			
		$2p_{1/2}$	Co^{3+}	794.58			
			Co^{2+}	796.00			
	Mn	$2p_{3/2}$	Mn^{3+}	641.31		0.77	
			Mn^{4+}	643.34			
		$2p_{1/2}$	Mn^{3+}	652.93			
			Mn^{4+}	655.19			
	O 1s		$O_{lattice}$	529.33			2.39
			O_{ad}	530.40			
			H_2O_{ad}	532.44			
		C 1s		284.3			

3. Conclusions

In this study, we presented a hybrid catalyst as a bifunctional catalyst for both ORR and OER. To synthesize the hybrid catalyst, 18 nm Co_3O_4 nanoparticles were uniformly deposited over the entire surface of a ~70 nm LSM nano-scaffold through simple electrospray and infiltration. The hybrid catalysts exhibited comparable OER and ORR performances compared to the benchmarked commercial catalysts. In particular, the hybrid catalysts followed a four-electron pathway, meaning a very effective path to the ORR. Due to differences in the electronic configurations of the ions in Co_3O_4 and LSM, the concentration of Co^{3+} and electron-rich phases were increased in the hybrid catalyst, contributing to the enhancement in the catalytic activity of ORR and OER.

4. Materials and Methods

$La_{0.5}Sr_{0.5}MnO_{3-\delta}$ nanoparticles were synthesized via electrospray using a precursor solution consisting of $La(NO_3)_3 \cdot 6H_2O$ (99+%, Sigma Aldrich Co.), $Sr(NO_3)_2$ (99+%, Sigma Aldrich Co.), $Mn(NO_3)_2$ (98%, Sigma Aldrich Co.), poly-vinylpyrrolidone (PVP, avg. molecular weight ~1,300,000 by LS, Sigma-Aldrich Co.), and N,N-dimethylformamide (DMF, Alfa Aesar Co.). The precursor solution was prepared by mixing the metal nitrates in DMF solvent with continuous stirring. After adding the PVP to the resulting solution, it was stirred for 12 h at 60 °C. The as-prepared LSM/PVP solution was filled into a plastic syringe with a capillary tip (D = 0.5 mm), electrically connected to a high-voltage power supply from Korea switching Co. The needle tip and aluminum collector were connected to the anode and cathode of the power supply, respectively, and were placed horizontally (Figure 2a). The applied voltage was 15 kV, and the distance between the tip and aluminum foil collector was 14 cm. The collected precursor was peeled off and heated at 850 °C for 4 h at a rate of 2 °C min^{-1} in air.

For the infiltration process, the precursor solution was prepared by dissolving $Co(NO_3)_2 \cdot 6H_2O$ (98+ %, Sigma Aldrich Co.) in ethanol, which was penetrated into the LSM powder (prepared via electrospray). The resultant product was calcinated at 450 °C for 15 min in air. These penetration–calcination steps were repeated until the loading amount reached 20 and 25 wt.%. The loading amounts of Co_3O_4 was chosen with the consideration of interfacial effects, investigated in a previous study [12]. The samples with 20 and 25 wt.% Co_3O_4 are designated hereafter as LSM@Co_3O_4-20 and LSM@Co_3O_4-25, respectively. The infiltrated LSM was sintered at 600 °C for 4 h at a ramping rate of 2 °C min^{-1} in ambient air to obtain Co_3O_4. The loading amount was calculated using Equation (3):

$$\text{Loading amount of } Co_3O_4 \text{ (wt.\%)} = (W_{LSM@Co3O4} - W_{LSM})/W_{LSM} \times 100 \quad (3)$$

where W_{LSM} is the weight of the LSM before infiltration and $W_{LSM@Co3O4}$ is the weight of LSM@Co_3O_4 after sintering at 600 °C.

X-ray diffraction (XRD, Bruker D8 Advance) was performed in the range of $20° < 2\theta < 60°$ for structural analyses. Scanning electron microscopy (SEM) (Nova SEM) was conducted to observe the microstructures of LSM and Co_3O_4@LSM. X-ray photoelectron spectroscopy (XPS) data were acquired with ESCALAB 250XI (Thermo Fisher Scientific) with a monochromated Al-Kα (ultraviolet He1, He2) X-ray source. To evaluate the electrocatalytic activity for ORR and OER, rotating ring-disk electrode (RRDE) experiments were performed with the prepared ink composed of 20 mg of sample, 0.9 mL of solvent (ethanol: isopropyl alcohol = 1:1), and 0.1 mL of 5 wt.% Nafion solution (Aldrich, 274704). The well-dispersed catalyst ink (5 mL) was spread on the pre-polished glassy carbon (GC) disk electrode (0.1256 cm^2). The electrochemical performances with an RRDE-3A rotating disk electrode system were recorded with a Biologic VMP3. A Pt wire, Hg/HgO electrode filled with 1 M NaOH, and 0.1 M KOH solution saturated with oxygen were respectively used as the counter electrode, reference electrode, and electrolyte. To estimate the amount of peroxide from the disk electrode during the ORR, +0.4 V was applied to the ring electrode.

To convert the potential values from vs. Hg/HgO to vs. the reversible hydrogen electrode (RHE), the difference was measured in a cell using Pt wires for the working and counter electrodes under H_2-saturated electrolyte 0.1 M KOH with an Hg/HgO reference electrode. The open-circuit potential was -0.89 V at 1 mV s^{-1} and the relationship between Hg/HgO and RHE was represented as:

$$E_{Hg/HgO} + 0.89 \text{ V} = E_{RHE}. \tag{4}$$

Author Contributions: S.K.: Conceptualization, Methodology, Data curation, Visualization, Investigation, Writing. G.K.: Conceptualization, Supervision, Reviewing and Editing. A.M.: Supervision, Reviewing and Editing, Funding acquisition. All authors have read and agreed to the published version of the manuscript.

Funding: Support from the Welch Foundation, grant F-1254, is gratefully acknowledged.

Institutional Review Board Statement: Not applicable.

Informed Consent Statement: Not applicable.

Data Availability Statement: The data presented in this study are available in manuscript.

Conflicts of Interest: The authors declare no conflict of interest.

References

1. Goodenough, J.B.; Kim, Y. Challenges for rechargeable batteries. *J. Power Sources* **2011**, *196*, 6688–6694. [CrossRef]
2. Fabbri, E.; Bi, L.; Pergolesi, D.; Traversa, E. Towards the next generation of solid oxide fuel cells operating below 600 °C with chemically stable proton-conducting electrolytes. *Adv. Mater.* **2012**, *24*, 195–208. [CrossRef]
3. Manthiram, A. A reflection on lithium-ion battery cathode chemistry. *Nat. Commun.* **2020**, *11*, 1550. [CrossRef] [PubMed]
4. Girishkumar, G.; McCloskey, B.; Luntz, A.C.; Swanson, S.; Wilcke, W. Lithium-air battery: Promise and challenges. *J. Phys. Chem. Lett.* **2010**, *1*, 2193–2203. [CrossRef]
5. Lee, J.S.; Kim, S.T.; Cao, R.; Choi, N.S.; Liu, M.; Lee, K.T.; Cho, J. Metal-air batteries with high energy density: Li-air versus Zn-air. *Adv. Energy Mater.* **2011**, *1*, 34–50. [CrossRef]
6. Cheng, F.; Chen, J. Metal–air batteries: From oxygen reduction electrochemistry to cathode catalysts. *Chem. Soc. Rev.* **2012**, *41*, 2172. [CrossRef] [PubMed]
7. Ju, Y.-W.W.; Yoo, S.; Kim, C.; Kim, S.; Jeon, I.-Y.Y.; Shin, J.; Baek, J.-B.B.; Kim, G. Fe@N-Graphene Nanoplatelet-Embedded Carbon Nanofibers as Efficient Electrocatalysts for Oxygen Reduction Reaction. *Adv. Sci.* **2015**, *3*, 1500205. [CrossRef]
8. Li, Y.; Gong, M.; Liang, Y.; Feng, J.; Kim, J.E.; Wang, H.; Hong, G.; Zhang, B.; Dai, H. Advanced zinc-air batteries based on high-performance hybrid electrocatalysts. *Nat. Commun.* **2013**, *4*, 1805. [CrossRef]
9. Agarwal, S.; Yu, X.; Manthiram, A. A pair of metal organic framework (MOF)-derived oxygen reduction reaction (ORR) and oxygen evolution reaction (OER) catalysts for zinc-air batteries. *Mater. Today Energy* **2020**, *16*, 100405. [CrossRef]
10. Jung, J.-I.; Park, S.; Kim, M.-G.; Cho, J. Tunable Internal and Surface Structures of the Bifunctional Oxygen Perovskite Catalysts. *Adv. Energy Mater.* **2015**, *5*, 1501560. [CrossRef]
11. Zhu, Y.; Zhou, W.; Yu, J.; Chen, Y.; Liu, M.; Shao, Z. Enhancing Electrocatalytic Activity of Perovskite Oxides by Tuning Cation Deficiency for Oxygen Reduction and Evolution Reactions. *Chem. Mater.* **2016**, *28*, 1691–1697. [CrossRef]
12. Kim, S.; Kwon, O.; Kim, C.; Gwon, O.; Jeong, H.Y.; Kim, K.-H.; Shin, J.; Kim, G. Strategy for Enhancing Interfacial Effect of Bifunctional Electrocatalyst: Infiltration of Cobalt Nanooxide on Perovskite. *Adv. Mater. Interfaces* **2018**, *5*, 1800123. [CrossRef]
13. Lu, Q.; Guo, Y.; Mao, P.; Liao, K.; Zou, X.; Dai, J.; Tan, P.; Ran, R.; Zhou, W.; Ni, M.; et al. Rich atomic interfaces between sub-1 nm RuOx clusters and porous Co_3O_4 nanosheets boost oxygen electrocatalysis bifunctionality for advanced Zn-air batteries. *Energy Storage Mater.* **2020**, *32*, 20–29. [CrossRef]
14. Fujiwara, N.; Nagai, T.; Ioroi, T.; Arai, H.; Ogumi, Z. Bifunctional electrocatalysts of lanthanum-based perovskite oxide with Sb-doped SnO_2 for oxygen reduction and evolution reactions. *J. Power Sources* **2020**, *451*, 227736. [CrossRef]
15. Sadighi, Z.; Huang, J.; Qin, L.; Yao, S.; Cui, J.; Kim, J.K. Positive role of oxygen vacancy in electrochemical performance of $CoMn_2O_4$ cathodes for Li-O_2 batteries. *J. Power Sources* **2017**, *365*, 134–147. [CrossRef]
16. Suntivich, J.; May, K.J.; Gasteiger, H.A.; Goodenough, J.B.; Shao-horn, Y. A Perovskite Oxide Optimized for Molecular Orbital Principles. *Science* **2011**, *334*, 1383–1385. [CrossRef] [PubMed]
17. Liang, Y.; Li, Y.; Wang, H.; Zhou, J.; Wang, J.; Regier, T.; Dai, H. Co_3O_4 Nanocrystals on Graphene as a Synergistic Catalyst for Oxygen Reduction Reaction. *Nat. Mater.* **2011**, *10*, 780–786. [CrossRef]
18. Kim, S.W.; Kim, H.; Yoon, K.J.; Lee, J.H.; Kim, B.K.; Choi, W.; Lee, J.H.; Hong, J. Reactions and mass transport in high temperature co-electrolysis of steam/CO_2 mixtures for syngas production. *J. Power Sources* **2015**, *280*, 630–639. [CrossRef]

19. Bu, Y.; Gwon, O.; Nam, G.; Jang, H.; Kim, S.; Zhong, Q.; Cho, J.; Kim, G. A Highly Efficient and Robust Cation Ordered Perovskite Oxides as a Bi-Functional Catalyst for Rechargeable Zinc-Air Batteries. *ACS Nano* **2017**, *11*, 11594–11601. [CrossRef]
20. Behnken, J.; Yu, M.; Deng, X.; Tüysüz, H.; Harms, C.; Dyck, A.; Wittstock, G. Oxygen Reduction Reaction Activity of Mesostructured Cobalt-Based Metal Oxides Studied with the Cavity-Microelectrode Technique. *ChemElectroChem* **2019**, *6*, 3460–3467. [CrossRef]
21. Hong, W.T.; Risch, M.; Stoerzinger, K.A.; Grimaud, A.; Suntivich, J.; Shao-Horn, Y. Toward the rational design of non-precious transition metal oxides for oxygen electrocatalysis. *Energy Environ. Sci.* **2015**, *8*, 1404–1427. [CrossRef]
22. Hu, J.; Shi, L.; Liu, Q.; Huang, H.; Jiao, T. Improved oxygen reduction activity on silver-modified LaMnO3-graphene via shortens the conduction path of adsorbed oxygen. *RSC Adv.* **2015**, *5*, 92096–92106. [CrossRef]
23. Wang, W.; Geng, J.; Kuai, L.; Li, M.; Geng, B. Porous Mn_2O_3: A Low-Cost Electrocatalyst for Oxygen Reduction Reaction in Alkaline Media with Comparable Activity to Pt/C. *Chem. A Eur. J.* **2016**, *22*, 9909–9913. [CrossRef] [PubMed]
24. Lambert, T.N.; Vigil, J.A.; White, S.E.; Delker, C.J.; Davis, D.J.; Kelly, M.; Brumbach, M.T.; Rodriguez, M.A.; Swartzentruber, B.S. Understanding the Effects of Cationic Dopants on α-MnO_2 Oxygen Reduction Reaction Electrocatalysis. *J. Phys. Chem. C* **2017**, *121*, 2789–2797. [CrossRef]
25. Chowdhury, A.D.; Agnihotri, N.; Sen, P.; De, A. Conducting $CoMn_2O_4$—PEDOT nanocomposites as catalyst in oxygen reduction reaction. *Electrochim. Acta* **2014**, *118*, 81–87. [CrossRef]
26. Dai, L.; Sun, Q.; Guo, J.; Cheng, J.; Xu, X.; Guo, H.; Li, D.; Chen, L.; Si, P.; Lou, J.; et al. Mesoporous Mn_2O_3 rods as a highly efficient catalyst for Li-O_2 battery. *J. Power Sources* **2019**, *435*, 226833. [CrossRef]
27. Sun, C.; Li, F.; Ma, C.; Wang, Y.; Ren, Y.; Yang, W.; Ma, Z.; Li, J.; Chen, Y.; Kim, Y.; et al. Graphene–Co_3O_4 nanocomposite as an efficient bifunctional catalyst for lithium–air batteries. *J. Mater. Chem. A* **2014**, *2*, 7188. [CrossRef]
28. Lee, D.U.; Park, H.W.; Park, M.G.; Ismayilov, V.; Chen, Z. Synergistic bifunctional catalyst design based on perovskite oxide nanoparticles and intertwined carbon nanotubes for rechargeable zinc-air battery applications. *ACS Appl. Mater. Interfaces* **2015**, *7*, 902–910. [CrossRef]
29. Zhao, X.; Li, F.; Wang, R.; Seo, J.M.; Choi, H.J.; Jung, S.M.; Mahmood, J.; Jeon, I.Y.; Baek, J.B. Controlled Fabrication of Hierarchically Structured Nitrogen-Doped Carbon Nanotubes as a Highly Active Bifunctional Oxygen Electrocatalyst. *Adv. Funct. Mater.* **2017**, *27*, 1–9. [CrossRef]
30. Yang, W.; Salim, J.; Ma, C.; Ma, Z.; Sun, C.; Li, J.; Chen, L.; Kim, Y. Flowerlike Co_3O_4 microspheres loaded with copper nanoparticle as an efficient bifunctional catalyst for lithium-air batteries. *Electrochem. Commun.* **2013**, *28*, 13–16. [CrossRef]
31. Chen, W.F.; Sasaki, K.; Ma, C.; Frenkel, A.I.; Marinkovic, N.; Muckerman, J.T.; Zhu, Y.; Adzic, R.R. Hydrogen-evolution catalysts based on non-noble metal nickel-molybdenum nitride nanosheets. *Angew. Chem. Int. Ed.* **2012**, *51*, 6131–6135. [CrossRef] [PubMed]
32. Zhang, H.; Qiao, H.; Wang, H.-Y.; Zhou, N.; Chen, J.; Tang, Y.-G.; Li, J.; Huang, C. Nickel Cobalt Oxide/carbon Nanotubes Hybrid as a High-performance Electrocatalyst for Metal/air Battery. *Nanoscale* **2014**, *6*, 10235–10242. [CrossRef]
33. Augustyn, V.; Simon, P.; Dunn, B. Pseudocapacitive oxide materials for high-rate electrochemical energy storage. *Energy Environ. Sci.* **2014**, *7*, 1597–1614. [CrossRef]
34. Balaya, P. Size effects and nanostructured materials for energy applications. *Energy Environ. Sci.* **2008**, *1*, 645–654. [CrossRef]
35. Wang, R.; Lang, J.; Liu, Y.; Lin, Z.; Yan, X. Ultra-small, size-controlled $Ni(OH)_2$ nanoparticles: Elucidating the relationship between particle size and electrochemical performance for advanced energy storage devices. *NPG Asia Mater.* **2015**, *7*, e183. [CrossRef]

Article

Syntheses and Characterization of Novel Perovskite-Type LaScO$_3$-Based Lithium Ionic Conductors

Guowei Zhao [1], Kota Suzuki [2,3], Masaaki Hirayama [2] and Ryoji Kanno [1,2,*]

[1] All-Solid-State Battery Unit, Institute of Innovation Research, Tokyo Institute of Technology, 4259 Nagatsuta, Midori-ku, Yokohama 226-8502, Japan; zhao.g.w@echem.titech.ac.jp
[2] Department of Chemical Science and Engineering, School of Materials and Chemical Technology, Tokyo Institute of Technology, 4259 Nagatsuta, Midori-ku, Yokohama 226-8502, Japan; suzuki.k.bf@m.titech.ac.jp (K.S.); hirayama@echem.titech.ac.jp (M.H.)
[3] Precursory Research for Embryonic Science and Technology (PRESTO), Japan Science and Technology Agency (JST), 4-1-8 Honcho, Kawaguchi-shi, Saitama 332-0012, Japan
* Correspondence: kanno@echem.titech.ac.jp; Tel.: +81-45-924-5401

Citation: Zhao, G.; Suzuki, K.; Hirayama, M.; Kanno, R. Syntheses and Characterization of Novel Perovskite-Type LaScO$_3$-Based Lithium Ionic Conductors. *Molecules* 2021, 26, 299. https://doi.org/10.3390/molecules26020299

Academic Editors: Stephane Jobic, Claude Delmas and Myung-Hwan Whangbo

Received: 14 December 2020
Accepted: 6 January 2021
Published: 8 January 2021

Publisher's Note: MDPI stays neutral with regard to jurisdictional claims in published maps and institutional affiliations.

Copyright: © 2021 by the authors. Licensee MDPI, Basel, Switzerland. This article is an open access article distributed under the terms and conditions of the Creative Commons Attribution (CC BY) license (https://creativecommons.org/licenses/by/4.0/).

Abstract: Perovskite-type lithium ionic conductors were explored in the (Li$_x$La$_{1-x/3}$)ScO$_3$ system following their syntheses via a high-pressure solid-state reaction. Phase identification indicated that a solid solution with a perovskite-type structure was formed in the range $0 \leq x < 0.6$. When $x = 0.45$, (Li$_{0.45}$La$_{0.85}$)ScO$_3$ exhibited the highest ionic conductivity and a low activation energy. Increasing the loading of lithium as an ionic diffusion carrier expanded the unit cell volume and contributed to the higher ionic conductivity and lower activation energy. Cations with higher oxidation numbers were introduced into the A/B sites to improve the ionic conductivity. Ce^{4+} and Zr^{4+} or Nb^{5+} dopants partially substituted the A-site (La/Li) and B-site Sc, respectively. Although B-site doping produced a lower ionic conductivity, A-site Ce^{4+} doping improved the conductive properties. A perovskite-type single phase was obtained for (Li$_{0.45}$La$_{0.78}$Ce$_{0.05}$)ScO$_3$ upon Ce^{4+} doping, providing a higher ionic conductivity than (Li$_{0.45}$La$_{0.85}$)ScO$_3$. Compositional analysis and crystal-structure refinement of (Li$_{0.45}$La$_{0.85}$)ScO$_3$ and (Li$_{0.45}$La$_{0.78}$Ce$_{0.05}$)ScO$_3$ revealed increased lithium contents and expansion of the unit cell upon Ce^{4+} co-doping. The highest ionic conductivity of 1.1×10^{-3} S cm^{-1} at 623 K was confirmed for (Li$_{0.4}$Ce$_{0.15}$La$_{0.67}$)ScO$_3$, which is more than one order of magnitude higher than that of the (Li$_x$La$_{1-x/3}$)ScO$_3$ system.

Keywords: lithium ionic conductor; perovskite structure; solid electrolyte; oxide

1. Introduction

All-solid-state lithium batteries have recently received considerable attention as safer, stable, compact, and reliable energy storage devices [1,2]. However, suitable lithium-based solid electrolytes are required for their fabrication [3–6]. In the past few decades, a range of crystalline and amorphous materials have been examined to prepare solid electrolytes [7–9], which can be roughly divided into two groups, namely sulfides [1,10–12] and oxides [13–16]. However, many of the sulfide-based compounds are unstable under air and react with the Li metal electrode [17]. In contrast, the oxide-based group is more stable and easily prepared. As a result, the latter is relatively convenient for battery fabrication in a dry air atmosphere [6].

In the oxide group, Li-La-Ti-O perovskite [13,18–21] is considered a promising candidate for use as a battery electrolyte due to its high ionic conductivity at room temperature ($>10^{-3}$ S cm^{-1}) [13]. However, high temperatures (>1000 °C) are required for its preparation and it suffers from electrochemical stability issues due to the reduction of the reducible element Ti^{4+} to Ti^{3+} during the electrochemical process, which gives rise to undesirable electronic conduction properties and ultimately affects its practical application [20]. Although lithium ionic conductors developed in this perovskite system have mainly been prepared based on the introduction of lithium vacancies at the *A*-site [13,22], the development of lithium ionic conductors through the introduction of lithium interstitials in the perovskite-type system has generally not been considered.

We herein focus on $LaScO_3$ [23] as a mother structure for solid electrolytes to investigate the novel perovskite system. We previously considered that the *B* site cation of Sc^{3+} would be expected to show a higher resistance to electrochemical reduction than Ti^{4+} [24]. In addition, it was thought that the low valent Li^+ cation could partially substitute the high valent La^{3+} cation, which could result in the generation of structural defects due to charge compensation, with examples including lithium-ion interstitials that may act as ion carriers for ion diffusion. Therefore, we herein report the development of lithium ionic conductors in the $LaScO_3$-based perovskite system. Initially, different ratios of lithium are introduced into $LaScO_3$ according to the chemical formula ($Li_xLa_{1-x/3}$)ScO_3, and their ionic conductivities are evaluated. To improve the ionic conductivity of ($Li_xLa_{1-x/3}$)ScO_3, further modification is carried out by introducing additional structural defects. As the dopant, Ce^{4+} and Zr^{4+} or Nb^{5+} are selected to partially substitute the *A*- or *B*- sites of ($Li_xLa_{1-x/3}$)ScO_3, and perovskite-type lithium-excess lithium ionic conductors are developed in a step-by-step manner. Finally, the composition exhibiting the highest conductivity is identified in the prepared $LaScO_3$ perovskite-based materials.

2. Results and Discussion

2.1. Syntheses and Ionic Conductivities of ($Li_xLa_{1-x/3}$)ScO_3

The X-ray diffraction (XRD) patterns of ($Li_xLa_{1-x/3}$)ScO_3 (x = 0–0.60) are shown in Figure 1a,b. The diffraction peaks of the main phase were identified to correspond to the orthorhombic *Pnma* $LaScO_3$ phase [25]. For the composition where x = 0.6, additional peaks attributed to the impurity ($LiScO_2$) phase were also observed, indicating that the solid solution limit with the $LaScO_3$ structure is $x < 0.6$. The peak corresponding to the 111 diffraction peak at ~24.5° did not show a clear shift upon increasing the x value (i.e., the Li^+ content), although the lattice volume increased slightly (Figure 1c). This indicates that a solid solution formed through lithium doping in the x range of 0.15–0.45. The absence of a significant diffraction peak shifts may be attributed to the introduced lithium ions located at the interstitial positions, which are not sufficient to affect the lattice change. The total content of La and Li at *A*-site exceeds 1 when $x > 0$, which indicates that a portion of the lithium ions may be located at the interstitial sites in addition to the *A*-sites within the structure of $LaScO_3$. These results clearly indicate that lithium-excess perovskite-type materials were successfully formed via this high-pressure synthesis route.

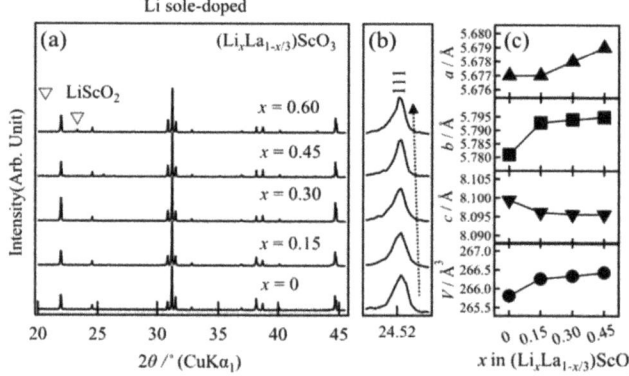

Figure 1. (**a**,**b**) X-ray diffraction patterns and (**c**) dependence of the lattice parameters on the composition of ($Li_xLa_{1-x/3}$)ScO_3 (x = 0, 0.15, 0.30, 0.45). Lattice parameter data for $LaScO_3$ (x = 0) were taken from the literature [25].

To determine the ionic conductivities of the samples, alternating current (a.c.) impedance measurements were conducted. The total resistance, R, was determined by calculating the intercept value of the semicircular plot with the x-axis. The ionic conductivities were calculated using $\sigma = d/(R \times A)$, where R is the resistance, d is the thickness, and A is the area of the pellet. It should be noted that the a.c. impedance could not be measured for the samples at temperatures < 523 K. Therefore, to compare these ionic conductivities with the actual measured experimental data, the ionic conductivity obtained at the intermediate temperature (623 K) of the test temperature interval was chosen. Figure 2a,b show the typical complex impedance spectra of all sintered compounds at 623 K. These spectra are composed of a semicircle at higher frequencies and a spike at lower frequencies, which correspond to the contributions of the bulk/grain boundary and electrode resistances, respectively. A semicircle with a capacitance value on the order of 10^{-11} F indicates a total resistance that consists of a mixed contribution (sum of the bulk and grain boundaries). Since the component separation for the bulk and the grain boundary resistance could not be obtained, the total conductivity was calculated from the semicircle as a total resistance.

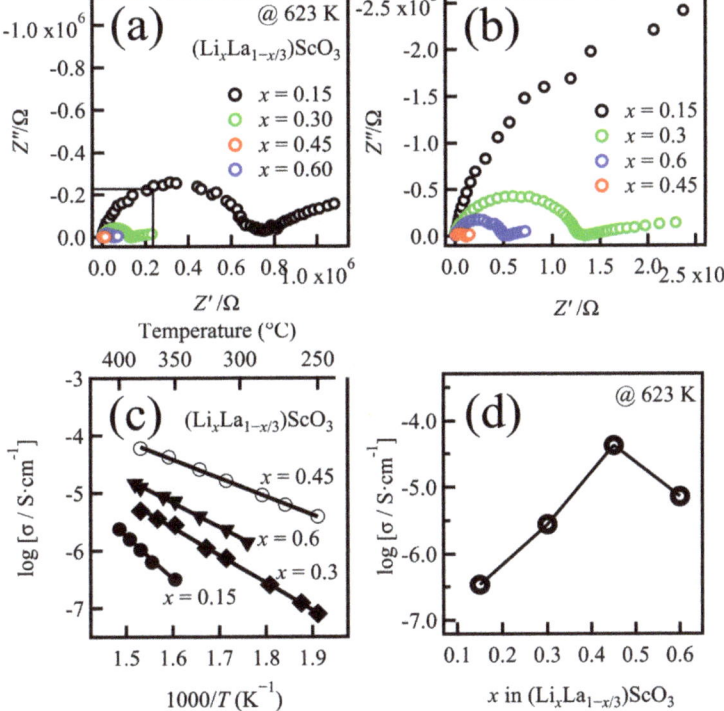

Figure 2. (a,b) Impedance plots obtained at 623 K, (c) temperature dependence of the conductivity, and (d) variation in the conductivity at 623 K based on the composition of $(Li_xLa_{1-x/3})ScO_3$ (x = 0.15, 0.30, 0.45, 0.60).

Figure 2c shows the temperature dependences of the conductivities, while Figure 2d displays the conductivity at 623 K as a function of the composition (x). Table 1 summarizes the ionic conductivities measured at 623 K and the activation energies E_a for the compounds. As indicated, the conductivity continuously increased upon increasing x from 0 to 0.45, and then slightly decreased where $x > 0.45$. The highest conductivity was observed for the composition with x = 0.45, yielding a value of 4.2×10^{-5} S cm^{-1} at 623 K with a low activation energy of 61 kJ mol^{-1}. The decrease in conductivity for the sample where

$x = 0.60$ was likely due to the formation of the impurity phase $LiScO_2$, which exhibits a relatively low ionic conductivity at the same temperature ($\sim 10^{-9}$ S cm^{-1} at 623 K) [26]. As the sample where $x = 0.45$ exhibits the highest ionic conductivity, $(Li_{0.45}La_{0.85})ScO_3$ was determined to be the optimal initial composition for examining the co-doping systems containing Ce^{4+}, Zr^{4+}, and Nb^{5+}.

Table 1. Ionic conductivities at 623 K and activation energies of the obtained $(Li_xLa_{1-x/3})ScO_3$.

Formula	Nominal Composition	Identified Phases	σ_{total}/S cm^{-1}	E_a/kJ mol^{-1}
$(Li_xLa_{1-x/3})ScO_3$	$LaScO_3$	$LaScO_3$-type	—	—
	$(Li_{0.15}La_{0.95})ScO_3$	$LaScO_3$-type	3.4×10^{-7}	143.0 ± 6.8
	$(Li_{0.30}La_{0.90})ScO_3$	$LaScO_3$-type	2.8×10^{-6}	91.6 ± 1.7
	$(Li_{0.45}La_{0.85})ScO_3$	$LaScO_3$-type	4.2×10^{-5}	61.0 ± 0.9
	$(Li_{0.60}La_{0.80})ScO_3$	$LaScO_3$-type, $LaLiO_2$	7.2×10^{-6}	80.7 ± 2.2

2.2. Syntheses and Ionic Conductivities of $(Li_xLa_{1-x/3})ScO_3$ co-doped with Ce^{4+}, Zr^{4+}, and Nb^{5+}

For the A-site doping system, two material search directions were examined. The first was La^{3+} substitution by Ce^{4+} with fixing of the lithium composition according to the formula $(Li_xLa_{0.87-4y/3}Ce_y)ScO_3$, where $x = 0.35$ or 0.40. The XRD patterns of the obtained samples are shown in Figure 3. Although the $LaScO_3$ phase accounted for the main diffraction peaks, an additional peak derived from the CeO_2 impurity was observed at $\sim 28°$. The intensity of this peak increased upon increasing the y value (i.e., the amount of Ce doping), while the lattice parameters increased slightly at the same time, as shown in Figure 3c. These results indicate that compositional and/or structural changes occurred depending on the y value, although the mono-phasic perovskite-type solid solution was not obtained.

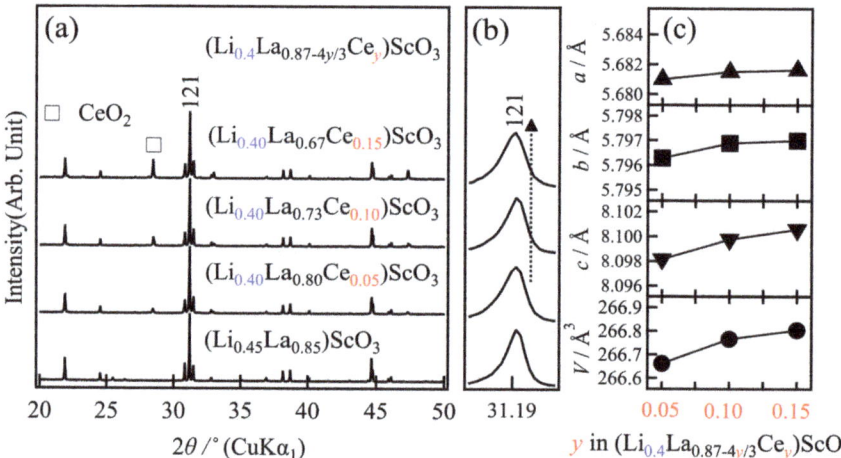

Figure 3. Variation in (**a**,**b**) the X-ray diffraction patterns, and (**c**) the lattice parameters based on the composition of the as-prepared $(Li_{0.4}La_{0.87-4y/3}Ce_y)ScO_3$ ($y = 0.05, 0.10, 0.15$).

The second direction of A-site doping was conducted using a fixed Ce content. In this case, charge neutrality was maintained using the La:Li ratio, giving $(Li_xLa_{0.933-x/3}Ce_y)ScO_3$, where $y = 0.05$ was the target composition. The XRD patterns of the obtained products are shown in Figure 4. The sample of $(Li_{0.45}La_{0.78}Ce_{0.05})ScO_3$ where $x = 0.45$ exhibited the mono-phasic characteristic of the $LaScO_3$ perovskite, while other compositions showed impurity formation. Even in this case, the small changes in the lattice parameters, which indicate compositional and/or structural changes, were confirmed (Figure 4c).

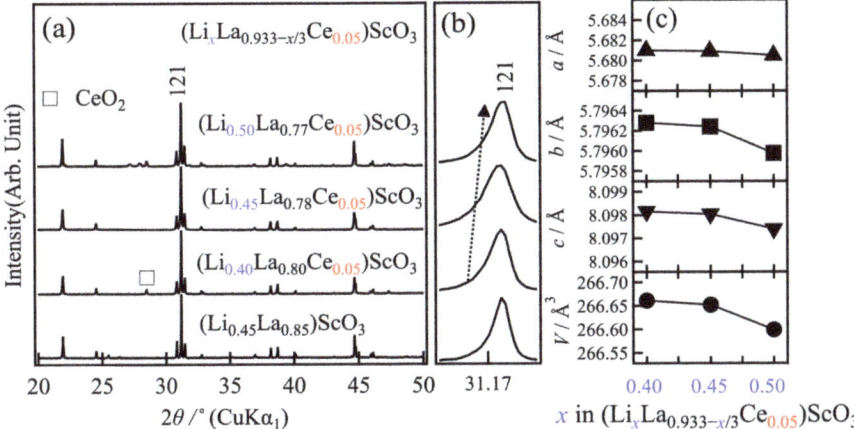

Figure 4. Variation in (**a**,**b**) the X-ray diffraction patterns, and (**c**) the lattice parameters based on the composition of the as-prepared (Li$_x$La$_{0.933-x/3}$Ce$_{0.05}$)ScO$_3$ (x = 0.4, 0.45, and 0.50).

B-site doping using Zr^{4+} or Nb^{5+} was subsequently examined according to the formulae (Li$_{0.45}$La$_{0.85-z/3}$)(Sc$_{1-z}$Zr$_z$)O$_3$ and (Li$_{0.45}$La$_{0.85-2z/3}$)(Sc$_{1-z}$Nb$_z$)O$_3$. As shown in Figures S1 and S2, evident impurity (LiLaO$_2$) formation was confirmed even with a small amount of doping (z = 0.05) for both dopants. Therefore, B-site doping reduces the lithium content in the perovskite-phase due to part of the lithium source being consumed by the impurity. As indicated in Table 2, where the obtained phases of all doping systems are summarized, the synthetic conditions employed herein did not result in a wide range of single-phase solid solutions upon A- or B- site co-doping into the (Li, La)ScO$_3$ system. As a result, (Li$_{0.45}$La$_{0.78}$Ce$_{0.05}$)ScO$_3$ was found to be the singular example of a pure LaScO$_3$-type phase considering all examined compositions.

Table 2. Dependence of the ionic conductivities at 623 K and the activation energies on the Ce^{4+}, Zr^{4+}, and Nb^{5+} co-doped (Li$_x$La$_{1-x/3}$)ScO$_3$ compositions.

Chemical Formula	Composition	Identified Phases	σ_{total}/S cm^{-1}	E_a/kJ mol^{-1}
(Li$_x$La$_{1-x/3}$)ScO$_3$	(Li$_{0.45}$La$_{0.85}$)ScO$_3$	LaScO$_3$-type	4.2 × 10^{-5}	61.0 ± 0.9
(Li$_x$La$_{1-x/3-4y/3}$Ce$_y$)ScO$_3$	x = 0.35 (fixed), y = 0.05 (Li$_{0.35}$La$_{0.82}$Ce$_{0.05}$)ScO$_3$	LaScO$_3$-type, CeO$_2$, unknown	2.1 × 10^{-5}	88.9 ± 0.7
	x = 0.35 (fixed), y = 0.10 (Li$_{0.35}$La$_{0.75}$Ce$_{0.10}$)ScO$_3$	LaScO$_3$-type, CeO$_2$	2.4 × 10^{-5}	87.9 ± 0.3
	x = 0.40 (fixed), y = 0.10 (Li$_{0.40}$La$_{0.73}$Ce$_{0.10}$)ScO$_3$	LaScO$_3$-type, CeO$_2$	8.8 × 10^{-4}	75.5 ± 0.6
	x = 0.40 (fixed), y = 0.15 (Li$_{0.40}$La$_{0.67}$Ce$_{0.15}$)ScO$_3$	LaScO$_3$-type, CeO$_2$	1.1 × 10^{-3}	75.4 ± 0.1
	x = 0.40, y = 0.05 (fixed) (Li$_{0.50}$La$_{0.77}$Ce$_{0.05}$)ScO$_3$	LaScO$_3$-type, CeO$_2$	5.5 × 10^{-5}	78.2 ± 2.5
	x = 0.45, y = 0.05 (fixed) (Li$_{0.45}$La$_{0.78}$Ce$_{0.05}$)ScO$_3$	LaScO$_3$-type	1.9 × 10^{-4}	82.9 ± 2.1
	x = 0.50, y = 0.05 (fixed) (Li$_{0.50}$La$_{0.72}$Ce$_{0.10}$)ScO$_3$	LaScO$_3$-type, CeO$_2$, unknown	1.3 × 10^{-5}	86.5 ± 0.3

Table 2. Cont.

Chemical Formula	Composition	Identified Phases	σ_{total}/S cm^{-1}	E_a/kJ mol^{-1}
$(Li_{0.45}La_{0.85-0.33z})(Sc_{1-z}Zr_z)O_3$	$(Li_{0.45}La_{0.833})(Sc_{0.95}Zr_{0.05})O_3$	LaScO$_3$-type, LiLaO$_2$	7.36×10^{-6}	77.9 ± 2.0
	$(Li_{0.45}La_{0.817})(Sc_{0.90}Zr_{0.10})O_3$	LaScO$_3$-type, LiLaO$_2$	8.03×10^{-6}	75.6 ± 1.8
$(Li_{0.45}La_{0.85-0.67z})(Sc_{1-z}Nb_z)O_3$	$(Li_{0.45}La_{0.817})(Sc_{0.95}Nb_{0.05})O_3$	LaScO$_3$-type, LiLaO$_2$	3.67×10^{-6}	81.1 ± 0.9
	$(Li_{0.45}La_{0.783})(Sc_{0.90}Nb_{0.10})O_3$	LaScO$_3$-type, LiLaO$_2$	9.50×10^{-6}	82.0 ± 1.9

All the co-doped samples were then subjected to a.c. impedance measurements; Figure 5a shows the typical a.c. complex impedance spectra of the sample with the nominal composition $(Li_{0.40}La_{0.67}Ce_{0.15})ScO_3$ ($x = 0.4$, $y = 0.15$) recorded at a range of temperatures. The conductivities of the samples were calculated following the same manner as that employed for the non-doped $(Li_xLa_{1-x/3})ScO_3$ system. Thus, Figure 5b presents the Arrhenius plots showing the effect of temperature on the conductivity for the representative samples of $(Li_xLa_{1-x/3-4y/3}Ce_y)ScO_3$. Since it was not possible to measure the a.c. impedance for every sample at 373 K, the representative conductivities at a higher temperature of 623 K were examined for comparison; the representative conductivities at 623 K and the activation energies for the co-doped samples are summarized in Table 2.

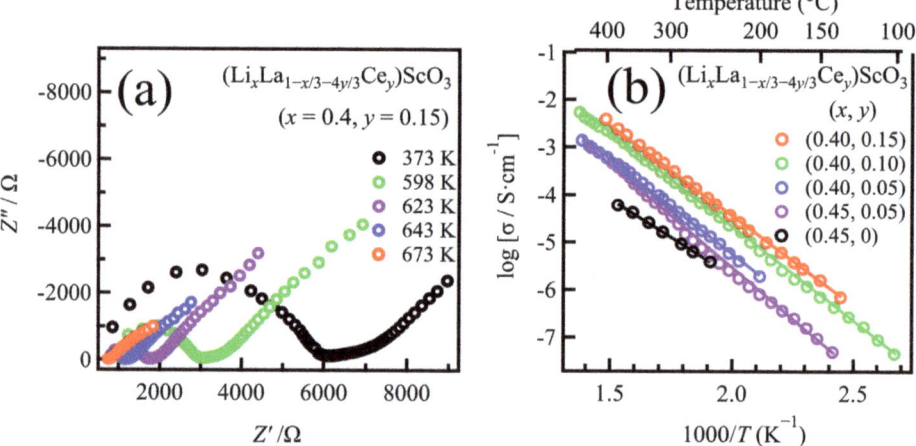

Figure 5. (a) Impedance plots of the representative co-doped $(Li_xLa_{1-x/3})ScO_3$ at various temperatures, and (b) temperature dependence of the conductivity of the representative samples in $(Li_xLa_{1-x/3-4y/3}Ce_y)ScO_3$.

For the Ce^{4+} co-doped system, a number of compositions were found to exhibit higher conductivities than the $(Li_xLa_{1-x/3})ScO_3$ system. For example, $(Li_{0.45}La_{0.78}Ce_{0.05})ScO_3$, which shows single-phase characteristics, presented a relatively high ionic conductivity of 1.9×10^{-4} S cm^{-1} at 623 K. This value is nearly five times higher than that of $(Li_{0.45}La_{0.85})ScO_3$ at the same temperature. The highest ionic conductivity (i.e., 1.1×10^{-3} S cm^{-1} at 623 K) was obtained for the Ce^{4+} co-doped $(Li_{0.4}La_{0.67}Ce_{0.15})ScO_3$, even though it contains an impurity. As noted above, a.c. impedance data could be obtained for all the Ce^{4+} co-doped samples at a temperature of 373 K, and below 523 K, it was also not possible to obtain these data for $(Li_xLa_{1-x/3})ScO_3$, which demonstrates that these Ce^{4+} co-doped samples have a higher ionic conductivity than that of $(Li_xLa_{1-x/3})ScO_3$. In contrast, the activation energy was increased upon Ce^{4+} co-doping (e.g., 75.2 kJ mol^{-1} for $(Li_{0.4}La_{0.67}Ce_{0.15})ScO_3$). The enhancement in the ionic conductivity at the same temperature could, therefore, originate from the increase in the pre-exponential factor (σ_0) in the Arrhenius equation: $\sigma T = \sigma_0 \exp(-E_a/kT)$ [5,27]. Since σ_0 is related to the intrinsic number of carriers and defects, Ce^{4+} co-doping may increase the number of charge carriers, which in turn contributes to ion migration.

The temperature dependences of the conductivities for the $(Li_{0.45}La_{0.85-z/3})(Sc_{1-z}Zr_z)O_3$, $(Li_{0.45}La_{0.83-1.33n})(Sc_{0.95}Zr_{0.05})O_3$, and $(Li_{0.45}La_{0.85-2z/3})(Sc_{1-z}Nb_z)O_3$ samples are shown in Figure S3 (Supporting Information). More specifically, no improvement in the conductivity was observed for the Zr^{4+} and Nb^{5+} co-doped systems, and relatively high activation energies were confirmed. The lack of an increase in the ionic conductivity may be due to impurities that consume part of the lithium source in the co-doped samples, thereby decreasing the content of mobile lithium ions. The increase observed for the $(Li_xLa_{1-x/3})ScO_3$ system was achieved by A-site Ce^{4+} co-doping, while the presence of M at the B-site was found to have no effect on the conductivity.

To investigate the electronic contribution to the total conductivity, direct current (d.c.) potentiostatic polarization measurements were performed for $(Li_{0.4}La_{0.67}Ce_{0.15})ScO_3$, which exhibited the highest conductivity of the examined samples. The steady-state current at specific constant voltages of the symmetrical cell Au | $(Li_{0.4}La_{0.67}Ce_{0.15})ScO_3$ | Au was evaluated by varying the applied voltage from 0.5 to 1.2 V at 623 K (Figure S4); a rapid current decay was observed shortly following a voltage application. From the slope of the I vs. V plot, the electronic conductivity of $(Li_{0.4}La_{0.67}Ce_{0.15})ScO_3$ was determined to be 7.6×10^{-7} S cm^{-1}, indicating that electron transport is negligible in $(Li_{0.4}La_{0.67}Ce_{0.15})ScO_3$ because the calculated value is three orders of magnitude lower than the total conductivity estimated by the a.c. impedance measurements. The conductivity observed for the Ce^{4+}-doped $(Li_{0.4}La_{0.67}Ce_{0.15})ScO_3$ was therefore attributed mainly to lithium ion diffusion.

These results therefore indicated that a small amount of Ce^{4+} co-doping enhanced the ionic conductivity. However, the critical factors involved in determining the observed high ionic conductivities remained unclear, and so crystal structure analysis and chemical composition analysis were subsequently carried out.

2.3. Crystal Structure Analysis

$(Li_{0.45}La_{0.85})ScO_3$ ($x = 0.45$) and $(Li_{0.45}La_{0.78}Ce_{0.05})ScO_3$ ($x = 0.45$, $y = 0.05$) were taken as representative highly-conductive samples for the doped and co-doped systems, respectively. These samples were also found to show single-phase characteristics, as discussed previously in the context of XRD phase identification. The crystal structures of these samples were then determined by the Rietveld analysis of synchrotron X-ray diffraction data obtained at 298 K. An orthorhombic perovskite-type structural (LaScO$_3$) [25] with *Pnma* symmetry was applied as an initial model for analysis. The atomic positions for the structure refinement models were as follows: La at the 4c sites, Sc at the 4a sites, O at the 4a and 8d sites.

During the structural refinement, some Li$^+$ was assumed to be present in the A-sites (La^{3+}), and the occupancy factor of the normal-site lithium (A-site) was constrained as 1−g(La) for both samples. The lithium present at the interstitial sites was not refined since X-rays are not sensitive to the light element lithium. The lattice constants, atomic isotropic displacement parameters (B_{iso}), fractional atomic coordinates, atomic occupancies, background parameters, ratio, zero-shift, and profile parameters were then refined. It is worth mentioning that the occupancy of Ce^{4+} was not considered as a refinement parameter because it possesses the same number of electrons as La^{3+}, which prevents its identification by X-ray analysis. Therefore, the composition of La and Ce in $(Li_{0.45}La_{0.78}Ce_{0.05})ScO_3$ was estimated by the combination of the La amount obtained by the Rietveld refinement and the La/Ce ratio obtained from inductively coupled plasma atomic emission spectroscopy (ICP-AES) analysis (see Table S1, Supporting Information).

Figure 6 shows the Rietveld refinement patterns, which are also presented in Tables 3 and 4. All diffraction peaks of the two samples were assigned to the orthorhombic perovskite structure. The obtained reliability factors (S and R_{wp}) confirmed that the proposed structural model for the diffraction data was reasonable.

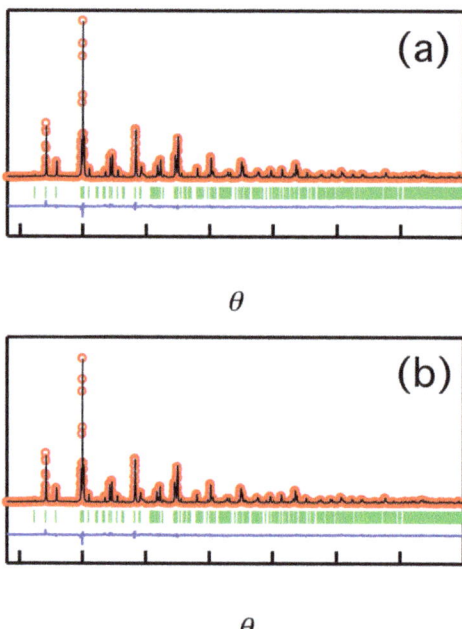

Figure 6. (Color online) Synchrotron Rietveld refinement patterns for $(Li_{0.45}La_{0.85})ScO_3$ (**a**) and $(Li_{0.45}La_{0.78}Ce_{0.05})ScO_3$ (**b**) at 298 K. Red: observed intensities; black: calculated intensities; and blue: difference plots. Green vertical markers denote the positions of the Bragg reflections of the space group *Pnma* (62) (perovskite structure).

Table 3. Crystallographic parameters of the synchrotron diffraction data for $(Li_{0.45}La_{0.85})ScO_3$ refined by Rietveld analysis.

Atom	Site	g	x	y	z	B_{iso}/Å^2
La	4c	0.938(2)	0.95637(10)	0.25	0.01062(13)	0.235(9)
Li	4c	=1−g(La)	=x (La)	=y (La)	=z (La)	=B (La)
Sc	4b	1.0	0.5	0	0	0.908(7)
O1	4c	1.0	0.53871(1)	0.25	0.90136(4)	0.788(6)
O2	8d	1.0	0.29065(7)	0.94553(6)	0.70253(7)	0.848 (3)

a = 5.67921(5) Å, b = 5.79488(4) Å and c = 8.09558(6) Å, V = 266.4281(36); R_{wp} = 8.638, S = 1.899, R_b = 6.080, R_f = 4.965.

Table 4. Crystallographic parameters of the synchrotron diffraction data for $(Li_{0.45}La_{0.78}Ce_{0.05})ScO_3$ refined by Rietveld analysis

Atom	Site	g	x	y	z	B_{iso}/ Å^2
La	4c	0.8212(5)	0.95622(4)	0.25	0.00892(6)	0.178(3)
Li	4c	=1−g(La)	=x (La)	=y (La)	=z (La)	=B (La)
Sc	4b	1.0	0.5	0	0	1.209(4)
O1	4c	1.0	0.53153(4)	0.25	0.90757(5)	0.699(3)
O2	8d	1.0	0.2951(6)	0.9419(3)	0.70075(3)	1.825(4)

a = 5.68088(6) Å, b = 5.79624(6) Å and c = 8.09803(9) Å, V = 266.6500(48); R_{wp} = 8.850, S = 2.315, R_b = 5.343, R_f = 3.752.

The refined composition of $(Li_{0.45}La_{0.85})ScO_3$ was calculated to be $Li_{0.062}La_{0.938}ScO_3$. Subsequently, the interstitial lithium, which cannot be detected by Rietveld analysis, was considered to satisfy the charge neutrality in the sample. Thus, the composition of $(Li_{0.45}La_{0.85})ScO_3$ was determined to be $Li_{0.186}La_{0.938}ScO_3$, and this value is comparable to that obtained by ICP-AES (Table S1). These analyses revealed that the practical amount of lithium is less than half of the nominal composition of $(Li_{0.45}La_{0.85})ScO_3$, which was attributed to the reaction of lithium reaction with the platinum capsule during the synthetic procedure [28].

The composition of the Ce^{4+} co-doped sample was also determined using the above process. From the Rietveld analysis, $Li_{0.1788}La_{0.8212}ScO_3$ was obtained, whereby the La composition was divided into La and Ce using the La/Ce ratio obtained from the ICP-AES results. Considering the charge neutrality, the composition of $Li_{0.493}La_{0.777}Ce_{0.044}ScO_3$ was determined for the co-doped $(Li_{0.45}La_{0.78}Ce_{0.05})ScO_3$. This value is comparable to both the nominal and ICP-AES results (Table S1).

The significant difference in the lithium amount between the doped and co-doped samples was clarified based on structural and compositional analysis results. More specifically, only 41% of the lithium present in the nominal composition was incorporated into the doped case sample.

In contrast, a nearly nominal composition was confirmed in the Ce^{4+} co-doped case. Therefore, the observed enhancement in the ionic conductivity upon co-doping was attributed mainly to the abundant lithium (carrier) number in the crystal structure despite a low doping level of 0.05. This assumption is consistent with the change in activation energy observed upon co-doping, which indicated an increase in the number of charge carriers.

The lattice parameters of the compound $(Li_{0.45}La_{0.85})ScO_3$ were determined to be $a = 5.67921(5)$, $b = 5.79488(4)$, $c = 8.09558(6)$ Å, and $V = 266.4281(36)$ Å^3, while those of $(Li_{0.45}La_{0.78}Ce_{0.05})ScO_3$ were $a = 5.68088(6)$ Å, $b = 5.79624(6)$ Å, $c = 8.09803(9)$ Å, and $V = 266.6500(48)$ Å^3. Compared with $LaScO_3$ [25] ($a = 5.6803$, $b = 5.7907$, $c = 8.0945$ Å, and $V = 266.252$ Å^3), the cell volumes were expanded by +0.07% and +0.15%, respectively. The co-doped $(Li_{0.45}La_{0.78}Ce_{0.05})ScO_3$ presented a larger unit cell size than $(Li_{0.45}La_{0.85})ScO_3$. Such an expansion of the cell volume may also contribute to the higher ionic conductivity of $(Li_{0.45}La_{0.78}Ce_{0.05})ScO_3$, because a larger unit cell size implies larger channels for lithium ion diffusion.

Again, the highest ionic conductivity (i.e., 1.1×10^{-3} S cm^{-1} at 623 K) for the Ce^{4+} co-doped $(Li_{0.4}La_{0.67}Ce_{0.15})ScO_3$ was almost two orders of magnitude higher than that of $(Li_{0.45}La_{0.85})ScO_3$ at the same temperature. The increase in the lithium content as an ionic diffusion carrier, in addition to the expansion of unit cell volume, was thereby considered to contribute synergistically to the higher ionic conductivity of the Ce^{4+} co-doped system. Determination of the interstitial lithium sites and the La/Ce distribution in the structure through neutron diffraction studies will be conducted in future studies.

3. Materials and Methods

Target materials with the general formula $(Li_xLa_{1-x/3})ScO_3$ and those containing Ce^{4+}, Zr^{4+}, and Nb^{5+} doping were synthesized via a solid-state reaction using the previously reported high-pressure method [28,29]. The starting materials were La_2O_3 (Kanto Chemical Co., Inc., Tokyo, Japan, $\geq$99.99% purity), Li_2O_2 (Kojundo Chemical Laboratory Co. Ltd., Sakado, Saitama Pre., Japan, 99% purity), Sc_2O_3 (Alfa Aesar, Thermo Fisher Scientific, Waltham, MA, USA, $\geq$99.9% purity), CeO_2 (Wako Pure Chemical Industries, Ltd., Osaka, Japan, $\geq$99.9% purity), ZrO_2 (Wako Pure Chemical Industries, Ltd., $\geq$99.9% purity), and Nb_2O_5 (Wako Pure Chemical Industries, Ltd., Osaka, Japan, $\geq$99.9% purity). Generally, the perovskite system can be synthesized at high temperatures (>1423 K) in an open ambient pressure atmosphere, which often results in lithium evaporation [22]. In this work, the mixed samples were encapsulated in a platinum capsule and subjected to high-pressure synthesis at 2 GPa and 1173–1373 K for 30 min, then quenched by cooling to room temperature. This high-pressure route ensured the incorporation of an accurate amount of Li, in addition to accelerating the reaction.

The synthesized products were characterized by XRD (Rigaku Smart Lab, and Miniflex 300, Rigaku, Tokyo, Japan) using Cu-$K\alpha$ radiation under an air atmosphere. Diffraction data were collected using a 0.01° step interval and scanning over the range of 10–50°. The lattice constants were determined using Si (SRM640d) as an internal standard for calibration. Synchrotron XRD patterns were recorded at 298 K using the BL02B2 and BL19B2 beamlines at the SPring-8 facility, operating with a wavelength of 0.5 Å. Structural parameters were refined by the Rietveld method using the RIETAN-FP program [30]. Rietveld analyses were conducted to refine the structural parameters using the diffraction data collected at each 0.01° step width over a 2θ range of 10–70°. Chemical analysis was performed by ICP-AES (ICPS-8100, Shimadzu, Tokyo, Japan).

The ionic conductivity of each sample was evaluated by a.c. impedance spectroscopy using a frequency response analyzer (Solartron 1260, AMETEK Scientific Instruments, Berwyn, PA, USA). A gold paste was painted onto each side of the sample as a blocking electrode. The pellets (~3 mm diameter, ~1 mm thickness) were then heated at 300 °C for 30 min under an argon atmosphere to obtain dry samples for carrying out the measurements. The samples were placed under an argon flow, and the measurements were performed between 298 and 673 K at frequencies ranging from 0.1 Hz to 3 MHz with an applied voltage of 20–100 mV.

The electrical conductivity of each sample was measured using the Hebb-Wagner polarization method [31]. The voltage, which was controlled using a potentiostat electrochemical interface (Solartron 1287, AMETEK Scientific Instruments, Berwyn, PA, USA), was applied from 0.1 to 1.2 V. The resulting current obtained due to the electronic contribution was evaluated.

4. Conclusions

Lithium ionic conductors were developed in the $(Li_xLa_{1-x/3})ScO_3$ system with a perovskite structure via a high-pressure solid-state reaction. The resulting novel lithium-excess perovskite-type materials exhibited solid solution formation where $0 \leq x < 0.60$. The highest ionic conductivity was observed for $(Li_{0.45}La_{0.85})ScO_3$, yielding a conductivity of 4.2×10^{-5} S cm^{-1} at 623 K with a low activation energy of 61 kJ mol^{-1}. Further improvement in the ionic conductivity of $(Li_{0.45}La_{0.85})ScO_3$ was examined by introducing aliovalent cations; Ce^{4+} and Zr^{4+} or Nb^{5+} were selected to substitute the A-site La^{3+} and B-site Sc^{3+}, respectively. Interestingly, improvements in the ionic conductivity of $(Li_xLa_{1-x/3})ScO_3$ were achieved only upon Ce^{4+} co-doping, with the presence of Zr^{4+} and Nb^{5+} having no such effect. A mono-phasic sample was obtained for the co-doped $(Li_{0.45}La_{0.78}Ce_{0.05})ScO_3$ system, which exhibited a higher ionic conductivity compared to $(Li_{0.45}La_{0.85})ScO_3$. Furthermore, an increase in the activation energy indicated that a rise in the pre-exponential factor (σ_0) might contribute to the enhanced conductivity. Compositional analysis and crystal structure refinement revealed the real compositions of $(Li_{0.45}La_{0.85})ScO_3$ and $(Li_{0.45}La_{0.78}Ce_{0.05})ScO_3$ to be $Li_{0.186}La_{0.938}ScO_3$ and $Li_{0.493}La_{0.777}Ce_{0.044}ScO_3$, respectively. Our results also indicated that an increase in the lithium amount and expansion of the unit cell volume could contribute to the higher ionic conductivity of $(Li_{0.45}La_{0.78}Ce_{0.05})ScO_3$ (i.e., 1.9×10^{-4} S cm^{-1} at 623 K). Indeed, even a small amount of Ce^{4+} co-doping had a sufficient effect on the ionic conductivity since it can significantly modify the number of charge carriers in the developed $(Li_xLa_{1-x/3})ScO_3$-based materials. Although the ionic conductivity of the obtained materials in this study is not yet very high, these materials are among the rarely studied lithium ion conductors with a perovskite structure consisting of a single-valent metal element at the B-site. Moreover, the lower valency of Sc^{3+} at B-site suggested that greater number of lithium interstitials could be generated compared to in the case of Ti^{4+} at the B-site, thereby rendering a higher carrier doping possibility. The obtained materials could be considered for use as template materials to further improve ionic conductivities until they have been further developed to approach a high lithium ion conductivity at low temperatures for practical use. These results are expected to contribute to the development of superior lithium-based solid electrolytes for application in all-solid-state lithium batteries, and the ongoing material search will be expected to expand the varieties of materials available for use in battery applications.

Supplementary Materials: The following are available online. Figure S1: (a) X-ray diffraction patterns, (b) observed shifts of the selected reflections, and (c) variation in the lattice parameters with the composition of $(Li_{0.45}La_{0.85-0.33z})(Sc_{1-z}Zr_z)O_3$ ($y = 0$ and 0.1). Figure S2: (a) X-ray diffraction patterns, (b) observed shifts of the selected reflections, and (c) variation in the lattice parameters with the composition of $(Li_{0.45}La_{0.85-0.67z})(Sc_{1-z}Nb_z)O_3$ ($z = 0$ and 0.1). Figure S3: Temperature dependence of the sample conductivity in (a) $(Li_{0.45}La_{0.85-z/3})(Sc_{1-z}Zr_z)O_3$ and $(Li_{0.45}La_{0.83-1.33n})(Sc_{0.95}Zr_{0.05})O_3$, and (b) $(Li_{0.45}La_{0.85-2z/3})(Sc_{1-z}Nb_z)O_3$. Figure S4: Steady-state current as a function of the applied voltage at 623 K in a dry Ar atmosphere for $(Li_{0.4}La_{0.67}Ce_{0.15})ScO_3$. Table S1: Chemical compositions of $(Li_{0.45}La_{0.85})ScO_3$ and $(Li_{0.45}La_{0.78}Ce_{0.05})ScO_3$.

Author Contributions: Conceptualization, R.K., K.S. and M.H.; validation, G.Z., and K.S.; formal analysis, G.Z. and K.S.; investigation, G.Z. and K.S.; resources, M.H., and R.K.; data curation, G.Z. and K.S.; writing—original draft preparation, G.Z.; writing—review and editing, K.S., M.H., and R.K; visualization, G.Z. and K.S.; supervision, M.H., and R.K.; project administration, M.H., and R.K. All authors have read and agreed to the published version of the manuscript.

Funding: This research was funded by ALCA-SPRING from the Japan Science and Technology Agency (JST) and a Grant-in-Aid for Scientific Research (S) of the Japan Society for the Promotion of Science (JSPS), grant number JPMJAL1301 and 17H06145, respectively.

Institutional Review Board Statement: Not applicable.

Informed Consent Statement: Not available.

Data Availability Statement: The data presented in this study are available in supplementary material.

Acknowledgments: This work was partly supported by ALCA-SPRING from the Japan Science and Technology Agency (JST) Grant Number JPMJAL1301 and a Grant-in-Aid for Scientific Research (S) (No. 17H06145) of the Japan Society for the Promotion of Science (JSPS). The synchrotron radiation experiments were performed at the SPring-8 facility with the approval of the Japan Synchrotron Radiation Research Institute (JASRI) (Proposal nos. 2016B1778 and 2019B1745).

Conflicts of Interest: The authors declare no conflict of interest.

Sample Availability: Not available.

References

1. Kamaya, N.; Homma, K.; Yamakawa, Y.; Hirayama, M.; Kanno, R.; Yonemura, M.; Kamiyama, T.; Kato, Y.; Hama, S.; Kawamoto, K.; et al. A lithium superionic conductor. *Nat. Mater.* **2011**, *10*, 682–686. [CrossRef] [PubMed]
2. Kato, Y.; Hori, S.; Saito, T.; Suzuki, K.; Hirayama, M.; Mitsui, A.; Yonemura, M.; Iba, H.; Kanno, R. High-power all-solid-state batteries using sulfide superionic conductors. *Nat. Energy* **2016**, *1*, 1–7.
3. Ma, C.; Chen, K.; Liang, C.; Nan, C.-W.; Ishikawa, R.; Morea, K.; Chi, M. Atomic-scale origin of the large grain-boundary resistance in perovskite Li-ion-conducting solid electrolytes. *Energy Environ. Sci.* **2014**, *7*, 1638–1642. [CrossRef]
4. Cao, C.; Li, Z.-B.; Wang, X.-L.; Zhao, X.-B.; Han, W.-Q. Recent advances in inorganic solid electrolytes for lithium batteries. *Front. Energy Res.* **2014**, *2*, 1–10.
5. Bachman, J.C.; Muy, S.; Grimaud, A.; Chang, H.-H.; Pour, N.; Lux, S.F.; Paschos, O.; Maglia, F.; Lupart, S.; Lamp, P.; et al. Inorganic solid-state electrolytes for lithium batteries: Mechanisms and properties governing ion conduction. *Chem. Rev.* **2016**, *116*, 140–162. [CrossRef]
6. Sun, C.; Liu, J.; Gong, Y.; Wilkinson, D.P.; Zhang, J. Recent advances in all-solid-state rechargeable lithium batteries. *Nano Energy* **2017**, *33*, 363–386. [CrossRef]
7. Van den Broek, J.; Afyon, S.; Rupp, J.L.M. Interface-engineered all-solid-state li-ion batteries based on garnet-type fast Li$^+$ conductors. *Adv. Energy Mater.* **2016**, *6*, 1–11.
8. Li, X.; Liang, J.; Yang, X.; Adair, K.R.; Wang, C.; Zhao, F.; Sun, X. Progress and perspectives on halide lithium conductors for all-solid-state lithium batteries. *Energy Environ. Sci.* **2020**, *13*, 1429–1461. [CrossRef]
9. Zhao, Q.; Stalin, S.; Zhao, C.-Z.; Archer, L.A. Designing solid-state electrolytes for safe, energy-dense batteries. *Nat. Rev. Mater.* **2020**, *1*, 229–252. [CrossRef]

10. Hori, S.; Suzuki, K.; Hirayama, M.; Kato, Y.; Saito, T.; Yonemura, M.; Kanno, R. Synthesis, structure, and ionic conductivity of solid solution, $Li_{10+\delta}M_{1+\delta}P_{2-\delta}S_{12}$ (M = Si, Sn). *Faraday Discuss.* **2014**, *176*, 83–94. [CrossRef]
11. Inoue, Y.; Suzuki, K.; Matsui, N.; Hirayama, M.; Kanno, R. Synthesis and structure of novel lithium-ion conductor $Li_7Ge_3PS_{12}$. *J. Solid State Chem.* **2017**, *246*, 334–340. [CrossRef]
12. Hori, S.; Suzuki, K.; Hirayama, M.; Kato, Y.; Kanno, R. Lithium superionic conductor $Li_{9.42}Si_{1.02}P_{2.1}S_{9.96}O_{2.04}$ with $Li_{10}GeP_2S_{12}$-type structure in the $Li_2S–P_2S_5–SiO_2$ pseudoternary system: Synthesis, electrochemical properties, and structure–composition relationships. *Front. Energy Res.* **2016**, *4*, 1–10. [CrossRef]
13. Inaguma, Y.; Liquan, C.; Itoh, M.; Nakamura, T. High ionic conductivity in lithium lanthanum titanate. *Solid State Commun.* **1993**, *86*, 689–693. [CrossRef]
14. Hallopeau, L.; Bregiroux, D.; Rousse, G.; Portehault, D.; Stevens, P.; Toussaint, G.; Laberty-Robert, C. Microwave-assisted reactive sintering and lithium ion conductivity of $Li_{1.3}Al_{0.3}Ti_{1.7}(PO_4)_3$ solid electrolyte. *J. Power Sources* **2017**, *378*, 48–52. [CrossRef]
15. Kuwano, J.; West, A.R. New Li^+ ion conductors in the system, Li_4GeO_4-Li_3VO_4. *Mater. Res. Bull.* **1980**, *15*, 1661–1667. [CrossRef]
16. Alpen, U.V.; Bell, M.F.; Wichelhaus, W. Ionic conductivity of $Li_{14}Zn(GeO_4)$ (Lisicon). *Electrochim. Acta* **1978**, *23*, 1395–1397. [CrossRef]
17. Wenzel, S.; Randau, S.; Leichtweiß, T.; Weber, D.A.; Sann, J.; Zeier, W.G.; Janek, J. Direct observation of the interfacial instability of the fast ionic conductor $Li_{10}GeP_2S_{12}$ at the lithium metal anode. *Chem. Mater.* **2016**, *28*, 2400–2407. [CrossRef]
18. Itoh, M.; Inaguma, Y.; Jung, W.-H.; Chen, L.; Nakamura, T. High lithium ion conductivity in the perovskite-type compounds. *Solid State Ion.* **1994**, *70*, 203–207. [CrossRef]
19. García-Martin, S.; García-Alvarado, F.; Robertson, A.D.; West, A.R.; Alario-Franco, M.A. Microstructural study of the Li+ ion substituted perovskites $Li_{0.5-3x}Nd_{0.5+x}TiO_3$. *J. Solid State Chem.* **1997**, *128*, 97–101. [CrossRef]
20. Birke, P.; Scharner, S.; Huggins, R.A.; Weppner, W. Electrolytic stability limit and rapid lithium insertion in the fast-ion-conducting $Li_{0.29}La_{0.57}TiO_3$ perovskite-type compound. *J. Electrochem. Soc.* **1997**, *144*, L167–L169. [CrossRef]
21. Kawakami, Y.; Ikuta, H.; Wakihara, M. Ionic conduction of lithium for perovskite type compounds, $(Li_{0.05}La_{0.317})_{1-x}Sr_{0.5x}NbO_3$, $(Li_{0.1}La_{0.3})_{1-x}Sr_{0.5x}NbO_3$ and $(Li_{0.25}La_{0.25})_{1-x}M_{0.5x}NbO_3$ (M = Ca and Sr). *J. Solid State Electrochem.* **1998**, *110*, 187–192. [CrossRef]
22. Stramare, S.; Thangadura, V.; Weppner, W. Lithium lanthanum titanates: A review. *Chem. Mater.* **2003**, *15*, 3974–3990. [CrossRef]
23. Bhalla, A.S.; Guo, R.; Roy, R. The perovskite structure—A review of its role in ceramic science and technology. *Mater. Res. Innov.* **2000**, *4*, 3–26. [CrossRef]
24. Hellstrom, E.E.; Van Gool, W. Li ion conduction in Li_2ZrO_3, Li_4ZrO_4, and $LiScO_2$. *Solid State Ion.* **1981**, *2*, 59–64. [CrossRef]
25. Liferovich, R.P.; Mitchell, R.H. A structural study of ternary lanthanide orthoscandate perovskites. *J. Solid State Chem.* **2004**, *177*, 2188–2197. [CrossRef]
26. Zhao, G.; Muhammad, I.; Suzuki, K.; Hirayama, M.; Kanno, R. Synthesis, crystal structure, and the ionic conductivity of new lithium ion conductors, M-doped $LiScO_2$ (M = Zr, Nb, Ta). *Mater. Trans.* **2016**, *57*, 1370–1373. [CrossRef]
27. Muy, S.; Bachman, J.C.; Chang, H.-H.; Giordano, L.; Maglia, F.; Lupart, S.; Lamp, P.; Zeier, W.G.; Shao-Horn, Y. Lithium conductivity and Meyer-Neldel rule in Li_3PO_4-Li_3VO_4-Li_4GeO_4 lithium superionic conductors. *Chem. Mater.* **2018**, *30*, 5573–5582. [CrossRef]
28. Mizuno, Y.; Matsuda, Y.; Suzuki, K.; Yonemura, M.; Hirayama, M.; Kanno, R. Synthesis, crystal structure, and electrochemical properties of $Li_{1.2+x}Mn_{0.3}Co_{0.2}Ni_{0.3}O_2$ (x > 0) for lithium-ion battery cathodes. *Electrochemistry* **2015**, *83*, 820–823. [CrossRef]
29. Phraewphiphat, T.; Iqbal, M.; Suzuki, K.; Hirayama, M.; Kanno, R. Synthesis and Lithium-Ion Conductivity of $LiSrB_2O_6F$ (B = Nb^{5+}, Ta^{5+}) with a Pyrochlore Structure. *J. Jpn. Soc. Powder Powder Metall.* **2018**, *65*, 26–33. [CrossRef]
30. Izumi, F.; Momma, K. Three-dimensional visualization in powder diffraction. *Solid State Phenom.* **2007**, *130*, 15–20. [CrossRef]
31. Neudecker, B.-J.; Weppner, W. Li_9SiAlO_8: A lithium ion electrolyte for voltages above 5.4 V. *J. Electrochem. Soc.* **1996**, *143*, 2198–2203. [CrossRef]

Article

Electronic and Magnetic Structures of New Interstitial Boron Sub-Oxides B$_{12}$O$_2$:X (X = B, C, N, O)

Samir F. Matar [1,*] and Jean Etourneau [2]

[1] Lebanese German University, Sahel-Alma, Jounieh 1200, Lebanon
[2] University of Bordeaux, ICMCB-CNRS, 33600 Pessac, France; jean.etourneau@icmcb.cnrs.fr
* Correspondence: s.matar@lgu.edu.lb; Tel.: +00-961-81-225-180

Abstract: The boron-rich boron sub-oxide rhombohedral B$_6$O considered in B$_{12}$O$_2$ full formulation has a large O-O spacing of ~3 Å and a central vacant position that can receive interstitial atoms X, forming a central O-X-O alignment in the dodecaboron cage as observed in well-known triatomic B$_{12}$ compounds as B$_{12}${C-C-C}, B$_{12}${N-B-N}, etc. Plane wave density functional theory (DFT) based calculations of unrestricted geometry relaxation of B$_{12}${O-X-O}, X = B, C, N, and O let one identify new ternary sub-oxides, all found cohesive while showing different d(X-O) distances ranging from d(B-O) = 1.95 Å down to d(O-O) = 1.73 Å with intermediate d(C-O) = 1.88 Å. The different magnitudes were assigned to the chemical affinities of X-inserts versus host oxygen with the increasing development of X-O bonding along the series with larger cohesive B$_{12}${O-O-O}. From the atom projected charge density, B presents none, while significant magnitudes are shown on C and N, the latter developing bonding with terminal oxygen atoms especially N. The presence of unpaired valence electrons leaves nonbonding charge density on X = C, N interstitial compounds, which, besides the relative isolation of the central C and N lead to the onset of magnetic moments: M(C) = 1.9 µ$_B$, and M(N) = 1 µ$_B$ in a ferromagnetic ground state. Atom-resolved assessments are provided with the magnetic charge density and electron localization function electron localization function (ELF) projections on one hand and the site and spin projected density of states and the chemical bonding based on the overlap integral S$_{ij}$ within the COOP criterion, on the other hand.

Keywords: *p*-magnetism; boron sub-oxide; interstitial atoms; DFT; DOS; ELF; charge density plots

Citation: Matar, S.F.; Etourneau, J. Electronic and Magnetic Structures of New Interstitial Boron Sub-Oxides B$_{12}$O$_2$:X (X = B, C, N, O). *Molecules* **2021**, *26*, 123. https://doi.org/10.3390/molecules26010123

Academic Editor: Stephane Jobic
Received: 6 December 2020
Accepted: 21 December 2020
Published: 29 December 2020

Publisher's Note: MDPI stays neutral with regard to jurisdictional claims in published maps and institutional affiliations.

Copyright: © 2020 by the authors. Licensee MDPI, Basel, Switzerland. This article is an open access article distributed under the terms and conditions of the Creative Commons Attribution (CC BY) license (https://creativecommons.org/licenses/by/4.0/).

1. Introduction

The sesquioxide B$_2$O$_3$ is the best known boron oxide and its hydration leads to the well-known boric acid B(OH)$_3$, which has many uses in the medical sector as an antibacterial and in chemistry as a pH buffer. However, more relevant to solid state chemistry is its use in combination with boron to build the B$_2$O$_3$-B phase diagram [1]. Boron rich compounds were identified as B$_6$O [2] characterized by a structure resembling the simplest form of boron, α-B$_{12}$ (space group $R\overline{3}m$, N°166) [3]. A small amount of oxygen sub-stoichiometry was identified by Olofsson and Lundström [4] who claimed that larger oxygen content can be attained at pressures above ambient conditions. For the sake of complete review, the B$_2$O composition with trigonal structure was proposed as an unsymmetrical analog of carbon by Endo et al. [5], but its structure as a carbon derived one was deemed as unstable by density functional theory (DFT) total energy calculations [6]. The quantum theoretical DFT framework [7,8] is also used in the present investigative work. The B$_{12}$ structure depicted in Figure 1a in both rhombohedral (1 formula unit FU) and hexagonal (3 FU) settings, shows a remarkable vacant space surrounded by 6 B1, that can be occupied by one, two, and/or three interstitial atoms leading to a large family of compounds (cf. [9] for a review on boron with enumerated families). B$_6$O has the same space group as α-B$_{12}$ and it can be expressed as B$_{12}$O$_2$ in fully stoichiometric formulation that we consider here as the host of foreign insertion elements (vide infra). In the structure shown in Figure 1b, oxygen atoms occupy the two-fold Wyckoff positions (2c) in the vicinity of

B1 while the B2 atoms are pushed farther along the diagonal towards the rhombohedron corners as with respect to their positions in B_{12}. In the hexagonal setting, oxygen atoms align along $c_{hex.}$. Concomitantly, the rhombohedral angle $\alpha_{rh.} = 63.2°$ is enlarged with respect to its magnitude in B_{12} where $\alpha_{rh.} = 58°$, i.e., there is an opening of the rhombohedron to receive the interstitials. A relevant feature resulting from the 3B1-O bonding is the large spacing between the two oxygen atoms, amounting to ~3 Å, highlighting their isolation from each other on one hand and offering the central void at the body center defined by the Wyckoff position $1b$ ($\frac{1}{2}, \frac{1}{2}, \frac{1}{2}$) on the other hand. In the context of the present investigation, the central position is made to receive interstitials of light elements called X, leading to express B_6O as $B_{12}O_2$:X depicted in Figure 1c in both rhombohedral and hexagonal settings. Note that this formulation highlighting central triatomic linear alignments is identified in other compounds as the recently investigated $B_{13}N_2$ expressed as $B_{12}N_2$:B [10,11] as well as $B_{12}C_2$:X (X = B, C) [12,13]. Consequently, the present paper has the purpose of presenting investigation results of original compounds $B_{12}O_2$:X considering herein a series of neighboring p-interstitial light elements: B, C, N, and O studied within the DFT. The original compounds belonging to the large family of α-B_{12} based chemical compounds will be shown to be cohesive and possessing particular electronic properties as well as magnetic ones for some of them. They are proposed to further broaden the scope of boron research in chemistry and physics. For the sake of clear and simple presentation, the results are presented for 1 FU within the rhombohedral setting.

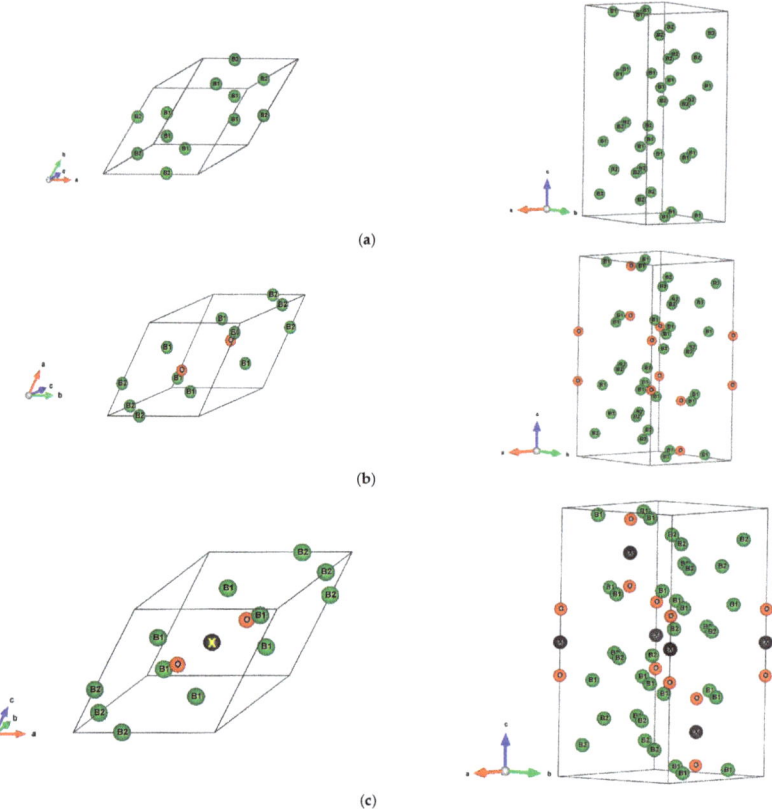

Figure 1. (color). Representation of the B_{12}-based structures in rhombohedral (left) and hexagonal (right) settings, with 1 and 3 formula units (FU) respectively. (a) α-B_{12}, (b) $B_{12}O_2$, (c) Proposed $B_{12}O_2$:X, with X representing a generic interstitial at the central void (X = B, C, N, and O).

2. Computational Framework

The search for the ground state structure and energy was carried out within DFT based calculations using plane-wave code Vienna Ab initio Simulation Package VASP [14,15] with the projector augmented wave (PAW) method [15,16] for the atomic potentials with all valence states especially in regard of the light elements B, C, N, and O. The exchange-correlation XC effects within DFT were considered with the generalized gradient approximation (GGA) [17]. This XC scheme was preferred to the homogeneous gas-based local density approximation (LDA), one which led in preliminary calculations to underestimated structure parameters, indeed, LDA is known to be over-binding [18]. The conjugate-gradient algorithm [19] was used in this computational scheme to relax the atoms onto the ground state. The tetrahedron method with Blöchl et al. corrections [20] and Methfessel–Paxton [21] scheme was applied for both geometry relaxation and total energy calculations. Brillouin-zone (BZ) integrals were approximated using a special **k**-point sampling of Monkhorst and Pack [22]. The optimization of the structural parameters was performed until the forces on the atoms were less than 0.02 eV/Å and all stress components were below 0.003 eV/Å^3. The calculations converged at an energy cut-off of 500 eV for the plane-wave basis set concerning the **k**-point integration with a starting mesh of $6 \times 6 \times 6$ up to $12 \times 12 \times 12$ for best convergence and relaxation to zero strains. Calculations are firstly carried out considering total spins configuration, pertaining to non-spin polarized (NSP) configurations. In a further step, due to the paramagnetic character observed from the valence electron count (VEC, see below), spin-polarized SP calculations were carried out.

Properties related to electron localization were obtained from real-space projections of the electron localization function (ELF) according to Becke and Edgecomb [23] and Savin et al. [24] as based on the kinetic energy in which the Pauli Exclusion Principle is included. ELF is a normalized function, i.e., 0 < ELF < 1, ranging from no localization for 0 (blue zones) and full localization for ELF = 1 (red zones) and free-electron gas behavior corresponds to ELF = $\frac{1}{2}$ (green zones), cf. Figure 2.

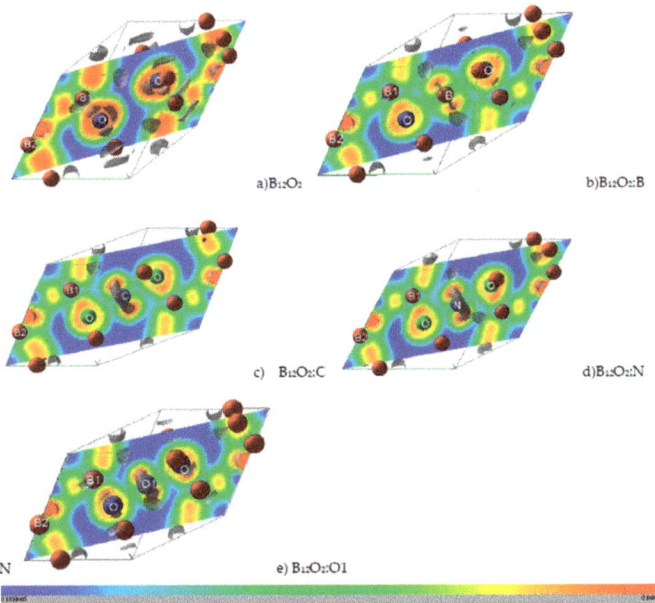

Figure 2. (color). ELF slice planes along diagonal plane. (**a**) $B_{12}O_2$, (**b**) $B_{12}O_2$:B, (**c**) $B_{12}O_2$:C, (**d**) $B_{12}O_2$:N, and (**e**) $B_{12}O_2$:O or $B_{12}O_3$. The ruler shows the color code with blue, green, and red corresponding respectively to 0 (no electron localization), $\frac{1}{2}$ (free electron like), and 1 (full electron localization). Grey volumes depict 3D ELF, especially for showing the non-bonded electrons around central inserted atoms: C, N, and O.

In the post-treatment process of the ground state electronic structures, the total charge densities "CHGCAR", as well as the magnetic charge density "CHGCAR_magn" are illustrated. The latter is computed if spin polarized SP calculations identify a magnetic solution, versus nonmagnetic NSP configuration.

From the geometry of the ground state structures NSP and SP, the detailed electronic site and spin projected density of states (PDOS) were obtained within DFT using full potential augmented spherical wave (ASW) method [25] and the GGA for the XC effects [17]. Also, the properties of chemical bonding are qualitatively assessed within ASW based on overlap matrix (S_{ij}) with the crystal orbital overlap population (COOP) criterion [26]. In short hand notation, the COOP's are the S_{ij}—weighted density of states (DOS). Positive, negative and zero COOP magnitudes (cf. Section 4) correspond to bonding, anti-bonding and non-bonding interactions.

In the minimal ASW basis set, the outermost shells were chosen to represent the valence states and the matrix elements. They were constructed using partial waves up to $l_{max} = 1$ for B, O, and X p-elements interstitials. In the most electronegative element, oxygen's low energy lying 2s states were omitted from the valence basis set in the DOS projection for the sake of clarity. Self-consistency was achieved when charge transfers and energy changes between two successive cycles were: $\Delta Q < 10^{-8}$ and $\Delta E < 10^{-6}$ eV, respectively. The Brillouin-zone integrations were performed using the linear tetrahedron method within the irreducible rhombohedral wedge following Blöchl et al. scheme [20].

3. Results from Energy Calculations

3.1. Trends of Cohesive Energies

Firstly, we examined the B_6O based compounds for their respective cohesive energies obtained considering total spin configuration within unconstrained, parameter-free, successive self-consistent sets of calculations at an increasing number of k-points. The structural results are shown in Table 1 and they will be discussed in the next section, focusing here on the cohesive energies deducted from subtracting the constituents' atomic energies, averaged per atom to enable comparison between the different stoichiometries. The atomic constituents' atomic energies in eV are as follows: E(B) = −5.56; E(C) = −6.48; E(N) = −5.2; and E(O) = −3.14.

Pure $rh.\alpha$-B_{12} was identified with E_{coh}./at. = −1.16 eV. Comparatively, $B_{12}O_2$ (B_6O) was found with a larger cohesive energy: E_{coh}./at. = −1.97 eV, explained by the selective bonding of B1 with the two oxygen atoms, the B1-O bond introducing an iono-covalent character.

Upon insertion with B, C, and N the cohesive energies: E_{coh}./at.($B_{12}O_2$:B) = −1.280 eV; E_{coh}./at.($B_{12}O_2$:C) = −1.266 eV and E_{coh}./at.($B_{12}O_2$:N) = −1.234 eV are all intermediate between α-B_{12} and $B_{12}O_2$. While they all remain within a close range of magnitudes, their decrease from B to C, and N is likely due to the decreasing difference of electronegativity χ between O and X: $\chi_O = 3.44$, $\chi_B = 2.04$, $\chi_C = 2.55$, and $\chi_N = 3.04$. This progressive decrease of the ionic character $\Delta\chi$(X-O) is concomitant with a slight decrease of the cohesive energy. However, the relatively large decrease of the cohesive energy from pristine $B_{12}O_2$ to $B_{12}O_2$:X is assessed through the competitive B1-O versus X-O bonding along with the decrease of d(X-O) within the series as shown in Table 1.

An exception is nevertheless observed upon inserting oxygen O1 at the Wyckoff position $1b$ ($\frac{1}{2},\frac{1}{2},\frac{1}{2}$) with a resulting E_{coh}./at.($B_{12}O_2$:O) = −1.56 eV, a larger magnitude than all other hetero-inserted (X $\neq$ O) compounds but lower than pristine $B_{12}O_2$. Based on this observation, besides the fact that B_{12} is further stabilized through engaging iono-covalent bonds of one of the two B substructures with oxygen, namely B1–O, the insertion of additional central oxygen O1 is unfavorable versus $B_{12}O_2$ because it involves a competitive O1-O bonding versus B1-O on which the B_6O structure is based, and hence weakening it. Indeed, the shortest X-O distance is observed for $B_{12}O_3$ with d(O1-O) = 1.73 Å (cf. Table 1), leading to suggest that O-O-O should be favored over O-X-O, independently of the chemical nature of X as observed from the respective cohesive energies.

Table 1. Crystal data of $B_{12}O_2$:X family. Space group $R\bar{3}m$, N°166. Rhombohedral settings. Void for interstitial atoms X: (1b) $\frac{1}{2}$, $\frac{1}{2}$, $\frac{1}{2}$ (cf. Figure 1c). Shortest distances are presented in Å unit (1Å = 10^{-10} m).

(a) B_6O or $B_{12}O_2$. Experimental [2] and (calculated) crystal parameters. a_{rh} = 5.15 (5.13) Å; α = 63.04° (63.19)°.

Atom	Wyckoff	x	y	z
B_1	6h	0.800 (0.804)	0.323 (0.323)	x
B_2	6h	0.347 (0.333)	0.002 (0.003)	y
O	2c	0.376 (0.377)	x	x

d(O-O) = 3.06 (3.046); d(B1-O) = 1.49 (1.489); d(B1-B2) = 1.78 (1.82).

(b) $B_{12}O_2$:B. Calculated crystal parameters; B at (1b) $\frac{1}{2}$, $\frac{1}{2}$, $\frac{1}{2}$. a_{rh} = 5.32 Å; α = 63.04°.

Atom	Wyckoff	x	y	z
B_1	6h	0.806	0.321	x
B_2	6h	0.333	0.003	y
O	2c	0.348	x	x

d(O-O) = 3.06 (3.046); d(B1-O) = 1.49 (1.489); d(B1-B2) = 1.78 (1.82).

(c) $B_{12}O_2$:C calculated crystal parameters (non-spin polarized (NSP)/SP close values); C at (1b) $\frac{1}{2}$, $\frac{1}{2}$, $\frac{1}{2}$. a_{rh} = 5.30 Å; α = 61.97°.

Atom	Wyckoff	x	y	z
B_1	6h	0.805	0.323	x
B_2	6h	0.333	0.004	y
O	2c	0.352	x	x

d(C-O) = 1.88; d(B1-O) = 1.52; d(B1-B2) = 1.83.

(d) $B_{12}O_2$:N calculated crystal parameters (NSP/SP close values); N at (1b) $\frac{1}{2}$, $\frac{1}{2}$, $\frac{1}{2}$. a_{rh} = 5.26 Å; α = 62.42°.

Atom	Wyckoff	x	y	z
B_1	6h	0.804	0.321	x
B_2	6h	0.333	0.003	y
O	2c	0.357	x	x

d(N-O) = 1.81, d(B1-O) = 1.52; d(B1-B2) = 1.73.

(e) $B_{12}O_2$:O (Also expressed as $B_{12}O_3$). Calculated crystal parameters O1 at (1b) $\frac{1}{2}$, $\frac{1}{2}$, $\frac{1}{2}$; a_{rh} = 5.25 Å; α = 62.55°.

Atom	Wyckoff	x	y	z
B_1	6h	0.804	0.320	x
B_2	6h	0.332	0.003	y
O	2c	0.363	x	x

d(O1-O) = 1.73; d(B1-O) = 1.525; d(B1-B2) = 1.81.

Another argument can be evoked, pertaining to the fact that linear O_3 is an unfavorable geometry both in the solid state and the isolated molecule. The angle O-O1-O amounts to 116.8°, smaller than the one identified by X-ray powder diffraction of solid ozone, amounting to 123.2° [27]. Both magnitudes are quite far from 180° characterizing a linear arrangement as in azide N_3^- anion or in CO_2, both with VEC = 16 electrons. Therefore, aligning three oxygen atoms in B_{12} host cavity (cf. Figure 1) is found less favorable energetically than the presence of only two oxygen atoms attached to one of the boron substructures, namely B1, arriving at $B_{12}O_2$ (or 2 B_6O). Nevertheless, $B_{12}O_3$ is found largely cohesive. Such a favorable energy situation can be addressed through the bonding of the two end oxygens with one of the boron substructures B1 in view of their respective electronegativities: χ_O = 3.44 versus χ_B = 2.04.

The cohesive energies hierarchies are then chemically assessed in absolute magnitudes as: $|E_{coh.}/at.(B_{12})| < |E_{coh.}/at.(B_{12}O_3)| < |E_{coh.}/at.(B_{12}O_2)|$, on one hand and $|E_{coh.}/at.(B_{12}XO_2)|$ (X = B,C,N)$| < |E_{coh.}/at.(B_{12}O_3)|$ on the other hand.

Lastly, the relatively $E_{coh.}/at.(B_{12}O_3)$ large magnitude suggests the potential formation of this sub-oxide under optimized pressure/temperature conditions within the B_2O_3-B phase diagram [1]. Further emphasizing this relevant hypothesis, we calculated the cohesive energy of experimental $B_{12}C_3$ (B_4C) also well known for use as an abrasive [13] and found a close magnitude of $E_{coh.}/at.(B_{12}C_3) = -1.57$ eV/at. This provides further support to our hypothesis of inserting X at the rhombohedron center on one hand and particularly oxygen with the potential existence of $B_{12}O_3$ on the other hand.

3.2. Trends of Valence Electron Count VEC

The valence electron count evoked above can help the further assessment of the electronic behavior in a preliminary step. For B, C, N, and O, VEC = 3, 4, 5, and 6, respectively. Then VEC(B_{12}) = 36 (12 × 3). Focusing on the sub-oxides, VEC($B_{12}O_2$) = 36 + 12 = 48 for $B_{12}O_2$ on one hand and VEC($B_{12}O_3$) = 36 + 18 = 54 on the other hand, expecting closed shells. For illustration, the electronic density of states (DOS) projected over each constituent (PDOS) in Section 4 exhibits a small energy band gap of 1 eV between the filled valence band (VB) and the empty conduction band (CB) for both binary sub-oxides.

Regarding the interstitial $B_{12}O_2$:X, the atomic constituents: B($2s^2$, $2p^1$), C($2s^2$, $2p^2$), N($2s^2$, $2p^3$), and O($2s^2$,$2p^4$) add on 3, 4, 5, and 6 valence electrons leading to VEC = 51, 52, 53, and 54, respectively, expecting a paramagnetic behavior for B and N as well as for the carbon case where the 2 p electrons remain unpaired.

3.3. Geometry Optimization Results

For $B_{12}O_2$, Table 1a presents the experimental and (calculated) structure results. A relatively good agreement is obtained for the lattice constants and the atomic positions for B1 and B2 substructures. This is also observed for the interatomic distances, the shortest one being B1-O at 1.49 Å.

Upon insertion of X at the body center the changes (Table 1b–e) are little for the B1 and B2 substructures but larger for $a_{rh.}$ and oxygen coordinate x_O which increase regularly on the one hand and for the d(X-O) which decreases from 1.95 Å for X = B down to 1.81 Å for X = N, along with the slight decrease of the cohesive energy discussed above, on the other hand.

The exception for the largely cohesive $B_{12}O_3$ was discussed above in relation with its potential existence. NSP calculations were complemented with SP ones especially for X = B, C, and N. Only C and N showed a magnetically polarized ground state with $\Delta E(SP - NSP) = -0.53$ and -0.34 eV, respectively, from the total energy calculations in the NSP and SP configurations and the subsequent development of a magnetic moment of 1.9 μ_B and 1 μ_B—Bohr magneton—on C and N, respectively (vide infra). The reason why only one electron polarizes on nitrogen, is likely due to the bonding of the other two with oxygen of pristine $B_{12}O_2$ as illustrated by the strongest bonding between X and O observed relatively for N-O (cf. Section 4 illustrations) leaving a single non bonded electron and showing a significantly large density of states at the Fermi level, detailed in Section 4.

3.4. Electron Localization ELF Representations

Further illustration regarding the localization of the electrons and the corresponding bonding is obtained from the 2D (diagonal plane) and 3D ELF isosurfaces representing strong localization domains shown in grey volumes in Figure 2. As introduced in the Computation Section, ELF is a normalized function with $0 \leq$ ELF ≤ 1. For the 2D ELF slice planes the ruler indicates the color code with 0 indicating zero localization with blue zones, ELF = $\frac{1}{2}$ for free electron like behavior with green zones and ELF = 1 for full localization with red zones.

Figure 2a shows the ELF of $B_{12}O_2$ with a remarkable feature of a large blue zone between the two oxygen atoms surrounded by red ELF of strong localization. The two oxygens feature isolated grey volumes of non-bonded electrons pointing towards the central void, and they are expected to be involved with the X-O bonding. In Figure 2b, B in $B_{12}O_2$:B at the center fills the electron gap between the two oxygens while showing little localization around it. The introduction of the other X's stresses the feature of larger localization and the occurrence of an increasingly large 3D grey torus around X (cf. Figure 2c–e) corresponding to non-bonding electrons. The emerging picture from the ELF representations is that besides showing the binding between O and B1 substructure, the presence of the X interstitial leads to the formation of a "3B1-O-X-O-3B1"-like linear complex along the rhombohedron diagonal, or along c if a hexagonal setup is considered (cf. Figure 1c).

3.5. Total and Magnetic Charge Densities

The charge density resulting from the self-consistent calculations for all $B_{12}O_2$:X are illustrated in Figure 3. The charge density projected onto the atomic constituents shows its prevalence on oxygen in the neighborhood of B1 substructure with no charge density on electron deficient boron, also observed in the central boron within $B_{12}O_2$:B. From X = C to O, there appears a charge density on X with an increasing size along with the increase of the number of valence electron, i.e., from four up to six.

Considering the VEC numbers of the different compositions and starting from the non-spin polarized NSP calculations, SP calculations were carried out by accounting for two spin channels, i.e., spin up ↑ and spin down ↓. The magnetic configuration was favored in the case of C and N providing a magnetic ground state and a magnetization $M(B_{12}O_2C) = 1.9\ \mu_B$ and $M(B_{12}O_2N) = 1\ \mu_B$. The atom projected magnetic charge density is shown in Figure 4 in the form of a torus identified only on the central interstitial atom, i.e., C and N with a larger volume on C proportionally to the twice-larger moment magnitude.

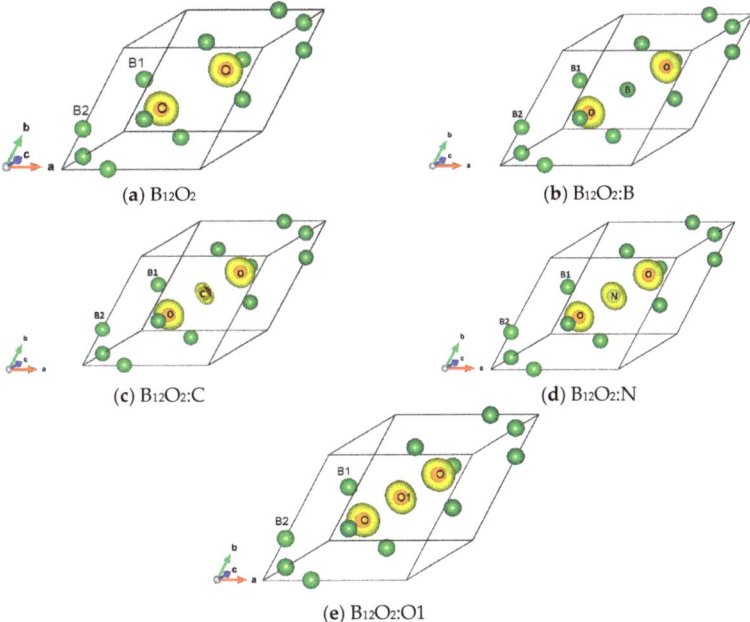

(a) $B_{12}O_2$ (b) $B_{12}O_2$:B

(c) $B_{12}O_2$:C   (d) $B_{12}O_2$:N

(e) $B_{12}O_2$:O1

Figure 3. (color). Charge density projected onto the atomic constituents in (a) $B_{12}O_2$, (b) $B_{12}O_2$:B, (c) $B_{12}O_2$:C, (d) $B_{12}O_2$:N, and (e) $B_{12}O_2$:O1.

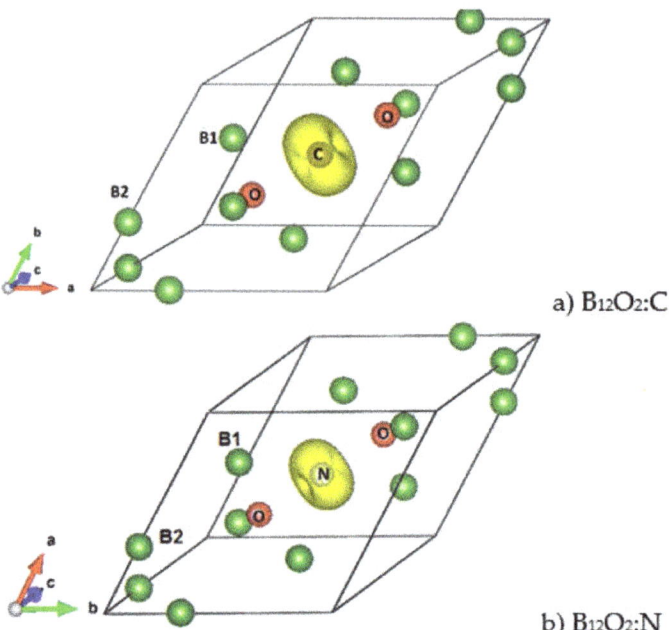

Figure 4. (color). Magnetic charge density in the two magnetically stable ternary compounds $B_{12}O_2$:C (a) and $B_{12}O_2$:N (b) exhibited by a torus centered on C and N and corresponding respectively to 1.9 μ_B and 1 μ_B magnetic moments.

4. Electronic Structure and Bonding

The specific role of each chemical constituent in $B_{12}O_2$:X can be assigned based on the projection of the electronic density of states DOS and the COOP in the two magnetic configurations, NSP and SP, respectively. Using the data in Table 1, the calculations were carried out within the full potential augmented spherical wave (ASW) method [25,26] using the GGA gradient functional for the DFT XC effects [17]. The plots (in color) are shown with highlighting of the interstitial atoms partial DOS (PDOS) in red, then, the host ones subsequently.

4.1. NSP Calculations

Reporting firstly on the two binary sub-oxides, $B_{12}O_2$ and $B_{12}O_3$, Figure 5 shows the PDOS. Along the *x*-axis the energy zero is with respect to the top of the valence band VB E_V because both compounds are insulating with a small band gap of 1 eV for both binary oxides, in accordance with the even VEC count discussed above. In $B_{12}O_2$ the O and B1 PDOS skylines show resemblance oppositely to B2 which is far from O, at -16 eV there can be seen the B1(s) PDOS with negligible contribution from O—the B1 and O s states being at much lower energy (in Figure 5 (bottom)). Similar features are observed for B2 in $B_{12}O_3$ as with the B2(s) DOS at -16 eV. Also, a larger intensity O-PDOS is observed likely due to the charge density from O1 interstitial. In Figure 5, the conduction band shows a sharp peak corresponding to the O1-O interaction.

The COOPs shown on the right hand side further highlight the interactions between the chemical species. In $B_{12}O_2$ iono-covalent B1-O interaction exhibits a larger intensity than B1-B2. However, at the top of the VB it shows anti-bonding intensity due to the B1 bonding with O, competitive of the B1 bonding with the other boron substructure, B2. The COOP panel of $B_{12}O_2$ shows further features pertaining to the larger B1-O bonding following the PDOS larger magnitude of O. B1-B2 bonding is also prevailing with positive COOP intensities throughout the VB, thus ensuring the stability of the boron host structure

B_{12}. The negative COOP's near the top of the VB are due to the competitive bonding of O with O1 versus B2.

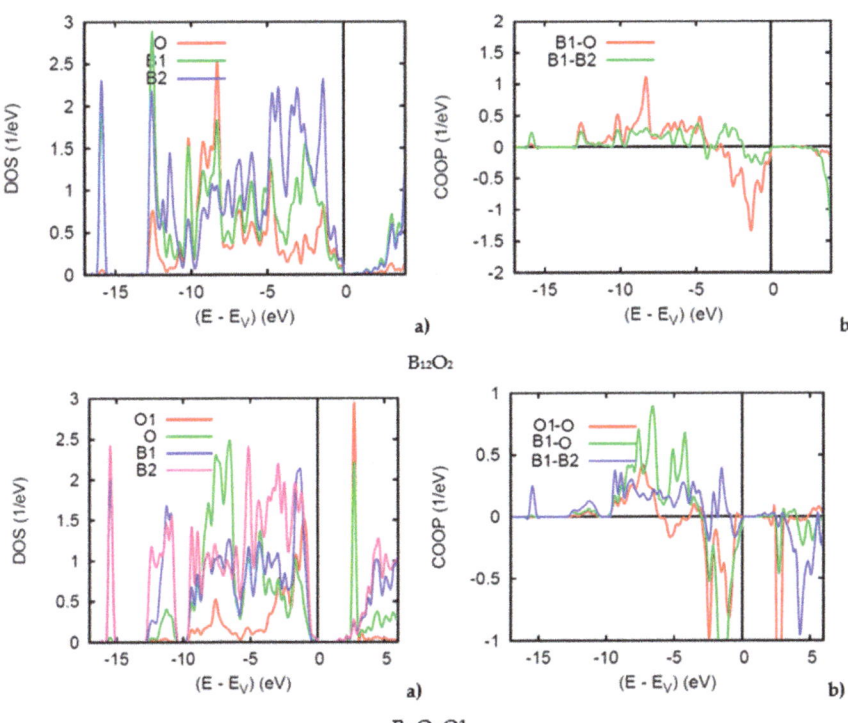

Figure 5. (color). (a) Site projected DOS and (b) chemical bonding from the COOP criterion of overlap valence populations in $B_{12}O_2$ (top) and $B_{12}O_2$:O1 (bottom).

The DOS and COOP panels in Figure 6 exhibit a loss of the band gap observed in the binary sub-oxides and the energy reference is now with respect to the Fermi level E_F. While other PDOS features are similar to the above discussion, we focus on the DOS at E_F: $n(E_F)$ in such spin degenerate NSP calculations. All three ternary compounds show a relatively large $n(E_F)$ due to the interstitial atoms, B, C, and N which bring 3, 4, and 5 extra valence electrons, but more specifically 1, 2, and 3 p electrons, respectively. In $B_{12}O_2$:B, B bringing one p electron shows a slight perturbation of the pristine $B_{12}O_2$ DOS of Figure 5a with a small $n(E_F)$ intensity from one p electron. More significant $n(E_F)$ intensities are observed for $B_{12}O_2$:C and $B_{12}O_2$:N (Figure 6). Such high PDOSs are significant of electronic system instability in such total spins configuration assessed in the framework of the Stoner theory of band ferromagnetism explained in the textbook of Peter Mohn [28] and detailed in a case study in ref. [29]. Then the NSP configuration is not the ground state one and further SP calculations are needed. The VASP calculations above have actually shown that the SP configuration was the ground state with the onset of finite moments on C and N.

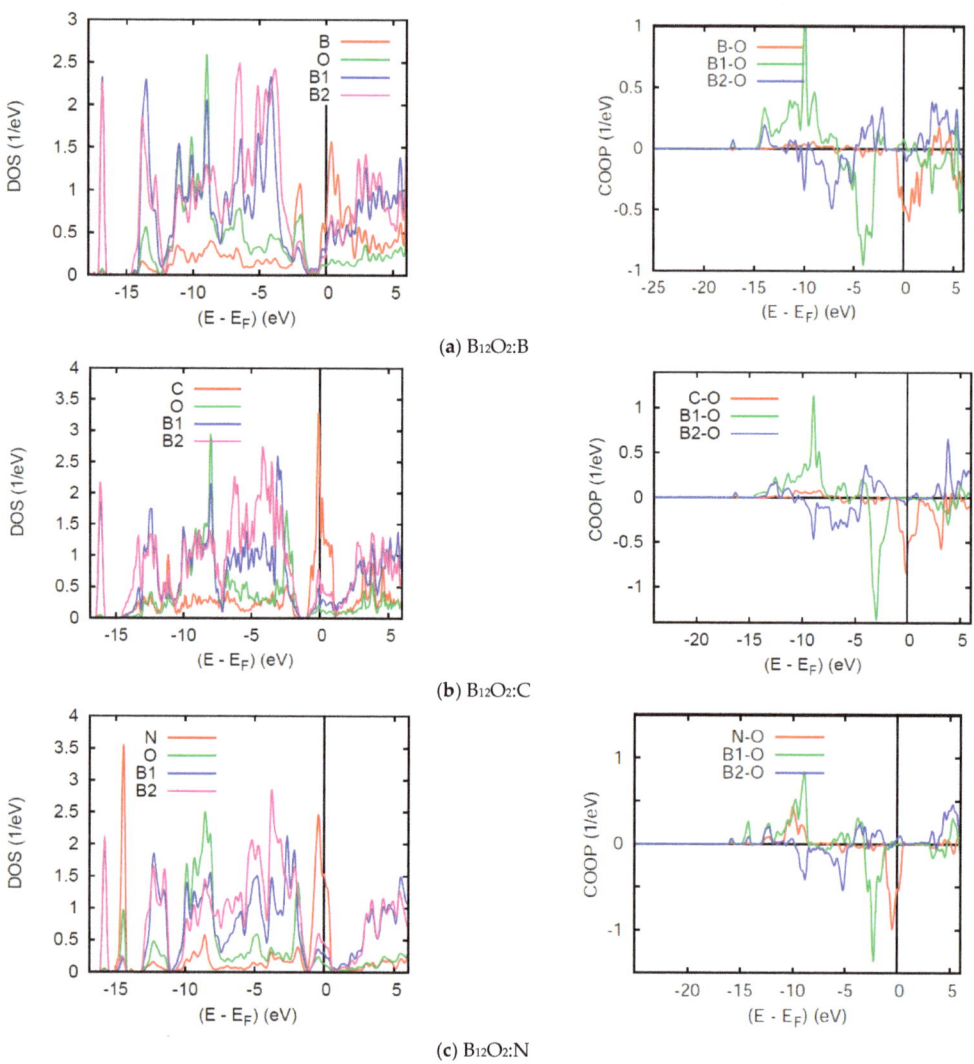

Figure 6. (color). $B_{12}O_2$:X. Site projected DOS (left) and chemical bonding from the COOP criterion (right) for (**a**) X = B; (**b**) X = C, and (**c**) X = N.

The right hand side COOP panels in Figure 6 show an increasing X-O COOP intensity along the series together with prevailing B1-O bonding and the competitive anti-bonding COOP near the top of the VB. Note that the large $n(E_F)$ correspond to anti-bonding states and stress further the electronic instability of the system in NSP configuration, meaning that the electronic system should stabilize upon accounting for two spin channels, majority spins (↑) and minority spins (↓) in SP calculations. Lastly, while negligible B-O and C-O COOPs are observed around −10 eV within the VB, relatively large COOP intensity is observed for N-O with a peak following the B1-O COOP shape. This implies a larger number of p- electrons from N involved with the bonding, namely two of them leaving one electron to spin polarize as shown below.

4.2. SP Calculations

SP calculations of $B_{12}O_2$:B did not lead to the onset of magnetization, in agreement with the VASP calculations. Expectedly, from the large C and N $n(E_F)$, both $B_{12}O_2$:C and $B_{12}O_2$:N led to magnetic solutions with a stable magnetization of 1.9 and 1 μ_B, respectively. From the FP-ASW calculations using Table 1 data, the SP-NSP energy differences are: $\Delta E(SP - NSP) = -0.62$ eV for $B_{12}O_2$:C and -0.37 for $B_{12}O_2$:N. These magnitudes are expectedly different from those obtained with the different method VASP (Section 3) where $\Delta E(SP - NSP) = -0.53$ and -0.34 eV, respectively, but they remain within range, thus confirming the trend towards SP ground state.

While the two p electrons of carbon polarize due to sufficient localization, nitrogen has only one electron carrying the magnetic moment as assessed in the paragraph above and in Sections 3.3 and 4.1.

Figure 7 shows the SP DOS in two subpanels for majority spins (↑) and minority spins (↓), named after their content of larger and smaller electron numbers as one may observe upon visual inspection of the energy downshift of the former (↑) and the energy up-shift for the latter (↓). However, such energy shift affects mainly the $n(E_F)$ states leading to a nearly insulating $B_{12}O_2$:C with very small $n(E_F)$ for both spin channels ↑, ↓ and non-negligible $n_\downarrow(E_F)$ in $B_{12}O_2$:N (half-filled band) which can be qualified as a half metallic ferromagnet. It needs to be highlighted here that such magnetic features are not unique in as far as they were also observed for transition metal ions as Cr in the rare transition metal oxide ferromagnet CrO_2 which exhibits a half-metallic ferromagnetic ground state computed by us within DFT back in 1992 [30]. However, it was discussed earlier in a phenomenological schematic in 1973, by John B. Goodenough, in his text book on transition metal oxides [31].

The SP COOPs reflect the SP-DOS mainly through the energy shifts of the UP ↑ and DOWN (DN ↓) spin COOP shown on the right hand side panels of Figure 7. The main observation is the lowering of the anti-bonding COOP at E_F.

Lastly, accounting for a possible anti-ferromagnetic state, complementary calculations were done through a doubling of the unit cell leading to two magnetic sub-cells with the first one considered as SPIN-UP and the second as SPIN-DOWN. The calculations led to an increase in the total energy and a decrease in magnetization, thus confirming the ferromagnetic ground state.

Nevertheless, it needs to be stressed that the calculations within DFT are at zero Kelvin implicitly. Since $\Delta G = \Delta H - T.\Delta S$, then $T.\Delta S = 0$ and the free energy ΔG corresponds to the enthalpy ΔH. Experimentally, the thermal effects are likely to play a significant role in the magnetic properties, such as the passage from ferromagnetic order to paramagnetic disorder state.

Syntheses efforts and subsequent measurements at low temperatures are likely to bring further assessment and clarification to the observations and results reported herein.

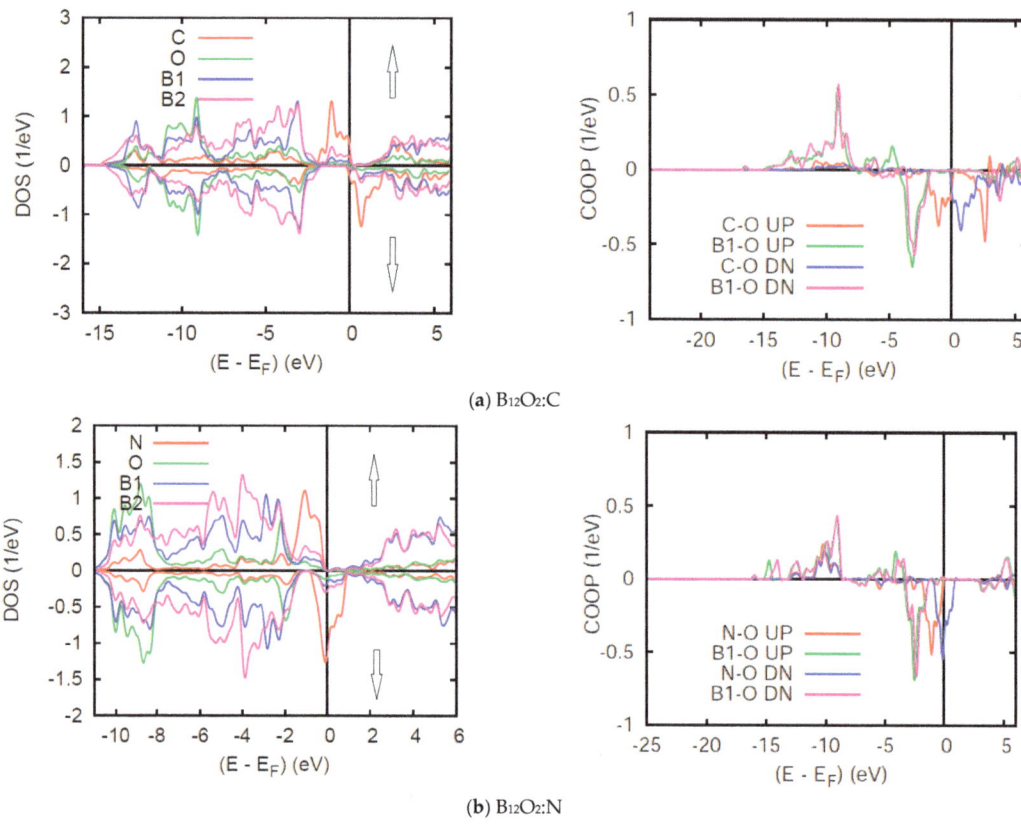

Figure 7. (color). $B_{12}O_2$:X. Site and spin projected DOS and COOP in (**a**) X = C, and (**b**) X = N corresponding to 1.9 and 1 μ_B, respectively.

5. Conclusions

Based on experimental observations of $rh.$-$B_{12}O_2$, its expression as $B_{12}\{O:X:O\}$ lets one identify potential linear tri-atomic arrangement of atoms where X stands for a vacant position alike other compounds as $B_{12}C_3$ expressed as $B_{12}\{C$-C-$C\}$, better known as B_4C. With X interstitials belonging to the 1st period, $rh.$-$B_{12}O_2$:X (X = B, C, N, O) form a new family of ternary boron sub-oxides that we investigated within quantum theoretical DFT based on two complementary methods, namely a plane-wave one allowing for geometry optimizations and a second one, the augmented spherical wave (ASW) for detailed atom projected DOS and bonding assessments. The resulting compounds were found cohesive, albeit with smaller magnitude than pristine $B_{12}O_2$. The largest cohesion magnitude within the family was found for $B_{12}\{O$-O-$O\}$ or $B_{12}O_3$ translating the possibility of its synthesis under P and T conditions, alike the well-known abrasive B_4C. $B_{12}O_2$ and $B_{12}O_3$ were found semi-conducting with a small band gap of 1 eV. Oppositely, the other $B_{12}O_2$:X (X = B, C, and N) were found metallic-like from the finite density of state at the Fermi level: $n(E_F)$. Large $n(E_F)$ magnitudes were identified for C and N in non-spin-polarized NSP calculations. Upon allowing for spin polarization SP, the onset of magnetization was found with $M(B_{12}O_2C)$ = 1.9 μ_B and $M(B_{12}O_2N)$ = 1 μ_B. Their ground state was identified energetically to be ferromagnetic versus anti-ferromagnetic configuration. The energy and crystal chemistry numerical observations were further illustrated with total and magnetic charge density projections, as well as with the electron localization function (ELF) mapping and overlap population based COOP. In view of the relevant electronic and potential magnetic properties, such compounds could be prepared. Especially with the $E_{coh.}/at.(B_{12}O_3)$ large magnitude (1.56 eV) relatively to the other $B_{12}O_3$:X compositions but close to $E_{coh.}/at.(B_4C) = 1.57$ eV, it can be suggested the potential synthesis of $B_{12}O_3$ under specific pressure/temperature conditions within the B_2O_3-B phase diagram. Such P,T experiments using Flash Spark Plasma SPS [32] sintering are underway at the ICMCB-CNRS-University of Bordeaux.

Author Contributions: Conceptualization, S.F.M. Methodology, S.F.M. and J.E. Software, S.F.M. Validation, S.F.M. Formal analysis, S.F.M. and J.E. Investigation, S.F.M. Resources, S.F.M. Data curation S.F.M. Writing—original draft preparation, S.F.M. and J.E. Writing—review and editing, S.F.M. and J.E. Visualization, S.F.M. and J.E. Supervision S.F.M. and J.E. Project administration, S.F.M. and J.E. All authors have read and agreed to the published version of the manuscript.

Funding: This research received no external funding.

Data Availability Statement: Data can be made available upon written request to the corresponding author and with a proper justification.

Acknowledgments: Drawing of Figure 1, Figure 3, and Figure 4 were done using VESTA software, free for academics [33]. Parts of the calculations were done on MCIA-University of Bordeaux Cluster-Computers.

Conflicts of Interest: The authors declare no conflict of interest.

References

1. Turkevich, V.Z.; Turkevich, D.V.; Solozhenko, V.L. Phase Diagram of the B-B_2O_3 System at Pressures to 24 GPa. *J. Superhard Mater.* **2016**, *38*, 216–218. [CrossRef]
2. Blomgren, H.; Lundstroem, T.; Okada, S. Structure refinement of the boron sub-oxide B_6O by the Rietveld methods. *AIP Conf. Proc.* **1991**, *231*, 197.
3. Decker, B.F.; Kasper, J.S. The crystal structure of a simple rhombohedral form of boron. *Acta Cryst.* **1959**, *12*, 503–506. [CrossRef]
4. Olofsson, M.; Lundström, T. Synthesis and structure of non-stoichiometric B_6O. *J. Alloys Compd.* **1997**, *257*, 91–95. [CrossRef]
5. Endo, T.; Sato, T.; Shimada, M. High-pressure synthesis of B_2O with diamond-like structure. *J. Mater. Sci. Lett.* **1987**, *6*, 683. [CrossRef]
6. Grumbach, M.P.; Sankey, O.F.; McMillan, P.F. Properties of B_2O: An unsymmetrical analog of carbon. *Phys. Rev. B* **1995**, *52*, 15807. [CrossRef] [PubMed]
7. Hohenberg, P.; Kohn, W. Inhomogeneous Electron Gas. *Phys. Rev.* **1964**, *136*, 864. [CrossRef]
8. Kohn, W.; Sham, L.J. Self-Consistent Equations Including Exchange and Correlation Effects. *Phys. Rev.* **1965**, *140*, 1133. [CrossRef]
9. Etourneau, J.; Matar, S.F. *rh.*-B_{12} as host of interstitial atoms: Review of a large family with illustrative study of $B_{12}\{CN_2\}$ from first-principles. *Prog. Solid State Chem.* **2020**, 100296. [CrossRef]
10. Kurakevych, O.O.; Solozhenko, V.L. Rhombohedral boron subnitride, $B_{13}N_2$, by X-ray powder diffraction. *Acta Cryst. C* **2007**, *63*, i80–i82. [CrossRef]
11. Matar, S.F.; Solozhenko, V.L. First-principles studies of the electronic and magnetic structures and bonding properties of boron sub-nitride $B_{13}N_2$. *J. Solid State Chem.* **2020**, 121840. [CrossRef]
12. Kirfel, A.; Gupta, A.; Will, G. The Nature of the Chemical Bonding in Boron Carbide $B_{13}C_2$. I. Structure and Refinement. *Acta Cryst.* **1979**, *35*, 1052–1059. [CrossRef]
13. Clark, H.K.; Hoard, J.L. The Crystal Structure of Boron Carbide. *J. Am. Chem. Soc.* **1943**, *65*, 2115–2119. [CrossRef]
14. Kresse, G.; Furthmüller, J. Efficient iterative schemes for ab initio total-energy calculations using a plane-wave basis set. *Phys. Rev.* **1996**, *54*, 11169. [CrossRef]
15. Kresse, G.; Joubert, J. From ultrasoft pseudopotentials to the projector augmented wave. *Phys. Rev.* **1999**, *59*, 1758–1775. [CrossRef]
16. Blochl, P.E. Projector augmented wave method. *Phys. Rev. B* **1994**, *50*, 17953–17979. [CrossRef]
17. Perdew, J.; Burke, K.; Ernzerhof, M. Generalized Gradient Approximation Made Simple. *Phys. Rev. Lett.* **1996**, *77*, 3865–3868. [CrossRef]
18. Ceperley, D.M.; Alder, B.J. Ground State of the Electron Gas by a Stochastic Method. *Phys. Rev. Lett.* **1980**, *45*, 566–569. [CrossRef]
19. Press, W.H.; Flannery, B.P.; Teukolsky, S.A.; Vetterling, W.T. *Numerical Recipes*, 2nd ed.; Cambridge University Press: New York, NY, USA, 1986.
20. Blöchl, P.E.; Jepsen, O.; Anderson, O.K. Improved tetrahedron method for Brillouin-zone integrations. *Phys. Rev.* **1994**, *49*, 16223–16226. [CrossRef]
21. Methfessel, M.; Paxton, A.T. High-precision sampling for Brillouin-zone integration in metals. *Phys. Rev.* **1989**, *40*, 3616–3621. [CrossRef]
22. Monkhorst, H.J.; Pack, J.D. Special k-points for Brillouin Zone integration. *Phys. Rev.* **1976**, *13*, 5188–5192. [CrossRef]
23. Becke, D.; Edgecombe, K.E. A simple measure of electron localization. *J. Chem. Phys.* **1990**, *92*, 5397–5401. [CrossRef]

24. Savin, A.; Jepsen, O.; Flad, J.; Andersen, O.K.; Preuss, H.; von Schneering, H.G. The ELF perspective of chemical bonding. *Angew. Chem. Int. Ed.* **1992**, *31*, 187–191. [CrossRef]
25. Eyert, V. Basic notions, and applications of the augmented spherical wave ASW method. *Int. J. Quantum Chem.* **2000**, *77*, 1007–1012. [CrossRef]
26. Hoffmann, R. Chemistry and Physics Meet in the Solid State. *Angew. Chem. Int. Ed.* **1987**, *26*, 846–878. [CrossRef]
27. Marx, R.; Iberson, R.M. Powder diffraction study on solid ozone. *Solid State Sci.* **2001**, *3*, 195–202. [CrossRef]
28. Mohn, P. *Magnetism in the solid state: An introduction*; Cardona, M., Fulde, P., von Klitzing, K., Merlin, R., Queisser, H.J., Störmer, H., Eds.; Springer: Berlin/Heidelberg, Germany, 2003.
29. Matar, S.F.; Mavromaras, A. Band magnetism in A_2T_2Sn (A = Ce, U.; T = Ni, Pd) from local spin density functional calculations. *J. Solid State Chem.* **2000**, *149*, 449. [CrossRef]
30. Matar, S.F.; Demazeau, G.; Sticht, J.; Eyert, V.; Kübler, J. Étude de la structure électronique et magnétique de CrO_2. *J. Phys. I* **1992**, *2*, 315–328. [CrossRef]
31. Goodenough, J.B.; Casalot, A.; Hagenmuller, P. Les Oxydes des Métaux de Transition. Monographies de Chimie Minérale. Gauthiers-Villars: Paris, France, 1973.
32. Manière, C.h.; Lee, G.; Olevsky, E.O. All-Materials-Inclusive Flash Spark Plasma Sintering. *Sci. Rep.* **2017**, *7*, 15071. [CrossRef]
33. Momma, K.; Izumi, F. VESTA 3 for three-dimensional visualization of crystal, volumetric and morphology data. *J. Appl. Crystallogr.* **2011**, *44*, 1272–1276. [CrossRef]

Commentary

John B. Goodenough's Role in Solid State Chemistry Community: A Thrilling Scientific Tale Told by a French Chemist [†]

Michel Pouchard

CNRS, University Bordeaux INP, ICMCB, UMR 5026, F-33600 Pessac, France; michel.pouchard@icmcb.cnrs.fr
† In September 1964 John B. Goodenough has been the "seed" for the future development of the Solid State Community.

Academic Editor: Paul A. Maggard
Received: 3 December 2020; Accepted: 16 December 2020; Published: 21 December 2020

Abstract: In this tribute to John B. Goodenough I will describe how John's talk on the metal-to-nonmetal transition of vanadium oxide VO_2, presented at the Bordeaux Conference (September 1964) attended by inorganic chemists, metallurgists, crystallographers, thermodynamicists and physicists, provided a pioneering vision of interdisciplinary research to come. John gave a complete description of the paradigm on how the physical properties of a solid depend on its structure and bonding, by employing the chemical notions as local distortions and interatomic distances as well as the physics notions such as band width and the Hubbard on-site repulsion U. I will illustrate how inspiring John's ideas were, by discussing the research examples of my own research group in the sixties-seventies. The fundamental approach of John B. Goodenough to Solid State Chemistry, leading particularly to lithium battery applications, is at the heart of the 2019 Nobel Prize awarded to John.

Keywords: structure; bonding; physical properties; collective or localized electrons; exchange integral

1. Introduction

Tuesday morning, 8 October 2019, my cell phone received a fantastic news that John B. Goodenough was awarded the Nobel Prize, the first one given to solid state chemistry. This was a joyful moment for all of his friends! Suddenly, I realized how profoundly a cell phone in my hand, a very common object these days, changed our everyday life thanks to the fast progress of research and technology in lithium batteries. On reflection, Pasteur's famous words came to my mind: "There are science and applications of science, which are so intimately bound together than the tree and the fruits that it bears . . . " In this Special Issue dedicated to John, I would like to reminisce on the role played by John in the early development of solid state chemistry, the tree that can bear so many different fruits in the future.

2. The Beginnings of Solid State Chemistry

In the early 20th century, inorganic chemistry was essentially concerned with the discovery of new elements such as the noble gases (W. Ramsay), fluorine (H. Moissan), rare-earths (G. Urbain), and the radioactive elements, radium and polonium, (the Curie's). These studies led to three Nobel prizes between 1904 and 1911. After the First World War, following the discovery of X-ray diffraction for the determination of crystal atomic structures (Laue, Bragg, Ewald), inorganic chemistry started dealing with more and more complex systems such as the inorganic core (Fe, Co or Mg) in biological molecules such like hemoglobin and metalloporphyrins (Perutz, Nobel Prize 1962). Simultaneously, quantum mechanics provided the foundation for the description of electronic structure and chemical bonding (De Broglie, Dirac, Heitler, Hund, Hückel, Mulliken). As early as 1939, Ewans elaborated a structural chemistry, which enables one to deduce chemical and physical properties from X-rays atomic structures. In the same year, Pauling published his famous textbook, "The nature of the chemical bond and the structure of molecules and crystals", establishing the valence-bond approach to chemical bonding, for which he received the Nobel Prize in 1954. The molecular orbital approach to chemical bonding, an alternative to the valence bond approach, was subsequently developed by Mulliken, who received the Nobel Prize in 1966.

In the early 20th century, inorganic chemistry in France was carried out by three forerunners: C. Friedel, H. Moissan and H. Le Chatelier, later by Friedel's students, G. Urbain and A. Chretien, by Moissan's students, P. Lebeau and L. Hackspill, and by Le Chatelier's student, G. Chaudron. They held prestigious Chairs at The Sorbonne, College de France and Engineering School institutions. At the end of The Second World War, inorganic chemistry was divided essentially in two areas, chemistry of solutions led by A. Chretien and metallurgy led by G. Chaudron; they disagreed totally on their visions of chemistry concerning the use of physical methods and applications. Chretien's students, P. Hagenmuller and A. Deschanvres, and Chaudron's student, R. Collongues, attempted to find some common grounds. As early as 1948, Chaudron had organized a series of International Symposia on the Reactivity of Solids. This series was rather limited in terms of topics and international participations, and gradually disappeared.

In Germany, inorganic solid state chemistry was most recognized for its successes in synthesis, for example, in the area of transition metals of high oxidation states by W. Klemm and in Zintl-type phases between alloys and normal compounds. However, collaborations with physicists and application attempts remained poorly developed. In USA, solid state inorganic chemistry was essentially developed in the industrial laboratories such as Dupont, Exxon, IBM and General Electrics, strongly motivated by the applications and materials science. There were high-level studies in thermodynamics and non-stoichiometry by J. S. Anderson in United Kingdom, and those in extended defects of non-metallic solids by A. Magneli and S. Anderson in Sweden and by D. Wadsley in Australia, following the pioneering works in this defects in solid field (Frenkel, Schottky and Wagner).

3. Bordeaux, September 1964:"Crystallization" of a Solid State Chemistry Community

In Spring 1960, Paul Hagenmuller organized, with his group, a long visit to the main inorganic chemistry laboratories in Germany (Klemm and Schäfer in Munster, Scholder in Karlsruhe, Glemser in Göttingen, Rabenau in Aachen). Four years later, he invited them as well as many foreign colleagues mostly from USA to Bordeaux for an important international symposium on the "transition metal oxides in solid state". The scientific community was broadly chosen: physicists (P. Aigrain as an expert of semiconductors, F. Bertaut representing L. Néel, the future French Nobel Prize winner in physics, C. Guillaud etc.), French and German inorganic chemists, a large group of metallurgists (G. Chaudron, J. Bénard, P. Lacombe, A. Michel etc.), a few specialists of thermodynamics, crystallographers (S. Anderson, G. Blasse, E.W. Görter etc.) as well as materials chemists of principally from USA (J. B. Goodenough, A. Wold, R. Roy).

In front of this mixed Assembly John Goodenough gave an unforgettable lecture on the metal/non-metal transition in vanadium oxide, VO_2 (Figure 1). This phase transition had been observed earlier, just before Second World War, by the metallurgist A. Michel, who attended the symposium. John first introduced the small range of composition homogeneity for this oxide ($VO_{2-\delta}$), close to the Magneli-oxide series $V_nO_{(2n-1)}$, with crystallographic shear planes beautifully imaged by high-resolution electron microscopy.

Figure 1. J.B. Goodenough at the Bordeaux-Congress of 1964 presenting the metal-non-metal transition (structural as electronic) in the vanadium di-oxide VO_2.

For the specialists of thermodynamics, John recalled the temperature vs. oxygen partial pressure conditions for the H_2/H_2O or CO/CO_2 gas-phase mixture needed for the synthesis, related to the famous Chaudron's diagram. Secondly, John turned to the crystallographers, showing the crystal symmetry breaking at the transition, from tetragonal P4/mnm to monoclinic P2$_1$/c, then to the chemists, showing how the VO_6 octahedron undergoes a large distortion in the monoclinic form. To the physicists, he spoke the band structure language, with the band splitting along the distortion from a $\pi^*(t_{2g})$ triplet into an occupied singlet b_{2g}^1 and an empty doublet e_g^0. He underlined the periodicity doubling along the c axis, as expected in this Peierls-type transition, following the $\pi^*(b_{2g}^1)$ band splitting into two split narrow bands, of which only the lowest one is doubly occupied ($b_{2g}^{2'}$). John showed to the chemists how this splitting modifies the electron distribution along the chain of vanadium atoms, increasing the electron density and bonding character in the V-V pairs, decreasing the density (of antibonding character) between the vanadium pairs. Everybody understood that the different descriptions correspond actually to the same physical reality, but a different picture emerged for different eyes, with everybody finding one's own place. As for myself, working at that moment on the vanadium oxide bronzes, John's talk was a revelation, a guideline for my future as a scientist. I often recall this magic moment. When growing single crystals by cooling a crucible filled with many components, one often finds crystallization into separate binary, sometimes ternary, compounds. However, introducing an appropriate seed induces the formation of a beautiful multi-component single crystal, in which all components find their own and desired places in a perfectly ordered three-dimensional arrangement. *In September 1964, John Goodenough was such a seed for the future development of the solid state community.*

4. Solid State Chemistry and Localized Versus Delocalized Electrons

In the early 1950s, transition-metal oxides were believed to possess strong ionic bonding and hence are insulators that can be described by crystal-field theory and by magnetic super-exchange or Zener-type double-exchange. However, the band structure description of insulating oxides (e.g., NiO) found that they possess partially-filled bands, hence predicting that they are metals in contradiction to experiment. In the late 1950s, sodium tungsten bronzes Na_xWO_3, discovered one century ago, were found to be a metal. Similarly, the mixed-valent manganite $La_{(1-x)}Sr_x MnO_3$ was found to be a ferromagnetic metal. In fact, two opposite theories existed for the description of the outer electrons of solids: the crystal-field and band theories. The crystal-field theory describes localized electrons at discrete atomic positions with a long residence time τ. With a short residence time τ, each electron feels the periodic potential of the crystal and becomes delocalized (collective). The latter is the case for s and p electrons, but the former for 4f electrons. d electrons lie in-between the two, localized in some crystals but delocalized in others. At the beginning of the sixties John explored this difference, especially for the AMO_3 perovskite and related structural classes of solids. These studies led to many review papers, book chapters/series and more specific articles [1–11].

There are two fundamental parameters to consider in understanding the metal vs. insulator behaviors of solids; the transfer energy b^{ca} between cationic and anionic orbitals and the on-site Hubbard U parameter separating the formal valences between the +m and +(m-1) oxidation states of a cation. For the free ion, U can be expressed as the energy difference between the successive ionization energies $EI^{m+1} - EI^m$ corrected by the electron-hole Coulomb attraction. The competition between b^{ca} and U, or the b^{ca}/U ratio, sets the boundary for the localized vs. delocalized character of the outer d-electrons. Besides, both parameters depend on the covalency: b^{ca} increases with covalency (and so does the band width), while U decreases with covalency (so does the Racah B parameter of the molecular chemists). Electrons become delocalized above a critical value b_c^{ca} and metallicity sets in just later (b_m), when the occupied and unoccupied states (bands) overlap. Thus, in the early 1960s, J. B. Goodenough proposed a phenomenological diagram, $b^{ca} = f(Z, S)$, to classify the transition metal oxides of perovskite structure with localized vs. delocalized d-electrons in terms of the atomic number Z and spin S of the transition metal ion.

The AMO_3 perovskite structure consists of corner-sharing MO_6 octahedra, involving thus a cubic close-packing arrangement of oxygen atoms with one quarter of which replaced by the large A cation at the body-center positions. With the ideal ionic radii R_A, R_M and R_O of A, M and O that allow the cations and anions to be in contact for the optimal Coulomb interactions, the geometric tolerance factor t,

$$t = \frac{(R_A + R_O)}{\sqrt{2}(R_M + R_O)},$$

becomes 1. Many parameters can modify the values of ionic radii:

(i) Coordination number, temperature, pressure, including chemical pressure, i.e., substitution with an ion of different radius.

(ii) Spin equilibrium or transition. For example, a d^6 cation with three possible configurations, low-spin t^6e^0, high-spin t^4e^2, and sometimes intermediate-spin t^5e^1, in a Jahn-Teller distorted site or in metallic compounds ($t^5\sigma^{*1}$), as discussed in details by John in his articles on LaCoO$_3$ and LaNiO$_3$ (9,10).

(iii) Charge disproportionation, as observed by Takano et al. [12] in CaFeO$_3$ synthesized under oxygen high-pressure (Fe^{4+} with t^3e^1, giving rise to Fe^{5+} with t^3e^0 plus Fe^{3+} with t^3e^2) and extensively discussed by John.

(iv) Insulator to metal transition, often associated with a large decrease in the cell volume, as found for hole-doped divalent cuprates.

Let us consider one example illustrating the point (i): silica and silicates have corner-sharing SiO$_4$ arrays at ambient pressure. Under geological pressure in the deep mantle, the (R_M/R_O) ratio strongly increases due to oxygen polarizability and compressibility; for the same number of inner electrons [10], the volume of the oxide ion is approximately thirty times greater than silicon. In these conditions, silicon becomes octahedrally coordinated in the ilmenite and perovskite structures. Conversely, a cation such as Fe^{2+} becomes smaller under pressure by changing its configuration from t^4e^2 into low-spin t^6e^0. As a consequence, the atomic and electronic structures strongly influence all properties of solids.

AMO$_3$ oxides may also adopt the hexagonal close-packing of oxygen, in which face-sharing octahedral chains develop along the c axis (t > 1). This less dense hexagonal form transforms naturally into the cubic form under pressure. In several general articles, John discussed such transitions through successive well-ordered face- centered cubic and hexagonal close-packing sequences, giving rise to various intermediate phases named "poly-types", such as 4-H or 9-H in the oxides BaTiO$_3$ and BaRuO$_3$, respectively, and in AMCl$_3$ chlorides. These early studies allowed John to classify structures on the basis of inter-cationic Coulomb repulsions (charge, distance) related to the covalent mixing and the structural pattern of corner-, edge-, or face-sharing. For edge-sharing entities, he defined a critical value (R_C) separating the localized and collective electrons. An important consequence concerns the MO$_2$ series: with the earliest 3d elements (low-covalent mixing and highest charge), only the rutile structure with two common edges is allowed. The six common edges of the 2D layered brucite-type structure was an important challenge for the electrode-battery materials in sodium cobalt bronzes by Fouassier et al. [13], Delmas et al. [14], and finally Tarascon et al. [15] due to reduced cationic charges in the mixed-valent bronzes.

For the t values smaller than unity, the twelve coordination site A of cubic AMO$_3$ oxides is too roomy for the mid and late rare-earth cations. To accommodate the small cations at the A site, the pristine cubic cell adjusts by a cooperative rotation of the MO$_6$ octahedra along the [111] axis for a R3m rhombohedral symmetry, and further along the [110] axis for an orthorhombic Pbnm symmetry. The bending angle in M-O-M bonds decreases smoothly with t to preserve the A-O contacts, typically until an eight-coordination environment is achieved. A few examples of the evolutions along the 3d series with A = Ln^{3+} or Sr^{2+} and Ca^{2+} are:

- For the largest M^{3+} cation (Ti) of $t_{2g}^1 e_g^0$ configuration with $b^{ca} < b_c < b_m$, only the orthorhombic Pbnm structure is accessible. The M-O-M bending increases along with the lanthanide series contraction, the band width decreases, and the band gap increases from 0.2 eV for La up to 1.2 eV for Y.
- For smaller cations (Cr^{3+} and Fe^{3+}), the increase in b^{ca} is not significant enough to overcome the effect of U, and the strong spin exchange interactions make these compounds antiferromagnetic insulators.
- For the intermediate lanthanum manganite LaMnO$_3$ and its high-spin $t_{2g}^3 e_g^1$ configuration, one can predict either an insulating Jahn-Teller distorted situation as well as a metallic state ($t^3\sigma^{*1}$ for b^{ca} overcoming b_m). The former configuration was accepted early, but not totally understood until John's enlighting explanation [7] based on the alternating ($x^2 - y^2$) and (z^2) atomic orbitals along both [100] and [010] axes, in agreement with the unusual ratio c/a < 1. Besides,

such orbital ordering was among the most spectacular illustrations of the Goodenough-Kanamori rules for ferromagnetic super-exchange. At this point, it is interesting to discuss briefly the case of Sr ferrates containing Fe^{4+} ions with the same d^4 configuration. b^{ca} is larger for Fe^{4+} than for Mn^{3+} cations, allowing the $t^3\sigma^{*1}$ metallic configuration for Fe in $SrFeO_3$. However, with the smaller Ca cation and the associated orthorhombic distortion, as well as with the decrease in b^{ca} and band width, a metallic state is not allowed anymore for $CaFeO_3$. As mentioned above, a new type of ordering occurs at low temperature, with a partial disproportionation of two Fe^{4+} into $Fe^{5+} + Fe^{3+}$. This interesting new behavior, observed by ^{57}Fe Mossbauer spectroscopy, was largely discussed by John and Mikio Takano [16]. A few years later, we were able to confirm this partial disproportionation by comparing their data to ours in La_2LiFeO_6 (with a much larger negative value of the isomer-shift $\delta = -0.41$ mm.s^{-1}) [17].

- $LaNiO_3$ (R3c) was the only rare-earth nickelate known up to the seventies. Its metallicity was proven only in 1965 by Goodenough and Racah [9], which results from the low-spin $t_{2g}^6 e_g^1$ configuration and $b^{ca} > b_m$, which induces the involvement of the $t_{2g}^6 \sigma^{*1}$ configuration.

The AMO_3 perovskites allow anionic vacancies different from ordered defects of Magneli and Wadsley type in crystallographic shear planes and so do the ReO_3-type oxides, which are related to the AMO_3 perovskites except that the large central A site is empty. Of particular interest was John's description of the Brownmillerite calcium ferrite mineral $Ca_2Fe_2O_5$ (2), i.e., $CaFeO_{2.50}$, exhibiting a large number of vacancies compared with the pristine AMO_3 perovskite: half the Fe^{3+} ions occupy the distorted corner-sharing octahedra with the remaining half occupying the corner-sharing tetrahedra, and the planes containing these Fe^{3+} ions alternate. We will discover in the following how important this description by John was.

The contributions of John Goodenough to solid state chemistry are too numerous to summarize in a few lines. His contributions broadened the vision of every solid state chemist by providing them with the crystal-field description of the ionic solids, the molecular description of the chemical bond, and the physics view of the delocalized outer-electrons.

5. Solid State Chemistry, Goodenough's Heritage

Every solid state inorganic chemist of my generation (young researchers in the 1960s to 1970s) has been influenced, or more often inspired, by John's ideas, which were guidelines for explaining or predicting new properties for new materials. Since it would be unrealistic to give an exhaustive list of his ideas and concepts, I will limit my reminiscences to those that greatly influenced the thinking and development of my research group in the 1960s to 1970s.

5.1. Electrical Conductivity of Tungsten Bronzes and the Goodenough-Sienko Debate

For Mike Sienko, the metallic conductivity observed in tungsten oxide bronzes resulted from the direct t_{2g}–t_{2g} overlap of W 5d orbitals. In contrast, John proposed an indirect overlap through the $2p_\pi$ orbitals of oxygen, on the basis of the insulating character of the double perovskite Sr_2MgReO_6, (W^{5+} and Re^{6+} are isoelectronic with $5d^1$). However, other explanations are possible due to the tetragonal distortion in the latter as well as the different nature and size of all ions involved. In a contribution to this debate, J. P. Doumerc [18] replaced some oxygen atoms by fluorine, more electronegative than oxygen, which should decrease the conductivity in an indirect mechanism. He showed experimentally this to be the case, but it was argued not to be an absolute proof: actually, since the Anderson-type anionic disorder (O and F) would lead to a similar effect. This issue was resolved many years later to the satisfaction of both John and Mike! For Mike, for the lowest values of x in Na_xWO_3, because the Fermi level E_F lies close to the center of the Brillouin zone, where the mixing between the $t_{2g}(W)$ and $2p_\pi(O)$ orbitals is symmetry-forbidden. For John, for much larger x values (for which the bronze is fully metallic) because the Fermi level lies far away from the zone center, where the cationic and anionic orbital mixing takes place. We were relieved to find this friendly solution!

5.2. Ternary Tungsten Oxides MWO_4 (M = Al, Cr, Ga) of W^{5+} Ions

Following John's lecture in 1964, the metal/non-metal transition of VO_2 led to numerous developments in USA (IBM) and in France (Orsay, Bordeaux) generating new collaborations between physicists and chemists, and to the discovery of new intermediate phases such as Marezzio et al.'s [19]. In this chromium-doped monoclinic variety of VO_2 (C2/m), every second chains of V^{4+} ions have dimerized V^{4+} pairs, as in what would be an ordered phase including the high and low-temperature forms of VO_2. Another way to stabilize the W^{5+} ions is to pair them in one chain and replace the W^{5+} ions of the other chain with trivalent d^0 cations (e.g., Al^{3+}), d^{10} cations (e.g., Ga^{3+}) or even a spherical d^3 ion (e.g., Cr^{3+}). The small W-W pairs were found to be particularly stable up to 1000 °C [20].

5.3. Nickelates of Perovskite Structural Type $LnNiO_3$ (Ln = rare-earth, La-Lu, Y)

In 1964, $LaNiO_3$ was the only known rare-earth nickelate. Five years later, in Bordeaux, G. Demazeau began his PhD studies, introducing in our laboratory the Belt-type high-pressure synthesis facilities adapted to produce high oxygen pressure by decomposing $KClO_3$ in situ. Among a large number of transition metal oxides, which system should one study? Perovskite, of course, for their high stability under pressure! The nickelate system appeared also as an obvious choice, with immediate questions to raise. Does a complete series from La up to Y exist as discovered for Ti^{3+} under reducing conditions in Germany in the late sixties? Low-spin trivalent nickel is much smaller than trivalent titanium. Do all nickelates $LnNiO_3$ possess low-spin Ni^{3+}? Are they metallic? In 1972 we obtained all the answers [21,22]. The series was complete, from the rhombohedral and metallic phase with large La to the orthorhombic and insulating phase with smallest rare-earths Lu and Y. As predicted, the increased tilting of the NiO_6 octahedra together with the decrease in rare-earth size narrows the σ^{*1} band, with $b^{ca} < b_m$. In addition, an antiferromagnetic order appeared at low temperature. Twenty years later, at IBM San Jose, Torrance and Lacorre [23,24] were able to draw the first phase diagram of the insulator-to-metal transition temperature T_t and the Néel temperature T_N for the $LnNiO_3$ family. Today, such a diagram is found in solid state chemistry textbooks. A strontium-doped member of this nickelate family, $Nd_{1-x}Sr_xNiO_3$, reduced by H_2 in a layered thin-film oxide AMO_2, was recently found to be a superconductor, stimulating new reflections on the superconductivity mechanisms [25].

5.4. Long-Range Ordering of Planar Defects in Some Non-Stoichiometric Perovskite-Type Ferrites

To my knowledge, John Goodenough was the first to propose the description of $Ca_2Fe_2O_5$ as an ordered succession of corner-sharing octahedron layers and corner-sharing chains of tetrahedra, arranged also in parallel layers. Both coordination types, approximately octahedral and approximately tetrahedral, agree with the spherical character of high-spin Fe^{3+} (d^5, S = 5/2), insensitive to the crystal field symmetry. For such a high extent of oxygen vacancies (0.50 per AMO_3 formula unit) with all A sites occupied, any crystallographic shear mechanism can be envisaged. During his PhD studies in Bordeaux (1975), Grenier et al. imagined that all $AFeO_{3-\delta}$ non-stoichiometric compositions with $0 < \delta < 0.50$ can be described with the same type of layer alternation of octahedral (O) and tetrahedral (T) coordinations, with various sequences such as OOTOOT ($\delta = 0.33$), OOOTOOOT ($\delta = 0.25$), as well as inter-growth situations (OOTOTOOTOT, $\delta = 0.40$). J.C. Grenier synthesized the solid solutions between $CaFeO_{2.50}$ and $CaTiO_3$ as well as $LaFeO_3$, and confirmed the proposed non-stoichiometry-based structural model by X-ray diffraction, high-resolution electron microscopy and ^{57}Fe-Mössbauer spectroscopy [26–28]. Such a type of layered, ordered planar defects is associated with the electron configuration of the cation Fe^{3+} implying the octahedral and tetrahedral symmetries for the basic polyhedra. Ten years later, oxides of other d-electron configurations like d^4, d^7 and d^9 were prepared and studied. In particular, d^9 ions led to the cuprates family, based on layers of square-pyramidal (C_{4v}) and square-planar (D_{4h}) polyhedra arrangements of Cu^{2+} ion.

5.5. Trivalent Cuprates, a Precursor to High-T_C Superconductors

Back in 1970, trivalent cuprates of perovskite structure were not known, despite the similarity between Cu^{3+} and Ni^{2+}. However, a decade earlier, Klemm, and subsequently Hoppe, synthesized $MCuO_2$ phases (M = Na, K) with Cu^{3+} in square-planar coordination, which exhibit diamagnetism and insulating behaviors. During John's visit to Bordeaux for a few months in the late sixties, we had a lot of discussions with him about the influence of bond covalency on the band width and electron delocalization and also about the similarity between d^1 (one electron) and d^9 (one hole) systems. A strong covalent character meant the presence of highly destabilized states at the top of the filled 3d band, and hence a large energy stabilization by hole doping. This realization led to the associated question: is covalency strong enough to overcome the Jahn-Teller instability and induce metallicity? The previous success in synthesizing the $LnNiO_3$ series under high oxygen pressure was an argument for attempting the synthesis of new Cu^{3+} oxides. We chose three compositions: (i) $LaCuO_3$ for its expected 3D metallicity, (ii) $LaSrCuO_4$, the first member of the Ruddlesden-Popper series of 2D perovskite-related structures, so as to evaluate the Jahn-Teller vs. metallicity competition, and (iii) $La_2Li_{1/2}Cu_{1/2}O_4$ to evaluate the influence of Li chemical pressure, an additional effect to prepare the S = 0 low-spin d^8 configuration of Cu. All these synthesis attempts were successful with physical properties as predicted. These results were published in 1972-73 [29–31], with the participation of John Goodenough, and that of Sir Nevill Mott (for the last paper), who was awarded the Nobel Prize in physics a few years later (1977). Ten years later, these results were enriched by the study of Raveau et al. on mixed-valent Cu^{2+}/Cu^{3+} cuprates [32]. This work was followed by the discovery of superconductivity at 35 K in 1986 by Bednorz and Müller in this system [33], for which they received the Nobel Prize in 1987.

Before concluding on John's achievements during the seventies, I would like to mention another important areas in which fundamental science and applications are intimately linked: Na^+ cationic conductivity in phospho-silicate oxides with tunnel structures (NASICON: Natrium-Super-Ionic-CONductors), as well as O^{2-} anionic conductivity in $La_{1-x}Sr_xGaO_{3-y}$ gallates for solid oxide fuel cell (SOFC) applications, and solar energy simultaneous capture and storage by wavelength-selective photo-anodes, generating hydrogen from water by photo-electrolysis (Honda-type experiment).

6. Concluding Remarks

In summary, John Goodenough greatly influenced the thinking of the solid state chemists and played a crucial role in the early development of solid state chemistry. He recognized the interdisciplinary nature of solid state chemistry and clearly demonstrated the paradigm that, for understanding and getting useful applications out of solid state chemistry, it is crucial to know the interrelationship between structure, bonding and properties. I would like also to point out more recent contributions of physics (P.G. de Gennes, Nobel Prize in 1991) for the soft matter and those of theoretical chemistry (R. Hoffmann, Nobel Prize in 1981, and his active disciples) in enriching this paradigm.

Funding: This research received no external funding.

Acknowledgments: The author deeply thanks A. Villesuzanne and M. Whangbo for their critical reading of the text, and P. Teissier for sending his book" Une histoire de la chimie du solide" Hermann Editeurs, (2014).

Conflicts of Interest: The author declares no conflict of interest.

References

1. Goodenough, J.B. *Magnetism and the Chemical Bond*; Wiley Interscience: New York, NY, USA, 1963.
2. Goodenough, J.B.; Longo, J.M. Crystallographic and Magnetic Properties of Perovskite and Perovskite-Related Compounds. In *Magnetic and Other Properties of Oxides and Related Compounds Part A*; Landolt-Börnstein Tabellen; New Series III/4a; Hellwege, J.M., Hellwege, K.H., Eds.; Springer: Berlin, Germany, 1970.
3. Goodenough, J.B. Metallic oxides. In *Progress in Solid State Chemistry*; Reiss, H., Ed.; Pergamon Press: Oxford, UK, 1971; Volume 5.

4. Goodenough, J.B.; Kafalas, J.A.; Longo, J.M. *Preparative Methods in Solid State Chemistry*; Hagenmuller, P., Ed.; Academic Press: New York, NY, USA, 1972; Chapter 1.
5. Goodenough, J.B.; Cooper, S.L.; Egami, T.; Zhou, J.-S. Localized to Itinerant Electronic Transition in Perovskite Oxides. In *Structure and Bonding*; Goodenough, L.B., Ed.; Springer: Berlin, Germany, 2001; Volume 98.
6. Goodenough, J.B. Transition-metal oxides with metallic conductivity. *Bull. Soc. Chim. France* **1965**, *4*, 1200.
7. Goodenough, J.B. Theory of the Role of Covalence in the Perovskite-Type Manganites [La, M(II)]MnO$_3$. *Phys. Rev.* **1955**, *100*, 564–573. [CrossRef]
8. Goodenough, J.B.; Wold, A.; Arnott, R.J.; Menyuk, N. Relationship Between Crystal Symmetry and Magnetic Properties of Ionic Compounds Containing Mn^{3+}. *Phys. Rev.* **1961**, *124*, 373–384. [CrossRef]
9. Goodenough, J.B.; Raccah, P.M. Complex vs. Band Formation in Perovskite Oxides. *J. Appl. Phys.* **1965**, *36*, 1031. [CrossRef]
10. Senaris-Rodriguez, M.A.; Goodenough, J.B. LaCoO$_3$ Revisited. *J. Solid State Chem.* **1995**, *116*, 224–231. [CrossRef]
11. Zhou, J.S.; Goodenough, J.B.; Dabrowski, B. Probing the metal-insulator transition in Ni(III)-oxide perovskites. *Phys. Rev. B* **2000**, *61*, 4401. [CrossRef]
12. Takano, M.; Nakanishi, N.; Takeda, Y.; Naka, S.; Takada, T. Charge disproportionation in CaFeO$_3$ studied with the Mossbauer effect. *Mat. Res. Bull.* **1977**, *12*, 923–928. [CrossRef]
13. Fouassier, C.; Matejka, G.; Reau, J.M.; Hagenmuller, P. Sur de nouveaux bronzes oxygénés de formule Na$_x$CoO$_2$ (x1). Le système cobalt-oxygène-sodium. *J. Solid State Chem.* **1973**, *6*, 532–537. [CrossRef]
14. Braconnier, J.J.; Delmas, C.; Fouassier, C.; Hagenmuller, P. Comportement électrochimique des phases Na$_x$CoO$_2$. *Mat. Res. Bull.* **1980**, *15*, 1797–1804. [CrossRef]
15. Amatucci, G.G.; Tarascon, J.M.; Klein, L.C. CoO$_2$, the end number of the Li$_x$CoO$_2$ solid solution. *J. Electrochem. Soc.* **1996**, *143*, 1114–1123. [CrossRef]
16. Takano, M.; Nasu, S.; Abe, T.; Yamamoto, K.; Endo, S.; Takeda, K.; Goodenough, J.B. Pressure-induce high-spin to low-spin transition in CaFeO$_3$. *Phys. Rev. Lett.* **1991**, *67*, 3267–3270. [CrossRef]
17. Demazeau, G.; Buffat, B.; Menil, F.; Fournes, L.; Pouchard, M.; Dance, J.M.; Fabritchnyi, P.; Hagenmuller, P. Characterization of six-coordinated iron (V) in an oxide lattice. *Mat. Res. Bull.* **1981**, *16*, 1465–1472. [CrossRef]
18. Doumerc, J.P.; Pouchard, M. Une nouvelle famille de composés minéraux: Les bronzes oxyfluorés de tungstène NaxWO3-xFx. *C. R. Acad. Sci.* **1970**, *270*, 547.
19. Marezio, M.; Mc Whan, B.; Remeika, J.P.; Dernier, P.D. Structural Aspects of the Metal-Insulator Transition in Cr-Doped VO$_2$. *Phys. Rev. B* **1972**, *5*, 2541–2551. [CrossRef]
20. Doumerc, J.P.; Vlasse, M.; Pouchard, M.; Hagenmuller, P. Synthèse, croissance crystalline, propriétés structurales et physiques d'un nouveau tungstate +V d'aluminium AlWO$_4$. *J. Sol. State Chem.* **1975**, *14*, 144–151. [CrossRef]
21. Demazeau, G.; Marbeuf, A.; Pouchard, M.; Hagenmuller, P.; Goodenough, J.B. Synthesis, Structure and Magnetic Study of New Family of Oxygenated Nickel (Iii) Compounds. *C. R. Acad. Sci.* **1971**, *272*, 2163.
22. Demazeau, G.; Marbeuf, A.; Pouchard, M.; Turrel, S.; Hagenmuller, P. Sur une série de composés oxygènes du nickel trivalent derivés de la perovskite. *J. Sol. State Chem.* **1971**, *3*, 582–589. [CrossRef]
23. Torrance, J.B.; Lacorre, P.; Nazzal, A.I.; Ansaldo, E.J.; Niedermayer, C. Systematic study of insulator-metal transitions in perovskites RNiO$_3$ (R=Pr, Nd, Sm, Eu) due to closing of charge-transfer gap. *Phys. Rev. B* **1992**, *45*, 8209. [CrossRef] [PubMed]
24. Lacorre, P.; Torrance, J.B.; Pannetier, J.; Nazzal, A.I.; Wang, P.; Huang, T.C.; Cui, Y.; Hikita, Y.; Hwang, H.Y. Synthesis, crystal structure and properties of metallic PrNiO3: Comparison with metallic NdNiO3 and semiconducting SmNiO3. *J. Solid State Chem.* **1991**, *91*, 225–237. [CrossRef]
25. Li, D.; Lee, K.; Wang, B.Y.; Osada, M.; Crossley, S.; Lee, H.R.; Cui, Y.; Hikita, Y.; Hwang, H.Y. Superconductivity in an infinite-layer nickelate. *Nature* **2019**, *572*, 624–627. [CrossRef]
26. Grenier, J.C.; Menil, F.; Pouchard, M.; Hagenmuller, P. Caracterisation physico-chimique du ferrite de calcium et de lanthane Ca$_2$LaFe$_3$O$_8$. *Mat. Res. Bull.* **1977**, *12*, 79–85. [CrossRef]
27. Pouchard, M.; Grenier, J.C. Vacancy ordering in oxygen-deficient perovskite-related ferrites. *C. R. Acad. Sci.* **1977**, *284*, 311.
28. Grenier, J.C.; Pouchard, M.; Hagenmuller, P. *Structure and Bonding*; Reinen, D., Ed.; Springer: Berlin, Germany, 1981; Volume 47.
29. Demazeau, G.; Parent, C.; Pouchard, M.; Hagenmuller, P. Sur deux nouvelles phases oxygénées du cuivre trivalent: LaCuO$_3$ et La$_2$Li$_{0,50}$Cu$_{0,50}$O$_4$. *Mat. Res. Bull.* **1972**, *7*, 913–920. [CrossRef]

30. Goodenough, J.B.; Demazeau, G.; Pouchard, M.; Hagenmuller, P. Sur un Nouvelle Phase Oxygénée du Cuivre + III: SrLaCuO$_4$. *J. Sol. State Chem.* **1973**, *8*, 325–330. [CrossRef]
31. Goodenough, J.B.; Mott, N.F.; Pouchard, M.; Demazeau, G.; Hagenmuller, P. Etude comparative du comportement magnétique des phases LaNiO$_3$ et LaCuO$_3$. *Mat. Res. Bull.* **1973**, *8*, 647–655. [CrossRef]
32. Nguyen, N.; Studer, F.; Raveau, B. Oxydes ternaires de cuivre à valence mixte de type K2NiF4 déficitaires en oxygène: Evolution progressive d'un état semi-conducteur vers un état semi-métallique des oxydes La$_{2-x}$Sr$_x$CuO$_{4-x2+\delta}$. *J. Phys. Chem. Solids* **1983**, *44*, 389–400. [CrossRef]
33. Bednorz, J.G.; Müller, K.A. Possible high T_c superconductivity in the Ba-La-Cu-O system. *Z. Phys. B Condens. Matter* **1986**, *64*, 189–193. [CrossRef]

Publisher's Note: MDPI stays neutral with regard to jurisdictional claims in published maps and institutional affiliations.

© 2020 by the author. Licensee MDPI, Basel, Switzerland. This article is an open access article distributed under the terms and conditions of the Creative Commons Attribution (CC BY) license (http://creativecommons.org/licenses/by/4.0/).

MDPI
St. Alban-Anlage 66
4052 Basel
Switzerland
Tel. +41 61 683 77 34
Fax +41 61 302 89 18
www.mdpi.com

Molecules Editorial Office
E-mail: molecules@mdpi.com
www.mdpi.com/journal/molecules

www.ingramcontent.com/pod-product-compliance
Lightning Source LLC
LaVergne TN
LVHW070445100526
838202LV00014B/1669